Research Topics in Aerospace

For further volumes:
http://www.springer.com/series/8625

Martin Wiedemann · Michael Sinapius
Editors

Adaptive, Tolerant and Efficient Composite Structures

 Springer

Editors
Martin Wiedemann
Institute of Composite Structures
  and Adaptive Systems
German Aerospace Center DLR
Braunschweig
Germany

Michael Sinapius
Institute of Composite Structures
  and Adaptive Systems
German Aerospace Center DLR
Braunschweig
Germany

ISBN 978-3-642-44531-6   ISBN 978-3-642-29190-6 (eBook)
DOI 10.1007/978-3-642-29190-6
Springer Heidelberg New York Dordrecht London

Printed on acid-free paper

Springer is part of Springer Science+Business Media (www.springer.com)

Contents

Part II   Structural Mechanics

Part III   Composite Design

17   Compliant Aggregation of Functionalities 225

18   Boom Concept for Gossamer Deployable Space Structures...... 237

19   Local Metal Hybridization of Composite Bolted Joints......... 251

Part V   Adaptronics

30   Autonomous Composite Structures   375

31   Design of a Smart Leading Edge Device   381

**32   Experimental Investigation of an Active Twist Model Rotor
Blade Under Centrifugal Loads**   391

Chapter 1
Introduction

Martin Wiedemann and Michael Sinapius

Abstract Polymer composites offer the possibility for functional integration since the material is produced simultaneously with the product. The efficiency of composite structures raises through functional integration. The specific production processes of composites offer the possibility to improve and to integrate more functions thus making the structure more valuable. Passive functions can be improved by combination of different materials from nano to macro scale, i.e. strength, toughness, bearing strength, compression after impact properties or production tolerances. Active functions can be realized by smart materials, i.e. morphing, active vibration control, active structure acoustic control or structure health monitoring. The basis is a comprehensive understanding of materials, simulation, design methods, production technologies and adaptronics. These disciplines together deliver advanced lightweight solutions for applications ranging from mechanical engineering to vehicles, airframe and space structures along the complete process chain. The book provides basics as well as inspiring ideas for engineers working in the field of adaptive, tolerant and robust composite structures.

The existence of load carrying structures has been taken for granted up to the point that their function is hardly being perceived. But with regard to the efficient use of resources load carrying structures receive a new importance since they represent masses and thus their movement requires energy. In many cases load carrying structures constitute a significant part of the total mass of a technical product. Many functions of a technical product, however, add their own masses. If the load carrying structure is called a primary structure with respective primary masses, all

M. Wiedemann (✉) · M. Sinapius
Institute for Composite Structures Adaptive Systems, German Aerospace Center DLR,
Lilienthalplatz 7, 38108, Braunschweig, Germany
e-mail: martin.wiedemann@dlr.de

M. Wiedemann and M. Sinapius (eds.), *Adaptive, Tolerant and Efficient Composite Structures*, Research Topics in Aerospace, DOI: 10.1007/978-3-642-29190-6_1,
© Springer-Verlag Berlin Heidelberg 2013

the parts providing the other functions are secondary masses. These parts are for example motors, sensors, brackets, housings, cables, media tubes, interior panels. Although these functions finally constitute the technical product, they are of minor importance with respect to the global load carrying function of the complete product. Especially in passenger transport a considerable number of functions is needed to ensure the required safety and to provide the requested comfort.

The traditional approach of product development is characterized by the separation of functions: in a safe, reliable and stable way and without unwanted deflections and vibrations passive primary structures carry the accelerated masses of all the other functions and the operational loads caused by the product mission. The division of technical parts into those responsible for the load carrying and the others that provide further functions is caused by the separate development processes of single technical components which are combined to more complex subcomponents and finally to the global product at a later stage in the development process. With regard to an efficient use of energy and the maximum saving of resources it must be the aim to reduce the moving technical product mass to the absolute minimum. If the structure is facing only small operational loads this might be achieved by the substitution of heavier material by lighter ones. In other areas as for example electronics, the functions can be minimized as well as the power generators or batteries making the products lighter. But in the field of transport, energy and production, i.e. whenever high loads have to be carried by large technical machineries or vehicles other measures have to be taken.

Lightweight design gained importance first in aviation whereas its theoretical foundations date back more than 100 years to Maxwell [1] and Mitchell [2]. It was the lightweight design of the airframes which made it possible to realize technically and economically relevant flight distances. But at the beginning of the 20th century lightweight design was also applied beside aviation in constructions where a shortage of material occurred. Pioneering examples of lightweight architecture were provided by Sukov [3] who was the builder of large pylons and transmission towers in Russia.

The traditional lightweight design develops its potential when loads have to be carried over a certain distance with a minimum of material application. If F is a single load or p a load per unit length or q a shear load per unit length and l the distance over which the load has to be carried, a characteristic structure parameter can be defined as follows:

$$K \equiv F/l^2 \ oder \ K \equiv p/l \ oder \ K \equiv q/l$$

The smaller this characteristic structure parameter K is, i.e. the slender the structure is, the more importance lightweight design gains. The use of the characteristic structure parameter allows the optimization of lightweight structures as shown by Wiedemann [4].

Lightweight design could demonstrate its potential first in the area of lightweight metal alloys through the increasing differentiation of structural parts and their related loading and failure mechanism. Topology optimizations have first

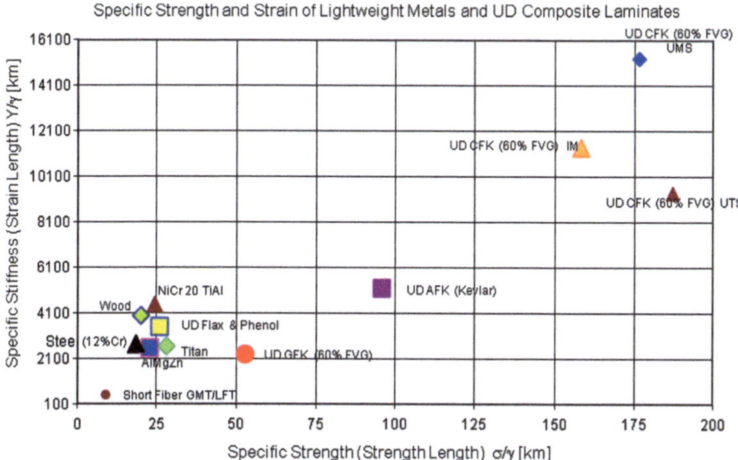

Fig. 1.1 Strength and stiffness of composite materials in comparison to metal alloys and bio materials

been applied successfully by analytical methods [5–8] in order to maximize the usage of the strength of the material.

Defined anisotropy with distinct strength or stiffness oriented in the same direction as the applied loads characterizes lightweight design solutions. Stringer stiffened plates, truss work structures with a maximum normal loading, sandwich structures with high bending stiffness in combination with low core material density are examples of the traditional lightweight design applied to many moving products today.

A further development in lightweight design took place by the introduction of composite materials which show a superior mechanical behaviour with regard to strength and stiffness compared to metal materials. Strength and stiffness are the most important mechanical properties that determine the required amount of structure material. If the yield strength is defined as σ, the Youngs modulus as Y and the material density as ρ, the specific strength R_σ and the specific stiffness R_Y can be taken for the comparison of different structure materials.

$$R_\sigma = \frac{\sigma}{\rho g}; \quad R_Y = \frac{Y}{\rho g}$$

The specific strength can be understood as the maximum length a material (e.g. a wire) can reach hanging down to the ground without a rupture caused by its own weight.

In Fig. 1.1 several composite materials, for example, are compared with lightweight metal materials. While the specific strength of the composites can be eight times higher than the one of lightweight metal alloys, the specific stiffness can reach up to four times higher values. In other words: if the structure has to sustain a certain load in strength a composite structure can be up to eight times lighter than the metal one. If the structure is sized by stiffness requirements, it can be up to four times lighter in composite materials. In practice in most of the cases this potential cannot be reached due to various other constraints, as for example the minimum skin thickness, local load introduction provisions and the possibilities for repair.

On the other hand in addition to the advantages in weight caused by the specific properties of composite materials the specific production process opens up new possibilities for the manufacturing of large complex structure components without extra joints and couplings which after all also cause additional weight.

This benefit in weight is being paid for by the price of the semi-finished products, i.e. fibers and resin, and by the consumption of time and energy and therefore costly production technologies.

Considering the increasing complexity of the material's characteristic parameters, i.e. the material anisotropy, the local complexity of load-material interaction as well as the need for the global analysis of complex structures, the analytical approaches have been replaced by numerical ones of which the finite element method (FEM) is the most popular one [9].

Today's research and the development of primary structures is aiming at achieving the maximum possible lightweight design in metal and—to an increasing extent—in polymer composite materials. Effort is being put into optimizing the structures in detail with respect to strength, stability, durability and damage tolerance by means of FEM.

The relatively high costs per part for lightweight composite structures are justified in aeronautics by the large snow ball effect in weight savings and have been accepted for a long time. But now the demand for composite material application in ground based transport systems (cars, trams, trains) is steadily increasing despite the higher costs and the smaller snow ball effects because of the growing demand to save energy in all technical fields.

The development in weight saving within primary structures, however, has been counter balanced in many cases by an increasing number of additional functions incorporated into the product with the consequence of a further weight increase. Sometimes this increase in weight is caused by more stringent safety requirements, sometimes by requested improvements in comfort. The total weight of the famous VW Golf, for example, has been increased due to safety and comfort requirements by 364 kg or 45% from 1974 to 2008 without increasing the seat capacity [10]. It shall be mentioned that this effect has partially been balanced by improving the efficiency of the motor and by reducing the fuel consumption by about 30%.

Even apart from this example it can be stated that for technical products the energy consumption has increased over the last decades despite the enormous improvements in lightweight design and the efficiency of the motors. This is mainly due to the increased number of functions incorporated into the technical products caused by the increasing safety and comfort requirements as already mentioned above.

A further step in the direction of a consequent lightweight design is the consideration of all these additional functions and their related material with the

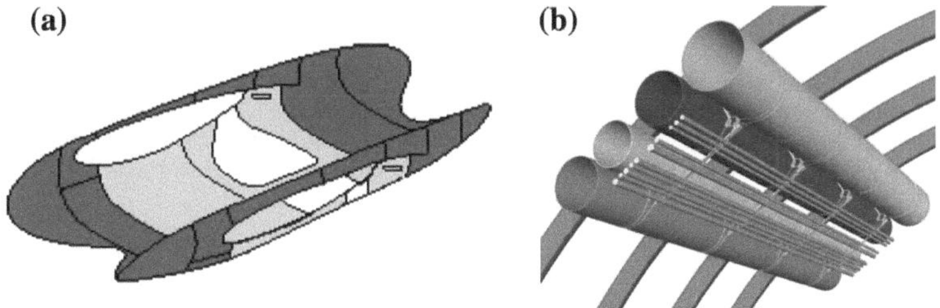

Fig. 1.2 a Secondary structure belly fairing A320. **b** Cabin piping

guiding idea that the more all the material which is required by further functions can be incorporated into the load carrying task of the primary structure, the more weight can be saved. Isolated product functions can be re-considered with respect to their capability of carrying a certain part of the load in order to further minimize the total weight of the product. In principle there are two approaches to be taken:

1. One can make distinct use of secondary masses to carry the product loads.
2. One can integrate as many of the additional functionalities as possible into the primary structure.

If secondary masses are used to carry the product loads, questions concerning the assembly of parts, regarding maintenance, inspectability, comfort, safety and often also certification have to be considered and answered. Secondary structures are frequently used as fairing and protective means in damage prone areas. The questions regarding their load carrying capability in case of damage have to be answered. Solutions have to be found to replace such damaged parts with reasonable effort. Secondary structures protect against surrounding conditions, for example passengers against low temperatures or pressures, against strong airflow or vibrations. The question is how many of these protective functions can be maintained, if the secondary structure is more integrated into the primary one. Load bearing joints must be designed without any tolerances and must be durable over the entire operational lifetime of the product but have to be repeatedly detachable at the same time. This requirement seems to be achievable. New assembly technologies are currently under development as for example detachable bonded joints.

Sometimes the use of secondary structures in the load carrying task of a primary structure requires the incorporation of a load-monitoring system [11] that indicates the potential loss of the load carrying capability. The design has to be made in such a way that this loss can be accepted temporarily which results in some limitations in the operational handling.

In the airplane fuselage for example a combination of the classic primary structure with the cabin lining and the design of a double shell structure could be a very efficient means to create a stiff structure against global and local bending. Large aerodynamic fairings, for example in the intersection between the fuselage and the wing, can be considered and designed so that they partially carry global loads, Fig. 1.2a. Similarly media routes possess the ability to release the primary

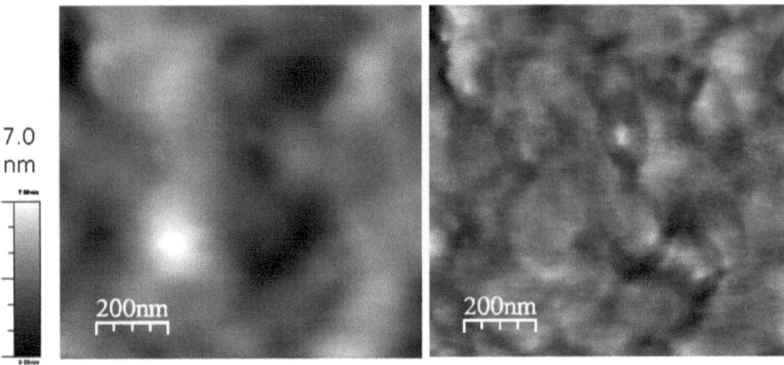

Fig. 1.3 Contrast of storage modulus of resin matrix reinforced by nanoparticles (*left* topography, *right* real part amplitude)

structure partially and therefore to contribute to the reduction of the total weight, Fig. 1.2b.

In car design the combination of primary and secondary structures is sometimes already realized, for example by the integration of the windshield front panel into the car structure.

The approach which the scientific articles of this book refer to is the second possible one, that is to integrate as many of the traditional non load carrying functions as possible into the primary structure itself and thus to use as much of the material as possible to carry the operational loads directly.

But the direct integration of further product functions into the primary structure causes a distortion in the concerned load paths. In many cases the integration of the function is possible through the combination of different materials that each have their specific properties and create a load carrying structure together. Already the composite structure is such a combination of materials. Composites integrate the high modulus and high strength fibers with the low strength and far less stiff resin system in order to finally stabilize the shape of the parts and keep the fibers in contour. Further materials can incorporate additional functions into the composite structure at different scales. On the smallest scale nano particles for example can improve the FST-properties (fire, smoke, toxicity) significantly as well as the compressive strength or fracture toughness of composites. An example of nano-technology research is given in Fig. 1.3 taken from Chap. 3 of this book. By scanning the surface of a nano particle reinforced specimen with an electron microscope the picture visualises the contrast of the complex storage modulus of integrated and functionalized nano particles made of Böhmite material. The stiffer areas are the brighter ones whereas in the surrounding area the grayer regions of the matrix material show a less stiff behaviour.

On the next scale level energy transforming functional materials like for example piezoelectric ceramics can bring sensory and actuator capabilities into the composite structure. But in many cases such combinations of functions cause local discontinuities in primary structural parameters as for example in density or in stiffness and strength.

A certain challenge in the design of composites with integrated functions is to ensure conformity between the material partners which are brought together in the primary structure and to minimize any detrimental effect with respect to the load carrying capability.

In the nineties of the last century the terms of active functional design, of function integration and the science of adaptronics were established [12–16]. In adaptronics the behaviour of load carrying structures is being influenced or adapted by means of energy transforming materials, often also called smart materials. Janocha [14] defines adaptronics as follows: Adaptronic structures are based upon a technology paradigm: the integration of actuators, sensors, and controls with a material or structural component.

Before the foundation of adaptronics many individual observations of special materials took place, sometimes more than four decades ago, like for example the piezoelectric behaviour of some ceramic materials [17] or the memory effect of some metal alloys [18]. In the eighties this collection of individual phenomena was more systematically analysed in the military laboratories of the US and the materials under consideration got the collective designation "smart materials" [19].

Piezoelectric ceramics for example can transform electric energy into mechanical energy and vice versa at a very high frequency, i.e. very fast. The effect is caused by the local polarization of the ceramic lattice structure, which is linked with a geometric asymmetry. The rhombohedral or tetragonal lattice structure can be oriented along an applied electric field with the effect of mechanical lengthening. Vice versa a mechanical deformation causes an electric voltage displacement that can be used to create a (small) flow of current.

Smart materials and the functions that can be realized with them are preferably integrated in composite structures which do not only possess superior mechanical properties in lightweight design but are manufactured in a specific production process. The combination of fibers and resin with the following curing process in a mould offers the advantage to integrate smart materials in the same shot. This has a great advantage over using metal lightweight structures, where the integration of either active or passive additional functions often requires an additional assembly step and extra mechanical fastening or bonding.

If piezoelectric ceramics can be brought into suitable conditions for integration, for example as piezoelectric patches, they can be perfectly integrated into a composite structure. These piezoelectric patches allow the integration of an active function into the composite primary structure. Figure 1.4 shows several possibilities how to integrate such patches and the effect of the various solutions to the mechanical parameters of strength, stiffness and yield strain.

The mechanical properties of the integration of the patch with the partial interruption of fibers (2nd to the left in the diagrams) but with a minimum of additional eccentricity are close to the properties of a later bonding of the patches to the undisturbed structure.

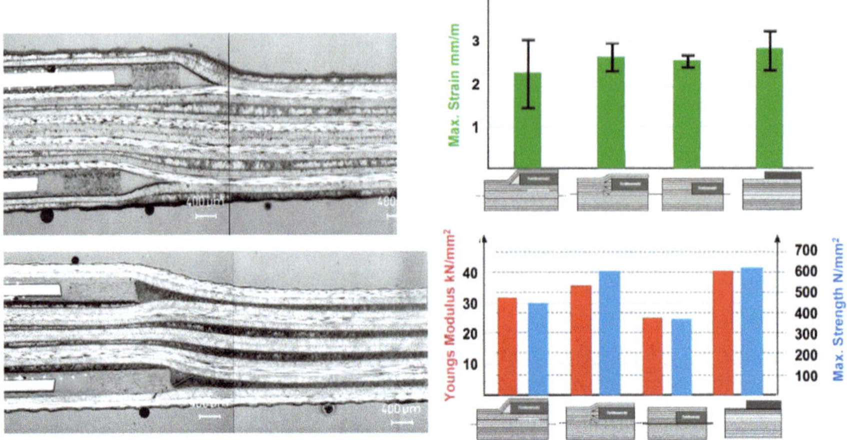

Fig. 1.4 Different types of the integration of piezoelectric foil into composite laminates and related mechanical properties

The potential of saving weight by the integration of additional functions into the primary structure works out considerably better in composite materials than into metal ones because of the integration process of the components fiber and resin.

As a consequence the disadvantage of the consumption of time and energy in the production process of composites can be turned into an advantage when it is used for the integration of a function at the same time. Functions or their carriers can now be integrated directly into the composite structure whereas they had to be applied to metal structures up to now within extra assembly steps and with brackets and fasteners which again add extra weight to the product.

The avoidance of extra production steps and assembly effort, the minimization of single parts and of the required time for assembly furthermore saves costs. In the ideal case this compensates for the extra costs caused by the pre-fabrics fiber and resin and the production of the composite structure.

The integration of functions into primary structures is often already realized, if passive functions are considered. Beside its mechanical properties each material owns further ones that are used implicitly or explicitly to ensure additional functions. Metals serve as a conductor for electrical grounding or for electro-magnetic shielding whereas composites with a relatively low thermal conductivity of the resin part can serve for thermal isolation and the high chemical resistivity helps to deal with liquid media. A well-known additional function of primary structures is the capability to absorb crash energy, which can be designed in both metal and composite parts.

Often passive functions can be integrated into the composite structures much better than in metal ones. The integration of media routes is possible as well as the integration of lighting or thermal functions, for example for local heating. Infor-mation routes can be incorporated to replace extra wiring and the possibilities of improvement which can be achieved by nano particles dispersed in the resin component have already been mentioned.

Protective layers can be incorporated and material properties varying in thickness direction can be realized in composites. A proper tolerance management can be used to ensure a required surface quality which is better than the one of metal structures due to the high ability to integrate parts and due to the higher stiffness leading to less deformation when being loaded. These features help to ensure a natural laminar flow which is much easier to realize and lighter than the alternative active flow control.

Further possibilities are offered by the integration of active functions into the composite primary structure as it can be done with smart materials. Active functions usually are those that help the structure to adapt itself to changing environmental conditions. Examples of such structures which are designed to adapt to changing operational conditions are the high lift devices on the wings of an airplane. These devices are used to adapt the lift and drag coefficients of the wing to the changing air speed and angle of attack during start, cruise and landing phases of the aircraft. Today the high lift devices are a combination of more or less rigid structural parts that are actuated by discrete mechanisms, e.g. links, drive shafts and hydraulic or electric motors. If the high lift devices can change their shape with the help of a structure integrated actuation—this capability is often called morphing—the discrete mechanism can at least partially be saved and thus some of the weight associated with them [20–22]. Wear and tear of the mechanical actuation systems might be reduced and—what is much more important—gapless surfaces can be realized that minimize the effect of stall and therefore minimize the acoustic emission and the loss of aerodynamic performance.

The contradiction between the need to create a strong and stiff structure with respect to the external load to be carried and the required flexibility for actuation can be overcome. An example of such a gapless, strong but also flexible structure that can be actuated actively, e.g. a morphing structure, will be given in Chap. 31 in this book.

To reduce the vibration of rotating structures or those excited by their surroundings in a large frequency range and to decouple them from the rest of the primary structure passive damper-spring combinations or active hydraulic dampers are used today. Also their additional masses and weight can be reduced partially if it can be achieved to integrate active vibration control and means for the reduction of vibrations directly into the primary structure as it has been shown for example by Schütze et al. [23]. Another example of the capabilities of structure integrated active vibration control is given in Chap. 33 within this book.

Structure vibrations in the higher frequency range cause sound emissions, which depending on their intensity lead to comfort, safety or even health problems for passengers in an aircraft or another vehicle. The protection against sound emission today is in almost all cases done by passive means with a remarkable additional weight, e.g. by damping materials.

If the structure can be built in such a way, that incorporated active functions suppress the sound emissions effectively, a lot of additional weight can be saved,

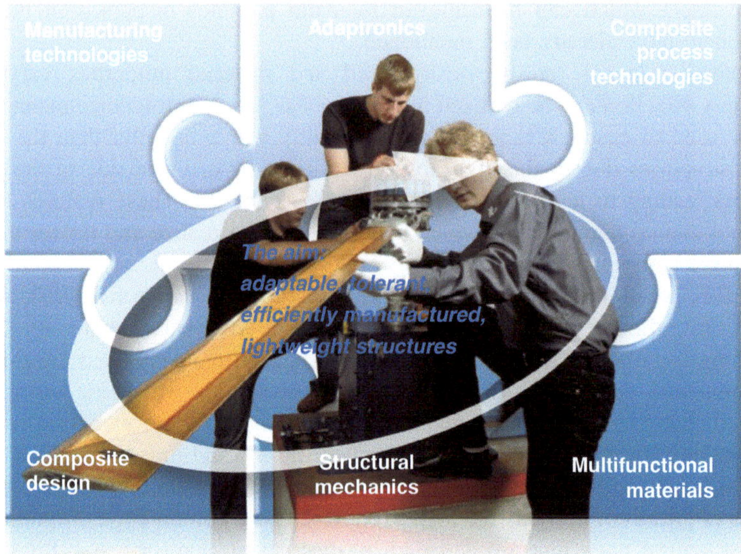

Fig. 1.5 Competencies required to realize adaptive, efficient and tolerant composite structures. *Example* adaptive composite rotor blade

as for example Weyer et al. [24] have demonstrated with a vehicle structure. Some more examples can be found in Chaps. 34 and 35.

In all of these examples the integration of functions aims to adapt the primary structure actively to the operational conditions the product is working in, i.e. to create adaptive composite structures.

An efficient integration of functions must deal with special challenges. The advantages of today's separation of functions and the combination of individual single components in one common product are the exchangeability and the possibility to replace single parts in case of malfunction. Individual components and functions have their individual failure probability and their individual reliability and operational lifetime. An efficient integration of functions is possible, if all the integrated functions possess the same reliability including the load carrying one to which they are exposed. To ensure a proper prediction of reliability and to develop the design of the primary structure with integrated functions a detailed knowledge of all the properties of the contributing constituents and their interactions is mandatory.

For the development of adaptive, efficient and tolerant composite structures a series of competencies have to be considered as being linked closely together, forming a multidisciplinary process chain. This process chain is elaborated on in detail within this book.

Figure 1.5 shows the competencies that are required to realize adaptive, efficient and tolerant composite structures and gives an example of the application of the smart materials mentioned above. A down scaled helicopter rotor blade is shown, which is actively twisted by applied piezoelectric patches in the upper and lower blade surface. More details about the active twist rotor are given in Chap. 32.

In the research field of **multifunctional materials** the basic properties of the composite constituents, e.g. fibers, resin systems and additives are investigated and their interactions are characterized in detail. In addition smart materials are explored and their potential is being analyzed with regard to the maximum efficient deployment in composite structures. The question raised repeatedly is how material properties found in the nano scale can be brought into a technically relevant macro scale and with what kind of treatment or technical process their capabilities can be used for the integration of functions. **Nano-Micro–Macro** is therefore the title of the initial chapter of the first part of this book which comprises chapters with exemplary results from research in the field of composite semi-finished components and smart materials.

The **structural mechanics** is the field of research that provides sizing methods for new composite structure design concepts and production technologies. In this domain also all simulation methods for the production processes are developed and the impact of manufacturing conditions on the final structure is evaluated, for example the so called spring-in effect, which is caused by the shrinkage of the resin during the curing process and the resulting induced stresses. The load carrying capability of the structure is assessed in structural mechanics. With the help of a detailed test pyramid and the associated validation-verification strategy the agreement of analytical and numerical prediction methods with the true structure behaviour is secured. Reliable design methods are mandatory, but the challenge is to ensure reliability—or robustness—when adding additional functions into the structure. There is the need to bring all of the elements of such a structure onto the same level of reliability. **Validation Approach for Robust Primary Structures** is therefore the title of the first chapter of the second part in which the following chapters will give a survey about sizing methods for composites as well as further examples for new methods developed for better simulation and testing. This part will also give some consideration to a closed process chain for simulation in the field of composite structures.

A central role within the whole development process of adaptive, efficient and tolerant composite structures is given to the research field of **composite design**. Here design solutions have to be developed that on the one hand satisfy the requirements of the load carrying primary structure but that on the other hand allow the aimed integration of functions. The answers must comply with the specific composite production constraints and ensure the required quality of the parts. The opportunities of composite structures in design and the boundary conditions to be observed for the successful integration of functions are described in the third part entitled **Compliant aggregation of functionalities**. The first chapter gives an overview of the questions to be answered and the challenges which the design is exposed when realizing adaptive and efficient composite structures.

Without a complete knowledge of the production technologies and the capability to control and steer the related processes, the efficient integration of functions into composite structures will not be successful. The domain of **composite**

technologies therefore is the one that deals with the comprehensive research in all aspects of the interaction of production process parameters with the quality of parts. At the same time a minimum of energy and time consumption in production is a must to ensure the competitiveness with metal structures. A maximum of process automation is aimed at in order to minimize the drawback of the high amount of manual work that is still common in the production of composite structures. A large variety of technologies exists and their specific process parameters are being explored with respect to their impact on the composite parts. Sensing measures are under development to monitor the processes on line. The concept behind is the development of self-controlled and furthermore self-correcting processes. This is why the first chapter of the fourth part is entitled **Self Controlled Composite Processing**. The chapters published in this part give examples of research results derived in the field of composite technologies in order to improve the control and the steering of the production process of efficient and tolerant composite structures.

Especially the integration of active functions into the composite structures is the research field of **adaptronics**. Here the concepts and the technologies are developed for the integration of active functions such as morphing, active vibration control (AVC), active acoustic structural control (ASAC) and structural health monitoring (SHM). The basics are investigated for the efficient realization of such functions with the help of smart materials. Steering means and control algorithms are being developed here as well as the adaptability to different operational conditions of the final product. The integration of active functions still requires an external power supply that makes the adaptive composite structures difficult to handle and requires a certain effort in joining the parts and the assembly of the primary structure. Some smart material cannot only transform electric into mechanical energy, but also harvest electric energy from mechanical deflections. With this capability structures can be built collecting the energy for their active functions directly from their mechanical loading. This is the idea behind the development of a new type of smart composite structures and therefore the title **Autonomous CFRP Structures** has been chosen for the first chapter of the fifth part. In the chapters collected in this part research results are being presented that give an inside view to various aspects of the integration of active functions into the composite structures.

This book is supposed to elucidate and to illustrate some results of research that demonstrate the conditions and the potentials of adaptive, efficient and tolerant composite structures. The large variety of possible applications and the numerous questions that need to be answered to their realization only allow a selection of research data, methods and results. This collection of chapters is meant to underline how the close cooperation and interaction between the technical disciplines of material development, structure mechanics, composite design, composite technologies and adaptronics can generate a new class of lighter primary structures which can be characterized as being adaptive and efficient by the maximum integration of active and passive functions.

These new composite structures with integrated additional functions are lighter than traditional primary structures with a comparable scope of functions. The relatively high costs of their semi-finished materials can be compensated by the high degree of integration of single parts and the reduced effort in assembly.

It is our conviction that the interdisciplinary research as outlined in this book is the chance to move on towards a resource efficient mobility using adaptive, efficient and tolerant composite structures.

References

1. Maxwell, C.: Scientific Papers II. Cambridge University Press, Cambridge (1869)
2. Michell, A.G.M.: The limit of economy of materials in frame structures. Philos. Mag. Ser. **8**, 589–597 (1904)
3. Suchov, V.G.: Die Kunst der sparsamen Konstruktion, bearbeitet von R. Graefe. Deutsche Verlags-Anstalt, Stuttgart (1990)
4. Wiedemann, J.: Leichtbau 2: Konstruktion. Springer-Verlag, Berlin-Heidelberg-New York (1996)
5. Hertel, H.: Leichtbau. Springer-Verlag, Berlin (1960). (Reprint 1980)
6. Timoshenko, S.P., Woinowsky-Krieger, S.: Theory of Plates and Shells. McGraw-Hill, NY (1959)
7. Pflüger, A.: Stabilitätsprobleme der Elastostatik. Springer-Verlag, Berlin (1975)
8. Kossira, H.: Grundlagen des Leichtbaus. Springer Verlag, Berlin-Heidelberg, NY (1996)
9. Knothe, K., Wessels, H.: Finite Element. Springer Verlag, Berlin-Heidelberg, NY (1991)
10. Drechsler, K.: CFK technologie im Automobilbau—Was man von anderen Märkten lernen kann. CCe.V. automotive symposium, Neckarsulm (2010)
11. Giurgiutiu, V., et al.: Active sensors for health monitoring of aging aerospace structures. University of South Carolina, Mechanical Engineering Department, SC 29208
12. Elspass, W.J., Flemming, M.: Aktive Funktionsbauweisen. Springer-Verlag, Berlin (1998)
13. Neumann, D.: Bausteine "Intelligenter Technik von Morgen—Funktionswerkstoffe in der Adaptronik. Wissenschaftliche Buchgesellschaft Darmstadt (1995)
14. Janocha, H., et al.: Adaptronics and Smart Structures. Springer, Berlin-Heidelberg, NY (1999). (ISBN 3-540-61484-2)
15. Jenditza, D., et al.: Technischer Einsatz Neuer Aktoren. Expert Verlag, Renningen-Malmsheim (1998). (ISBN 3-8169-1589-2)
16. Guran, A., et al.: Structronic Systems: Smart Structures, Devices and Systems. World Scientific, Singapore, New Jersey, London, Hong Kong (1998). (ISBN 981-02-2955-0)
17. Jaffe, B., Roth, R.S., Marzullo, S.: Piezoelectreic properties of lead zirconate-lead titanate solid solutiuon ceramics. J. Apll. Phys. **25**, 809–810 (1954)
18. Bank, R.: Shape Memory Effects in Alloys, p. 537. Plenum, NY (1997)
19. Thomson, B.S., Gandhi, M.V.: Smart Materials and Structures Technologies. An Intelligence Report. Technomic Publishing Company, Lancaster (1990)
20. Lu, K.J., Kota, S.: Design of compliant mechanisms for morphing structural shapes. J. Intell. Mater. Syst. Struct. **14**(6), 379–391 (2003)
21. Nagel, B., Monner, H.P., Breitbach, E.: Aerolastic tailoring transsonischer tragflü-gel auf basis anisotroper und aktiver strukturen. Deutscher Luft- und Raumfahrt-kongress, Dresden (2004)
22. Campanile, L.F., Carli, V., Sachau, D.: Adaptive wing model for wind channel tests. In: Paper presented at the RTO AVT symposium on "active control technology for enhanced

performance operational capabilities of military aircraft, land vehicles and sea vehicles", Braunschweig, Germany, and published in RTO MP-051, 8–11 May 2000
23. Schütze, R., Goetting, H.C., Breitbach, E., Grützmacher, T.: Lightweight engine mounting based on adaptive CFRP struts for active vibration suppression. Aerosp. Sci. Technol. **6**, 381–390 (1998)
24. Tom, W., Elmar, B., Olaf H.: Self-tuning active electromechanical absorbers for tonal noise reduction of a car roof. In: Proceedings of inter-noise 2007 Istanbul, the 36th International Congress & Exhibition on Noise Control Engineering (Ber.-Nr:in07_220), Inter-Noise 2007, 2007-08-28–2007-08-31, Istanbul (Turkey) (2007)

Part I
Multifunctional Materials

Chapter 2
Nano-Micro-Macro

Peter Wierach

Abstract New materials with superior properties are the basis to exceed existing technological barriers and to explore new fields of application. Especially composites as multiphase materials offer the possibility to influence their properties or to add even new functionalities by a proper choice and combination of the different phases. In this context it is of particular importance to understand the interactions between the different material phases. This includes for example the effect of nanoscale additives in resins as well as the effect of microscopic manufacturing defects, like pores, on the macroscopic material properties. A systematic material design is only possible if cause and effect on the different material scales is well understood. Nanotechnology gives the opportunity to manipulate the structure of materials on a level, allowing to realize properties and functionalities that can't be achieved with conventional methods. Beside the improvement of mechanical, thermal, optical and electrical properties, the incorporation of new "smart materials" on a technical relevant scale is in the focus of our research.

2.1 Looking at Composite Materials at Different Scales

In literature it is common to distinguish four characteristics scales to describe the hierarchical constitution of materials [1]:

- The nanoscopic scale with a characteristic length of 10^{-9} m or a few nanometers,
- the microscopic scale with a characteristic length of 10^{-6} m or a few micrometers,

P. Wierach (✉)
DLR, Institute of Composite Structures and Adaptive Systems,
Lilienthalplatz 7, 38108, Braunschweig, Germany
e-mail: peter.wierach@dlr.de

M. Wiedemann and M. Sinapius (eds.), *Adaptive, Tolerant and Efficient Composite Structures*, Research Topics in Aerospace, DOI: 10.1007/978-3-642-29190-6_2,
© Springer-Verlag Berlin Heidelberg 2013

- the mesoscopic scale with a characteristic length of 10^{-4} m or hundreds of micrometers,
- the macroscopic scale with a characteristic length of 10^{-2} m or centimetres and more.

An engineer, designing a fiber composite structure, is interested in properties describing the macroscopic behaviour of a certain material. Especially mechanical properties like stiffness and strength are required for the sizing of structures. Typically these properties are experimentally determined on a coupon level. A fiber composite material is build up by several layers of fiber materials oriented in different directions to carry the occurring loads. This mesoscopic scale describes the fiber architecture of the composite material. Also larger defects, like fiber undulations and pores, are in this order of magnitude. On a microscopic scale we look at a single fiber and the surrounding resin. A carbon fiber for example has a typical diameter of 7 μm. The behaviour of materials is finally determined by their smallest constituents, the atoms and molecules, and the way they interact among each other. On this scale also nano-sized particles and their interactions with the polymer are addressed.

2.2 Improving Fiber Composite Materials with Nanoscaled Particles

The modification of polymers with particles is a well known technology to adjust their properties. Depending on the particle size, shape, mechanical properties etc. attributes like viscosity, toughness, hardness, flammability can be affected to name a few. In the last decade the use of nano scaled particles came more and more into focus. At least one spatial direction of the filler material has to be in the order of nanometers to be designated as nano filler material. Although nano fillers have been used before, the current research aims to understand the effects that are related by adding particles of this scale to allow a systematic material development. There are several reasons why it is interesting to look at nanoscaled particles.

Due to the very small size of the individual particles the interphases between the particles and the surrounding polymer are extreme high. As an example the interphase area in a cubical volume of 1 mm^3 that is filled with 10% spherical particles, having a diameter of 1 μm, is 600 mm^2. In comparison to that the interphase area of particles with a diameter of 1 nm in the same volume with the same particle volume content is 600.000 mm^2. As a result a large portion of the polymer accumulates in the polymer/particle interphase. Assuming a good bonding between the particles and the polymer a relatively small fraction of particles can have a great effect on the properties of the polymer.

For fiber composite materials, where not only the neat resin is considered, nanoscaled particles have another important advantage, which is related to the

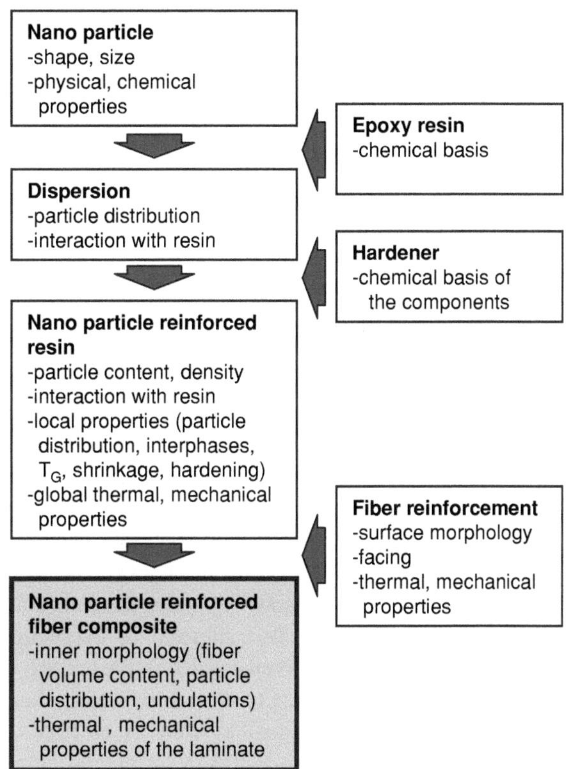

Fig. 2.1 Schematic drawing of parameters and influencing variables for nano particle modified fiber composites [3]

manufacturing process of fiber composites. Currently most high quality continuous fiber reinforced composites (fiber volume content around 60%) are manufactured with the prepreg method. Due to rising production costs, liquid resin infusion (LRI) methods are a promising cost effective alternative [2]. Quite a few different LRI processes have been developed so far. In all processes the liquid resin is infiltrated into the dry fiber material at process specific temperature- and pressure profiles. If a particle modified resin is used, filter effects can occur during the infiltration process, leading to particle density gradients in the composite material. With nanoscaled particles these effects can be minimized enabling a homogeneous particle distribution even in the consolidated fiber composite material.

Another issue in this context is the viscosity of the modified resin. LRI processes require a low viscosity of the resin. Otherwise the flow resistance during the infiltration process is to high leading to not infiltrated areas. Also in this case nanoscaled particles offer the chance to avoid this problem since only small particle volume contents are needed to get an effect.

Figure 2.1 illustrates the principle way from the nano particle to the reinforced fiber composite material and lists the most important parameters [3]. It is obvious that a great number of variables influence the fabrication process of the nano particle reinforced fiber composite material.

2.3 Smart Material Systems

The ideal basic material for "smart structures" possesses both actuator and sensor characteristics as well as load-bearing capabilities. Unfortunately, there is yet no existing material known that unites all of these characteristics in a sufficient manner. The assembly of a multifunctional material system is therefore realized by combining actuator and sensor materials with a primarily load-bearing component. Fiber composites are well-suited for the purpose of assembling a smart structure, since the active components can be inserted during the production process and, thus, become an inherent part of the structure. Additionally, the specific requirements and characteristics of active materials can be accounted for by the several production options for composites such as selection of the fiber/matrix material and the layup of the structure. Hence, the smart structure is designed and manufactured as a whole with embedded smart components including its supporting infrastructure of, e.g., lead wires, electrodes and terminals for power supply.

Sensors and actuators based on multifunctional materials are a substantial component of smart composite structures. Such multifunctional materials also called "smart" or "intelligent", are energy converters or transducers that respond in a technically usable manner to an external stimulus. The most widely employed types respond to an electric, thermal or magnetic field with a change in their mechanical properties. Well-known representatives are piezoelectric materials (load/deformation response to an electric field), shape memory alloys (temperature dependent load/deformation) as well as electro- and magnetorheological fluids (influence of shear transmission in an electrical or magnetic field respectively). Typically, the underlying actuation mechanism is caused by a microscopic reconfiguration in the material and operates in both directions. A change in the mechanical characteristics due to external loads can be detected and, thus, allows for sensor use also.

The reliable subsequent treatment and structural integration of the usually very sensitive materials is however, connected with some complexity and risk. The goal of our research activities is therefore to develop multifunctional material systems to enable the setup of reliable and reproducible smart structures that can be successfully applied in industrial production.

2.4 Integration of Smart Materials on a Macroscopic Level

Piezoceramic materials are very often used to fabricate smart composite structures. Figure 2.2 shows typical piezoceramic materials used for this purpose. The main reasons for the popularity of piezoceramics are their ability to operate at high frequencies and their stiffness, typically 60 GPa, matching the stiffness of many composite materials. The latter is of special importance for the usage as actuator. In the past quite a few investigations have been made to study the integration of

Fig. 2.2 Macroscopic piezoceramic fibers ($\varnothing$200 µm) and plates ($50 \times 30 \times 0.2$ mm^3) that are typically used for the fabrication of smart composite structures

piezoceramic components into composite materials [4, 5]. In the following an exemplary case will be presented to discuss some issues related to the integration of components of this size into composite materials [6].

Here a piezocomposite with a monolithic piezoceramic plate with dimensions of $50 \times 25 \times 0.2$ mm is considered (also referred to as transducer). The external dimensions of the piezocomposite transducer, including insulation and electrical contacts, were $58 \times 29 \times 0.47$ mm^3. At this point the detailed transducer design is not of special interest. The transducers were integrated into a 16 layer laminate with quasi isotropic ply lay up $[(0/+45/90/-45)_2]_S$. The specimens were manufactured using the Differential Pressure Resin Transfer Moulding Process (DP-RTM) with unidirectional fiber material having an area weight of 125 g/m^2 (type UD-CST 125/300 from SGL). With a fiber volume content of 60% this result in a theoretical thickness of 0.12 mm for each ply. The fiber material was infiltrated with the three component resin system LY556/HY917/DY070 at 6 bar and 120°C. To enable a symmetrical specimen design two transducers were embedded in opposite positions.

Three different integration configurations were investigated as depicted in Fig. 2.3 According to the thickness of the transducers, in configuration (a) four layers of the prepreg material were cut out to accommodate the transducer. In configuration (b) no plies were cut out resulting in a thickened specimen and resin rich pockets at the edges of the transducer. A partial cut out leaving the plies in 0° direction intact was realized in configuration (c). As a reference additional specimen with bonded transducers were manufactured.

Figure 2.4 shows the final specimen design. The specimens were loaded in a tensile testing machine until failure. To measure the specimen deformation, strain gauges were attached inside (section 1) and outside (section 2) the integration area. An exemplary test result with a specimen configuration with no cut out (configuration b) is shown in Fig. 2.5. The stress was calculated with respect to the cross section area in each section and the test load. The gradient of the stress/strain curve describes the Young's modulus in the considered cross section. In addition to that the sensor signal of the piezoelectric transducer was measured during the test (red curve). Discontinuities in the sensor signal would indicate a crack in the

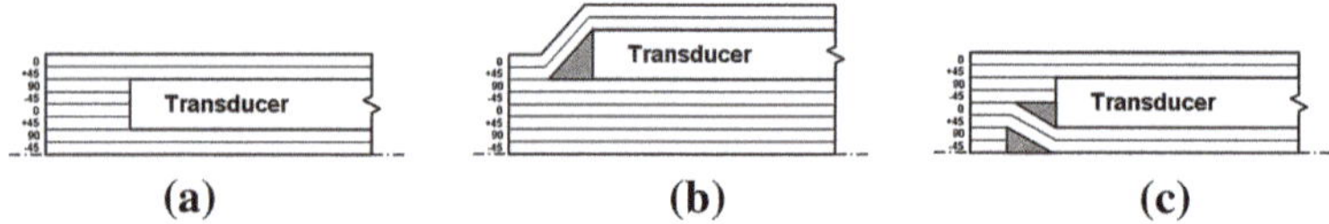

(a)   **(b)**   **(c)**

Fig. 2.3 Different integration configurations for macroscopic piezoelectric transducers. **a** Complete cut out. **b** No cut out. **c** Partial cut out

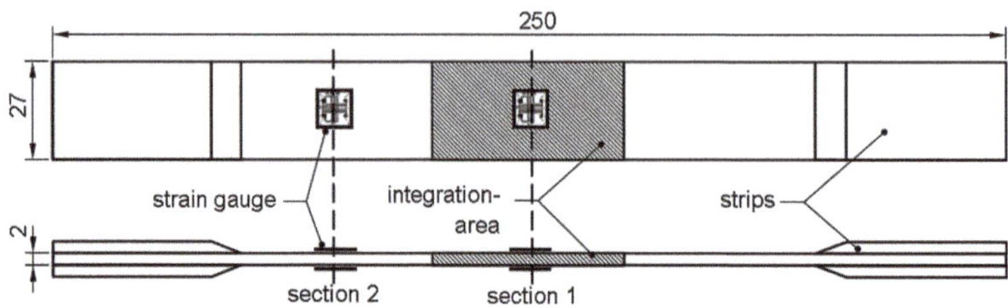

Fig. 2.4 Specimen design with embedded piezoelectric transducers

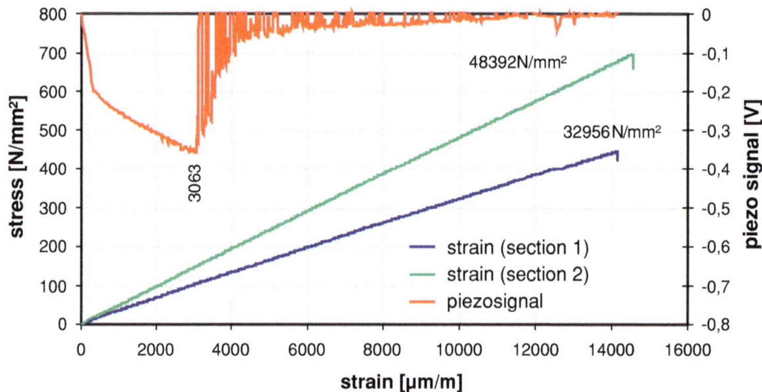

Fig. 2.5 Exemplary result of a tensile test with a specimen of configuration (b) no cut out

ceramic. Looking at the sensor signal in Fig. 2.5, first cracks are initiated at a strain of approximately 0.3% whereas the failure of the complete structure occurs at much higher strain levels. The piezoceramic fragments still generate signals but the performance is decreased. Nevertheless in comparison to very brittle plain ceramics, the strength of the embedded ceramic material is improved. Reasons for the higher strength are a mechanical pre compression in the ceramic due to higher thermal shrinkage of the composite material when cooled down from the curing temperature and the stoppage of crack propagation due to the embedding in a more ductile material.

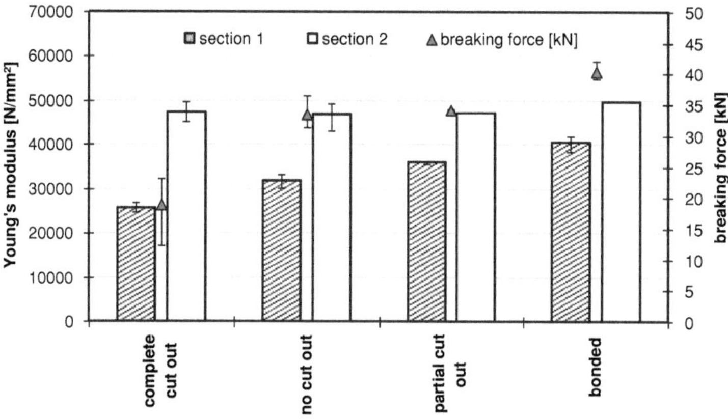

Fig. 2.6 Results of the tensile test with integrated piezoceramic transducer

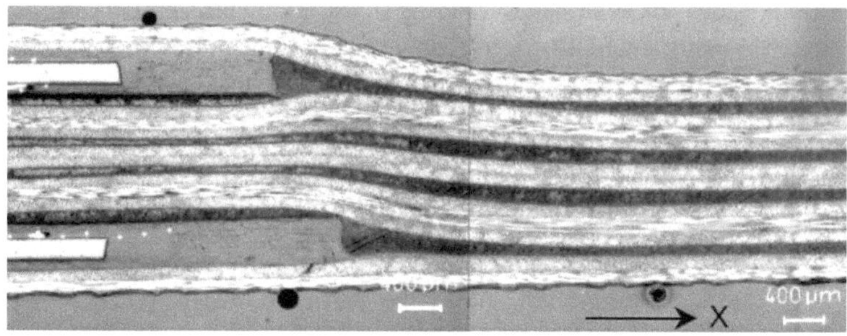

Fig. 2.7 Micrograph of a specimen cross section an embedded piezoelectric transducers (configuration with no cut out)

A summary of all test results is given in Fig. 2.6. At least 6 specimens were investigated for each configuration. The diagram depicts the breaking force and the Young's Modulus for each cross section. With respect to the reference specimen, for all configurations a clear reduction of the breaking force and Young's Modulus is observed for the cross section with the integrated transducer. As expected configuration (a) with a complete cut out shows the worst results. Best results are achieved with configuration (c) with a partial cut out. Even the configuration with no cut out is worse. Due to the large increase in thickness of the laminate in the integration area, large resin rich pockets are formed, resulting in high stress peaks in this area (Fig. 2.7). During loading the initiation of cracks was observed in the resin pockets actually.

The results show that an integration of smart materials of this macroscopic size in a monolithic composite material is most likely to have a negative effect on the strength of the material. In this example only the tension properties have been considered. It can be expected, that the impact on compression and fatigue properties is even worse. A structural conformable, and in most cases applied configuration, is to bond transducers of this size onto the surface of the material to form the actual multifunctional material system. A good example for a structural conformable integration in this context is to bond the active components on the inside of the face sheets of a sandwich material.

2.5 Integration of Smart Materials on a Micro- and Nanoscopic Level

In the example described before the integration of smart materials into fiber composites was done by relatively large devices on a macroscopic level. A much tighter and more structural conformable way to integrate new functionalities into fiber composite materials can be expected, if the functional components have a similar or even smaller size than the constituent parts of the composite. One possible solution is the use of micro- or even nanoscopic particles. First steps in this direction have been made by using magnetostrictive materials. Some preliminary results of this research will be used to highlight the potential of this approach [7, 8].

Its outstanding magnetostrictive properties make Terfenol-D particles suitable for the use as sensor particles in smart composite materials. In this study the Terfenol particles were purchased from ETREMA (USA) and were received in batches with very different particle sizes ranging from a few micrometers up to 300 μm. To investigate the effect of particle size it is necessary to fractionate Terfenol-D particles. Figure 2.8 shows the particle composition of the original batch and after a sieving process to select particles with a size below 20 μm.

A homogeneous dispersion of the particles in the epoxy resin is fundamental to fabricate specimen with a reproducible quality. Master batches of epoxy resin with Terfenol-D particles were produced with a triple roll mill and with a basket mill. The resin used was the three component system LY556, HY917, HY070. To investigate the influence of particle size and particle content several master batches with different compositions were produced.

Since Terfenol-D has a very high density of 9.25 g/cm^3, especially in comparison to epoxy resins (1.2 g/cm^3), the manufacturing of specimen with a uniform particle distribution is a very challenging task. To avoid sedimentation a suitable solution is to use a magnetic field during the curing procedure. By this measure the Terfenol-D particle form chain structures oriented along the magnetic flux lines (Fig. 2.9).

The epoxy-Terfenol-D composite coupons were produced using a small casting mould. During casting two hard ferrite magnets were placed at the outer surfaces of the mould to generate a magnetic field (Fig. 2.10). To get the magnetic field as homogeneous as possible, all parts of the casting mould, including assembly screws, were made of aluminium. During casting the mould was heated to 80°C. The resin plate was pre-cured for 4 h at 80°C and post-cured for 4 h at 120°C.

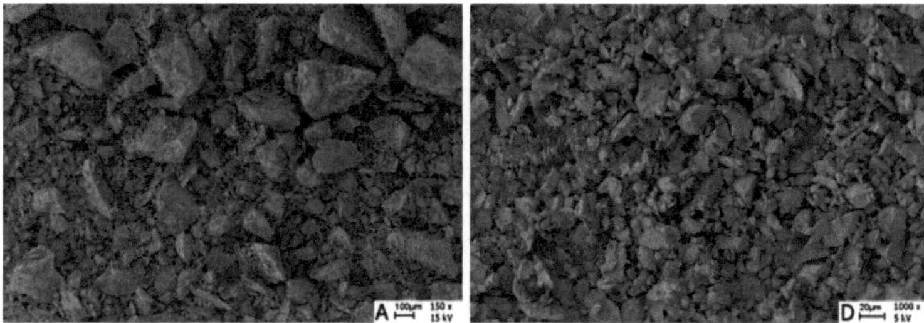

Fig. 2.8 SEM images of Terfonal particles (*left* original batch; *right* particles after sieving with a size <20 μm)

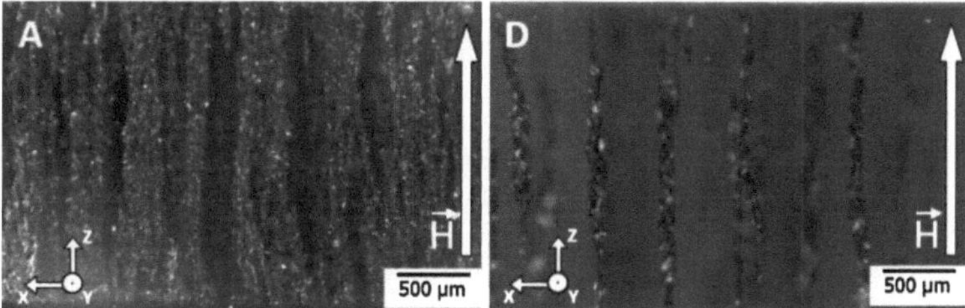

Fig. 2.9 Micrographs of specimen cross sections with Terfenol-D particles of different size and particle content aligned along magnetic flux lines (*left* particle size 15 μm and 20 wt%; *right* particle size <20)

After curing coupons with dimensions of $90 \times 10 \times 2$ mm^3 were cut to size from the plate.

To verify the sensory effect of the Terfenol-D/epoxy composite the coupons were loaded in a universal testing machine. The change of the magnetic flux density against the mechanical stress was measured using a hall-effect sensor, which was applied above the center of the coupon. First measurements showed only little changes in the magnetic flux density. This was significantly improved after performing a post magnetization process with 1.5 T for 30 s using a customized electromagnet. A summary of the test results is shown in Fig. 2.11.

All specimens showed reproducible changes of the magnetic flux density depending on the applied mechanical stress. Besides the particle distribution the particle size as well as the particle concentration had an influence on the magnetic flux density change. Maximum changes were measured with higher particle concentration and lower particle size.

Tough a lot of open question remain the first results are very promising, offering numerous possibilities for future research activities. Up to now only modified neat resins have been investigated. The challenges to integrate them into fiber composites still have to be explored. It is also very interesting to look at alternative smart materials exploiting other physical effects.

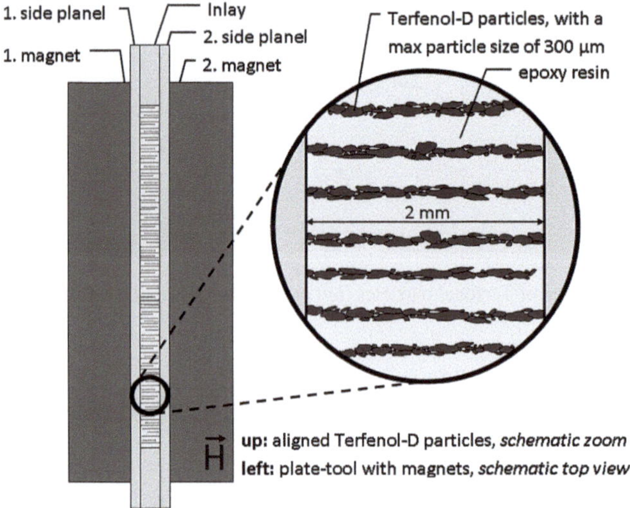

Fig. 2.10 Schematic illustration of the plate tool with magnets and enlarged view of particle alignment

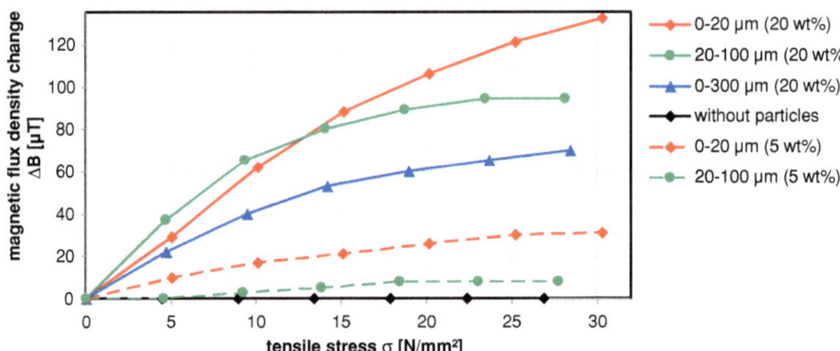

Fig. 2.11 Plots of the measured magnetic flux density change under tensile load

2.6 Summary and Conclusion

The following articles reflect a part of the research work that is dealing with the behavior of composite materials in combination with smart materials on different scales and how this knowledge can be used to enhance the performance by improving mechanical properties, by facilitating the production process of parts and by adding new functionalities.

The improvement of mechanical properties of fiber composites by adding nanoscaled particles materials will be addressed in two articles. Different particle types (epoxy-silica and boehmites) are investigated. Especially matrix dominated properties of fiber composite laminates like shear strength, compression strength and damage tolerance are considered.

Another article discusses the use of carbon nanotubes (CNT's) as a new superior actuator material for smart composite structures. The goal is to transfer both the extraordinary mechanical properties of CNT's and the possibility to actuate them with very low voltages on a macroscopic composite material.

Piezocomposites and biological inspired piezoceramic honeycomb actuators, as reliable, robust and advanced possibilities to form smart material systems with macroscopic components are presented in two further articles.

References

1. Lu, G., Kaxiras, E.: Overview of multiscale simulations of materials. Handb. Theor. Comput. Nanotechnol. **10**, 1–33 (2005)
2. Herbeck, L., Kleineberg, M.: Single line injection–a successful liquid resin infusion technology for commercial aircraft applications. In: International symposium on composite manufacturing for aircraft structures, Flevoland, Netherlands (2002)
3. Arlt, C.: Wirkungsweisen nanoskaliger Böhmite in einem Polymer und seinem Kohlenstofffaserverbund unter Druckbelastung. Dissertation, Otto von Guericke Universität Magdeburg (2011)
4. Dürr, J.K.: Integration von Piezokeramiken in Faserverbundstrukturen. Dissertation, Universität Stuttgart (2006)
5. Shukla, D.R., Vizzini, A.J.: Interlacing for improved performance of laminates with embedded devices. J. Smart Mater. Struct. **6**, 225–229 (1996)
6. Wierach, P.: Entwicklung multifunktionaler Werkstoffsysteme mit piezokeramischen Folien im Leitprojekt Adaptronik. Adaptronic Congress, Wolfsburg, Germany (2003)
7. Kubicka, M.: Untersuchung von magnetostriktiven Partikeln zur Detektion von Eigenspannungen in CFK-Polymermatrices. Diplomarbeit, Universität Magdeburg (2011)
8. Kubicka, M., Mahrholz, T., Sinapius, M.: Magnetostrictive properties of epoxy resins modified with Terfenol-D particles for detection of internal stress in CFRP—Part 1: materials and Processes. J. Mater. Sci. submitted (2012)

Chapter 3
Piezocomposite Transducers for Adaptive Structures

Peter Wierach

Abstract Low profile actuators are a basic technology for smart structures. Bonded on surfaces or embedded in composite structures they work as actuators and sensors to control the structural behaviour. The simplest types are based on thin piezoceramic plates (typical thickness 200 μm) provided with surface electrodes to operate in the lateral d_{31}-mode. This type of actuator is able to generate strains of 500 μm/m. To achieve higher deformations it is necessary to use the d_{33}-effect. The difficulty is to generate the necessary in-plane electrical field. A common solution is the use of interdigitated electrodes consisting of two comb like electrodes with opposite polarity that are placed on the surface of the piezoceramic material. Known as Active Fiber Composites (AFC's) or Macro Fiber Composites (MFC's) these kinds of actuators can produce strains of 1,600 μm/m. The drawback of interdigitated surface electrodes is a very high driving voltage of up to 1,500 V. A promising concept to overcome this drawback is presented. It is based on the use of multilayer technology for low profile actuators. Within these actuators the electrodes are incorporated in the piezoelectric material during the sintering process as very thin layers with little impact on the actuator stiffness. This allows a significant reduction of the electrode distance and therefore also a reduction of the driving voltage. To utilize the multilayer technology for low profile actuators, standard multilayer stacks are diced into thin plates. In this configuration the electrodes are not only on the surface of the piezoelectric material but cover the whole cross section. In a second step these plates are embedded into a polymer to build a piezo-composite. Without the mechanical stabilization of the surrounding polymer the handling of the fragile multilayer

P. Wierach (✉)
DLR, Institute of Composite Structures and Adaptive Systems,
Lilienthalplatz 7, 38108, Braunschweig, Germany
e-mail: peter.wierach@dlr.de

M. Wiedemann and M. Sinapius (eds.), *Adaptive, Tolerant and Efficient Composite Structures*, Research Topics in Aerospace, DOI: 10.1007/978-3-642-29190-6_3,

plate would be extremely difficult or nearly impossible. Several prototypes have been build and achieved an active strain of 1,200 µm/m at a voltage of 200 V. Using other materials an active strain of 1,600 µm/m is possible.

3.1 Piezocomposite Technology

Actuators and sensors on the basis of smart materials are essential parts of adaptive systems. As integral structural components these materials can also provide load bearing capabilities at the best. Due to several advantages, piezoceramic materials are most commonly used as smart materials for adaptive systems. The reasons for the popularity of piezoceramic materials are on the one side of technical nature but there are also some "practical" reasons. On the technical side it's especially their ability to operate at high frequencies and the high stiffness of the material (typical 60 GPa), what is of special importance when used as actuators. In addition to that it is relatively easy to activate the material by a simple electrical field. The "practical" reasons are primarily the good availability and the reliable quality of the material. Piezoceramic materials are produced on an industrial level and a growing number of manufactures offer different types and shapes. Also the price is quite moderate in comparison to more exotic materials. The main disadvantage of piezoceramic materials is their inherent brittleness. Whereas it is no problem to apply high compression loads, tensile loads must be avoided at any time. Therefore the processing and structural integration of this sensitive material has to be done very carefully to avoid damages.

In adaptive systems, piezoceramic actuators are used to reduce, to generate or to detect deformations or vibrations with a special focus on distributed actuation and sensing systems. This means that the actuator forces are not transmitted at two discrete points but in-plane, usually by a laminar actuator, which is connected to the structure by a bonding layer. Especially for light weight and thin walled constructions this concept offers some advantages since no massive bearings are necessary. Because very thin laminar actuators are needed for this purpose, the brittleness of the piezoceramic material becomes even more serious. An appropriate solution for this problem is the use of so called "piezocomposites". Piezocomposites are a combination of piezoceramic materials with ductile polymers to form a robust and easy to use actuator and/or sensor. Especially their susceptibility to damage is significantly reduced by this measure. By the arrangement of the piezoceramic material the properties of the composite (e.g. stiffness or damping) can be specifically adjusted. Components like electrodes, electrical contacts or insulators can also be embedded into the composite. Generally the embedding is done at the curing temperature of the polymer (typical 120–180°C). Because of the different coefficients of thermal expansion (CTE) of the polymer in comparison to the piezoceramic material and due to the shrinking of the polymer during curing, the piezo material is provided with a beneficial mechanical pre-compression. This pre-compression allows applying (limited)

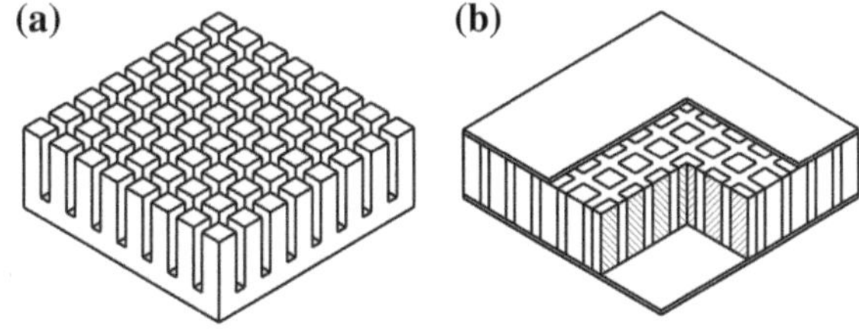

Fig. 3.1 Schematic drawings of a 1–3 piezocomposite for ultrasonic applications manufactured with the dice and fill process; (**a**) ceramic block after dicing; (**b**) piezocomposite with electrodes and filled with polymer

tensile loads to the actuator during operation. The following list summarizes the most important advantages of piezocomposites:

- Protection of the brittle piezoceramic material
- improved handling
- defined and easy to use electrical connectors
- electrical insulation
- pre-compression of the piezoceramic allowing to apply tensile loads
- reduction of stress peaks due to the soft surrounding material hence reduced crack propagation, larger passive deformations and improved lifetime
- possibility to realize complex architectures.

It should be mentioned, that the idea of piezo-composites is not totally new. Traditional applications for piezo-composites are ultrasonic transducers for medical diagnostic, health monitoring or sonar applications (Fig. 3.1). In comparison to monolithic devices piezo-composite transducers can be designed to have optimized acoustic impedance or to decouple radial oscillations.

3.2 State of the Art for Piezocomposite Transducers for Adaptive Structures

Considering the three piezoelectric effects

- longitudinal effect or d_{33}-effect,
- transversal or d_{31}-effect, and
- shear or d_{15}-effect,

in adaptive structures only the longitudinal and the transversal effect are used for technical relevant piezocomposites. More recently some research was published on actuators based on the shear-effect, but it seems that they only show limited advantages in comparison to extension mode actuators [1, 2].

The arrangement of the electrodes determines which effect will be used. The simplest configuration can be realized by using the d_{31}-effect. In this case the

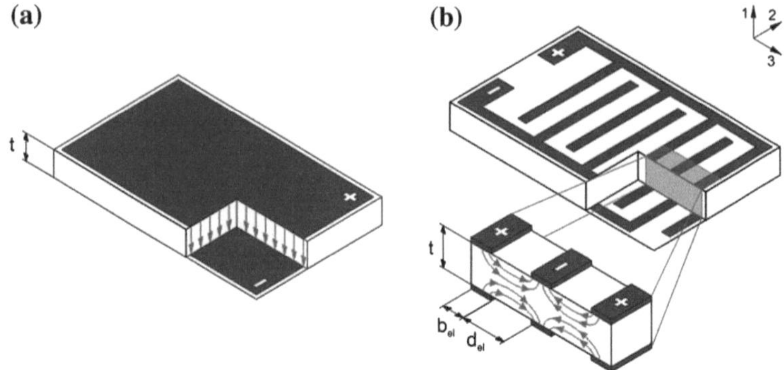

Fig. 3.2 Field distribution in different electrode configurations; (**a**) d_{31}-actuator; (**b**) d_{33}-actuator with interdigitated surface electrodes

in-plane contraction of the piezoceramic material is used when a positive electrical field is applied in perpendicular direction through the thickness of the piezoceramic plate. Thus the piezoceramic plate is provided with very thin layers of conductive material (a few μm) to build uniform surface electrodes. The electrical field is generated homogenously between these electrodes (Fig. 3.2a). The thickness of the piezoceramic plate defines the distance between the upper and lower electrode and therefore the voltage that is needed to generate a certain electrical field. With a usual plate thickness of 0.2 mm a voltage of 200 V is necessary to generate an electrical field of 1 kV/mm.

Up to three times higher deformations can be achieved with the d_{33}-effect. In this case the electrical field and the effective deformation have the same direction. Applying a positive field will result in an expansion of the piezoceramic device in the direction of the field and in a contraction perpendicular to this direction. The challenge is the generation of an in-plane electrical field. A feasible technical solution is the use of interdigitated electrodes. In this configuration the electrodes are made of two comb-like electrodes with opposite polarity, which are applied on the surface of the piezoceramic material. The electrical field is generated between the fingers of the electrode and penetrates the piezoceramic material as well. Due to this special design the electrical field is not very homogenous (Fig. 3.2b). This has a direct impact on the minimal electrode distance and hence on the operation voltage. If the distance between the electrode fingers is too small in comparison to the thickness of the piezoceramic material, the electrical field can not sufficiently penetrate the piezoceramic material and the efficiency of the actuator is reduced. In addition to that, the areas below the electrode fingers do not contribute to the actuation strain. If the electrode distance is reduced, the number of electrode fingers increases and also the "dead" areas below the electrodes. This can only partly be compensated by very thin electrode fingers. Besides technical limitations in producing very thin electrode fingers, such a configuration will also cause very high electrical field gradients in the vicinity of the electrodes. These high gradients are leading to high mechanical loads in the piezoceramic material, having an impact on lifetime and durability. A suitable electrode distance for a piezoceramic device with a thickness of 0.2 mm is between 0.5 and 1 mm. In this case, without considering the field inhomogeneity, a voltage of 500–1,000 V is necessary to generate an electrical field of 1 kV/mm.

Meanwhile several types of low profile piezocomposites for adaptive structures have been developed (and partly commercialized) and are shortly described in Table 3.1. For piezo composites with interdigitated electrodes, fiber like architectures that are referred to as Active Fiber Composites (AFC) turned out to be advantageous in comparison to monolithic designs. Cracks caused by the inhomogeneity of the electrical field will propagate through monolithic plates whereas these cracks are stopped at each interface between the polymer and the piezoceramic fiber. Besides this, fiber based actuators allow a directed actuation what is of advantage for some applications. A negative aspect of fibers with a circular cross section is that there is only a very small contact area between the electrode fingers and the piezoceramic fiber. Thus the penetration of the electrical field is aggravated resulting in even higher operation voltages or less performance. An improvement is the usage of fibers with rectangular cross sections to reduce the dielectric loss.

An essential drawback of fiber based composites is the very labour intensive manufacturing process. Up to know it is primarily handwork to place many single fibers close to another. This causes quality problems resulting in deviations of the actuator characteristics. Also the production and the following sintering process of PZT fibers are very cost intensive. An alternative manufacturing process uses commercially available PZT-wafers that are cut into ribbons. In this case the wafer is placed on a tacky film and cut with a wafer saw that is common within the production of silicon based integrated circuits. With this automated process the rectangular fibers or ribbons are aligned exactly in parallel. This type of piezocomposite is called Macro Fiber Composite (MFC). The last two examples in Table 3.1 refer to more exotic configuration, which haven't achieved a wider acceptance yet. Piezoceramic tubes or hollow fibers, which are provided with surface electrodes on the in- and outside allow a directed actuation in d_{31}-mode operation. They also offer the possibility to incorporate additional passive fiber materials like glass- or carbon fibers to improve their mechanical strength. The idea behind piezocomposites with very thin fibers is to improve the mechanical stability by assuming that the propagation of cracks can be reduced more efficiently. The problem is that due to the dielectric losses of the surrounding polymer the actuator performance is very poor. Possible applications are thin sensor sheets with minimal impact on the host structure.

The properties of the various designs differ with regard to voltage range, size, exact maximum strain and force. But at least it is the piezoceramic material and the operation mode that determines the performance. Therefore it is reasonable to look at some exemplary properties of d_{31}- and d_{33}-piezo composites that were derived using Macro Fiber Composite actuators from Smart Material GmbH. Some important properties are summarized in Table 3.2.

Table 3.1 Overview of state of the art low profile piezocomposites

Name	Type	Cross section	Description
QuickPack	d_{33} d_{31}		A monolithic piezoceramic plate with uniform surface electrodes bonded between polyimide films with etched circuit paths to contact the electrodes. To operate in the d_{33}-mode a bare plate is bonded between polyimide films with etched Interdigitated electrodes [3, 4]
FlexPatch	d_{31}		A monolithic plate with surface electrodes glued between a polyimide films by melting the film up to 325°C. Stripes of nickel are applied on the surface to contact the electrodes [5, 6]
DuraAct	d_{31}		A monolithic plate with surface electrodes embedded by resin transfer moulding. Flexible fabrics of conductive material (e.g. copper) completely cover the electrode to provide a reliable and damage tolerant electrical contact [7–10]
PowerAct	d_{33}		A bare monolithic piezoceramic plate provided with creases by laser cutting (not intersected). Interdigitated electrodes are applied by gluing the plates between polyimide films with etched circuit paths [11]
Active fiber composite (AFC)	d_{33}		Piezoceramic fibers with circular cross section embedded in a polymer with interdigitated electrodes etched or screen printed on a polyimide- or polyester-film. Designs by means of fibers with rectangular cross-section are used to improve the performance [12–17]
Macro fiber composite (MFC)	d_{33} d_{31}		A bare monolithic plate is diced to ribbons and glued between polyimide films with etched interdigitated electrodes. Configurations using diced plates with surface electrodes also allow operation in d_{31}-mode [18–21]
Piezoceramic tubes	d_{31}		Piezoelectric tubes (or hollow fibers) with typical outer diameters of 800 μm and inner diameters of 400 μm, provided with inner and outer surface electrodes to operate in the d_{31}-mode are embedded in a polymer to allow an anisotropic actuation [22, 23]
Thin fibers	d_{33}		Very thin fibers ($\varnothing < 50$ μm) are embedded in a polymer to form a sheet with an arbitrary fiber distribution. Interdigitated electrodes are directly applied to the polymer with a special screen printing technique [24]

Table 3.2 Typical properties of state of the art low profile piezocomposites [25]

		d_{33}	d_{31}
Operation voltage	U_{max} [V]	1,500	360
	U_{min} [V]	−500	−60
Capacity	[nF/cm^2]	0.42	4.5
Charge constant	[pm/V]	460	−370
Strain/volt	[μm/m/V]	0.7–0.9	−2
Max. strain	[μm/m]	1,600	500
Charge/strain	[pC/ppm]	1,670	3,250

Fig. 3.3 Comparison of d_{31}-(MFC type P2) and d_{33} actuators (MFC type P1) over a voltage range of 0–400 V

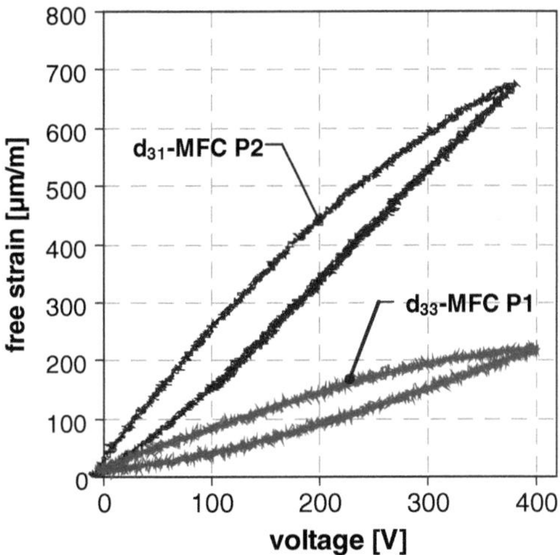

As mentioned before, for the price of very high voltages it is possible to generate much higher strains with d_{33}-actuators as with d_{31}-actuators. Hence it makes sense to look at the ratio of applied voltage and active strain. For state of the art piezo-composite actuators this ratio is much better for d_{31}-actuators. Figure 3.3 shows exemplary strain-voltage curves of a d_{31}-MFC (type P1) and a d_{33}-MFC (type P2) over a voltage range of 0–400 V and −500–1,500 V respectively. Both types of actuators had the same active area. For comparability reasons the absolute strain values of the d_{31}-actuator were plotted in the diagram. In this voltage range the superiority of the d_{31}-actuator is obvious. If the d_{33}-actuator is driven over its complete voltage range (−500–1,500 V) much higher strains can be achieved (Fig. 3.4).

Expectedly the capacity of the d_{31}-device is much higher, because the capacity is among others a function of the electrode distance. That means that the power, which is needed to drive a d_{33}-piezo composite is not inevitable higher. On the other side, if the device is used to generate energy (especially by means of energy

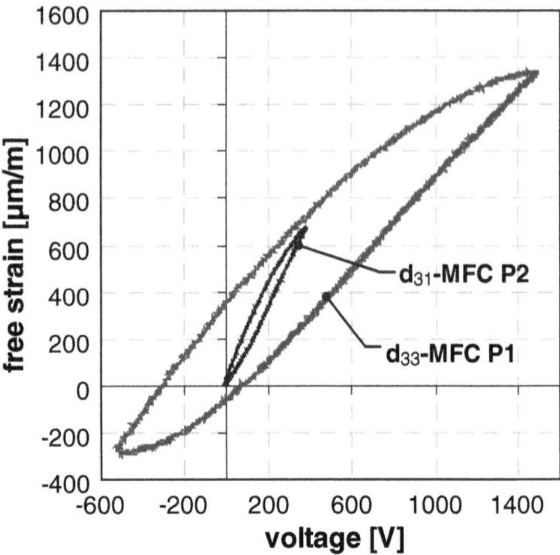

Fig. 3.4 Free strain of a d_{33}-actuator (MFC type P1) over a voltage range of −500–1,500 V

harvesting) it is beneficial to have a high capacity because more charge is generated when the transducer is deformed.

It becomes clear, that there is no perfect piezocomposite suitable for all purposes. The choice must always be made based on the given requirements and application scenarios.

3.3 A Modular Manufacturing Concept for Piezocomposites

Besides the operation voltage and the piezoceramic the complete material composition of a piezocomposite is of great importance. With respect to the variety of possible applications for piezocomposites the use of standardized solutions is very often not feasible. The goal was to develop new piezocomposites with improved performance parameters that can easily be adapted to different requirements. This requires the possibility to access every component of the piezocomposite to enable a material specific selection of the components to get a compatible material system:

- Piezoceramic material
- design and material of the electrodes
- shape of the piezoceramic
- insulation material
- surface quality
- design of electric contacts.

Figure 3.5 shows the principle design of the developed piezocomposite. For the development of the manufacturing technology, standard piezoceramic wafers provided with uniformly electroded surfaces to operate in the lateral d_{31}-mode and

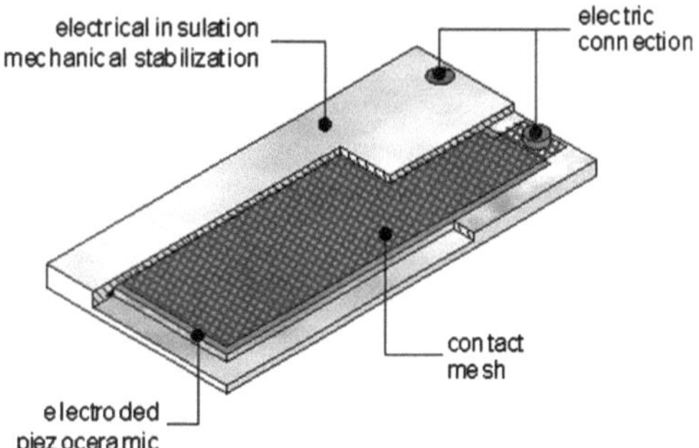

Fig. 3.5 Design of the pre-encapsulated patch

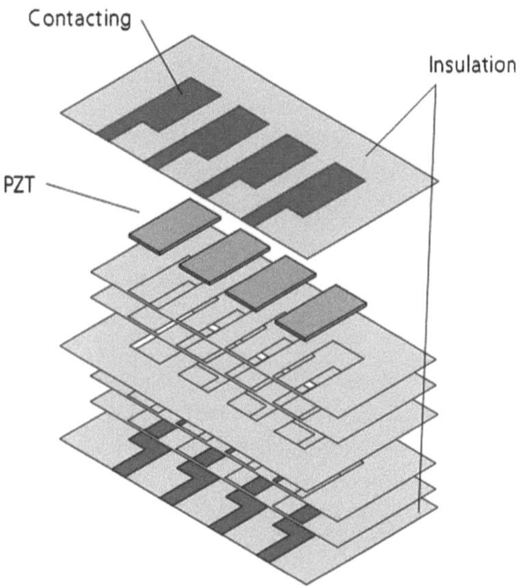

Fig. 3.6 Manufacturing of the encapsulated PZT-patches

a typical size of $50 \times 30 \times 2$ mm^3 were used. Good experiences have been made with plates from PI-Ceramic (PIC 255) and CeramTec (SONOX P53), but it is also possible to use any other piezoelectric material.

Figure 3.6 shows the manufacturing procedure for the new type of actuators. Considering the productivity of the process a set of piezocomposites is manufactured in one step and cut to size afterwards (Fig 3.4). The piezoceramic is embedded between thin layers of insulating fiber material (d < 0.05 mm) and layers with contacting structures. The contacting layer is made of a copper mesh (wire diameter 0.03 mm) or a metalized polyester fleece having the shape of the piezoceramic wafer. In case of a break the patch will still work because the contacting covers the whole electrode of the piezoceramic so that the broken pieces stay in the electrical field where they can be controlled.

Table 3.3 typical material parameters to calculate the pre-compression in a piezocomposite with RTM 6 resin and PIC 255 ceramic

Curing temperature	T	180	[°C]
Temperature difference to room temperature	ΔT	160	[K]
CTE insulation	α_{ISO}	64	[10^{-6}/K]
Young's modulus insulation	E_{ISO}	3	[GPa]
Thickness insulation	t_{ISO}	0.1–0.3	[mm]
CTE ceramic in 1-direction	α_{PZT}	5.8	[10^{-6}/K]
Young's modulus ceramic in 1-direction	E_{PZT}	62	[GPa]
Thickness ceramic	t_{PZT}	0.2	[mm]

The piezocomposites are produced with an improved RTM technology, the so-called DP-RTM (Differential Pressure—Resin Transfer Moulding) [26]. This guarantees an extreme high quality and reproducibility of the components. The fiber material is laid out in dry state which facilitates the positioning of the contact structures and ceramics. The DP-RTM procedure becomes especially interesting as it is not necessary to provide massive moulds since the clamping forces in the autoclave are created by the differential pressure. Thus, a simple sheet plate can serve as sub mould while a vacuum foil is applied for the upper mould. Fiber volume content and fluid rate can directly be controlled by adjusting differential pressure during the stages of injection. In order to minimize the weight of the active fiber composite a high fiber volume content is required. This can be achieved by increasing the differential pressure. Simultaneously, the increasing mechanical load on the ceramics has to be considered since it might result in mechanical damage of the brittle actuators.

Piezocomposites are cured at elevated temperatures. This result in a very beneficial pre-compression of the ceramic material since the coefficients of thermal expansion (CTE) of the resin and the insulating materials are higher than the CTE of the ceramic material. The pre-compression in the ceramic material can be calculated with Eq. (3.1).

$$\sigma th = \Delta T \left[\frac{2 E_{PZT} E_{ISO}\, t_{ISO} (\alpha_{ISO} - \alpha_{PZT})}{2 E_{ISO}\, t_{ISO} + E_{PZT}\, t_{PZT}} \right] \tag{3.1}$$

Using the material parameters of Table 3.3 the resulting pre-compression σth is 26 MPa for an insulation layer thickness of 0.1 mm and 73 MPa for an insulation layer thickness of 0.3 mm. In this case a resin (RTM6 from Hexcel Composites) that cures at 180° in combination with a PIC 255 ceramic from PI-Ceramic was chosen. It has to be noted that the young's modulus and the CTE of a polarized piezoceramic are depending on the polarization direction. Experimental measurements [27] confirmed these calculated results.

The materials and components of the piezocomposites had been selected and optimized with special regard to the integration into fiber composite structures but

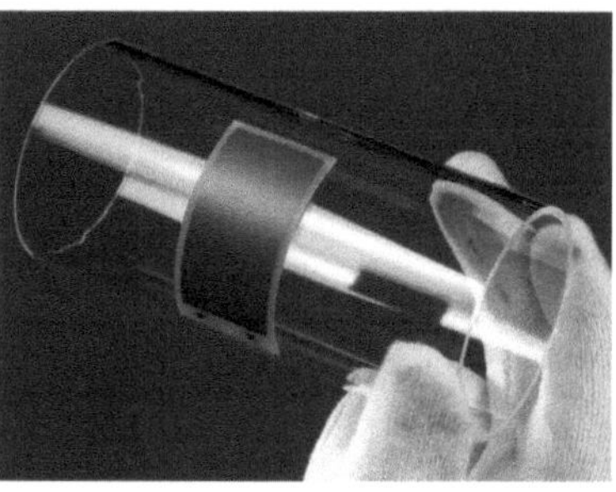

Fig. 3.7 Standard piezocomposite transducer ($54 \times 34 \times 0.4$ mm^2) with solder points as electrical contact bonded on a tube to demonstrate the flexibility of the piezocomposite

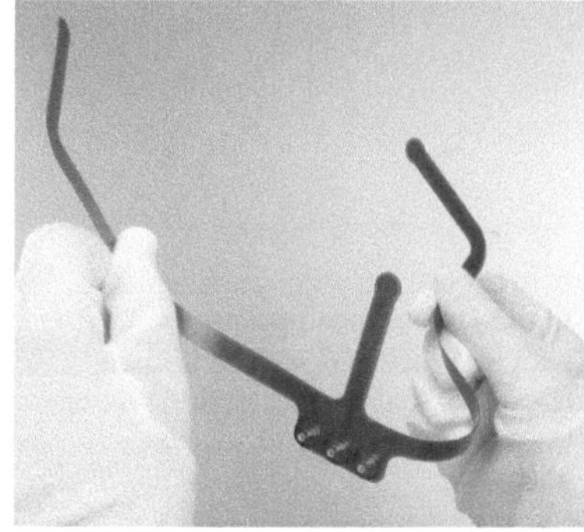

Fig. 3.8 Piezocomposite with three circular shaped piezoceramics and embedded BNC connectors as a sensor array for structural health monitoring (SHM)

they can also be attached on any surface. By default the electric contacts are realized as solder points (Fig. 3.7) but the use of insulated wires or any kind of plugs is also possible. Figure 3.8 shows a piezocomposite with special bayonet nut connectors (BNC) that were directly embedded into the composite. Due to the adaptability of the manufacturing process it is possible to produce patches of nearly any shape. This is very interesting for circular or curved structures where hexagonal or curved patches are most suitable. Figure 3.9 demonstrates the possibilities to manufacture customized patches. The separated pieces were cut into shape by laser and embedded into an array to form a complex piezocomposite. Besides piezoceramic plates it was also demonstrated that the manufacturing concept is applicable to fiber based piezocomposites as shown in Fig. 3.10.

3.4 Multilayer Piezocomposites

In many applications it is required to significantly reduce the operation voltage of piezoelectric actuators without reducing the active strain. The use of high voltages in technical systems is associated with some severe drawbacks. Besides harder official safety regulations there are insulation issues and high voltage electronic

Fig. 3.9 Piezocomposite with complex design to cover the surface of an adaptive satellite mirror for high precision shape control [28]

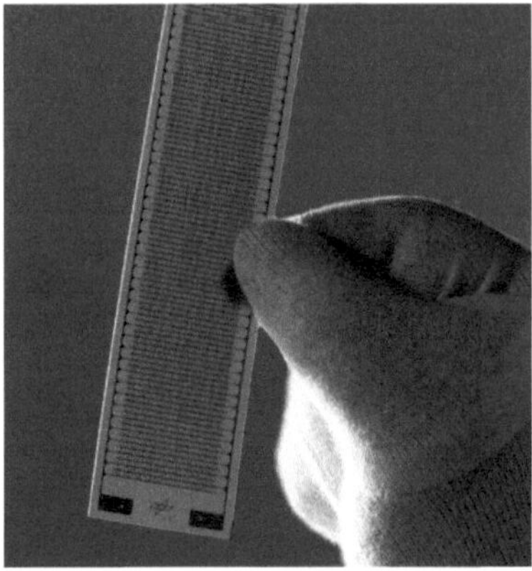

Fig. 3.10 Piezocomposite with interdigitated surface electrodes and piezoceramic fibers (200 μm fiber diameter)

components are usually more expensive. Also the acceptance of the user is may be lacking. The desire to reduce the operation voltage of piezoelectric actuators led to the development of multilayer stack actuators [29–31]. Conventional stack actuators are made of piezoceramic plates, which are glued together in a stacking sequence. To contact the electrodes, sheets of copper are also incorporated within the glue layer. The drawback of this design is the decreasing stiffness of the actuator with increasing length (or increasing number of glue layers). Also the operation voltage cannot be reduced significantly because this would mean a reduction of the piezoceramic plate thickness; hence an increased number of plates with even more glue layers. Also the manufacturing and handling of individual thin plates is difficult.

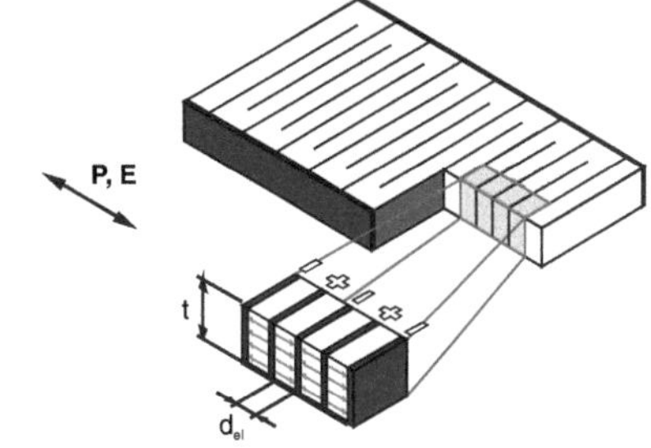

Fig. 3.11 Field distribution in a multilayer plate cut from a multilayer stack

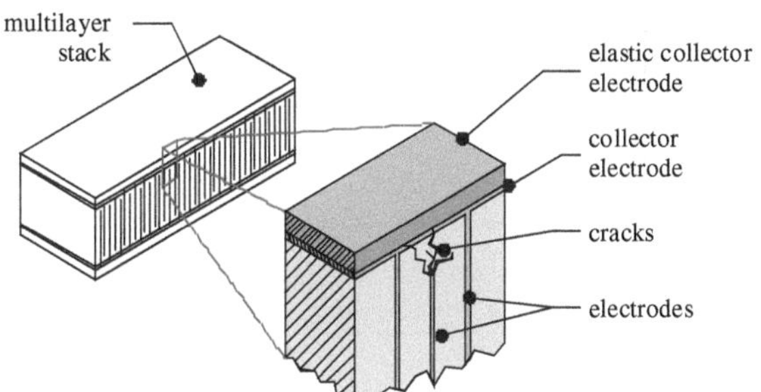

Fig. 3.12 Cracks caused by inhomogeneous electrical fields at the electrode tip

In the manufacturing process of multilayer stacks the electrodes are incorporated during the sintering process as very thin layers. The stack itself is a monolithic block with integrated electrode layers. Therefore their influence on stiffness and performance is very low. This allows a reduction of the distance between the electrodes, what leads to a significantly reduced operation voltage

To utilize the multilayer technology for low-profile piezocomposites a technology has been developed that allows to cut multilayer stacks into thin (0.2–0.3 mm) plates and to embed the fragile multilayer plates into a composite to form a robust and easy to use transducer. As depicted in Fig. 3.11 this design results in a homogenous field distribution over large areas of the piezoceramic material.

3.4.1 Manufacturing of Multilayer Piezocomposites

Starting point for the manufacturing of a multilayer piezocomposite is a commercial available multilayer stack. The dimensions of the stack configuration, which has been used for the technology development, are $18 \times 5 \times 5$ mm^3. The stack is provided with passive ceramic endplates, so that the actual active length of the stack is 15 mm with an electrode distance of 53 μm.

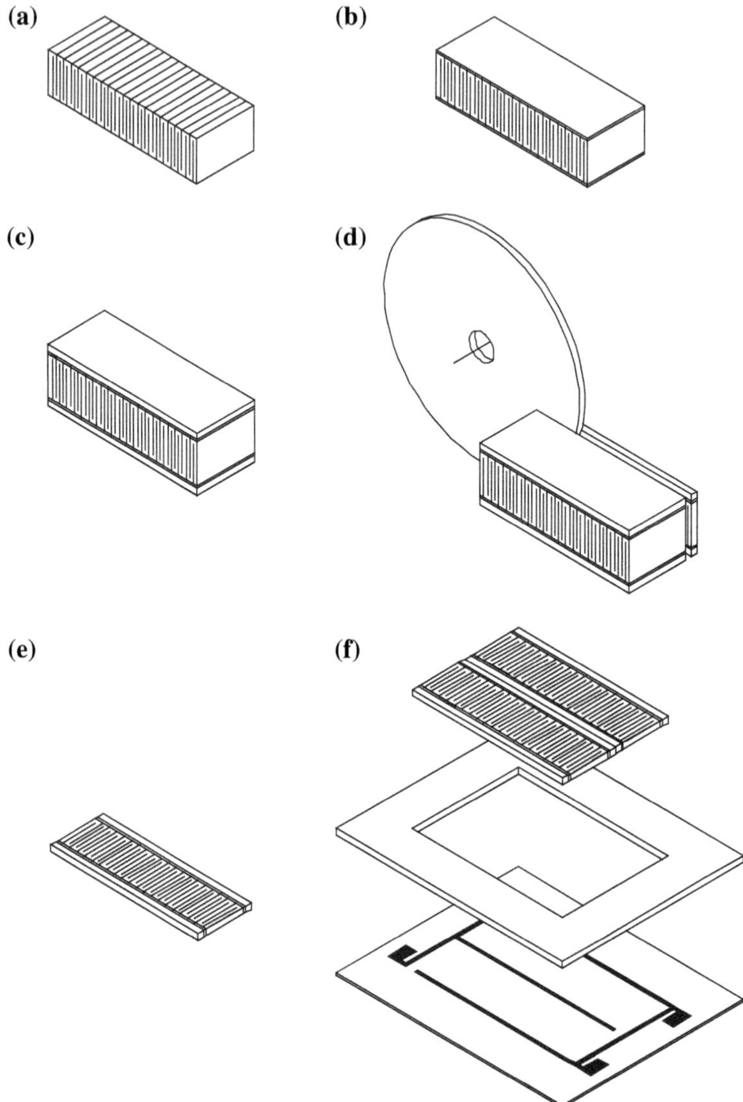

Fig. 3.13 Manufacturing steps; (**a**) multilayer stack; (**b**) collector electrode; (**c**) flexible collector electrode; (**d**) dicing; (**e**) multilayer plate; (**f**) packaging

In a first step the multilayer stack (Fig. 3.13a) is provided with a thin conductive collector electrode (Fig. 3.13b). External loads or loads that are generated during operation of the stack actuator can lead to cracks in the collector electrode (Fig. 3.12). Main causes for cracks are inhomogeneous electrical fields, which appear only at the tips of the electrodes. Because these cracks are limited to certain

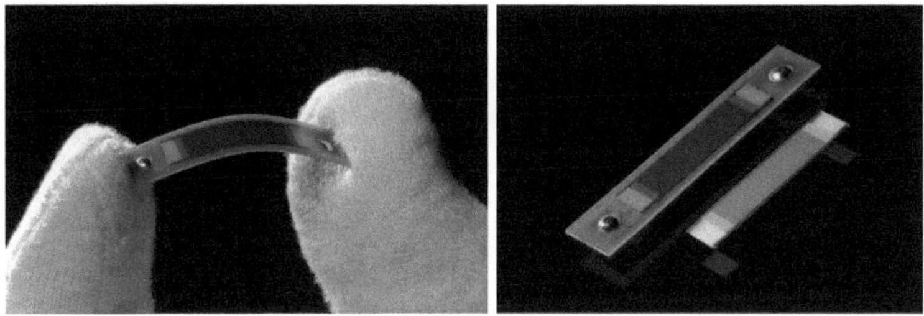

Fig. 3.14 Piezocomposites made of a stack with dimension of $18 \times 5 \times 5$ mm^3

regions they do not have a critical impact on the actuator performance, but they can cause cracks in the collector electrode as well. This would result in a partially or complete failure of the actuator. To compensate for this problem an additional elastic collector electrode is applied (Fig. 3.13c). The collector electrode is made of an elastic conductive fiber material to stop the propagation of the cracks. After the application of the elastic collector electrode thin plates with a thickness of 0.3 mm are cut from the stack (Fig. 3.13d, e).

The next step in the manufacturing process is the embedding of the brittle and sensitive multilayer plate in a polymer to form the actual composite. This is done using the modular manufacturing concept for piezocomposites as described before. The outer layers of the composite consist of thin polyimid films (25 µm) to guarantee a good electrical insulation. To fix the position of the multilayer plates during the manufacturing process, but also to enable the resin flow, a frame of non conductive fiber material is used. Preferably this frame is made of a polyester fleece with cut outs having the size of the multilayer plates. Because the dimensions of the multilayer plates are limited, an array of plates can be arranged in one composite to enlarge the active area. To increase the productivity of the process several composites are manufactured at once and separated afterwards.

Figure 3.14 shows exemplary piezocomposites that were build with the standard multilayer stack configuration. These composites have been used to characterize the performance of this new actuator configuration.

3.4.2 Free Strain of Multilayer Piezocomposites

A typical strain-voltage curve of the multilayer piezocomposite is plotted in Fig. 3.15. The strain of the multilayer piezocomposite was calculated using the results of the measured displacement with respect to an active length of 15 mm. With an electrode distance of 53 µm and a maximum voltage of 120 V, a maximum electrical field of 2.26 kV/mm was applied. All measurements were made applying a quasi static excitation of 0.1 Hz with a triangle wave form.

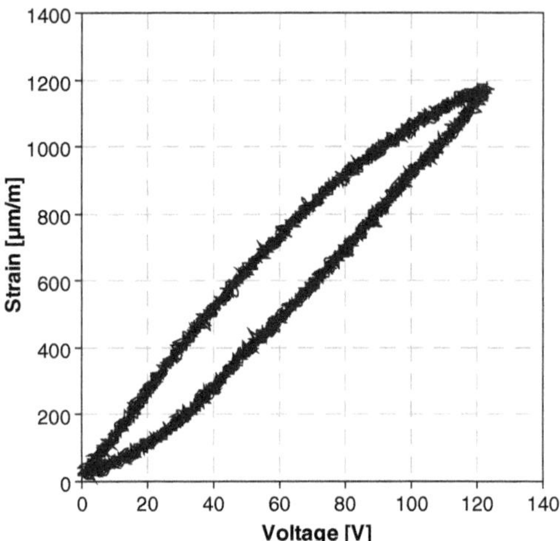

Fig. 3.15 Exemplary strain-voltage curve of a multilayer piezocomposite

Table 3.4 Results of the free strain measurement with a set of 12 specimens

Average free strain	[μm/m]	1,285
Standard deviation	[μm/m]	56.5
Standard deviation	[%]	4.4
Min	[μm/m]	1,193
Max	[μm/m]	1,380

Table 3.4 summarizes the results of the free strain measurement with a set of 12 specimens. The average active free strain that was measured with an operation voltage of 120 V was 1,285 μm/m. In comparison to state of the art d_{33}-piezo-composites with interdigitated surface electrodes, which need voltages of up to 1,500 V to achieve same active strain levels, this actuator has demonstrated that it is possible to drastically reduce the operation voltage of d_{33}-piezocomposites.

3.5 Summary and Conclusion

A considerable number of developments in the area of piezocomposites for adaptive systems demonstrate the relevance of this technology. This is also emphasized by an increasing number of commercial products in this context.

Based on fiber composite manufacturing processes a new technology was developed to build up reliable and easy to use piezocomposites. Due to their design these piezocomposites are characterized by a high damage tolerance and a mechanical pre-compression of the brittle piezoceramic material. It is also possible to realize very complex configurations that can be adapted in shape and material composition to meet the requirements of different application scenarios.

Fig. 3.16 Piezocomposite
with integrated acceleration
sensor

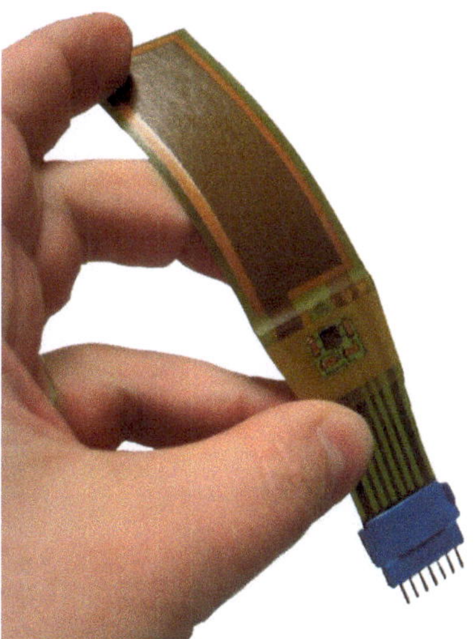

State of the art d_{33}-piezocomposites incorporate surface mounted interdigitated electrodes. The drawback of such a configuration is a very high operation voltage (typical 1,500 V). The utilization of multilayer technology for low profile piezocomposites allows a significant reduction of the operation voltage. A technology to fabricate multilayer piezocomposites is presented. It was demonstrated that an active strain of 1,285 μm/m can be achieved with an operation voltage of 120 V.

Following the basic idea of adaptive systems the potential of the piezocomposite technology is not yet fully explored. It is also feasible to integrate more functionalities than a piezoelectric transducer into the composite. An example is shown in Fig. 3.16. During the manufacturing process also a micro system (in this case an acceleration sensor) including all necessary circuits and electronics was integrated into a composite to form a combined actuator and sensor module. This module is useful for the control of plate vibrations, since it allows a vibration measurement perpendicular to the plate surface. A piezoceramic transducer alone would only measure the in-plane strain of the plate. Another example is the incorporation of aerodynamic flow sensors as part of an active flow control system [32].

References

1. Benjeddou, A., Deu, J.F.: Piezoelectric transverse shear actuation and sensing of plates. J. Intell. Mater. Syst. Struct. **12**(7), 435–449 (2001)
2. Wierach, P., Rienecker, R.: An experimental study on the use of piezoelectric shear actuators for smart structures. In: 18th international conference on adaptive structures and technologies, Ottawa, Canada (2007)
3. Lazarus, K. B.: Packaged strain actuator. United States Patent 5,656,882 (1997)

4. www.mide.com. Accessed 12 Dec 2011
5. Horner, G.: Piezoelectric composite device and method for making same. United States Patent Application 20020038990 (2002)
6. Horner, G.: Smart actuator research. Adaptronic Congress, Berlin (2001)
7. Wierach, P.: Entwicklung multifunktionaler Werkstoffsysteme mit piezokeramischen Folien im Leitprojekt Adaptronik. Adaptronic-Congress, Wolfsburg (2003)
8. Wierach, P.: Elektromechanisches Funktionsmodul. German Patent DE 10051784 C1 (2002)
9. http://www.invent-gmbh.de/s08x_duraact.htm. Accessed 12 Dec 2011
10. http://www.physikinstrumente.de/de/pdf/extra/ PI_Katalog_Piezokomposite_Flaechenwandler_DuraAct_Brosch.pdf. Accessed 12 Dec 2011
11. Master, B.P.: Laser machining of electroactive ceramics. United States Patent 6,337,465 (1993)
12. Bent, A.A.: Anisotropic actuation with piezoelectric fiber composites. J. Intell. Mater. Syst. Struct. **6**(3), 338–349 (1995)
13. Bent, A.A.: Piezoelectric fiber composites with interdigitated electrodes. J. Intell. Mater. Syst. Struct. **8**(11), 903–919 (1997)
14. Gentilman, R.: Enhanced performance active fiber composites. In: SPIE 10th symposium on smart structures and materials, San Diego, USA (2003)
15. http://www.matsysinc.com/products/transducers/. Accessed 12 Dec 2011
16. www.advancedcerametrics.com. Accessed 12 Dec 2011
17. Lammer, H.: Einsatz adaptiver Materialien und deren Wirkungen bei Sportgeräten an Beispiel Tennisschläger Is18 sowie Ski Ic 300 der Fa HEAD Sport. Adaptronic congress, Wolsburg, Germany (2003)
18. Wilkie, W.K.: Low-cost piezocomposite actuator for structural control applications. In: SPIE 7th international symposium on smart structures and meterials, Newport Beach, California, USA (2000)
19. Wilkie, W.K.: Free strain electromechanical characterization of the nasa macrofiber composite piezoceramic actuator. In: SPIE 8th symposium on smart structures and materials, Newport Beach, California, USA (2001)
20. Wilkie, W.K.: Method of fabricating a piezoelectric composite apparatus. United States Patent 6,629,341 (2003)
21. www.smart-material.com. Accessed 12 Dec 2011
22. Cannon, B.J.: Feasibility study of microfabrication by coextrusion of hollow fibres for active composites. J. Intell. Mater. Syst. Struct. **11**, 659–670 (2000)
23. Wierach, P.: Smart composites based on piezoceramic tubes. In: SPIE 9th symposium on smart structures and materials, San Diego, California, USA (2002)
24. Gesang, T.: Herstellung und Eigenschaften aktorischer Fasermodule. Adaptronic Congress, Wolfsburg, Germany (2003)
25. Schoenecker ,A., Roedig, T., Gebhardt, S., Keitel, U., Daue, T.: Piezocomposite transducers for smart structure applications. In: SPIE 12th symposium on smart structures and materials, San Diego, USA (2005)
26. Sigle, C.H., Kleineberg, M., Pabsch, A., Herrmann, A.S.: Das DP-RTM-Verfahren, eine Fertigungstechnologie zur wirtschaftlichen Herstellung hochwertiger Faserverbundbauteile. In: Brökel, K. (ed) 2. Workshop Konstruktionstechnik–Innovation–Konstruktion– Berechnung, pp. 403–414. Shaker Verlag, Herzogenrath (1998)
27. Wierach, P.: Entwicklung von Piezokompositen für Adaptive Systeme. Dissertation, Technische Universität Braunschweig (2009)
28. Wierach, P., Monner, H.P., Schönecker, A., Dürr, J.K.: Application specific design of adaptive structures with piezoceramic patch actuators. In: SPIE's 9th annual international symposium on smart structures and materials, San Diego, California, USA (2002)
29. Wierach, P.: Piezokeramischer Flächenaktuator und Verfahren zur Herstellung eines solchen. Patent Application DE 10 2006 0404 316 A1 (2006)

30. Wierach, P.: Low profile piezo actuators based on multilayer technology. In: 17th international conference on adaptive structures and technologies, Taipei, Taiwan (2006)
31. Grohmann, B., Maucher, C., Jänker, P., Wierach, P.: Embedded piezoceramic actuators for smart helicopter rotor blade. In: 16th AIAA/ASME/AHS adaptive structures conference, Schaumburg, Illinois, USA (2008)
32. Beutel, T., Leester-Schädel, M., Wierach, P., Sinapius, M., Büttgenbach, S.: Novel pressure sensor for aerospace purposes. Sens. Transducers J. **115**(4), 11–19 (2010)

Chapter 4
Nanoscaled Boehmites' Modes of Action in a Polymer and its Carbon Fiber Reinforced Plastic

Christine Arlt, Wibke Exner, Ulrich Riedel, Heinz Sturm
and Michael Sinapius

Abstract Laminates of carbon fiber reinforced plastic (CFRP), which are manufactured by injection technology, are reinforced with boehmite particles. This doping strengthens the laminates, whose original properties are weaker than those of prepregs. Besides the shear strength, compression strength and the damage tolerance, the mode of action of the nanoparticles in resin and in CFRP is also analyzed. It thereby reveals that the hydroxyl groups and even more a taurine modification of the boehmites' surface alter the elementary polymer morphology. Consequently a new flow and reaction comportment, lower glass transition temperatures and shrinkage, as well as a changed mechanical behavior occur. Due to a structural upgrading of the matrix (higher shear stiffness, reduced residual stress), a better fiber-matrix adhesion, and differing crack paths, the boehmite nanoparticles move the degradation barrier of the material to higher loadings, thus resulting in considerably upgraded new CFRP.

4.1 Challenges of Future Carbon Fiber Reinforced Plastics

Increasing ecological awareness as well as quality and safety demands, which are present e.g. in the aerospace and automotive sectors, lead to the need to use more sophisticated and more effective materials. Thereby cost efficiency is indispensable.

C. Arlt (✉) · W. Exner · U. Riedel · M. Sinapius
Institute for Composite Structures, German Aerospace Center DLR,
Lilienthalplatz 7, 38108, Braunschweig, Germany
e-mail: christine.arlt@dlr.de

H. Sturm
Bundesanstalt für Materialprüfung (BAM), Unter den Eichen 87, 12205, Berlin

M. Wiedemann and M. Sinapius (eds.), *Adaptive, Tolerant and Efficient Composite Structures*, Research Topics in Aerospace, DOI: 10.1007/978-3-642-29190-6_4,
© Springer-Verlag Berlin Heidelberg 2013

CFRPs have successfully been established in lightweight construction due to their excellent mechanical performance, while having a low specific weight. Currently the injection–and the prepreg (pre-impregnated fibers)—technology are the widest spread manufacturing techniques for processing high performance CFRPs. The injection of liquid resin into dry fibers has the advantages of lower manufacturing costs and higher potential of volume production. However, prepreg manufactured CFRPs still have the higher property level.

Typically, CFRP-structures are dimensioned via their fibers under external loading conditions. But it is well known, that laminate failure is often determined by the matrix behavior under miscellaneous loading conditions. Especially for laminates manufactured by injection technology the remaining residual stresses in the laminate and limited matrix properties hinder the development of these high performance materials to their full potential. Hence the idea of strengthening the injection matrix for CFRP is obvious.

One approach to progress is the incorporation of fillers. In numerous projects several materials as metal oxides, glasses or carbonates have been investigated [1 and there within]. Initially micro scaled particles were used for reinforcement. Problems occurred due to brittleness of the polymer, increased viscosity and filtration of the particles by the fiber. In order to overcome these problems nanoscaled fillers became the focus of research [2]. Improved chemical, physical and mechanical material properties have been realized by many researchers. The accomplished changes mostly depend on the material characteristics of the filler, the particle size and shape, the dispersion quality, the filler content and the particle–matrix interaction [3, 4]. By altering the nanoparticles surface molecules, the particle–matrix interaction can be tailored and therewith the polymer network be formed [5]. Various research results show an increase of mechanical performance with improved resin-particle bonding [6, 7]. Especially covalent bonding is named to enhance the load transfer and make the nanocomposites more resistant [6–10]. A limiting factor of the influence of a strong interphase is the network mobility. With a progress of cross linking density the network mobility decreases and the influence the particle-resin interphase becomes less influential [11]. For that reason general statements are difficult.

So, the challenge of this research is to figure out relationships between the particles' surface molecules, nano-polymer properties and the properties of corresponding CFRP.

4.2 Resin-Particle Interactions

For the reinforcement of CFRPs nanoscaled boehmite is used. The particles have a cubical shape and a primary size of 14 nm. Boehmite is a aluminum oxide hydroxide (γ-AlO(OH)). Therefore it posses a high number of hydroxyl groups at its surface, which allow a surface modification. In the presented work pure boehmite (HP14) and taurine modified particles (HP14T) are analyzed. According to literature

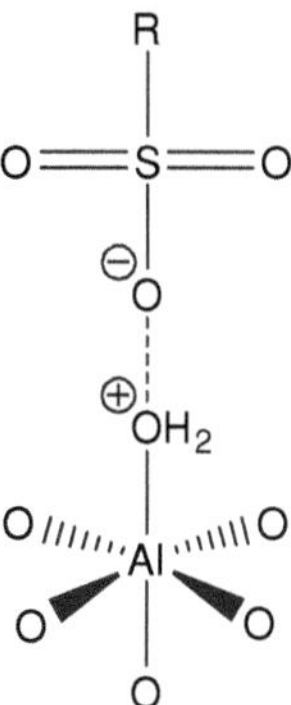

Fig. 4.1 Outer-sphere complex between sulfonat und boehmite (detail view of crystal) [12]

[12] the taurine molecule probably reacts in an acid base reaction with the boehmite and forms an outer-sphere complex. Figure 4.1 presents the formed complex.

By the means of BET measurements and elemental analysis the specific surface area and the atomic composition of the particles are tested. With the help of these results a surface coverage of 16% with taurine is calculated for the particles HP14T.

The processed particles were provided by the company Sasol Germany GmbH as a dry powder. In the dry state the particles are clustered into agglomerates and aggregates. To obtain single nanoscaled particles, the powder was dispersed into the resin. This was done with a three-roll mill (Exakt, 80E) for HP14 and with a bead mill for HP14T. In both machines the particles get separated by mechanical loads. To analyze particle size and distribution scanning election microscope (SEM) picture were made of the nanocomposites. In Fig. 4.2 two representative pictures are presented. The results show a fine and homogeneous particle distribution. Only a small number of minor agglomerates are found.

While processing the liquid nanocomposites a major difference in viscosity of both systems becomes obvious. Figure 4.3 shows the viscosity during cure at 80°C. Comparing the nanocomposite of pure boehmite with the pure resin, a significant increase in viscosity is observed. The taurine modified particles in contrast affect the flowability hardly.

Further experiments show for the masterbatch of HP14 a thixotropic behavior, while the masterbatch of HP14 only have a slight shear thinning. The different viscosity of both liquid systems indicates a different particle-resin interaction caused by the different surface properties. While the interaction of the pure boehmite and the resin seems strong, the modification with taurine weakens this interphase.

To verify this assumption both particles were processed with 4-tert-butylphenyl glycidyl ether. The 4-tert-butylphenyl glycidyl ether is used as a model substance for the resin. After the reaction with the model substance the particles got thoroughly washed and analysed by means of attenuated total reflection (ATR) and nuclear magnetic resonance spectroscopy (NMR). Results showed an opening of the epoxy group. A clear product could not be identified, but strong interactions between the particles and the resin can be assumed. Taurine molecules can not

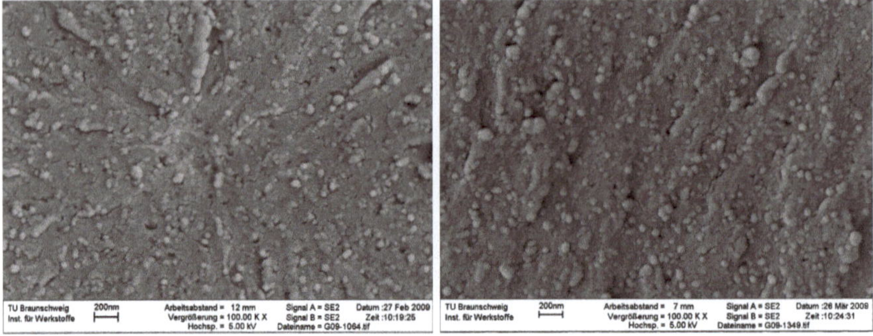

Fig. 4.2 SEM-pictures of cured nanocomposites with a filler content of 15 wt% of boehmite; *left* HP14; *right* HP14T

Fig. 4.3 Viscosity of nanocomposites with a filler content of 7.5 wt% during cure at 80°C

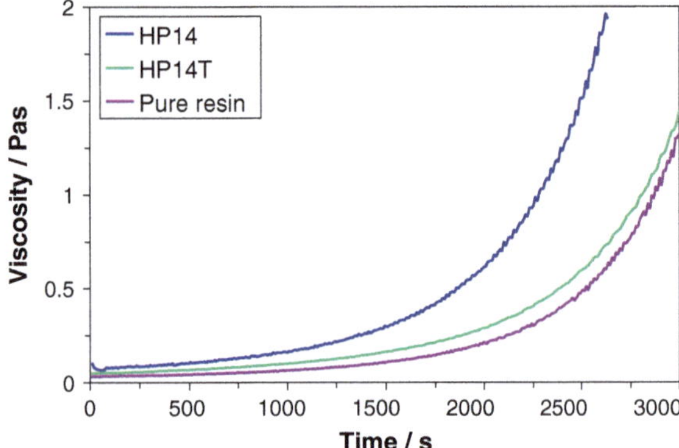

react with the resin. For that reason they act as a spacer, which results in significant lower viscosities.

The astonishing difference in flowing behavior caused by the modification is essential for the manufacturing of CFRPs by injection technology. Only with the particles HP14T this manufacturing method is realizable.

4.3 Particle-Polymer Interphases

The first approach of detecting changed particle-resin interactions and interphase strength by the altering the boehmites' surface molecules showed that taurine-modified particles seem to interact less strong with the epoxy than the unmodified analogon (viscosity profiles, Sect. 4.2, Fig. 4.3).

Now the resulting changes in the polymer network and therewith its Young's modulus especially in the close-up range of the particles HP14 and HP14T shall be considered.

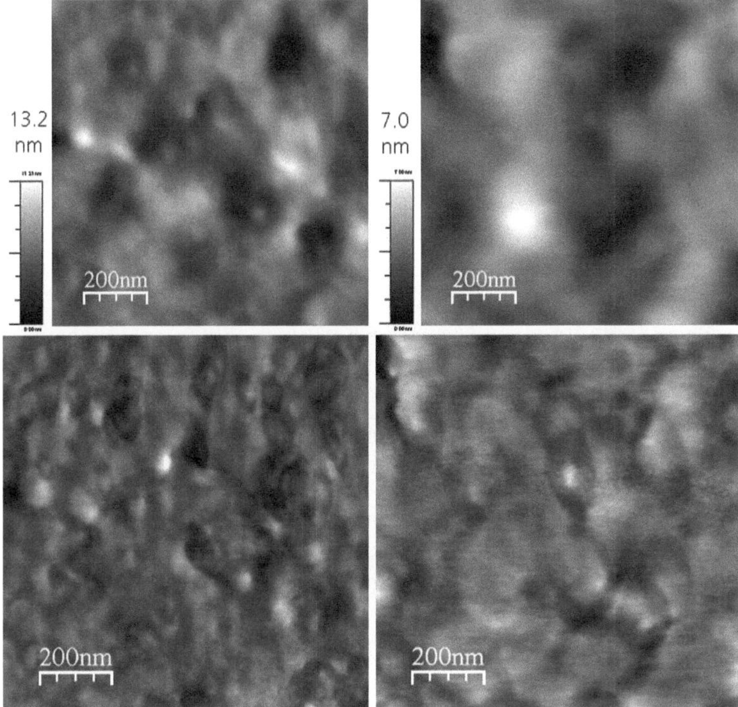

Fig. 4.4 FFM, excitation frequency 75 kHz: topographies (*top*) and pictures of real part amplitude (contrasts of storage modulus) (*bottom*) of nanocomposites; left: HP14 7.5 wt.-%: small "halo"; right: HP14T 15 wt%: big "halo"

For that reason, samples of cured nanocomposites are prepared using abrasive grinding paper followed by polishing using a diamond suspension. First inspection using Atomic Force Microscopy (AFM) operated in the Force Modulation Mode (FMM) showed partially contamination with a soft, viscous surface layer, which was gradually removed by a magnetically coupled plasma (air, $\sim$ 10-1 mbar).

Force Modulation Microscopy (FMM) is a variation of the AFM contact mode. The sample is mounted on a piezo (resonance frequency 6 MHz, $\Delta d/\Delta U \sim 2$ nm/ V, both for the unloaded case) and a sufficient drift range must be realized to suppress acoustic near field effects, which in our case was realized by the sample thickness itself (~ 1.5 cm). The actuator is excited with an AC voltage leading to a Δz of the surface of <2 nm. The resulting cantilever bending is analyzed by means of a synchronized lock-in amplifier. Soft sample sites deform under the contact load of the AFM tip leading to low cantilever vibration amplitude, very stiff contacts would lead to a high lever vibration. Such a stiff and loss-free contact allows tuning of the phase shift between excitation and response in a way that the loss (imaginary part) is close to zero. For the samples inspected, the real part amplitude, which can be equated with the local storage modulus within the sample, showed the most stable and reliable contrasts.

Figure 4.4 illustrates these stiffness alterations (bottom). For both nanocomposites' samples (HP14 and HP14T) dark "halos" surrounding light "dots" can be observed. As the lighter regions are the stiffer ones, it can be assumed, that the particles are surrounded by a clearly soft matrix. These soft regions are even softer than the remaining polymer.

After the viscosity analysis and the reactions with the model substance these pictures could be a further piece of the puzzle showing that both, the particles HP14, and the modified ones, HP14T, change the polymer network pattern in the close-up range around the particles. Thus a softer and less dense polymer, possibly a resin enriched zone, is formed. The close-up range extends up to 200 nm for HP14T and 100 nm for HP14 around each particle. While Wetzel [13] assumes a stronger network around his particles, Holst [14] and Solomko [15] on the other hand describe the contrary phenomenon, as can be found in this work: supramolecular structures which allow fewer cross-links and therewith a reduced network density compared to the unaffected matrix. At which moment these interphases are built is yet unclear. But it can be assumed that first bonds occur while dispersing the particle powder into the resin [5].

4.4 Selected Properties and the Nanocomposites' Particle-Network

The pictures of the FFM measurements give a clear indication towards the creation of soft zones around the incorporated particles. This understanding is supported by kinetic results. Nanocomposites with different filler contents were analyzed by the means of dynamic differential scanning calorimetry (DSC). For the particles HP14, as well as for the particles HP14T, a decreasing reaction enthalpy (released heat calculated per gram reactive resin) with increasing filler content was observed. This indicates the sterical hindrance of the cross linking [11]. At the surface of the particles further creation of the network is not possible and the total degree of cure decreases.

Another interesting result is shown by the DSC measurements. Comparing the reaction rates of the cure of the nanocomposites a striking observation can be made. The results show an increase of reaction rate by the incorporation of the pure boehmite particles, while the taurine modified ones decelerate the cure. In literature [16] a catalytic effect of boehmite on the ring opening of epoxies is described. This observation hints again towards the function of taurine as a spacer between particle and resin. The hindrance of direct contact blocks the catalytic effect of the boehmite surface. This observation clearly correlates with the results of the viscosity and the FFM-measurements.

The cured nanocomposites filled with different particle contents from 0 to 15 wt% were characterized by three-point bending (DIN EN ISO 178) and tension (DIN EN ISO 527) tests.

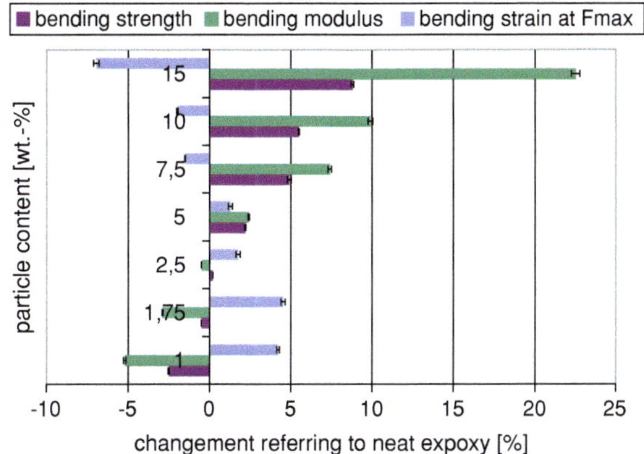

Fig. 4.5 Results of bending tests for nanocomposites filled with a varying content of HP14T

In Fig. 4.5 the flexural modulus, the bending strength and the strain at maximum force as results from the bending tests are depicted. With increasing filler content a decreasing strain at maximum force and increasing bending strength and flexural modulus related to the pure resin are observed. Surprisingly the nanocomposites with low filler content, in particular with 1 and 1.75 wt%, have the opposite effects as described.

In contrary to the bending results, no effect of the low filler contents could be observed for the tensile tests. The results show an increasing youngs modulus and tensile strength and a decreasing strain at maximum force with raising filler contents. For both tests no significant difference between the particles HP14 and HP14T occurred.

From the described results a model of the reinforcement of epoxy resins by nanoparticles is concluded. Nanoparticles create a soft zone in the matrix due to the sterical hindrance of the cure. This leads to increased network mobility. For that reason the nanocomposite becomes less stiff at low filler contents. At higher filler contents the strengthening of the particles becomes more and more dominant. The mechanical performance of the nanocomposite increases. Figure 4.6 shows a model of the described effects.

4.5 Conclusion: Nanoparticles' Mode of Action in CFRP

Incorporating up to 15 wt% surface modified boehemite nanoparticles into the polymer results in increased modulus, strength and fracture toughness of the nanoparticle reinforced polymers. Therefore, this can considerably enhance properties of CFRP. The shear strengths [interlaminar shear strength (ILSS) and ±45° shear strength (IPSS)], the compression strength (CS), their damage tolerance [compression after impact (CAI)] and the fiber-matrix adhesion [5], which are particularly limiting parameters for highly stressed aerospace materials, can be improved by up to 25%. Various effects are responsible:

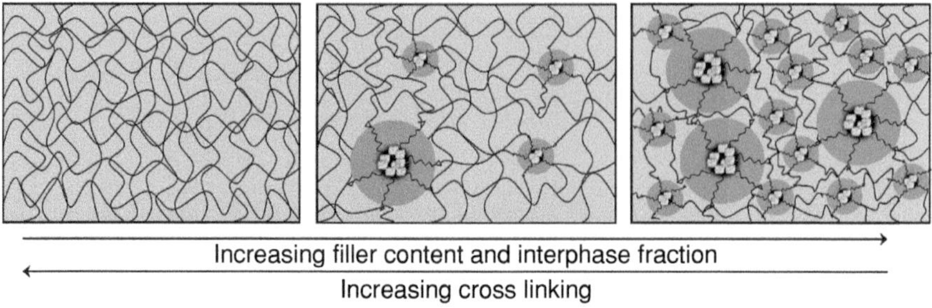

Fig. 4.6 Cross linking model of change in network by incorporation of nanofillers

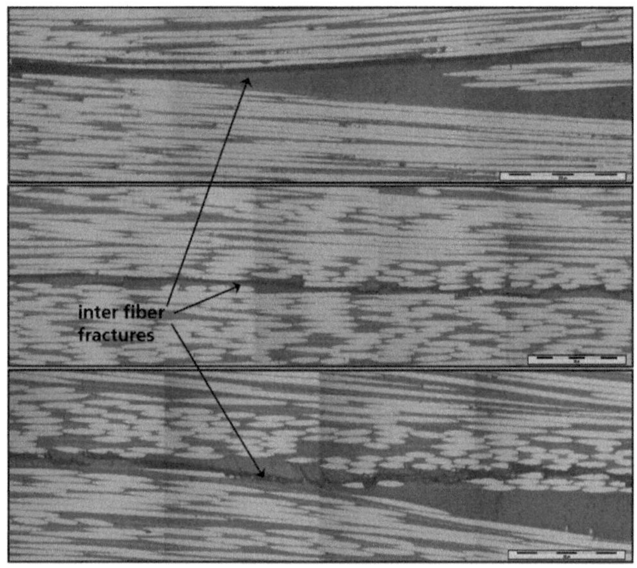

Fig. 4.7 Micrographs of ILSS-samples (HTA fibers, 59–61 vol.% fiber content, *top*: reference, *middle*: 7.5 wt% HP14T, *bottom*: 15 wt% HP14T): Differing crack paths in matrix (*dark*) between fibers (*light*). (Measuring bar: 200 μm)

The enhanced resin stiffness and strength prevent fiber buckling while

1. reduced initial stresses [5] lead to decelerated crack propagation and delamination growth, and, therefore, to a higher resistance against first ply failure and
2. very effective crack deflection, crack pinning and crack bridging processes [5, 17] (differing crack paths; Fig. 4.7) lead to a disproportionately enhanced interlaminar energy release rate (G_{1c}).
3. Furthermore with wisely chosen particle surface molecules, the resin-particle interactions can be adjusted. Also a new network pattern can be built, which, when combined with lower resin shrinkage, may lead to significantly higher fiber-matrix adhesion.

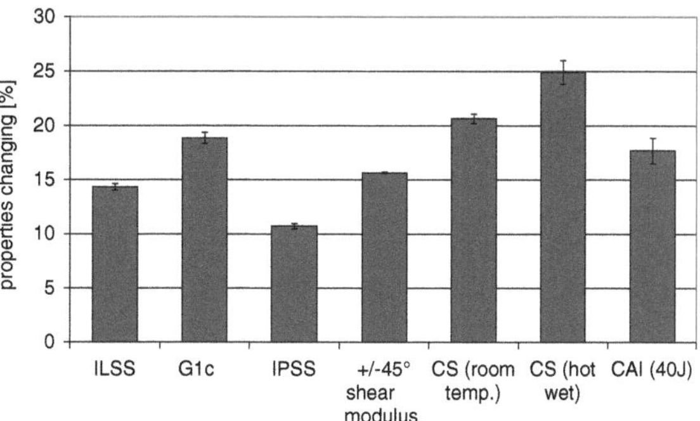

Fig. 4.8 Changed mechanical properties of boehmite nanoparticles reinforced CFRP compared to non-doped CFRP (all laminates: HTA fibers, 59–61 vol.% fiber content)

In summary all these attributes result in a considerably upgraded new CFRP (Fig. 4.8).

Acknowledgments The authors would like to thank the Helmholtz Association for the funding of the virtual institute 'nanotechnology in polymer composites'. Under its umbrella most of the research could be done. Furthermore the authors thank Sasol Germany GmbH for providing the particles and for the great cooperation all the years.

References

1. Fu, S., Feng, X., Lauke, B., Mai, Y.: Effects of particle size, particle/matrix interface adhesion and particle loading on mechanical properties of particulate-polymer composites. Compos. B **39**, 933–961 (2008)
2. Mahrholz, T., Stängle, J., Sinapius, M.: Quantitation of the reinforcement effect of silica nanoparticles in epoxy resins used in liquid composite moulding processes. Compos. A **40**, 235–243 (2009)
3. Schilde, C., Arlt, C., Kwade, A.: Influence of dispersing process in the production of nanoparticle composites. Chem. Ing. Tech. **81**(6), 775–783 (2009)
4. Shahid, N., Villate, R., Barron, A.: Chemically functionalized alumina nanoparticle effect on carbon fiber/epoxy composites. Compos. Sci. Technol. **65**, 2250–2258 (2005)
5. Arlt, C.: Nanoscaled boehmites' modes of action in a polymer and its carbon fiber reinforced plastic under compression load. Doctoral Thesis, Otto-von-Guericke-Universität Magdeburg (2011)
6. Hosseinpour, D., Guthrie, J., Berg, J., Stolarski, V.: The effect of interfacial contribution to the mechanical properties of automotive topcoats. Prog. Org. Coat. **54**, 1882–1887 (2005)
7. Shukla, D., Kaisisomayajula, S., Parameswaran, V.: Epoxy composites using functionalized alumina platelets as reinforcements. Compos. Sci. Technol. **68**, 3055–3063 (2008)
8. Guo, Z., Pereira, T., Choi, O., Wang, Y., Hahn, H.T.: Surface functionalized alumina nanopar-ticle filled polymeric nanocomposites with enhanced mechanical properties. J. Mater. Chem. **16**, 2800–2808 (2006)
9. Zunjarrao, S.C., Singh, R.P.: Characterization of the fracture behavior of epoxy reinforced with nanometer and micrometer sized aluminum particles. Compos. Sci. Technol. **66**, 2296–2305 (2006)
10. Mukherjee, N., Wavhal, D., Timmons, R.B.: Composites of plasma surface functionalized barium titanate nanoparticles covalently attached to epoxide matrices: synthesis and evaluation. Appl. Mater. Interfaces **2**(2), 397–407 (2010)

11. Putz, K.W., Palmeri, M.J., Cohn, R.B., Andrews, R., Brinson, L.C.: Effect of cross-link density on interphase creation in polymer nanocomposites. Macromolecules **41**, 6752–6756 (2008)
12. Karlsson, P.M., Baeza, A., Palmuvist, A.E.C., Holmberg, K.: Surfactant inhibition of aluminium pigments for waterborne printing inks. Corros. Sci. **50**, 2282–2287 (2008)
13. Wetzel, B.: Mechanische Eigenschaften von Nanoverbundwerkstoffen aus Epoxydharz und keramischen Nanopartikeln. Dissertation, Technische Universität Kaiserslautern (2006)
14. Holst, M.: Reaktionsschwindung von Epoxidharz-Systemen. Dissertation, Technische Universität Darm-stadt (2001)
15. Solomko, W.P.: Eine Modellvorstelleung über Polymer-Füllstoff-Systeme. Plaste Kautsch. **22**(11), 858–860 (1975)
16. Hoebbel, D., Nacken, M., Schmidt, H.: The effect of nanoscaled metal oxide sols on the structure and properties of glycidoxypropyltrimethoxysilane derived sols and gels. J. Sol-Gel. Sci. Technol. **19**, 305–309 (2000)
17. Fröhlich, J.: Nanostructured thermoset resins and nanocomposites containing hyperbranched blockcopolyether liquid rubbers and organophilic layered silicates. Doctoral Thesis, Universität Freiburg (2003)

Chapter 5
Advanced Flame Protection of CFRP Through Nanotechnology

Alexandra Kühn and Michael Sinapius

Abstract This chapter refers about the optimization of fire resistance of CFRP. For this optimization the most promising nano scaled additives are used and varied regarding the particle content, size and effect of flame retardancy. One major challenge is to optimize the particle dispersion and to determine the optimal particle concentrations in consideration of the effect of flame retardancy and the resulting material properties. Additionally a fire testing method has to be determined that resolves the potentially small differences in the used variations. Therefore standard fire and mechanical tests are used as well as a simple thermal material method, given with the thermo gravimetric analysis (TGA) including a single differential thermal analysis (SDTA) that also suits for a fast comparison of the materials fire properties. Hereby obtained results are compared and a related behaviour of the fire properties can be shown between standard fire tests and thermal material tests using the TGA-device.

5.1 Protection Against Fire

A major fire is normally defined as an independent spreading fire causing damages and health impairments to objects and people. The causes of a fire can be various. Especially regarding the safety of civil aircrafts the fire prevention is of particular importance. A fire scenario of great relevance is the so called Post Crash Fire.

A. Kühn (✉) · M. Sinapius
Institute of Composite Structures and Adaptive Systems,
German Aerospace Center DLR, Lilienthalplatz 7, 38108,
Braunschweig, Germany
e-mail: Alexandra.Kuehn@dlr.de

M. Sinapius
e-mail: michael.sinapius@dlr.de

M. Wiedemann and M. Sinapius (eds.), *Adaptive, Tolerant and Efficient Composite Structures*, Research Topics in Aerospace, DOI: 10.1007/978-3-642-29190-6_5,

A Post Crash Fire is a fire which can occur after a crash caused by leaking kerosene and an ignition source. In this case the mechanical stability has to be guaranteed until survivors and casualties have been evacuated. There is not only the risk of heavily loaded components falling down and injuring people but also the immediate danger through the release of great heat, smoke and toxic gases. Especially during the combustion of fiber composites containing polymer matrices many reaction products are released which can not be characterized as harmless. In addition to the strong emission of smoke and soot which probably block escape routes these are unhealthy gases and toxic fiber particles in the air breathed. While the term fire protection mostly describes concrete actions to extinguish a fire the flame retardancy directly influences the material itself so that a fire ideally can be prevented. One possibility is the usage of hardly or non flaming materials another option is to use additives within or at the component which effectively optimize the fire properties. The importance of flame retardants increases continuously [1, 2].

5.1.1 Flame Retardants for Fiber Composites

Though CFRP laminates are already widespread used and are still prevailing for further applications they show some critical drawbacks that prevent a more extensive use. Especially the low fire resistance of CFRP mainly due to the plastic matrix is a problem to be solved. Generally flame retardants are applied as coatings [3] that disadvantageously increase the components weight. An approach to prevent this disadvantage could be the modification of the polymeric matrices of CFRP by nano particles. This modification bears the opportunity of material integrated flame retardancy for CFRP manufactured with the injection technology. On the one hand the usage of nano particles is compulsory to prevent the particles from being filtered out by the fiber semi finished products with diameters of a few microns during the injection process. On the other hand nano scaled particles have a disproportionately higher surface than the same amount of larger particles. This bears the possibility of an even further enhancement of the effect of flame retardancy where for example a cooling effect is transmitted by the particles surface. In addition to the improvement of the flame properties of CFRP material integrated particles can lead to a structural reinforcement in case of a fire. Another demand for the daily use is that the mechanical properties of the materials like stiffness and strength should not be influenced negatively by the application of these modifications. These effects can not be reached by classical micro particles but due to nano particles as already shown in previous chapters of this book. Subsequently the integrated flame retardants used for the optimization of the flame properties of CFRP have to be nano scaled. For the selection of suitable particles based on nanotechnology the combustion process has to be regarded. A fire to get started and keep going needs the following prerequisites as pictured in Fig. 5.1 (left): heat, oxygen and burnable material. If these prerequisites are present the heat input initiates the pyrolysis. This is followed by an oxidation of the pyrolysis products

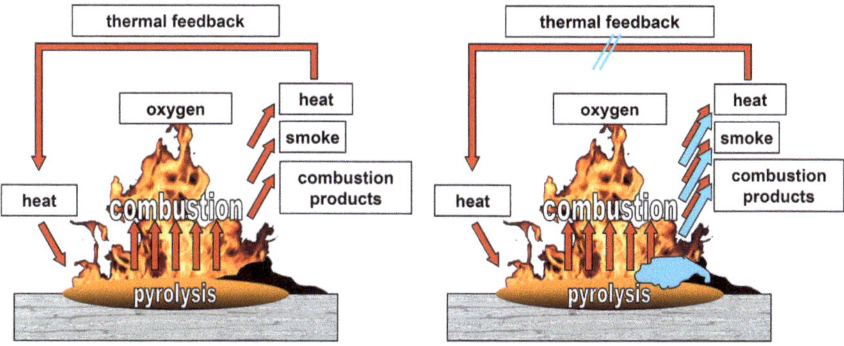

Fig. 5.1 Schematic combustion process (*left*); influence of flame retardants on combustion process (*right*) [1]

resulting in further heat, smoke and several combustion products. The heat thereby produced in turn results in a thermal feedback keeping the combustion going. Due to the oxidation of pyrolysis products a char is formed acting as a flame barrier.

The combustion process can be influenced by flame retardants as pictured in Fig. 5.1 (right) either by building an additional intumescing barrier or char preventing the oxygen from feeding the base of fire or by diluting the combustion products through inert gas or by actively cooling through for example the release of water and therefore preventing thermal feedback. Nanotechnology offers several particle systems which are already well examined in literature for the use as flame retardants. Suitable particle systems are carbon nano tubes (CNT) which function due to thermal conduction cooling down the base of fire, organoclays building an intumescing barrier layer and aluminium trihydroxide (ATH) and magnesium dihydroxide (MDH) known for splitting up water at definite temperatures and therefore cooling down and diluting the combustion products. All particle systems integrated into the matrix additionally reduce the burnable material by their embedded content. It depends on the dispersion quality if there is a further use due to the nano particles small size. Nano particles tend to agglomerate as described in the sections before so the use of dispersion devices is necessary to separate the agglomerates ideally up to the particles primary size. If possible several predefined agglomerate sizes can be produced to examine the influence of this value on the resulting effect of flame retardancy [1–5].

5.1.2 Fire Tests and Supplemental Characterizations

To evaluate the optimizations achieved by the modifications used in this research small scale fire tests as well as thermal tests were made. Due to these tests it is possible to determine whether the use of flame retardants is necessary or basically suitable. However, these coupon tests do not represent the real properties of a structural component exposed to fire. Therefore original scale components have to

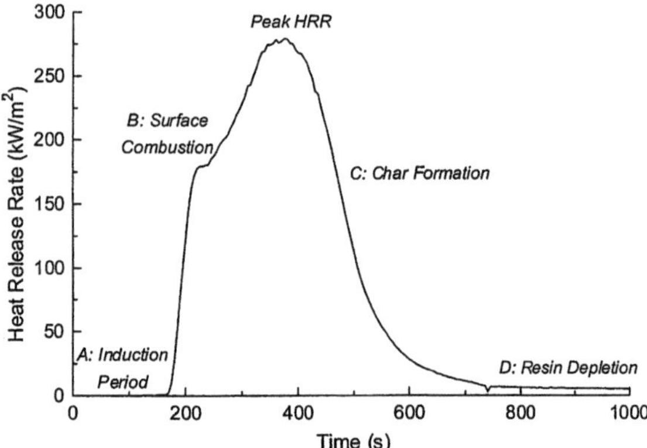

Fig. 5.2 Schematic picture of the heat release rate of a polymer [1]

be tested in a suitable experimental set-up with authentic loads in a realistic fire scenario. The major drawback of small scale coupon tests is that normally standardized burners are used and the experimental set-up does not meet the conditions of a real fire. The same applies to the release of the evaluated smoke and toxic gases giving no directly usable prediction about a realistic toxicity as for example the carbon monoxide concentration of a lab fire may significantly differ from the concentration resulting from a real fire. Though the deliverables of lab scale tests differ from these of a realistic fire scenario the results are of great importance for the estimation of the fire behaviour and toxicity of materials.

While the standard fire-smoke-toxicity (FST) tests qualitatively picture the behaviour of a fire especially in the case of fiber reinforced polymer composites, the most important quantitative value to be measured in a fire is the rate of the heat released representing the released energy which is schematically shown in the Fig. 5.2. Hereby each material has a specific character which can be measured using a cone calorimeter or an Ohio-State-University (OSU) calorimeter as widespread used and accepted standard methods.

Also important is the progression of the mass decomposition which also can be recorded by the cone calorimeter or using small scale tests performed with the thermo gravimetric analysis.

For this research flammability tests performed by a Bunsen burner according to AITM 2.0002B were chosen to visualize the optimization due to the modification of the polymer. The flammability test samples were 75×305 mm in size with a thickness of 2 mm and they were exposed to the Bunsen burner flame for 12 s. Afterwards the after burn duration was recorded and the burning notch length was measured.

For a comprehensive material characterization the quantity of heat produced at combustion was measured by a thermo gravimetric analysis TGA/SDTA 851 (Mettler Toledo, Germany) containing a single differential thermal analysis realized by the usage of a temperature sensor placed near the probe. The temperature of the combustion- or oven chamber is predetermined by the chosen temperature program. Knowing the probes temperature as well as the oven temperature the heat flow can be determined and consequently the quantity of heat. The weighing cell of the device records the probes weight during the whole measurement procedure. Therefore two important measured variables concerning the combustion process could be determined using the TGA. These measurements were conducted using a mixture of oxygen and nitrogen at a mixture ratio of 2:1 as a purge gas and a gas flow rate of 60 ml/min. The intension was to use a gas mixture with higher nitrogen content for a more realistic measurement environment. Unfortunately a mixture of both gases could not contain more than 20 ml/min of nitrogen. This is a limitation given by the device. Therefore a mixture of both gases could not simulate the conditions of air without further equipment. The temperature programme for this measurement was set to run from 200–600°C and the heating rate was 4 K/min to accelerate the measurement procedure. Samples with an initial weight of 25 mg were used.

The pyrolysis combustion flow calorimetric (PCFC) measurements were conducted using samples of 2 mg in size. The temperature programme for this measurement was set to run from 75–750°C and the heating rate was 60 K/minute. These measurements were conducted using a mixture of oxygen and nitrogen at a mixture ratio of 1:4 as in air and a gas flow rate of 100 ml/min.

The research was conducted with a Cone Calorimeter (Fire Test Technology, East Grinstead, UK) according to ISO 5660. The samples of the Cone Calorimeter were 100×100 mm in size with a sample thickness of 2 mm and were conditioned for at least 48 h at 23°C and 50% relative humidity before the measurement. The measurement was conducted with an irradiance of 50 kW/m^2 at a distance of 25 mm between the radiation source and the samples surface [6–12].

5.2 Materials and Methods

For the optimization of fire properties different types of nanoparticles are available as presented in the previous section. Some of these particles are commonly available as micron particles and are already widespread used. There are already several studies focussing on the effect of flame retardancy of CNTs and organoclays respectively nanoclays which can be found in literature [13–17]. The examinations of the fire properties of CFRP herein focus on the use of ATH-nanoparticles provided by the company Sasol [18].

5.2.1 Nanoparticles and Resins

The ATH particles by the company Sasol were delivered in several versions differing in particles size, surface modification and modification amount realized through the modification during different manufacturing process steps of the particles.

The resin used in this study is an epoxy resin which is partly well approved in aircraft industry. It is a three-component diglycidyl ether of bisphenol A (DGEBA) based epoxy resin with a T_G of 120°C (Araldite LY556 from Huntsman), the anhydride curing agent HY917 and the amine accelerator DY070. This resin system is cured for 4 h at 80°C and postcured for 4 h at 120°C.

5.2.2 Dispersion Process and Material Characterisation

Several dispersion devices were available for the examination as for example a dissolver, a torus mill, a ball mill and later during the last period of the examination a calender also usable for highly vicious systems containing high particle contents. During this process the particles provided by the company Sasol could be optimized themselves in respect of the surface modification with the aim of a better dispersion quality. As the result a modification of the particles surface with taurine showed the best dispersion quality. This optimization was accomplished by the company Sasol taking into account the results of the dispersion examination. To produce a particle-resin-masterbatch during the dispersion process only the resin component is used to avoid curing during this process step. For further processing the particle-resin-batch is mixed with the hardener and accelerator component. The dispersion quality is verified by the SEM. The ATH particle system by the company SASOL was used for a comprehensive material characterisation giving a detailed correlation between the particle content and the resulting reduction of the exothermal combustion heat. These tests were performed by the TGA as explained in the previous section [19].

5.3 Results and Discussion

5.3.1 Thermal Characterization of Very Small Scale Test Specimens

For the purpose of a comprehensive material characterization in addition to the nano particle modified polymer with particle contents ranging from 5 to 40% in steps of 5% the particle powder was measured as well solely in order to achieve the expected reduction of the combustion heat. Due to this knowledge the expected reduction of the combustion heat could be calculated for all measured particle contents. The results are given in Fig. 5.3.

It can be seen that the statistical spread of the measurement is comparatively high for very high particle contents as well as the neat resin specimens. For high particle contents this is presumably caused by a worse dispersion quality compared to lower particle contents. The huge statistical spread of the neat resin samples

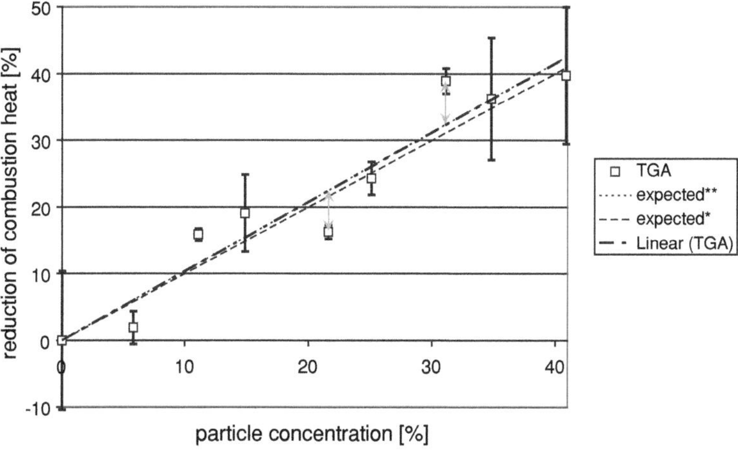

* calculated curve giving the expected reduction of the combustion heat induced
  by the reduction of the burnable material due to the particles

** calculated curve giving the expected reduction of the combustion heat induced
   by the reduction of the burnable material due to the particles and by the cooling
   effect of the particles

Fig. 5.3 Expected and measured reduction of combustion heat depending on the particle content
of ATH1

results from the used gas flow program which obviously was not appropriate for
the high combustion heat release peaks of the neat resin probes showing a very
discontinuously curve character. Two expected curve characteristics are shown in
the diagram. One expected* curve represents the reduction of the combustion heat
induced by the reduction of the burnable material due to the particles, the second
expected** curve also takes into account the cooling effect of the particles. The
trend line of the measured results hereby corresponds well to the expected** curve
characteristics which is not a surprise and certifies the test method. For some
particle contents the deviation from the expected** curve is significantly high as
for example regarding the particle concentrations of 20 and 30%. Where the
measured value is below the expected** curve the particles are not as efficient as
otherwise. This could result through a bad dispersion quality and many micro
scaled agglomerates. The minimum reduction of the combustion heat still
shouldn't be below the expected* curve resulting from the reduction of the
burnable material. For the statistical spread of the particle content was lower than
the deviation in the reduction of the combustion heat some effects of the particles
increasing the combustion heat of the sample have to be taken into account. This
could be an increase of the samples surface due to the particles. Where the
measured results surprisingly show a greater reduction of the combustion heat this
is supposed to be caused by the increased specific surface of the thoroughly
dispersed nano particles for the previously measured particle powder is agglom-
erated and therefore not nano scaled but micro scaled. This nano effect concerning

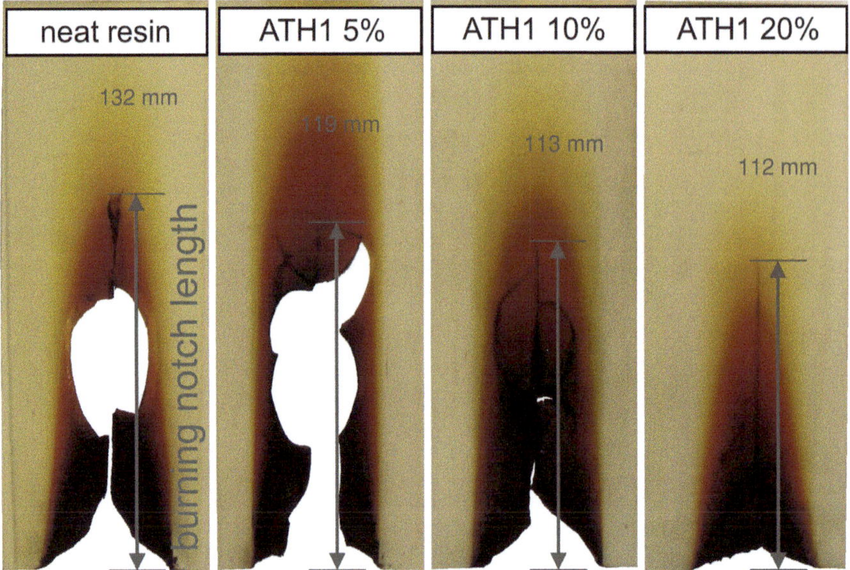

Fig. 5.4 Flammability test specimen with varying particle contents of ATH1

the flame retardancy will be researched and confirmed in further investigations. Obviously there is a linear correlation between the reduction of combustion heat and the particle content.

5.3.2 Comparison to Standard Fire Test Methods

The behaviour showing a linear correlation between the reduction of the fire properties and the particle content seen in the thermal material characterisation corresponds to the results of the flammability test. These tests prove the reduction of the burning notch length of ATH1-modified resin probes according to the increase of the particle content as shown in Fig. 5.4. Probes with an ATH1-Particle content of 20% achieved an improvement of 15% which is a little bit less than expected due to the modification and measured using the TGA.

The results of the examinations using the PCFC and the Cone calorimeter are pictured in Fig. 5.5 next to the results of the examinations presented in the previous section. The very small scale probes of the PCFC show similar results as the TGA test specimens. The trend leads to even higher reductions of the combustion heat. This even exceeds the expectations due to the measured cooling effect of the particles and supports the theory that there is an increased effect using nano particles instead of standard micro scaled particles. On the other hand the results of the examinations using the cone calorimeter show a contrary effect. Comparable to

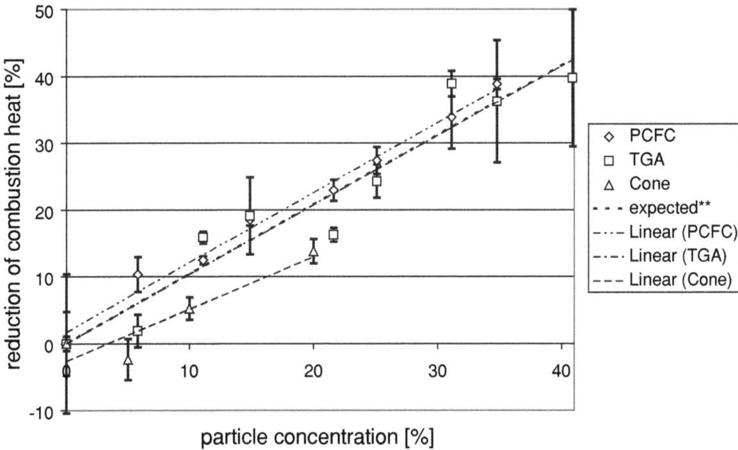

Fig. 5.5 Reduction of combustion heat depending on the particle content of ATH1 measured with PCFC, TGA and cone calorimeter

the trend of the flammability test a reduction of the combustion heat of only about 15% could be reached using a particle content of 20%. For low particle contents of about 5% even an increase of the combustion heat can be detected which supports the consideration of a flammability increasing effect due to the particles.

Very small scale tests seem to profit more from the modification of the polymer with nano scaled ATH than larger scale test configurations though a good trend of the results can be given using very small scale tests which makes them suitable for a comprehensive material characterisation.

Additionally it could be shown, that the nano scaled particles ATH1 by the company SASOL are well suitable as flame retardants for polymers used in CFRP.

References

1. Mouritz, A.P., Gibson, A.G.: Fire properties of polymer composite materials. Springer, Berlin (2006)
2. Quintiere, J.G., Walters, R.N., Crowley, S.: Flammability Properties of Aircraft Carbon-Fiber Structural Composite, Final Report, Federal Aviation Administration (2007)
3. Dodds, N., Gibson, A., Dewhurst, D., Davies, J.: Fire behaviour of composite laminates. Compos. Part A Appl. Sci. Manuf. **31**, 689–702 (2000)
4. Troitzsch, J., Becker, W., Bönold, H., Kaufmann, W., Prager, F.H., Rupprecht, H., Sand, H., Schiffer, H.-W.: International plastics flammability handbook: principles–regulations–testing and approval, 2nd edn. Carl Hanser Verlag, München (1990)
5. Ratna, D.: Epoxy Composites: Impact Resistance and Flame Retardancy, vol. 16, No. 5. Rapra Technology, Wales (2005)
6. Smith, E.E.: Heat Release rate calorimetry, fire technology fourth quarter (1996)

7. Babrauskas, V.: Development of the Cone Calorimeter-a Bench-Scale Heat Release Rate Apparatus Based on Oxygen Consumption. National bureau of standards, Washington. Fire Mater. **8**(2), (2004)
8. Lindholm, J., Brink, A., Hupa, M.: Cone Calorimeter—A Tool for Measuring Heat Release Rate, Åbo Akademi Process Chemistry Centre, Finland (2009)
9. Kandola, B.K., Horrocks, A.R., Padmore, K., Dalton, J., Owen, T.: Comparison of Cone and OSU Calorimetric Techniques to Assess the Flammability Behaviour of Fabrics used for Aircraft Interiors. Centre for Materials Research and Innovation, University of Bolton, UK. Fire Mater. **30**(4), 241–255 (2006)
10. Schartel, B., Bartholmai, M., Knoll, U.: Some comments on the use of cone calorimeter data. Polym. Degrad. Stab. **88**, 540–547 (2005)
11. Lyon, R.E., Walters, R.N.: Pyrolysis combustion flow calorimetry. J. Anal. Appl. Pyrolysis **71**, 27–46 (2004)
12. Morgan, A.B., Galaska, M.: Microcombustion calorimetry as a tool for screening flame retardancy in epoxy. Polym. Adv. Technol. **19**, 530–546 (2008)
13. Beyer, G.: Short communication: carbon nanotubes as flame retardants for polymers. Fire Mater. **26**, 291–293 (2002)
14. Schartel, B., Pötschke, P., Knoll, U., Abdel-Goad, M.: Fire behaviour of polyamide 6/multiwall carbon nanotube nanocomposites. Eur. Polymer J. **41**, 1061–1070 (2005)
15. Kashiwagi, T., Grulke, E., Hilding, J., Groth, K., Harris, R., Butler, K., Shields, J., Kharchenko, S., Douglas, J.: Thermal and flammability properties of polypropylene/carbon nanotube Nanocomposites. Polymer **45**, S 4227–4239 (2004)
16. Schartel, B., Bartholmai, M., Knoll, U.: Some comments on the main fire retardancy mechanisms in polymer nanocomposites. Polym. Adv. Technol. **17**, 772–777 (2006)
17. Beyer, G.: Filler blend of carbon nanotubes and organoclays with improved char as a new flame retardant system for polymers and cable applications. Fire Mater. **29**, 61–69 (2005)
18. Brasch, A., Erfinder; Sasol Germany GmbH, Patentinhaber. Verfahren zur Herstellung von Aluminiumtrihydraten mit hohem Porenvolumen. Deutsches Patent DE 103 32 776 B4 2009.04.09 (2009)
19. D. Könnicke, A. Kühn, T. Mahrholz, M. Sinapius: Polymer nanocomposites based on epoxy resin and ATH as a new flame retardant for CFRP: preparation and thermal characterization. J. Mater. Sci. **46**(21), 7046–7055 (2011)

Chapter 6
Fundamental Characterization of Epoxy-Silica Nanocomposites Used for the Manufacturing of Fiber Reinforced Composites

Thorsten Mahrholz and Michael Sinapius

Abstract Nanocomposites based on silica nanoparticles and high performance epoxy resins are investigated for their suitability as a new type of matrix for fiber-reinforced polymers (FRP) using injection technologies (LCM). The key focus is on the determination of the processing parameters at varying silica nanoparticle content. The homogeneous distribution of the nanoscaled silica in the epoxy matrix is proven by photon cross correlation spectroscopy (PCCS) and scanning electron microscopy (SEM) analysis. Depending on the silica content of the composite, its stiffness, strength and toughness can be increased significantly compared with the neat resin. The mechanical performance is discussed by failure mechanisms based on the analysis of the fracture surface morphology. Moreover, resin shrinkage and the thermal expansion are significantly reduced both important for lowering internal stress in FRP. The injectability of the nanocomposite for the purpose of lamination using the LCM technology is nearly unaffected. Epoxy-silica nanocomposites are now proven to be a new high performance polymer matrix for FRP structures manufactured by the low cost LCM techniques.

6.1 Introduction

Prepreg technology is currently the most widely used manufacturing technique for making high performance fiber composites (FRP). Despite good composite quality, this manufacturing technique carries the disadvantages of high manufacturing

T. Mahrholz (✉) · M. Sinapius
Institute for Composite Structures, German Aerospace Center DLR, Lilienthalplatz 7,
38108, Braunschweig, Germany
e-mail: thorsten.mahrholz@dlr.de

M. Wiedemann and M. Sinapius (eds.), *Adaptive, Tolerant and Efficient Composite Structures*, Research Topics in Aerospace, DOI: 10.1007/978-3-642-29190-6_6, © Springer-Verlag Berlin Heidelberg 2013

costs and minimal potential cost savings. Injection techniques (Liquid Composite Moulding—LCM), such as RTM, VARI, DP-RTM and SLI are alternative manufacturing techniques that have been developed in recent years [1–3] and are already in industrial application. Low manufacturing costs achieved through a combination of inexpensive resins and semi finished fiber materials are decisive advantages compared with the prepreg technology. However, the properties of high performance composites produced by LCM techniques have not yet reached the level of the Prepreg composites. This is essentially caused by the matrix shrinkage in the polymer system, which leads to internal stress reducing the material performance. Traditional microfillers which are investigated for reducing the matrix shrinkage lead to manufacturing problems i.e. increased viscosity and filtration effects and brittleness of the matrix. In order to avoid these problems so called nanocomposites i.e., thermosetting resins filled with nanoparticles (1–100 nm) are proposed.

In the present study nanoscaled silica particles produced in a sol–gel technique [4, 5] are focused on to eliminate main disadvantages of the LCM technology and, at the same time, to increase the material composite qualities. Espcially the thermal, mechanical and rheological properties of the nanocomposites are thoroughly studied. Moreover the influences of nanoparticles on the macroscopic properties of the polymer matrix are discussed with the fracture surface morphology as a reference. Finally the suitability of this new type of polymer matrix for manufacturing of fiber composite materials through injection technology is assessed. The long term objective of the investigation is to use nanoparticles for making tailored higher performance reaction resin available for the manufacturing of FRP by LCM techniques. The use of nanocomposites as a new type of polymer matrix will significantly improve the spectrum of properties of FRP, thus expanding the range of applications of fiber composite structures.

6.2 Materials and Preparation

The material used in this study is an epoxy resin, diglycidyl ether of bisphenol A (DGEBA) (Araldite LY556; Huntsman), cured by an anhydride curing agent, 4-methyl-1,2-cyclohexanedicarboxylic anhydride (Aradur HY917; Huntsman) and accelerated by an amine, 1-methyl-imidazole (DY070; Huntsman). The nanoparticle system is a commercially available spherical silica pre-dissolved in a bifunctional epoxy resin (similar to Araldite LY556). The master batch provided by Hanse-Chemie AG (Germany) consists of 40 wt% silica nanoparticles. These particles are manufactured by means of a sol–gel technique and grow directly in the polymer matrix [4, 6]. The primary particle size can be adjusted through quenching processes and is within a range of 8–50 nm (note of the producer). The particle surface was also modified by Hanse-Chemie AG with a reactive organic silane which allows a polymerisation directly with the resin matrix and prevents particle agglomeration.

Table 6.1 Thermal parameters of the prepared epoxy-silica nanocomposites

Sample	SiO_2 (wt%)	ρ^a (g/ cm^3)	ΔH^b (J/ g)	T_g^b (DSC) (°C)	T_g^c (DMA) (°C)	HDT^d (°C)	$\alpha_{T<Tg}^e$ (10^{-6}/K)	$\alpha_{T>Tg}^e$ (10^{-6}/K)
NA-0	0	1.209	384	130.8	126.1	116.8	62.0	184.0
NA-5	5	1.235	369	132.3	127.5	118.3	58.4	180.3
NA-15	15	1.286	324	127.4	126.2	116.4	54.7	171.1
NA-25	25	1.340	280	127.9	122.6	112.5	45.8	158.9

[a] Specific gravity measured for the cured samples at 25°C
[b] Specific reaction enthalpy (ΔH) and glass transition temperature (T_g) determined by DSC
[c] Glass transition temperature (T_g) determined by DMA (tanδ)
[d] Heat distortion temperature determined as onset temperature from DMA curves
[e] Coefficient of thermal expansion (CTE) at various temperature ranges

The influence of the nanoscaled silica concentration on the range of properties of the epoxy matrix (Araldite LY type) was investigated in a series of tests in order to determine the best concentration range for FRP production. Therefore the master batch of silica (40 wt%) was first diluted with the neat epoxy resin and afterwards well mixed with the hardener and accelerator. In each case the mix ratio of the epoxy resin (Araldite LY556), the hardener (Aradur HY917) and the accelerator (DY070) was 100: 90: 0.5 wt%. The silica concentration was adjusted within a range of 0–25 wt%. After a short de-gassing process the mixture was poured into a pre-heated aluminium mould. The modified resins were pre-cured at 80°C for 4 h and post-cured at 120°C for 4 h. Cured samples were then cut and ground for mechanical testing. The filled and unfilled matrix resins were characterised extensively in terms of their thermal, rheological and mechanical properties. More details can be taken from a former published paper [7]. An overview of the prepared nanocomposites as well as the corresponding thermo-mechanical parameters is given in Tables 6.1 and 6.2.

6.3 Characterization of the Nanocomposites

6.3.1 Analysis of the Nanoparticle Distribution

It is well known that the degree of dispersion of nanoparticles in a polymer matrix is a governing parameter which controls the final properties of the resulting nanocomposites. Only an extremely homogeneous particle distribution with the development of primary nanoparticles efficiently reinforces the polymer matrix. Thus, the quality of the dispersions manufactured as a mixture of the silica master batch, the neat epoxy resin, the anhydride hardener and accelerator are investigated for a broad range of silica content. The liquid and solid states are analyzed by means of photonen cross correlation spectroscopy (PCCS) and scanning electron microscopy (SEM), respectively. SEM images prove the homogeneous

Table 6.2 Mechanical parameters of the prepared epoxy-silica nanocomposites

Sample	SiO$_2$	Tension test			Flexural test		G$_{1c}^a$	K$_{1c}^a$
	(wt%)	E$_t$ (MPa)	σ_t (MPa)	ε_t (%)	E$_f$ (MPa)	σ_f (MPa)	(J/m^2)	(MPa m$^{1/2}$)
NA-0	0	3359	88	3.26	3484	157	77	0.517
NA-5	5	3526	91	4.51	3607	160	137	0.713
NA-15	15	3978	95	5.10	4048	165	177	0.865
NA-25	25	4555	98	4.40	4525	171	196	0.957

[a] Fracture toughness parameters: G$_{1c}$—critical energy release rate; K$_{1c}$—critical stress intensity factor

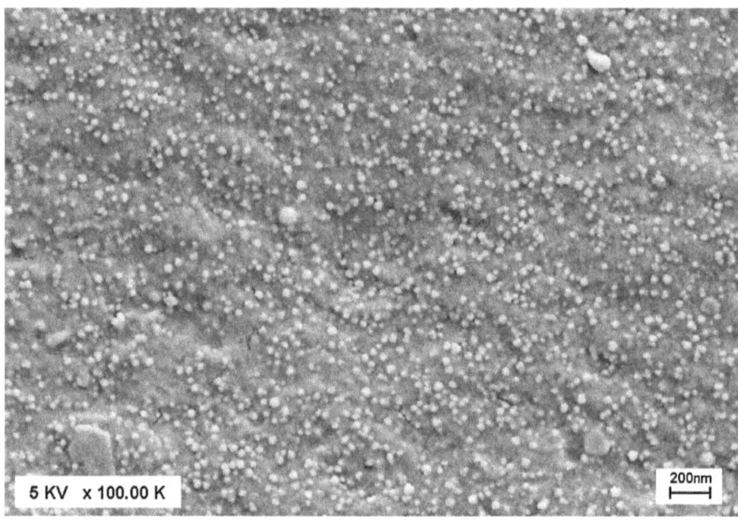

Fig. 6.1 SEM image of a nanocomposite based on silica and epoxy resin (25 wt% silica). Surface prepared by cryoscopic fracture

distribution of the spherical silica particles in the cured nanocomposite as well as the absence of any major agglomeration of particles (see Fig. 6.1). The PCCS demonstrates the very narrow and monomodal particle size distribution (2–50 nm) with an average size of 16 nm in the liquid resin (Fig. 6.2). It is evident that a good dispersion quality is retained in the nanocomposite from the liquid to the fully cured state and that primary particles are obtained.

6.3.2 Rheological Properties

The viscosity and resin gel-time are key factors for FRP structures manufactured by LCM. These factors determine the injectability of the resin and the fabric wetting process. Therefore, the flow behaviour of the nanocomposites are

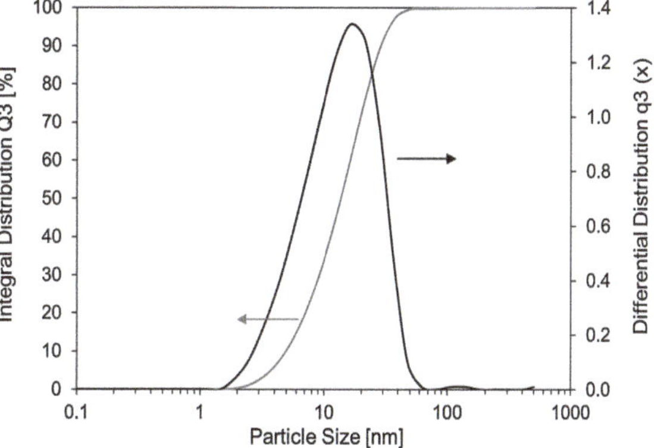

Fig. 6.2 Differential and integral particle size distribution of 25 wt% silica nanoparticles in an epoxy resin determined by PCCS

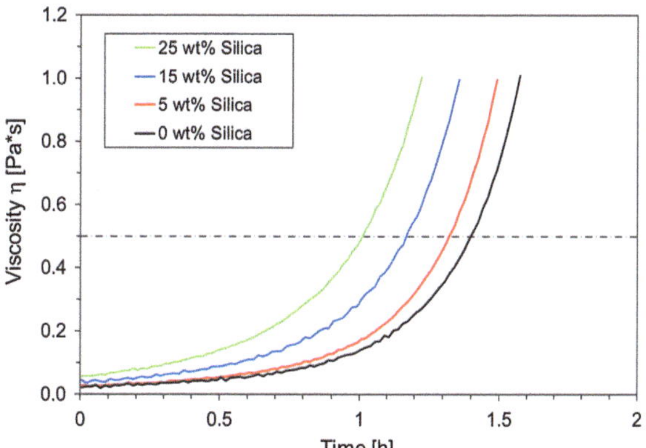

Fig. 6.3 Dynamic viscosity of epoxy-silica nanocomposites as function of process time and varying silica contents (0–25 wt%). Conditions: 80°C and constant shear rate of 4 Hz. Limiting viscosity number for LCM is 0.5 Pa*s

investigated in a series of rheological experiments, increasing the filler content up to 25 wt% silica at a typical injection temperature of 80°C for the neat epoxy resin used. The limiting viscosity number for injection technology is normally in the range of 500 mPa*s. The time required to reach that limit is defined as pot life with respect to the injection technology used. The corresponding isothermal viscosity curves are depicted in Fig. 6.3.

Increasing the filler content under the described conditions leads to a reduction in pot life from 1.4 to 1 h. DSC measurements have shown that this is not due to any catalytic effects [7]. It is likely that the observed reduction in gel-time can be explained by a pure physiochemical effect. As the curing reaction progresses, long polymer chains develop which interact extensively with the large surface of the nanoparticles. This causes a disproportional increase in viscosity compared with the reference resin. However, with regard to the filler content studied herein, the reduction in pot life as well as the small increase of the initial viscosity are

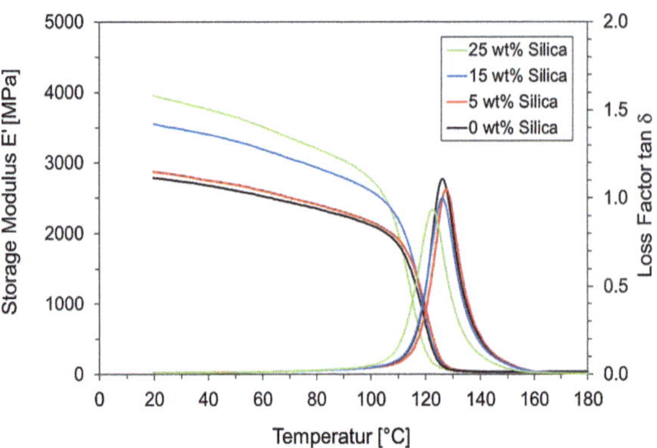

Fig. 6.4 Storage modulus and loss factor of the epoxy-silica nanocomposites as a function of temperature and degree of silica filling measured by DMA (tension mode)

acceptable and are still in the range of processability for the LCM technique. The injectability is retained in the modified resins.

6.3.3 Thermal Characterization by DMA and DSC

The heat distortion temperature (HDT) known as an important key driver for FRP was investigated for epoxy-silica nanocomposites in an extended temperature range measured by dynamic mechanical analysis (DMA). Here the HDT is defined as the onset temperature taken form the storage modulus E' by the tangent method. The glass transition temperature was analysed by the loss factor (tanδ). The results are summarized in Table 1 and are depicted in Fig. 6.4.

Obviously the DMA data reveal a high influence of the content of silica nanoparticles on the stiffness (E') of the modified epoxy resin at a temperature below T_g, i.e. in the glassy state at about 20–100°C, while the modulus becomes almost indendent of the silica content at the glass transition state (ca. 120°C). Consequently the material properties in the glassy state are considerably determined by the nanoparticles. In the range of the glass transition a nanoparticle effect is no more clearly detectable i.e. above the onset temperature E' seems to be predominantly determined by the pure matrix properties. On the other hand the slightly reduced loss factors (tanδ), indicating a lower damping behaviour, suggest an improved particle–matrix bonding [5] more discussed in Sect. 3.8. However, in terms of the nearly constant values of HDT the silica nanoparticles examined here do not enable an extension of the thermal field of applications.

Additionally, the glass transition temperature (T_g) was determined by DSC. The values of T_g decrease slightly by 2% from 131 to 128°C by increasing the filler content up to 25 wt% silica (Table 6.1). Obviously, the used silica nanoparticles have only a slight influence on T_g even at high degrees of filling. Interestingly the detected glass temperatures from the DSC show an amazingly good correlation with those ones from the DMA measurements (low shift of tanδ). This also

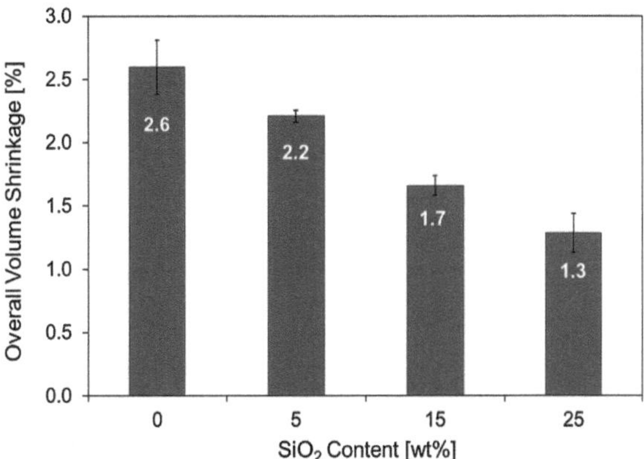

Fig. 6.5 Overall volume shrinkage of epoxy-silica nanocomposites at RT as a function of silica content

demonstrates that an important parameter for designing or dimensioning of FRP is approximately unchanged.

Recently, the reduction of T_g by nanoscaled silica was also reported by other authors, but they identified a more extensive decrease of T_g [8, 9]. Mostly the effects are attributed to a reduction of crosslinking density in the polymer matrix, i.e. the chemical and physical structure of the network is significantly affected by the presence of silica nanoparticles, e.g. by the preferential adsorption of curing agent and/or low molecular weight oligomers. It is assumed in literature that the particles are surrounded by a softer polymer shell and that the strength of the polymer matrix increases with increasing distance from the surface [10]. Such interphase layers between the nanoparticles and the matrix have properties differing from those of the main matrix. However, the development of a heterogeneous network by the silica nanoparticles examined in this study seems considerably less significant. Another reason might be that the epoxy resin in the master batch is not absolutely identical with the epoxy resin used for the thinning process. Differences in the average molecular weight of the pre-polymer DGEBA also induce lower T_g. Therefore it has to be concluded that the presence of silica nanoparticles increases the modulus dramatically whereas the glass transition temperature is only small affected. However the mechanisms are unknown and have to be identified by a detailed analysis of the polymer-particle interfaces.

6.3.4 Quantitation of Resin Shrinkage

Epoxy resins tend to build-up residual stresses induced by resin shrinkage in FRP structures [11]. Generally, high stress levels reduce the performance of FRP and should be prevented. Nanoparticles as fillers might play an important role in the reduction of resin shrinkage. Therefore the overall volume of shrinkage was

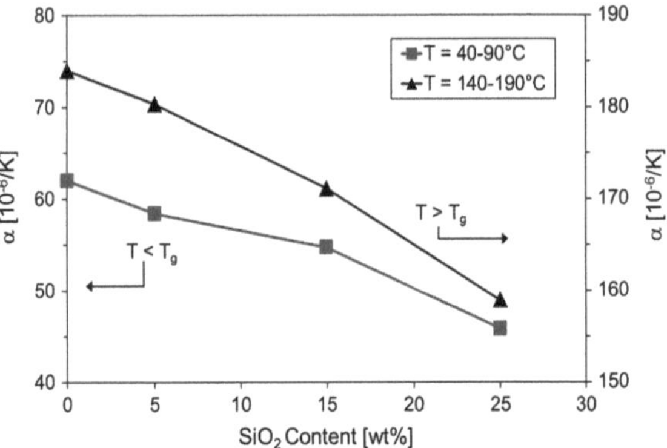

Fig. 6.6 Linear coefficient of thermal expansion (CTE) for epoxy-silica nanocomposites as a function of the temperature range and the degree of filling (standard deviation of CTE below 5%)

determined with density measurements. Figure 6.5 shows the effect of silica nanoparticles on the course of resin shrinkage. Interestingly, the shrinkage can be reduced by approx. 50% depending on the nanoparticle content. Both the substitution of resin with non-shrinkable nanoparticles and the less exothermic character of the curing process (Table 6.1) might lead to lower shrinkage stress and internal mechanical stress in the composite which improves the suitability of the FRP (e.g. higher tolerance to damage). It should be pointed out that silica nanoparticles also decrease the coefficient of thermal expansion (CTE) as shown below and improve thermal conductivity of the polymer matrix [7]. Both parameters also affect the shrinkage process.

6.3.5 Determination of Coefficients of Thermal Expansion

Using thermo-mechanical analysis (TMA), the linear coefficients of thermal expansion were determined dilatometrically for temperature ranges both below and above T_g. The results are presented in Fig. 6.6 and summarized in Table 6.1. Depending on silica content and the temperature range, a significant depression in the α-value of up to 30% is observed (form 62.0 to 45.8 $\times$ 10^{-6}/K). This effect can be attributed to the considerably smaller CTE of the silica nanoparticles ($\alpha = 0.5$–0.9×10^{-6}/K [12]) in comparison to the significantly higher CTE of neat polymer matrix ($\alpha_{40-90°C} = 62.0 \times 10^{-6}$/K). Moreover, the CTE might be also affected by the degree of strong interfacial properties of the nanoparticles with the epoxy resin as reported in [5]. Therefore, we can expect by the decreased CTE that resin shrinkage is reduced and the thermal stress in the composite will be smaller.

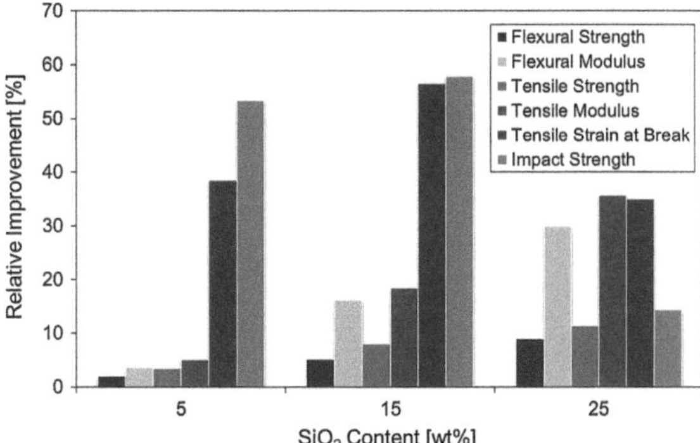

Fig. 6.7 Relative improvements of mechanical parameters for epoxy-silica nanocomposites in relation to the neat epoxy resin and as a function of various degrees of silica filling (G_{1c}: +155% and K_{1c}: +85%; both here not listed)

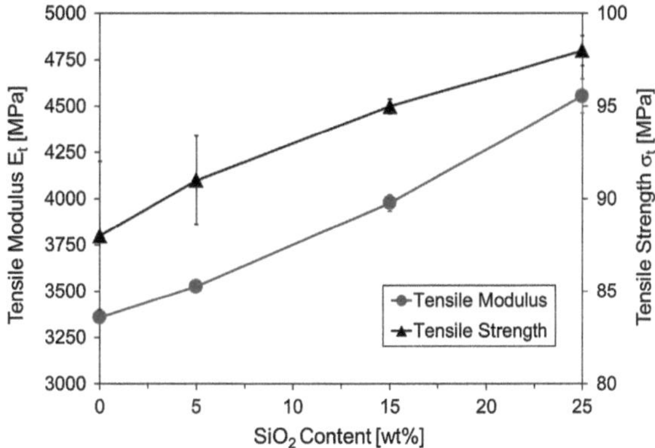

Fig. 6.8 Tensile properties of epoxy-silica nanocomposites as a function of the silica filler content

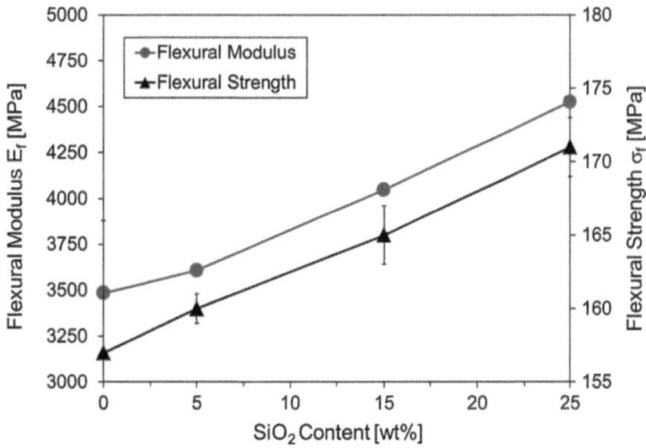

Fig. 6.9 Flexural properties of epoxy-silica nanocomposites as a function of the silica filler content

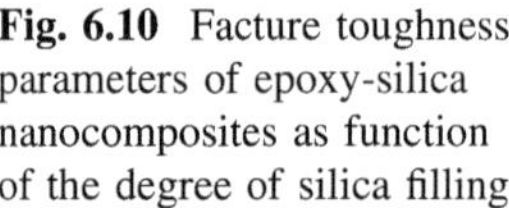

Fig. 6.10 Facture toughness parameters of epoxy-silica nanocomposites as function of the degree of silica filling

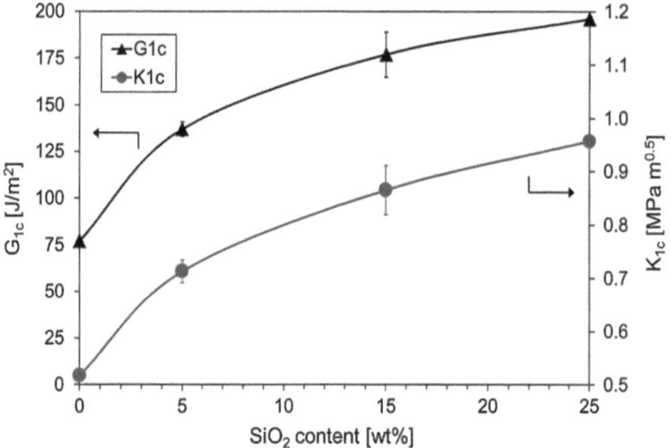

6.3.6 Static Mechanical Characterization

The improvements of the mechanical parameters of the epoxy resin by adding silica nanoparticles were investigated systematically. The absolute values are summarized in Table 6.2. The relative improvements of the mechanical properties of the nanocomposites as well as the course of absolute values for tensile, flexural and fracture toughness tests are depicted in Figs. 6.7, 6.8, 6.9, and 6.10.

The linear increase in stiffness and strength of the nanocomposites as the silica content increases is clearly distinguishable. The tensile modulus is increased by 36% and the flexural modulus by 30% at a silica content of 25 wt%. The tensile strength (ultimate) and flexural strength are increased by 11 and 9%, respectively. Interestingly, the toughness properties are also increased simultaneously as indicated by comparing the course of lines in Figs. 6.8, 6.9, and 6.10. The fracture toughness was determined by measuring the critical energy release rate (G_{1c}) and the critical stress intensity factor (K_{1c}) using compact tension test samples (CT). Both values quantify the fracture resistance at the process of crack propagation. Obviously the G_{1c} and K_{1c} values are improved systematically with an increased concentration of silica nanoparticles compared with the neat resin, about 155% for G_{1c} and 85% for K_{1c}. The remarkable toughening effect is also underlined by the course of the strain at break (ε (0% SiO$_2$) = 3.26%; ε (15% SiO$_2$) = 5.10%) and the impact strength published in [7]. The reasons for this toughness behaviour are discussed in detail in the following chapter having a closer look at the fracture topology given by SEM und AFM images.

Furthermore, it is evident that the mechanical parameters investigated here show no maximum for the curves, which means that the mechanical maximum for this material has not been reached yet and the optimum in therms of the silica content may not yet has been adjusted. On the other hand the matrix starts to become silgthly brittle above 15 wt% silica considering the strain at break. However, this small negative effect is more than compensated by the significant increase in the stiffness, strength and toughness such that the current optimum for the

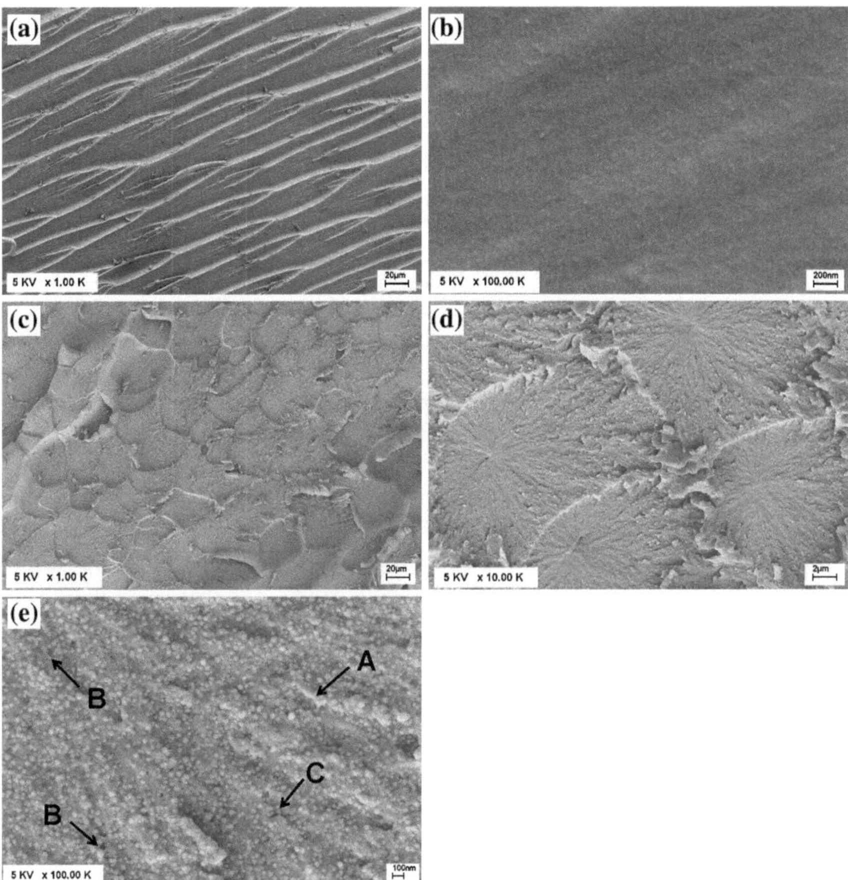

Fig. 6.11 SEM micrographs at various magnifications for the tensile fractured surface of (**a, b**) the neat epoxy resin and (**c–e**) the epoxy-silica nanocomposites filled with 25 wt% silica. *Arrows* indicate different fracture effects: (*A*) shear banding/shear lips, (*B*) formation of cavities, (*C*) de-bonding

nanocomposite was limited to a filler content of 25 wt% silica. A further increase in the concentration of nanoparticles results in more manufacturing problems, e.g. in reduction of pot life. The slight increase in matrix density is not addressed here because it does not interfere with the lightweight construction aspect.

In this way, the strength, stiffness and toughness of the polymer matrix can be improved simultaneously by using these hard, surface functionalized silica nanoparticles i.e. the stiffness-toughness paradox seems to be solved. This effect has already been observed with some other nanoparticle systems [13] and might be attributed to a good particle–matrix interaction. A comparable improvement of resin performance using micro-scaled fillers cannot be achieved, especially due to the much more marked brittleness effects and the reduced injectability of the resin.

6.3.7 Identification of Failure Mechanisms by Analysing the Fracture Surface Topology

The failure mechanism of the nanocomposites is qualitatively evaluated by SEM analysis on tensile fractured surfaces. Figure 6.11 shows SEM micrographs of the neat resin and the corresponding nanocomposite with 25 wt% silica at various magnifications.

The SEM micrograph of the neat resin shows a fracture surface at a low magnification (1000×) with river markings indicating the direction of crack propagation and rather smooth areas in between. Especially, the formation of smooth surfaces indicates the brittle behaviour of the matrix which is typical for epoxy resins. The corresponding theory of postulated failure mechanisms are summarized in the literature [14]. In contrast, the nanocomposites show a more detailed structured surface with many parabolic slices. The density of these slices also increases with increasing nanoparticle content. Although the nanoparticles are mostly homogeneously dispersed, it is assumed that the formation of slices is induced by areas with accumulated nanoparticles causing local centres of high stresses [15].

Moreover, the average roughness values (R_a) of the fracture surface was quantified by AFM measurements in the contact mode. Rastering of a 10×10 µm fracture areas reveals higher R_a values for the nanocomposite (189.5 nm) than for the neat resin (48.3 nm). Figure 6.12 illustrates the different topographic profiles of the investigated samples. Obviously nanoparticles generate greater fracture areas through the development of rougher surfaces, which allows them to resist crack propagation. In consequence, the fracture energy is effectively dissipated as illustrated by the increased values of fracture toughness (G_{1c}; K_{1c}) as well as strain at break (Fig. 6.10 and Table 6.2). This leads to the conclusion that the addition of nanoparticles to the polymer matrix significantly affects the roughness of fracture surface and, in consequence, the balance between tough and brittle behaviour of the matrix.

The higher magnification of SEM micrographs (100,000 ×) reveals the formation of primary nanoparticles covered by a thin polymer shell (interlayer phase). The core–shell structure is confirmed by the large particle diameter of 50 nm compared to the average nanoparticle diameter of 16 nm measured with PCCS. Obviously these nanoparticles generate highly efficient particle–matrix interactions which is also confirmed by the observed low degree of de-bonding and pull-out effects. Accordingly the nanoparticles are well embedded in the polymer matrix and the fracture behaviour is dominated by cohesive crack propagation in the surrounding matrix. As is reported in literature, the extension of the interphase

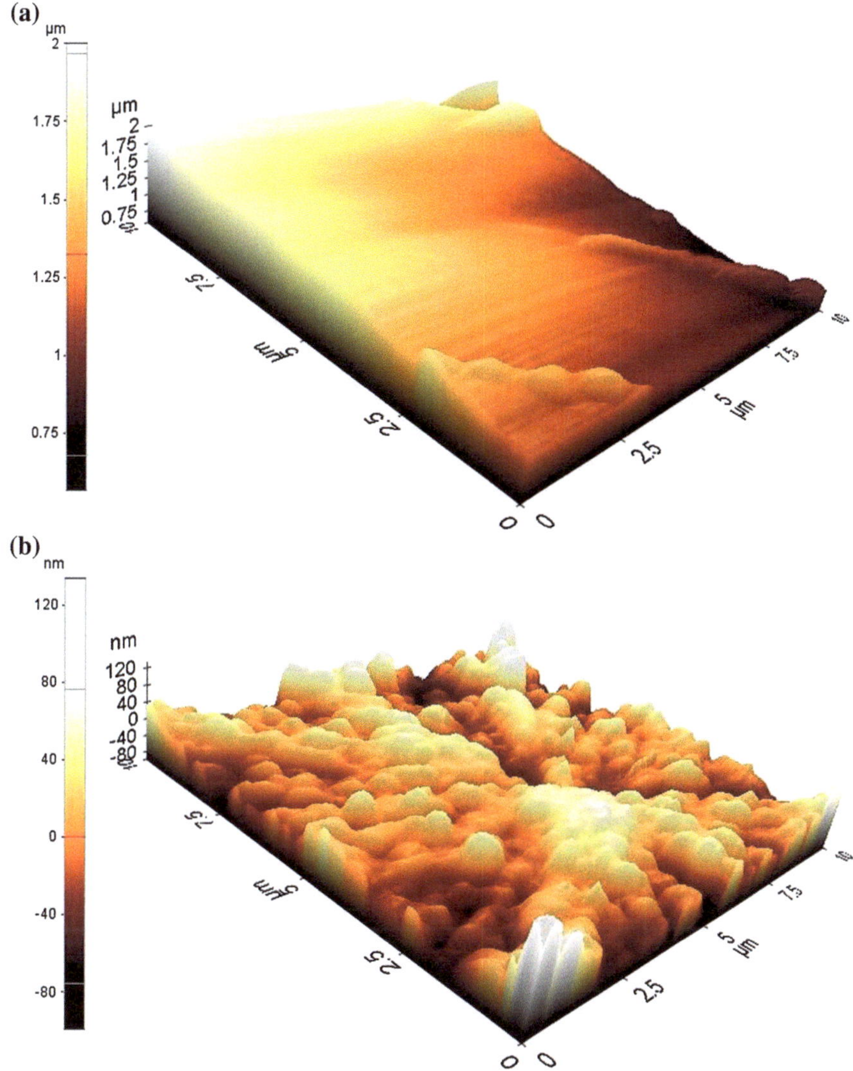

Fig. 6.12 AFM topographic profiles of the tensile fracture surface of **a** the neat epoxy resin and **b** the corresponding epoxy-silica nanocomposites with 25 wt% silica

is proportional to the particle size [16] and is proposed to be in the range of 2–50 nm [18]. Therefore it might be expected that high degrees of filling (low interparticle distance) lead to an intersection of the particle interlayers creating the development of a strong polymer network (the bulk polymer becomes an interlayer polymer). This interpretation corresponds fairly well with the increase of stiffness and strength as well as the improved toughening effect created with an increased silica content (Table 6.2). The influence on T_g is unexpected small.

It is noteworthy that the failure mechanism seems to be a superposition of various effects. The SEM micrographs suggest shear banding and crack deflection (see arrows in Fig. 6.11). In addition particle–matrix de-bonding (voiding) and pull-out effects of particles are indicated although to a lesser extent. A more detailed interpretation of the failure mechanism is not in the scope of this paper. The influence of the interphase on the mechanical characteristics of the nano-composite and a detailed analysis of the failure mechanism will be the focus of future research.

6.4 Conclusion

The paper presents the results of the investigation of epoxy-silica nanocomposites as a new type of matrix for manufacturing fiber-reinforced polymers (FRP) using injection technology (LCM). The stiffness, strength, and toughness of the composite can be significantly increased depending on the silica content compared with neat resin. These results correlate very well with the fracture surface analysis performed by means of SEM and AFM. The functionalized silica nanoparticles initiate highly profiled fracture surfaces which indicate better energy dissipation and results in increased matrix toughness. The development of core–shell structures (strong particle–matrix bonding) indicates the efficiency with which the nanoparticles are embedded in the matrix. The effective embedding might be responsible for the high levels of stiffness and strength. Moreover, the resin shrinkage and the thermal expansion coefficient are significantly reduced and the thermal conductivity increased. Therefore silica nanoparticles might entail lower shrinkage stress and also internal mechanical stress in the FRP which will extend their applicability, e.g. towards higher damage tolerance. Interestingly the silica nanoparticles seem to have slight influence on the glass transition temperature confirmed by DSC and DMA respectively. Moreover an optimum particle content of 25 wt% silica for the tested epoxy resin was identified also maintaining the injectability of these doped resins for their application in LCM processes. All in all epoxy-silica nanocomposites seem to be a new class of high performance polymer matrices for the manufacturing of FRP structures by low cost LCM techniques.

Acknowledgments The authors wish to thank Mr. C. Schilde (Institut of Particle Technology, Technical University Braunschweig, Germany) for AFM measurements and Dr. S. Sprenger (Hanse-Chemie AG; Geesthacht, Germany) for providing with the resin master batch.

References

1. Rudd, C.D., Long, A.C., Kendall, K.N., Mangin, C.G.E.: Liquid Moulding Technologies. Woodhead Publishing Limited, Cambridge (1998)
2. Kruckenberg, T.K., Paton, R. (eds.): Resin Transfer Moulding for Aerospace Structures. Kluwer Academic Publisher, Dordrecht (1998)
3. Parnas, R.S.: Liquid Composite Molding. Hanser Publishers, Munich (2000)
4. Adebahr, T., Roscher, C., Adam, J.: Reinforcing nano-particles in reactive resins. Eur. Coat. J. **4**, 144–149 (2001)
5. Kang, S., Hong, S., Choe, C., Park, M., Rim, S., Kim, J.: Preparation and characterization of epoxy composites filled with functionalized nanosilica particles obtained via sol-gel process. Polymer **42**, 879–887 (2001)
6. Rosso, P., Ye, L., Friedrich, K., Sprenger, S.: A toughened epoxy resin by silica nanoparticle reinforcement. J. Appl. Polym. Sci. **100**, 1849–1855 (2006)
7. Mahrholz, T., Stängle, J., Sinapius, M.: Quantitation of the reinforcement effect of silica nanoparticles in epoxy resins used in liquid composite moulding processes. Composites: Part A **40**, 235–243 (2009)
8. Preghenella, M., Pegoretti, A., Migliaresi, C.: Thermo-mechanical characterization of fumed silica-epoxy nanocomposites. Polymer **46**, 12065–12072 (2005)
9. Zhang, H., Zhang, Z., Friedrich, K., Eger, C.: Property improvements of in situ epoxy nanocomposites with reduced interparticle distance at high nanosilica content. Acta Mater. **54**, 1833–1842 (2006)
10. Hartwig, A., Sebald, M., Kleemeier, M.: Cross-linking of cationically polymerised epoxides by nanoparticles. Polymer **46**, 2029–2039 (2005)
11. Lange, J., Toll, S., Manson, J.: Residual stress build-up in thermoset films cured below their ultimate glass transition temperature. Polymer **38**(4), 809–815 (1997)
12. Lide, D.R. (ed.): CRC Handbook of Chemistry and Physics. CRC Press, Boca Raton (2007)
13. Wetzel, B., Rosso, P., Haupert, F., Friedrich, K.: Epoxy nanocomposites—Fracture and toughening mechanisms. Eng. Fract. Mech. **73**, 2375–2398 (2006)
14. Cantwell, W.J., Roulin-Moloney, A.C.: Fractography and failure mechanisms of unfilled and particulate filled epoxy resins. In: Roulin-Moloney, A.C. (ed.) Fractography and Failure Mechanisms of Polymers and Composites, pp. 233–290. Elsevier Applied Science, London and New-York (1989)
15. Zhang, H., Zhang, Z., Friedrich, K., Eger, C.: Property improvements of in situ epoxy nanocomposites with reduced interparticle distance at high nanosilica content. Acta Mater. **54**, 1833–1842 (2006)
16. Zhong, Y., Wang, J., Wu, Y.M., Huang, Z.P.: Effective moduli of particle-filled composite with inhomogeneous interphase: Part I + II. Compos. Sci. Technol. **64**, 1345–1362 (2004)
17. Nohales, A., Munoz-Espı, R., Felix, P., Gomez, C.M.: Sepiolite-reinforced epoxy nanocomposites: thermal, mechanical, and morphological behavior. J. Appl. Polym. Sci. **119**, 539–547 (2011)
18. Schadler, L.S.: Polymer-based and polymer-filled nanocomposites. In: Ajayan, P.M., Schadler, L.S., Braun, P.V. (eds.) Nanocomposites Science and Technology. Wiley-VCH, Weinheim (2003)

Chapter 7
Carbon Nanotube Actuation

Steffen Opitz, Sebastian Geier, Johannes Riemenschneider,
Hans Peter Monner and Michael Sinapius

Abstract The outstanding electrical and mechanical properties of single carbon
nanotubes (CNT) are the motivation for an intensive research in various fields of
application. The actuation effect constitutes the foundation for any application as a
multifunctional material and within the field of adaptronics. The effect is in the
majority of cases investigated by a CNT configuration of stochastically aligned
CNT, so-called bucky-paper, in an electrolytic environment. The chapter presents an
analytical model for a detailed understanding and investigation of the actuation
process. The complete description and parameterization of the model is documented.
Initial results from experiments with aligned CNT structures and the application of
solid electrolytes are presented.

7.1 Introduction

Carbon nanotubes were discovered in 1991 by Iijima [1]. Their excellent electrical and
mechanical properties in combination with the actuating effect which was first
observed by Baughman et al. in 1999 [2] are the reason why carbon nanotubes are
considered to be the actuating material of the future. There are different ways to
analyze the actuation. One setup to detect the effect is a bimorph bender [3, 4]. The
large deflection which can be achieved is a major advantage of this setup. A significant
disadvantage is the unknown influence of the passive structure. Because of this it is not
possible to gain reliable data about the actuating effect itself. The bending actuator

S. Opitz (✉) · S. Geier · J. Riemenschneider · H. P. Monner · M. Sinapius
Institute for Composite Structures, German Aerospace Center DLR,
Lilienthalplatz 7, 38108, Braunschweig, Germany
e-mail: Steffen.Opitz@dlr.de

M. Wiedemann and M. Sinapius (eds.), *Adaptive, Tolerant and Efficient Composite
Structures*, Research Topics in Aerospace, DOI: 10.1007/978-3-642-29190-6_7,
© Springer-Verlag Berlin Heidelberg 2013

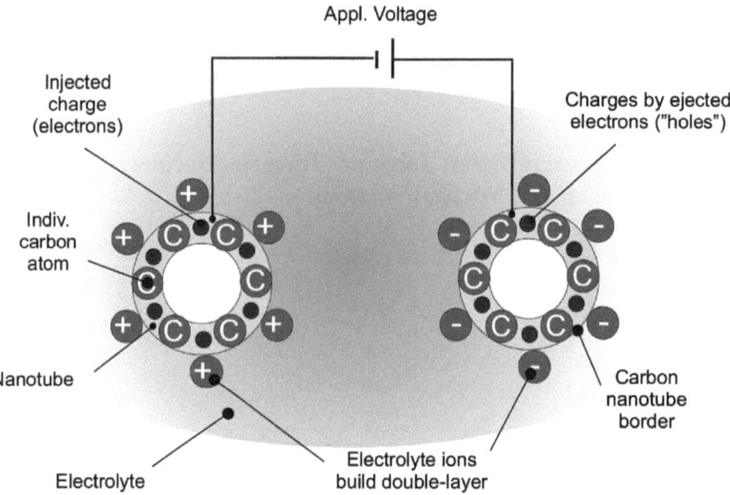

Fig. 7.1 Electrochemical double layer

setup is rather suitable to build up demonstrators than to obtain more information about the effect. An alternative attempt is to detect the strain of the bucky-paper directly [2, 5–11]. As a result the deflections are smaller but there is no disturbing influence of a passive structure. Hence the deflections are a direct consequence of the actuating effect. The investigation of the in-plain actuation effect enables the validation of analytical descriptions of the actuation effect. The chapter presents an analytical model, its experimental validation and initial results with solid electrolytes.

7.2 The Actuation Phenomenon of CNT

Today's standard configuration of Carbon Nanotube Actuators (CNA) is based on bucky-papers of CNT, which are extracted from aqueous CNT dispersion via vacuum filtration [12]. This structure provides the possibility to examine the effect of actuation but as the single carbon nanotubes are only connected by weak Van-der-Waals forces, this arrangement cannot benefit from the high elastic modulus of single carbon nanotubes. For actuation purpose the bucky-paper is put into an electrolyte being either liquid or solid. The bucky-paper as well as the electrolyte is conductive, although they differ in the mechanism of conduction. While the bucky-paper is electronically conducting the electrolyte is an ionic conductor. If the solid bucky-paper is brought in contact with the liquid electrolyte a charged phase interface is established (Fig. 7.1). Any excess charge on the electronic conductor is balanced by an ionic charge of equal magnitude and opposite sign on the solution side of the interface. This charge distribution is called the electrochemical double layer and causes the capacitive behavior of the system [13].

7.3 An Analytical Model for the Actuation Mechanism

The analytical model of a CNA is based on energy considerations. In the lower frequency range terms of kinetic energy may be disregarded. Then, the energy function is governed by the storage of potential energy in the mechanical as well as in the electrical system

$$E_{CNA} = \frac{Q^2}{2 \times C} - U_e \times Q + \frac{\kappa \times S^2}{2} - F_e \times S \tag{7.1}$$

where Q is the charge and S the elastic deformation of the CNT actuator. U_e is the external voltage and F_e is the external force, both being responsible for the work applied to the CNT. C is the system capacity and κ the stiffness. The latter is determined by the macroscopic CNT structure including its material property and its geometry. The disregard of a contribution of kinetic energy is based on the fact that the bucky-paper has a very small mass and that the velocities which have quadratic influence to the kinetic energy are low. Furthermore the system dissipates energy, mechanically as well as electrically. The energy dissipation may be expressed by a force and voltage, respectively which are related to the deformation velocity $\dot{S}$ in the sense of viscous damping and the current $\dot{Q}$.

$$\begin{aligned} \dot{S} &= -\gamma \times F \\ \dot{Q} &= -\sigma \times U \end{aligned} \tag{7.2}$$

Within this formulation the parameter γ is the inverse damping parameter and σ is the electrical conductivity. Equations (7.1) and (7.2), constitute the fundamentals of the system. The variation of the total energy (Eq. 7.1) leads to the force and the voltage outside the equilibrium.

$$\frac{dE_{CNA}}{dS} = F = -F_e + \kappa \times S + \frac{Q^2}{2} \frac{d}{dS} C^{-1} \tag{7.3}$$

$$\frac{dE_{CNA}}{dQ} = U = -U_e + \frac{Q}{C} \tag{7.4}$$

Because of the stochastically alignment of the single CNTs the bucky-paper has a quasi-isotropic material behavior. Thus a coupling between charge and displacement is not an option. An electro-mechanical coupling through Coulomb-forces is plausible. Due to this fact the last term in Eq. (7.3) respects those Coulomb-forces. The linearized dependency of this term of the mechanical system is:

$$C^{-1} = C_0^{-1} - \alpha \times S \tag{7.5}$$

where α is the coupling coefficient. This linear approach simplifies the dissipating forces of Eq. (7.3).

$$F = -F_e + \kappa \times S - \frac{Q^2}{2} \times \alpha \tag{7.6}$$

$$U = -U_e + \frac{Q}{C_0} - \alpha \times Q \times S \tag{7.7}$$

Incorporating those expressions into Eq. (7.2) yields:

$$\begin{aligned}
\dot{S} &= -\gamma \left(\kappa \times S - F_e - \frac{Q^2}{2} \times \alpha \right) \\
\dot{Q} &= -\sigma \left(\frac{Q}{C_0} - \alpha \times Q \times S - U_e \right)
\end{aligned} \tag{7.8}$$

These differential equations describe the actuation of CNTs in an electrolytic environment. The actuation is determined by five parameters

1. stiffness κ
2. damping parameter γ
3. capacity C_0
4. resistance R
5. Coulomb-coupling parameter α

which have to be determined experimentally.

7.4 Experimental Setup

The setup consists of a three-electrode measuring cell [14]. Within this cell the rectangular bucky-paper specimen is utilized as the working electrode. Because of the good electrical conductivity of the material the counter electrode can be build up the same way. The reference electrode is formed by a saturated Calomel electrode (Hg/Hg_2Cl_2, KCl (sat'd)).

These electrodes are driven by a potentiostat which controls the voltage drop between reference electrode and working electrode by adjusting the current between the counter electrode and the working electrode. An aqueous unimolar NaCl electrolyte solution was used for the experiments. During the measurement the electrodes are immersed into the NaCl solution. The electrolyte provides the mobile ions which are necessary for the actuating effect [15]. The bucky-paper specimen is clamped at the upper and lower end (Fig. 7.2), whereas the lower clamp incorporates the electrical contacting. This clamp is fixed. The upper clamp is used to apply a constant pre-stress. This is needed to avoid bending of the specimen. The elongation in in-plain direction of the bucky-paper due to the actuating effect can be measured by detecting the displacement of the upper clamp.

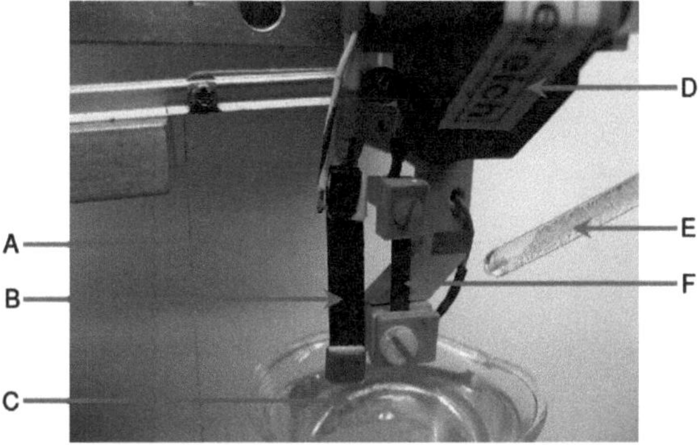

Fig. 7.2 Test setup; *A* string for pre-stress; *B* counter electrode; *C* beaker with electrolyte; *D* laser-triangulator; *E* reference electrode; *F* bucky-paper specimen

7.5 Parameterization

In a first step of system identification mechanism the coupling between mechanical and electrical system is disregarded, i.e. $\alpha = 0$, leading to two separate differential equations of first order.

$$\dot{S} = -\gamma(\kappa \times S - F_e)$$
$$\dot{Q} = -\sigma\left(\frac{Q}{C_0} - U_e\right) \tag{7.9}$$

These equations can also be written as:

$$\frac{1}{\gamma} \times \dot{S} + \kappa \times S = F_e$$
$$\frac{1}{\sigma} \times \dot{Q} + \frac{Q}{C_0} = R \times I + \frac{Q}{C_0} = U_e \tag{7.10}$$

The first expression characterizing the mechanical system constitutes the differential equation of forces for a parallel arrangement of a stiffness κ and a damping element $1/\gamma$. The second differential equation of Eq. (7.10) is obtained by applying Kirchhoff's voltage law to a series circuit of a resistance R and a capacity C_0 (RC-element). Since the force is constant the influence of the mechanical system to the electrical system can be disregarded under the condition that only frequencies above 0 Hz are analyzed.

7.6 Electrical System

An impedance spectroscopy is performed in order to identify the electrical parameters. A voltage sweep signal from $f = 0.02$ to 5 Hz with an amplitude of 0.25 V and an offset of 0.38 V is used for stimulation. The response of the current is recorded and the frequency response of the admittance is determined.

According to Eq. (7.10) the frequency response approximated by the model can be calculated as follows.

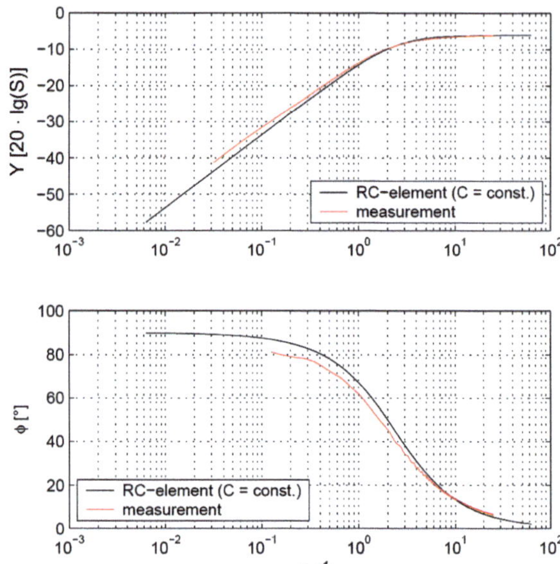

Fig. 7.3 Adapted frequency response of the admittance

$$R \times I + \frac{1}{C} \times \int I\, dt = U$$
$$\left(R + \frac{1}{j\omega \times C}\right) \times I = U$$
$$Y = \frac{I(j\omega)}{U(j\omega)} = \frac{1}{R + \frac{1}{j\omega \times C}} = \frac{C \times j\omega}{RC \times j\omega + 1} \tag{7.11}$$

Figure 7.3 compares the measured and calculated admittance. The logarithmic scaling of the axes provides the possibility to split the amplitude response of the admittance into two sections. For higher frequencies the capacitive behavior of the electrolytic double layer at the phase interface between solid electrode and electrolyte is negligible. Hence the amplitude response reaches a constant value which corresponds to the conductance of the system. The parameter of interest R is the inverse of this constant value. The ohmic behaviour of the system can also be seen in the phase response which approaches $0°$ (Fig. 7.3). In the range of low frequencies the systems characteristic is dominated by the capacity of the double layer. The amplitude response converges to a straight line with constant slope. The influence of the double layer on the phase response is visible through a constant value differing from $90°$. The capacity of the system can be determined via the barrier frequency ω_b and the already known resistance R.

$$\omega_b = \frac{1}{R \times C} \tag{7.12}$$

For low frequencies the frequency response of the series connection of a resistance and a capacity is characterized by a constant slope of 20 dB/decade for the amplitude and by a phase angle converging to 90°. This differs slightly from the measured characteristic of the system. Looking at the recorded responses of amplitude and phase it can be noticed that the constant slope is smaller than 20 dB/ decade and that the phase converges to a value lower than 90°. Nevertheless, the model incorporating the RC-circuit is capable to describe the behavior of the electrical system [14]. The observed deviations can be considered by a modification of the model. The capacity used to represent the electrical properties of the electrochemical double layer postulates an ideal plane electrode surface [16]. The real surface of the bucky-paper is rather porous. To activate the capacity deep inside the pores the ions have to transcend an additional electric resistance caused by the limited conductivity of the electrolyte in the pores [17]. A branched ladder network can characterize the electrochemical double layer even better than a simple RC-circuit [18, 19]. The frequency range extends as the number of parallel branches increases. This improvement is associated with an increase of the number of parameters which have to be determined. Fortunately there is an empirical observation that helps to decrease the number of parameters: the double layer capacity is proportional to R^{-n} [16]. The porous surface of the electrode can be modeled by a constant phase angle element (CPE) [18]. The impedance of such an element can be described using a fractional derivative. In the frequency domain the electrochemical double layer can be modeled using Eq. (7.13).

$$Z_{CPE} = \frac{1}{(j \times \omega)^n \times K} = R_{CPE}(\omega) + j \times X_{CPE}(\omega) \tag{7.13}$$

The developed model of the electrical system can be modified replacing the capacity by the constant phase element.

Next, the mechanical system and the actuating effect have to be examined to complete the description of the system. Furthermore the applicability of the model to the existing test setup has to be validated. The workspace of an electrolytic bucky-paper actuator lies within the electrochemical stability window $(U_{\min} < U < U_{\max})$ and can be divided into three regions depending on the potential which is applied [20].

The bucky-paper shows different electromechanical behavior within these regions. All experiments are made in regions II and III for the parameter identification. The excitation voltage is chosen above the voltage of minimum strain U_0.

7.7 Mechanical System

The stiffness κ and the damping parameter γ are the characteristic mechanical parameters of the system. Especially the stiffness of the system is directly determined by material properties of the bucky-paper. The material characteristics may change with the frequency due of the microscopic structure, which is characterized by stochastically aligned CNTs (Fig. 7.4).

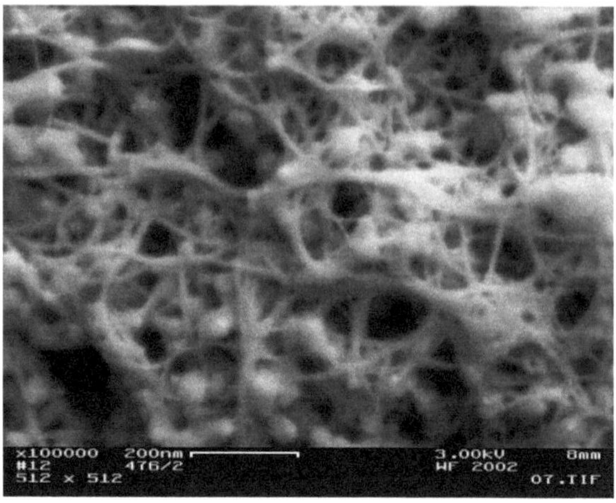

Fig. 7.4 REM picture of microscopic structure of bucky-paper

On the other hand the frequency range of interest is small (0–5 Hz) due to the appropriate working frequency of an electrolytic actuator. That is why it has to be verified whether the behavior of the mechanical system can be described with constant parameters or not. Additionally the influence of the pre-stress and the electrolyte to the specimen is examined. The stiffness is identified by an appropriate test-setup depicted in Fig. 7.5.

Dynamic tension tests are realized to determine the elastic modulus at different frequencies f and initial tensions σ_0. A dynamic load is applied to the bucky-paper (A) via an piezoelectric actuator (F). The generated force and resulting strain are measured by sensors (E, C). To examine the influence of the electrolyte the bucky-paper is wetted with a unimolar NaCl electrolyte. A voltage sweep applied on the piezoelectric actuator is used for the excitation of the system. The arising force and the displacement are measured simultaneously. Since the dimensions of the specimen are known, the Youngs modulus E of the material can be determined. Figure 7.6 shows the results of the investigation.

The results clearly show that the frequency does not have a big effect to the elastic modulus. Also the influence of the electrolyte is small. In contrast the pre-stress has an intense influence to the Youngs modulus. The higher the initial tension σ_0 the bigger is the elastic modulus E.

With the knowledge of the elastic modulus E and the dimensions of the specimen (width W, thickness T, length L) the stiffness parameter κ can be calculated.

$$\kappa = \frac{E \times W \times T}{L} \tag{7.14}$$

The second mechanical parameter of the system characterizes the energy dissipation through damping. There are several sources for damping in the test-setup, whereas the system stiffness is dominantly determined by the material properties of the bucky-paper. Hence the damping of the complete system cannot be estimated by the loss modulus of the bucky-paper. As a consequence the damping parameter

Fig. 7.5 Test setup for stiffness identification; *A* bucky-paper specimen; *B* steel frame; *C* laser-triangulator; *D* fixture; *E* force sensor; *F* piezoelectric actuator

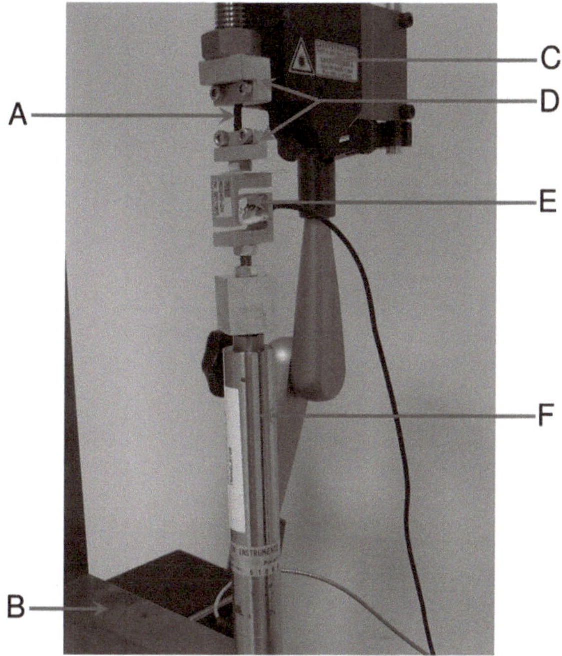

Fig. 7.6 Influence on the Youngs modulus

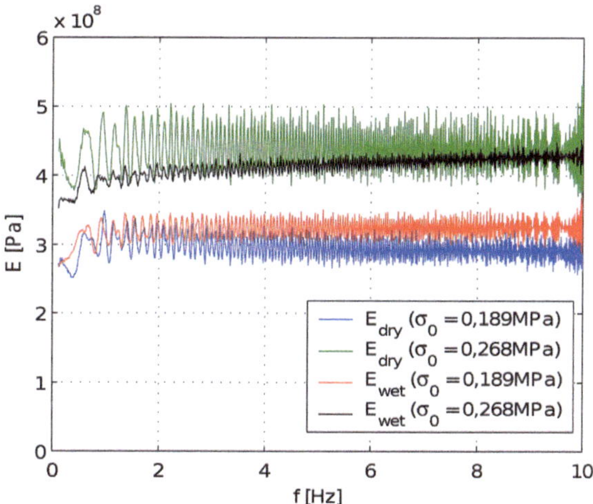

γ has to be determined using the complete setup. Ideally γ should be quantified from the frequency response function of the dynamic compliance which is described by the differential equation of the model.

$$\frac{1}{\gamma} \times \dot{S} + \kappa S = F_{act} + F_e \qquad (7.15)$$

Since only relative displacements are analyzed, the constant external force F_e can be disregarded. The displacement S and the actuation force F_{act} which causes the displacement have to be measured. The actuation force cannot be measured while the measurement of the displacement is directly possible. Using the knowledge of the electrical system and the model, the actuation force can be estimated from the charge signal which is obtained by integrating the current.

$$F_{act} = \frac{\alpha}{2} \times Q^2 \tag{7.16}$$

This equation postulates that the Coulomb-effect is the only actuating effect. The reason for this attempt is the observed actuation of the system. The value of the coupling parameter α must be known. The next section elucidates how this parameter can be determined and gives detailed information about the actuating effect. The velocity signal is required for calculating γ. The signals are transformed into the frequency domain, yielding a frequency- dependent damping coefficient

$$\gamma(j\omega) = \frac{|j\omega \times S(j\omega)|}{|F_{act}(j\omega) - \kappa \times S(j\omega)|} \tag{7.17}$$

There is no phase shift between velocity and damping force, as long as γ is defined as the parameter of a viscous damping. In consequence, the absolute values of force and velocity are taken for calculating γ.

7.8 Coupling Mechanism

The effect which links the mechanical and the electrical system is supposed to be caused by Coulomb-forces. These forces remain active in the system equilibrium. Due to the quadratic relationship between charge Q and Coulomb-force F_{act} this effect generates positive strains only. Since all former investigations showed that the effect does not disappear when the system approaches to the equilibrium state, the Coulomb-effect is identified as dominant [14, 20]. This assumption is confirmed by the parabolic characteristic of the static potential-strain curve (Fig. 7.7). Also the consideration of possible strains via Hartree–Fock-calculations shows that the Coulomb-effect dominates the strain generation [15]. In consequence the parameter α is determined to describe the actuation. Therefore the balance of forces in the static state is used. In the stationary state all damping forces disappear. Hence the Coulomb-forces are in equilibrium with the elastic forces of the bucky-paper which simplifies Eq. (7.15) to:

$$\kappa S = \frac{Q^2}{2} \times \alpha \tag{7.18}$$

The elastic forces can be calculated using the previously determined Youngs modulus and the measured displacement. The charge signal is needed in order to specify the value of α. Since a direct measurement of the charge is not possible,

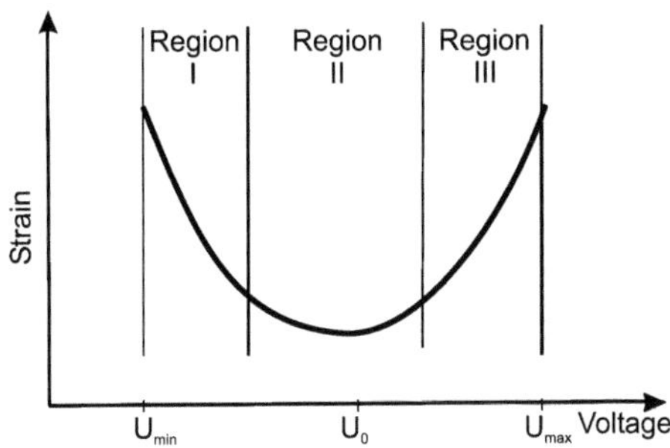

Fig. 7.7 Workspace of the carbon nanotube actuator

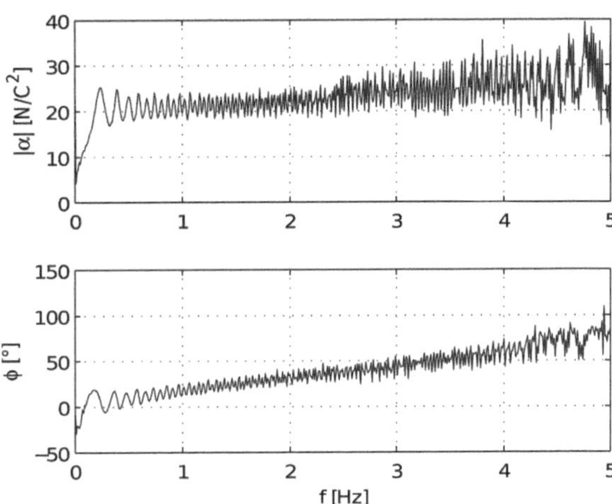

Fig. 7.8 Calculated frequency response of the coupling parameter α

the signal has to be calculated via integration of the current signal. The squared charge signal and the displacement signal are transformed into the frequency domain and the frequency spectrum of α can be determined by:

$$\alpha(j\omega) = \frac{2 \times \kappa \times S(j\omega)}{Q^2(j\omega)} \tag{7.19}$$

Experimentally the system is stimulated by a voltage-sweep from $f = 0.02$ to 5 Hz with an amplitude of $\hat{U} = 0.25\,\text{V}$ and an offset of $U_{off} = 0.45\,\text{V}$.

Figure 7.8 shows the calculated amplitude and the phase response of α. In the low frequency range there is no phase shift between squared charge signal and displacement. Postulating that the Coulomb-force is directly proportional to the squared charges this means that there is also no phase shift between force and strain. The value is determined from the low frequency range since damping forces are negligible small there. α is determined via this method to $\alpha = 21\,\frac{N}{C^2}$.

7.9 Validation of the Model

All system parameters, i.e. stiffness κ, damping parameter γ, capacity C_0, resistance R, and Coulomb-coupling parameter α are now available for a simulation of the actuation effect of a bucky-paper consisting of Carbon Nanotubes. The model has to be validated. Thus, the measured responses of the real system, the original model and the model incorporating the constant phase element (CPE) to different excitations will be compared. To facilitate the calculation some simplifications can be made. The order of magnitude of the voltages which are generated through the inverse effect can be estimated. This voltage ranges within some μV. Therefore the reaction from the mechanical system to the electrical system can be disregarded $(\alpha Q S = U_e)$. Two differential equations of first order are obtained which are only coupled in one direction.

$$\frac{1}{\gamma} \times \dot{S} + \kappa \times S = F_e + \frac{Q^2}{2} \times \alpha$$

$$R \times \dot{Q} + \frac{Q}{C_0} = U_e \tag{7.20}$$

$$R \times I + \frac{1}{C_0} \int I \, dt = U_e$$

The constant external force F_e can be disregarded. The mechanical system can be described consistently by the frequency response function of the dynamic compliance since the mechanical model has not to be changed.

$$\frac{S(j\omega)}{F_{act}(j\omega)} = \frac{1}{\kappa + \frac{1}{\gamma}j\omega} \tag{7.21}$$

The parameter identification reveals that a constant phase angle element can approximate the properties of the electrochemical double layer better than a simple capacity. Hence the model is extended and the modification is validated. The replacement of the impedance of the capacity by the impedance of a constant phase angle element results in the frequency response function of the modified model.

$$\left(R + \frac{1}{(j\omega)^n \times K} \right) \times I(j\omega) = U(j\omega)$$

$$Y_{\mathrm{mod}}(j\omega) = \frac{I(j\omega)}{U(j\omega)} = \frac{K \times (j\omega)^n}{R \times K \times (j\omega)^n + 1} \tag{7.22}$$

The modified response function $Y_{\mathrm{mod}}(j\omega)$ transfers into the original model if $n = 1$ with the condition that the values of the capacity C and the parameter of the CPE K are identical. As a result the modified model can be used for both calculations regarding the value of the parameter n.

The simulation is performed in the frequency domain, although the response of the system in the time domain is of interest. The following calculation sequence is used for the simulation of the system.

1. The signal of the exciting voltage is specified in the time domain where it can easily be adapted to the exciting signal of the experiment. Adjacent the signal is transformed into the frequency domain via Fourier transformation.
2. The obtained spectrum of the input voltage is multiplied with the frequency response function of the admittance resulting in the spectrum of the current.
3. The response of the electrical system is transformed into the time domain via inverse Fourier transformation in order to facilitate the comparison with the measured data of the current.
4. The charge signal is needed for the calculation of the actuation force which stimulates the mechanical system. Since a measurement of the charge is impossible, the characteristic is determined via integration of the current.
5. The charge signal is squared and multiplied with $0.5\,\alpha$ in order to obtain the exciting force.
6. The frequency spectrum of the force is multiplied with the frequency response function of the dynamic compliance yielding the displacement response which is consecutively transformed into the time domain.
7. The swelling of the bucky-paper in aqueous environment enforces the measurement of the relative displacement [20]. As a consequence the minimum of both, the measured and the calculated signal is set to zero.

Different types and frequencies of actuation signals are used in order to evaluate the quality of the simulation. Sine, square wave and ramp are used as periodic stimulation. Additionally, a sweep signal from $f = 0.01$ to 5 Hz is utilized to excite the CNA. Exemplary time responses are presented here. A cyclic voltammetry is a convenient and common procedure in order to get a general idea of the system behavior in the complete workspace of the actuator [2, 20, 21]. Hence the input signal for the cyclic voltammetry (a slow ascending and descending ramp) is used to check the developed model. Figure 7.9 depicts the time responses of the electrical and the mechanical system to the excitation with a frequency of $f = 0.01$ Hz, an amplitude of $U_{pp} = 1$ V and an offset of $U_{off} = 0.05$ V.

The model using the constant phase angle element provides a better approximation of the produced strains. Especially in the second half of the period the improvement is evident. The modification of the model by a CPE shows the biggest effect in the range of low frequencies of the admittance. Increasing the frequency is linked to a decrease of the amplitudes of the displacement caused by the inertia of the system. The best fit can be found in the range of 'low' signal frequencies. This trend is also identifiable looking at the other types of analyzed signals.

Generally the current response of the modified model has less deviation from the measurement than the response of the original model. The superior approximation of the properties of the electrochemical double layer by a constant phase angle element is evident in the range of low signal frequencies. For the examined

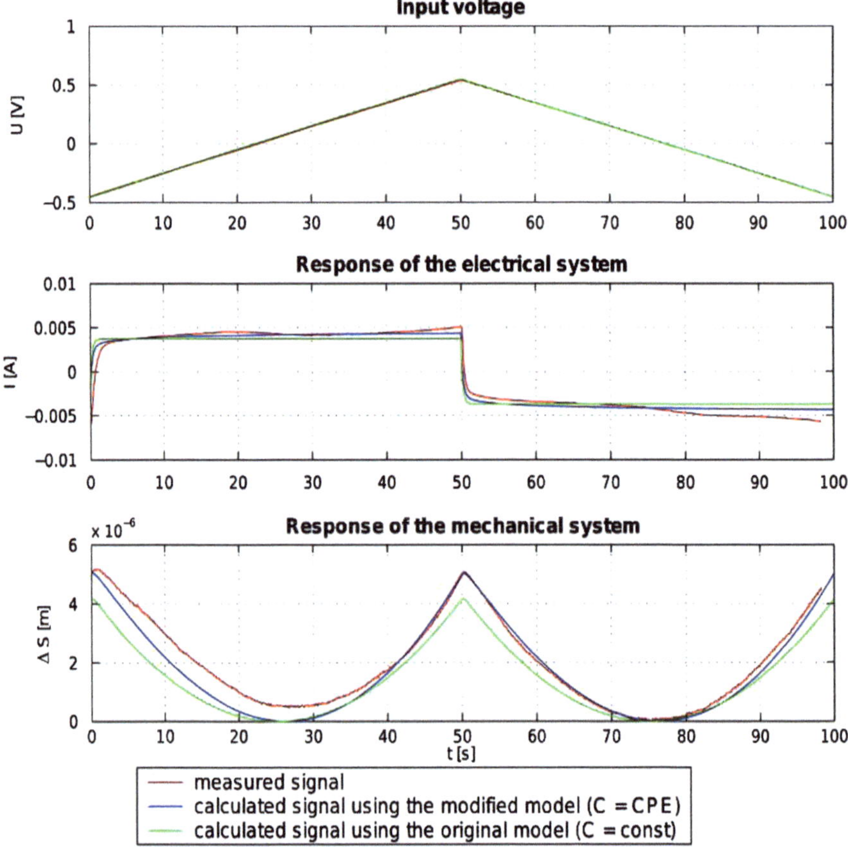

Fig. 7.9 Cyclic voltammetry $f = 0.01$ Hz

test signals both models can describe the behavior of the electrical system with average deviations of less than 10%.

The sweep signal can be used to excite the system within a well-defined frequency range and is suitable for a measurement with continuously varying frequency. Consequently the complete frequency range for the actuation of the bucky-paper specimen can be covered with a single measurement. Figure 7.10 shows the time domain responses to the sweep excitation.

Obviously the modified model using the CPE improves the amplitude response in the whole frequency range. The modified model reduces the relative deviations in the range from $f = 0.3$ to 1 Hz by up to almost 20%. The working frequency of the electrolytic actuator is below $f = 2$ Hz. Within this range the relative deviations of the frequency response are less than 10%.

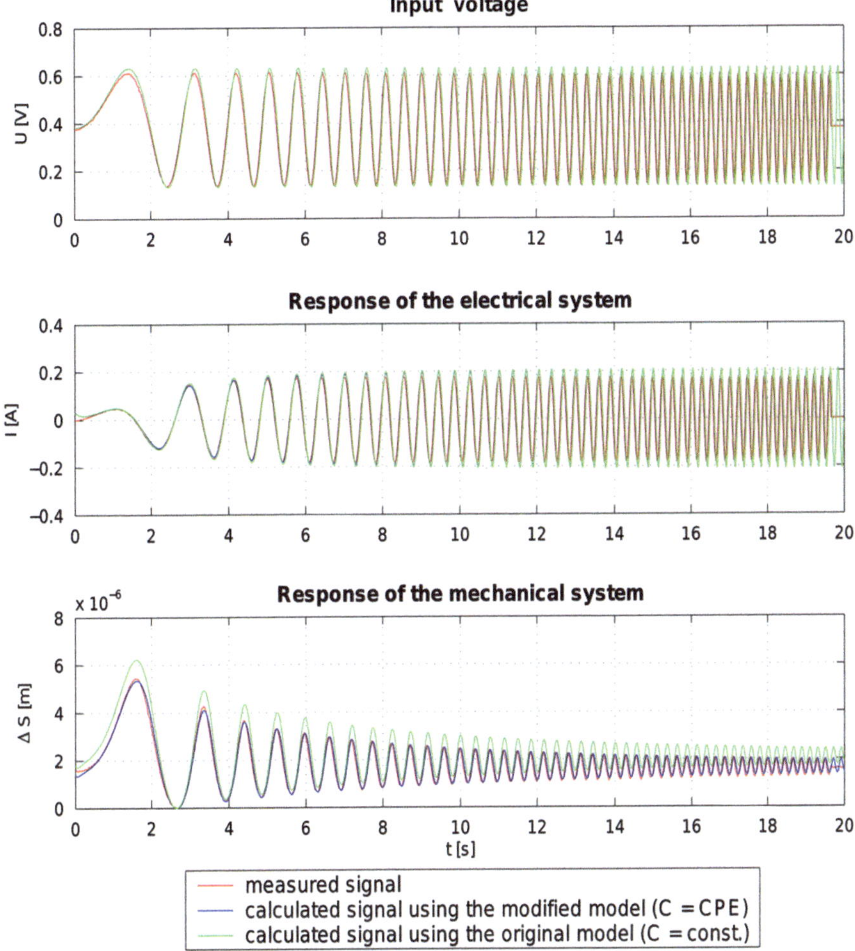

Fig. 7.10 Response of the system to a sweep excitation $f = 0\text{--}5$ Hz

7.10 Solid Electrolytes for CNT Based Actuators

A prerequisite for structural integration within a real application is a CNT-based actuator that is of solid state. Most of the already presented work refers to two-dimensional paper like CNT-architectures containing randomly oriented CNTs, called bucky-papers. The experiments described so far take place within liquid electrolytes. Even though this is not an option for real applications this experimental approach is helpful getting a fundamental understanding of the actuation mechanisms themselves.

This experimental case was investigated by several publications [22], focusing on the experimental actuator-characterization [6, 20] and the modeling of such a system [18]. In addition to intensive research, done on this topic, there is the need to insert an actuator into a smart system. A possible solution is the incorporation of solid electrolytes to form a hybrid actuator of electrolyte and CNT-structure. But there are just a few publications about their use so far [23, 24]. Among those most apply NAFION or FUMION as electrolyte. The use of such electrolytes has to be handled with care. Two aspects must be considered: only the use of an accurate test set-up, like it is the out-of-plain test rig [25], can compensate secondary deflection causing effects like mass transfer along the field lines within an electric field [26, 27] or thermal degradation. The second remarkable effect is based on the active behaviour of polymers themselves. It can be shown that thin sheets of Nafion show significant active strain using voltages up to ± 2 V [28].

Therefore accurate measurements are needed to analyze the active potential of the individual components. With this knowledge it is possible to make statements about their contribution to the global measured free stroke of the hybrid actuator.

7.11 Specimen Processing and Experimental Setup

Within the presented work three different CNT-based materials are used and processed as CNT-based electrode for hybrid actuators. On the one hand 2-dimensonal paper-like architectures of randomly oriented CNTs are manufactured by commercially available CNT-powder. The used multi-walled CNTs are provided by Bayer (Baytubes CP 105P) and single-walled CNTs are purchased from Thomas Swan (Elicarb PR0925). The papers are processed within a multistep high-pressure filtration process. On the other hand vertically aligned multi-walled CNTs [29] provided by the group of Schulte (TU Hamburg-Harburg) are investigated. These CNT-arrays are grown by chemical-vapor-deposition (CVD). Due to the processing time the CNT-array heights differ from 300 to 500 μm with CNT diameters of 20–50nm (Fig. 7.11)

The solid electrolyte, NAFION D2021 is provided by DuPont in the liquid state.

Using NAFION as coating material for CNT architectures revels that shrinkage is often critical for weak bucky-papers and for the alignment of CNT-arrays. Therefore thin sheets of NAFION are manufactured at first using air-brush technology. Afterwards the CNT-electrode and the NAFION-membrane are bonded to each other using liquid NAFION as adhesive. A post-experimental compression and storage under high humidity atmosphere is necessary for reproducible measurements.

The experiments for analyzing the active free strain of hybrid actuators and NAFION-sheets take place in an out-of-plain set-up (Fig. 7.12). The active material is clamped between two copper electrodes. The upper electrode has a small hole used for an optical device (MircoEpsilon Opto NCDT 2400-0.08) to measure directly the free stroke of the Au-coated surface of the active material. The used measurement chain is analogue to the test set-up used for analyzing the liquid electrolyte apart from the connected reference electrode and counter electrode. The reason therefore is the lack of space for positioning a reference electrode within the thin polymer film (less than 30 μm).

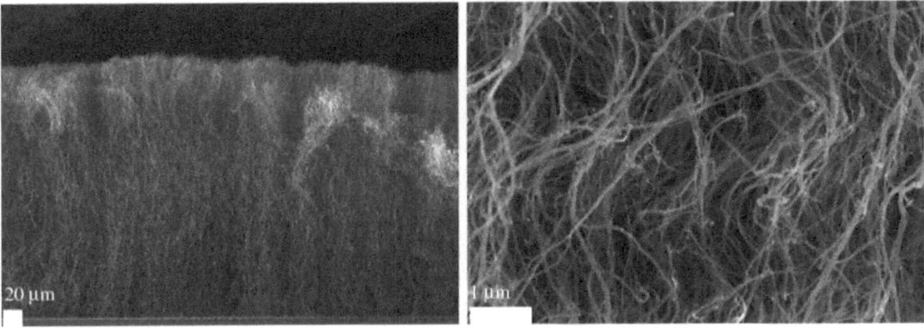

Fig. 7.11 Vertically aligned multi-walled CNTs called arrays

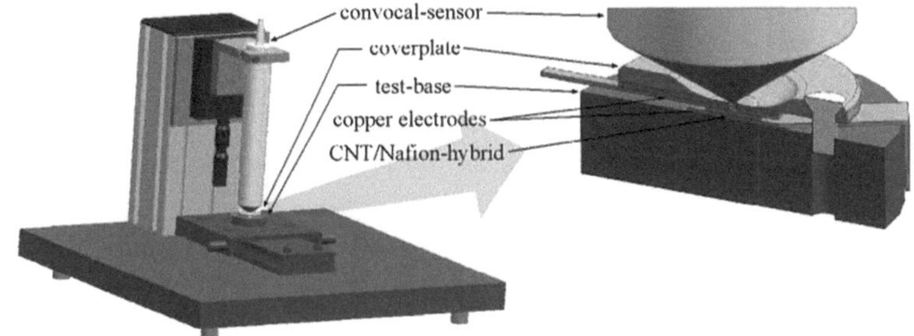

Fig. 7.12 Out-of-plain test set-up

7.12 Initial Test Results with Solid Electrolytes for CNA

Actuators of all material combinations can be successfully manufactured and tested. Nevertheless the shrinkage of NAFION due to its time-dependent water loss is crucial for the mostly weak CNT-based architectures like CNT-papers. Especially the bucky-papers containing single-walled CNTs are very weak. Most of these coated architectures are broken into small pieces due to their low internal stability. The number of tested samples is too small for making reliable statements. In contrast the multi-walled CNT based bucky-papers turn out to be more durable. An additional drawback is their water dependent reproducible active performance (Fig. 7.13).

Even the former highly vertically aligned CNT-arrays are afterwards disordered (Fig. 7.14). Two measures are necessary to profit from the more efficient strain generation of the aligned CNTs. The first is the increase of the CNT concentration within the actuator volume. The second is the increase of the stiffness of the architecture built up by the CNTs.

Beside the affected CNT-alignment after NAFION coating the manufactured hybrids also suffer from structural instability. During the drying the NAFION coating causes cracks and uneven surfaces. Therefor measurements become difficult with less reliable results. A solution for a more appropriate preparation of aligned CNTs in order to preserve their alignment during coating would be the

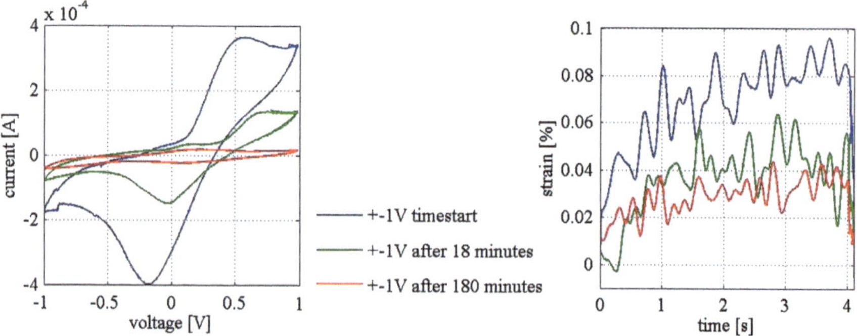

Fig. 7.13 Strain-degradation during testing

Fig. 7.14 Disordered multi-walled CNTs after coating and drying of NAFION

conversion of their orientation from vertical (standing) to horizontal (lying) by a rolling process [30]. A further improvement can be achieved by additional mechanical compression of the aligned CNT mat.

The multi-walled CNT-based papers are the only architecture within this test campaign whose stability enables a high number of successfully tested samples. However, the found relations of free strain and excitation voltage reveals no clear trend like it is found in the former mentioned experiments using liquid electrolytes. Due to the smaller surface area of multi-walled CNTs the measured free strain was only in the range of |0, 1%| at excitation voltages of +-1 V. Higher strain values (|0,3%|) can be reached using higher voltages up to +-3 V. According to the chemical stability window of aqueous electrolytes (+-1 V) the measured strain at +-3 V is related to irreversible chemical effects. For a better understanding of the active behavior of the hybrid-actuators, NAFION is tested as single sheets of differing thicknesses. The air-brush-technique enables to control the thickness of the sheets within 10 μm. The tests are carried out in the out-of-plain set-up as well. It can be found that all investigated sheets are almost inactive within the range of ±1 V (Fig. 7.15).

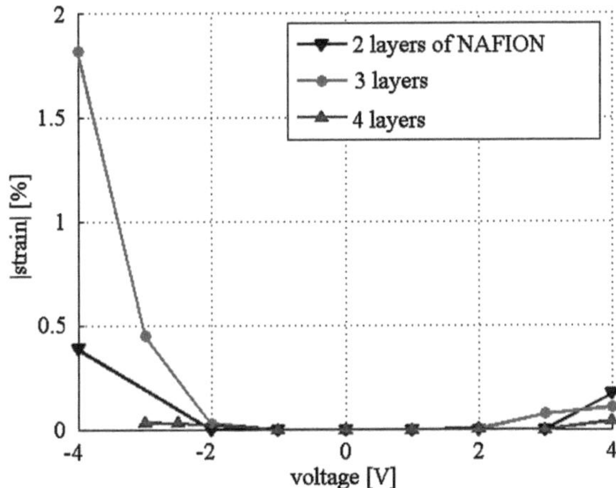

Fig. 7.15 NAFION-sheets are tested with the voltage-range of ±4 V

Table 7.1 Different characteristics of NAFION-sheets during strain tests

Number of layers	Thickness [μm]	Posttest weight loss [%]
2	49	0.43
3	52	0.23
4	70	0.15

The usage of the high strains generated at voltages beyond +-1 V is questionable in terms of reproducible active performance as irreversible chemical reactions take place. As an indicator for these irreversible reactions a weight analysis (see Table 7.1) of the sheets reveals a weight-loss inversely proportional to the sheet thickness. This records the trend of high water-loss of thin sheets due to their higher surface area per volume. Therefore an earlier actuator failure due to the lost ability of ion-movement can be expected using rather thin NAFION-films at high voltages.

7.13 Conclusion

The chapter examines the actuation mechanism of Carbon Nanotubes in an electrolytic environment. A model for a comprehensive and deepened understanding of the actuation mechanism is presented. The derived model is validated for a CNA configuration which is based on bucky-paper in a liquid electrolyte. One focus of this chapter is the determination of the parameters which are needed in the model. In the course of the parameter estimation it emerged that the properties of the electrochemical double layer could rather be approximated by a constant phase angle element than by a pure capacity. The application of the initial model based on a pure capacitor leads to parameters which vary with the frequency. The implementation of the fractional derivative of the constant phase angle element allows describing the electrical behavior with constant parameters. One additional parameter is needed for this improvement which is the exponent n of the constant phase angle element.

The results and deductions like the quadratic relationship between activation voltage and free stroke found in liquid electrolytes cannot be transferred to solid electrolytes with the same reproducibility. However, the measured strokes are similar in comparison to the liquid systems but the ionic conductive system NAFION is totally different and shows a strong dependency upon its humidity. Although different and more effective CNT-architectures as well as CNT types are used because of their alignment and active surface the measured free stroke is similar over all. Nevertheless hybrid actuators are built and tested. The next challenges are to capsule a hybrid of effective CNT-material (aligned CNT-paper) and to bond sheets of NAFION for more reproducible performance. The solid electrolyte NAFION is an available, handable and low-harmful solid ionomere with high industrial relevance (fuel cells). Until now no active behavior of NAFION itself is detected within the voltage range of interest for a carbon nanotube actuator avoiding its usage.

Acknowledgment The authors would like to thank the German Research Foundation (DFG) for financially supporting the research.

References

1. Iijima, S.: Helical microtubules of graphitic carbon. Nature **354**(6348), 56–58 (1991). (London, UK)
2. Baughman, R.H., Cui, C., Zakhidov, A.A., Iqbal, Z., Barisci, J.N., Spinks, G.M., Wallace, G.G., Mazzoldi, A., De Rossi, D., Rinzler, A.G., Jaschinski, O., Roth, S., Kerzesz, M.: Carbon nanotube actuators. Science **284**, 1340–1344 (1999)
3. Yeo-Heung, Y., Miskin, A., Kang, P., Jain, S., Narasimhadevara, S., Hurd, D., Shinde, V., Schulz, M.J., Shanov, V., He, P., Boerio, F.J., Shi, D., Srivinas, S.: Carbon nanofiber hybrid actuators: part I–liquid electrolyte-based. J. Intell. Mater. Syst. Struct. **17**(2), 107–116 (2006)
4. Yeo-Heung, Y., Miskin, A., Kang, P., Jain, S., Narasimhadevara, S., Hurd, D., Shinde, V., Schulz, M.J., Shanov, V., He, P., Boerio, F.J., Shi, D., Srivinas, S.: Carbon nanofiber hybrid actuators: part II—solid electrolyte-based. J. Intel. Mater. Syst. Struct. **17**(3), 191–197 (2006)
5. Barisci, J. N., Spinks, G.M., Gordon, G.W., John, D.M., Ray, H.B.: Increased actuation rate of electromechanical carbon nanotube actuators using potential pulses with resistance compensation. Smart Mater. Struct. **12**, 549–555 (2003)
6. Mazzoldi, A., De Rossi, D., Baughman, R.H.: Electro-mechanical behavior of carbon nanotube sheets in electrochemicial actuators. In: Bar-Cohen, Y. (ed.) SPIE, EA-PAD, vol. 3987 (2000)
7. Spinks, GM., Wallace, G.G., Baughman, R.H.: Carbon nanotubes actuators. In: Bar-Cohen, Y. (ed.) Electroactive Polymer (EAP) Actuator as Artificial Muscles, Book Chapter 8, vol. PM 136, pp. 223–246. SPIE Press Monograph (2001)
8. Spinks, G.M., Wallace, G.G., Carter, C.D., Zhou, D., Fifield, L.S., Kincaid, C.R., Baughman, R.H.: Conducting polymer, carbon nanotube, and hybrid actuator materials. In: Bar-Cohen, Y. (ed.) Smart Structures and Materials 2001: Electroactive Polymer Actuators and Devices, vol. 4329, pp. 199–208. SPIE (2001)

9. Spinks, G.M., Wallace, G.G., Lewis, T.W., Fifield, L.S., Dai, L., Baughman, R.H.: Electrochemically driven actuators from conducting polymers, hydrogels, and carbon nanotubes. In: Wilson, A.R., Asanuma, H. (ed.) Smart Materials, vol. 4234, pp. 223–231. SPIE (2001)
10. Tahhan, M., Van Truong, T., Spinks, G., Wallace, G.G.: Carbon nanotube and polyaniline composite actuators. Smart Mater. Struct. **12**, 626–632 (2003)
11. Spinks, G.M., Xi, B., Van Truong, T., Wallace, G.G.: Actuation behaviour of layered composites of polyaniline, carbon nanotubes and polypyrrole. Synth. Met. **151**(1), 8591 (2005)
12. Riemenschneider, J., Mahrholz, T., Mosch, J., Monner, H.P., Melcher, J.: Carbon nanotubes-smart material of the future: experimental investigation of the system response. ECCOMAS, (Jul 2005)
13. Schmickler, W.: Interfacial electrochemistry. Oxford University Press, Oxford (1996). (ISBN: 0195089324, 1996)
14. Riemenschneider, J., Temmen, H., Monner, H.P.: CNT based actuators: experimental and theoretical investigation of the in-plain strain generation. J. Nanosci. Nanotechnol. (2006)
15. Ghosh, S., Gadagkar, V., Sood, A.K.: Strains induced in carbon nanotubes due to the presence of ions: Ab initio restricted hatree-fock calculations. Chem. Phys. Lett. **406**, 10–14 (2005)
16. Brug, G.J., van den Eeden, A.L.G., Sluyters-Rehbach, M., Sluyters, J.H.: The analysis of electrode impedances complicated by the presence of a constantphase element. J. Electroanal. Chem. **176**, 275–295 (1984). March
17. Waidhas, M.: Grundlegende Technologie von Doppelschichtkondensatoren. In: Siemens ZVEI-Workshop, (Jan 2004)
18. Riemenschneider, J., Opitz, S., Sinapius, M., Monner, H.P.: Modeling of carbon nanotube actuators: part I {modeling and electrical properties. J. Intell. Mater. Syst. Struct. **20**(2), 245–250 (2009)
19. Riemenschneider, J., Opitz, S., Sinapius, M., Monner, H.P.: Modeling of carbon nanotube actuators: part II—mechanical properties, electro mechanical coupling and validation of the model. J. Intell. Mater. Syst. Struct. **20**(3), 253–263 (2009)
20. Riemenschneider, J., Mahrholz, T., Mosch, J., Monner, H.P., Melcher, J.: System response of nanotube based actuators. Mech. Adv. Mater. Struct. **14**(1), 57–65 (2007)
21. Barisci, J.N., Wallace, G.G., MacFarlane, D.R., Baughman, R.H.: Investigation of ionic liquids as electrolytes for carbon nanotube electrodes, vol. 6, issue 1, pp. 22–27 (2004)
22. Barisci, J.N., Wallace, G.G., Baughman, R.H.: Electrochemical studies of single-wall carbon nanotubes in aqueous solutions. J. Electroanal. Chem. **488**, 92–98 (2000)
23. Landi, B.J., Raffaelle, R.P., Heben, M.J., Alleman, J.L., VanDerveer, W., Gennett, T.: Single wall carbon nanotube-nafion composite actuators. Nano Lett. **2**(11), 1329–1332 (2002)
24. Landi, B.J., Raffaelle, R.P., Heben, M.J., Alleman, J.L., Van Derveer, W., Gennett, T.: Development and characterization of single wall carbon nanotube-nafion composite actuators. Mater. Sci. Eng. B **116**(3), 359–362 (2005)
25. Kosidlo, U., Eis, D.G., Hying, K., Haque, M.H., Kolaric, I.: Development of measurement set-up for electromechanical analysis of bucky-paper actuators. AZojono: J. Nanotechnol. Online **3**, 1–11 (2007)
26. Shahinpoor, M., Kim, K.J.: Ionic polymer-metal composites: I. fundamentals. Smart Mater. Struct. **10**(4), 819–833 (2001)
27. Shahinpoor, M., Kim, K.J.: Ionic polymer-metal composites: Iii. Modeling and simulation as biomimetic sensors, actuators, transducers, and artificial muscles. Smart Mater. Struct. **13**(6), 1362–1388 (2004)
28. Akle, B., Akle, E., Duncan, A., Lee, D.J., Wallmersperger, T.: Forced and free displacement characterization of inonic polymer transducers. Proc. SPIE **7287**, 72870N (2009)
29. Prehn, K., Adelung, R., Heinen, M., Nunes, S.P., Schulte, K.: Catalytically active cnt-polymer-membrane assemblies: From synthesis to application. J. Membr. Sci. **321**, 123–130 (2008)
30. Wang, D., Song, P., Liu, C., Wu, W., Fan, S.: Highly oriented carbon nanotube papersmade of aligned carbon nanotubes. Nanotechnology **19**(7), 1–6 (2008)

Chapter 8
Piezoceramic Honeycomb Actuators

Jörg Melcher

Abstract The success of active vibration control in adaptronics decisively depends on the performance of the actuators used as multifunctional structural elements. DLR's concept of impedance matched actuators proposes novel actuators in honeycomb design (Fig. 8.1). The partitions of their hexagonal cells are made of piezoceramic materials. The honeycomb geometry guarantees lightweight properties with nearly perfect load distributions in the plane perpendicular to the cell tubes. Furthermore there is a dynamic flexibility in this plane even when the static stiffness is very high. New fabrication methods are required to realize honeycomb actuators with very small cells, low wall thicknesses and perfect rounded corners. In order to fulfill the lightweight conditions self-organizing effects were used to guarantee optimal 3D geometries. In this context, a local thermal treatment of ceramic structures by employing lasers initiates self-organizing mechanisms via material flow. Experimental comparisons between piezoelectric honeycomb actuators and conventional monolithic actuators demonstrate a reduction of electrical power by a factor of 500. Honeycomb actuators seem to be the most promising candidate for technical applications that require low energy consumption.

8.1 Active Control of Mechanical Impedances

In recent years many multifunctional actuators and sensors have been developed in adaptronics for active control of noise and vibrations. Usually actuators are driven far away from their resonances to avoid control and design efforts. Being in a sharp

J. Melcher (✉)
DLR, Institute of Composite Structures and Adaptronics,
Lilienthalplatz 7, 38108, Braunschweig, Germany
e-mail: joerg.melcher@dlr.de.

M. Wiedemann and M. Sinapius (eds.), *Adaptive, Tolerant and Efficient Composite Structures*, Research Topics in Aerospace, DOI: 10.1007/978-3-642-29190-6_8,
© Springer-Verlag Berlin Heidelberg 2013

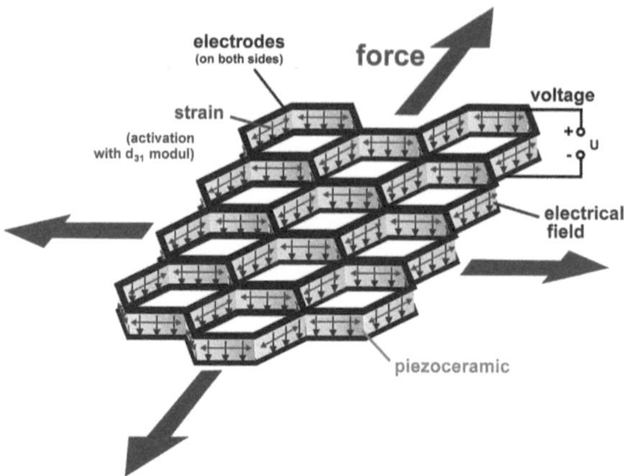

Fig. 8.1 Piezoceramic honeycomb actuator

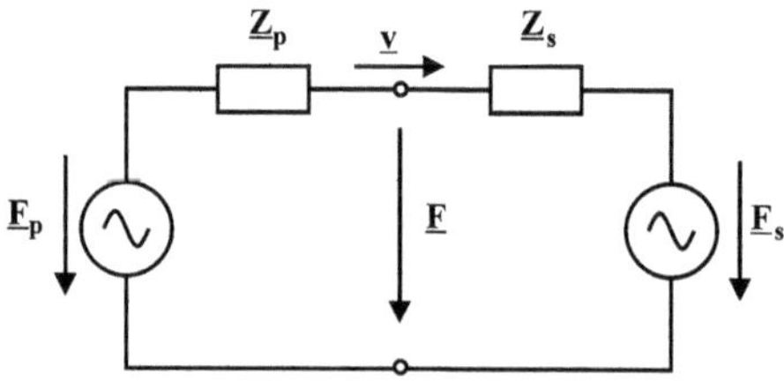

Fig. 8.2 Electrical analogue circuit of the structure-actuator junction

contrast to that DLR proposes actuator and sensor designs with structural conformity [1]. This strategy requires a simultaneous design of all components from the very first beginning leading to actuator constructions, which transfer function is optimally matched to the structural system. Consequently it is expected that the actuator characteristics are not linear any more, the dynamic characteristics of the transfer function are shaped.

Some research investigations used the impedance formulation in order to have a better understanding of the structural dynamics of systems with structurally integrated actuators and sensors [2, 3] For active vibration control a minimum of energy effort can be achieved by implementing multifunctional actuator elements with properly chosen mechanical input impedances [4, 5].

As long as the actuators do not support from a serial mass, e.g. from those of tuned mass dampers, but they have load carrying capacities as multifunctional elements, the junction point of structure and active system can be derived from a simple electrical equivalent circuit [6], see Fig. 8.2. Its left hand side represents the complex primary excitation $\underline{F}_p$ together with the dispersive mechanical input impedance $\underline{Z}_p = Z_p' + j \cdot Z_p''$, the right hand side the actuator impedance $\underline{Z}_s = Z_s' + j \cdot Z_s''$ and its complex generating force $\underline{F}_s$. The small circles are the electrical terminals corresponding to the junction. Using the force-voltage analogy, the voltage across the terminals corresponds to the force $\underline{F}$ between actuator and structure, while the current through the terminals is equivalent to the velocity $\underline{v}$ of the vibrating junction. For the general case where both the primary source as well as the active system are operated the junction values are

$$\underline{F} = \frac{\underline{Z}_s}{\underline{Z}_p + \underline{Z}_s} \cdot \underline{F}_p + \frac{\underline{Z}_p}{\underline{Z}_p + \underline{Z}_s} \cdot \underline{F}_s \tag{8.1}$$

and

$$\underline{v} = \frac{1}{\underline{Z}_p + \underline{Z}_s} \cdot \underline{F}_p - \frac{1}{\underline{Z}_p + \underline{Z}_s} \cdot \underline{F}_s. \tag{8.2}$$

Their quotient yields the mechanical impedance at the junction point:

$$\underline{Z} = \frac{\underline{F}}{\underline{v}} = \frac{\underline{F}_p}{\underline{F}_p - \underline{F}_s} \cdot \underline{Z}_s + \frac{\underline{F}_s}{\underline{F}_p - \underline{F}_s} \cdot \underline{Z}_p. \tag{8.3}$$

This impedance is the most significant physical parameter in an active control process. Depending on the particular requirements, the impedance $\underline{Z}$ is

- a "matched impedance" in case of an optimal active power absorption, whereas also "reactive impedances" are included compensating an existing mass or stiffness by adding a "negative mass" or a "negative stiffness",
- an "infinite impedance" in case of an optimal active vibration suppression, or
- a "zero impedance" in cases of a perfect active vibration isolation or an optimal force flow compensation.

The function of the power extraction depending from the secondary force $\underline{F}_s$ is given by the "power parabola", see Fig. 8.3:

$$\begin{aligned} P_w(\underline{F}_s) &= \tfrac{1}{4}\left(\underline{F} \cdot \underline{v}^* + \underline{F}^* \cdot \underline{v}\right) \\ &= \frac{1}{2 \cdot |\underline{Z}_p + \underline{Z}_s|^2} \cdot \left[\underline{Z}_s |\underline{F}_p|^2 - (\underline{Z}_p - \underline{Z}_s)\, \underline{F}_p \underline{F}_s^* - \underline{Z}_p |\underline{F}_s|^2 \right] \cdot \end{aligned} \tag{8.4}$$

The *optimal active power absorption* can be reached with the secondary force

$$\underline{F}_{s,matched} = \frac{\underline{Z}_p^* - \underline{Z}_s}{\underline{Z}_p^* + \underline{Z}_p} \cdot \underline{F}_p. \tag{8.5}$$

At the junction location this force causes the junction force and velocity

$$\underline{F} = \frac{\underline{Z}_p^*}{2 \cdot \underline{Z}_p'} \cdot \underline{F}_p \quad \text{and} \quad \underline{v} = \frac{1}{2 \cdot \underline{Z}_p'} \cdot \underline{F}_p \tag{8.6a}$$

and the matched impedance

$$\underline{Z} =: \underline{Z}_{matched} = \underline{Z}_s + \underbrace{\frac{\underline{F}_{s,matched}}{\underline{v}}}_{actively\ controlled\ part} = \underline{Z}_p^*. \tag{8.6b}$$

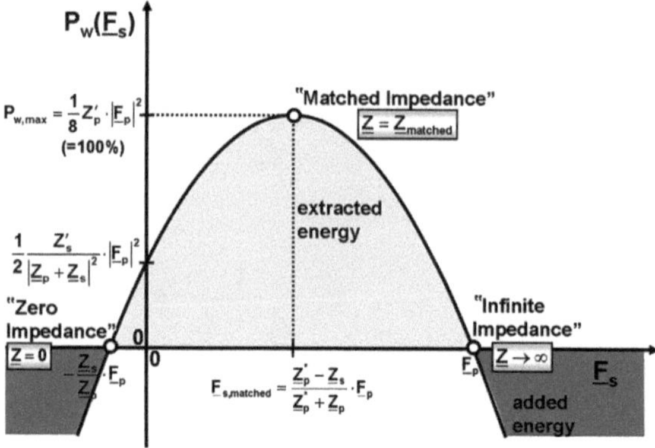

Fig. 8.3 Power extraction depending from active impedance adjustment (power parabola)

Generally the normalized power extraction by $\underline{Z}_s$ is given by

$$\frac{\underline{P}_W\left(Z'_s, Z''_s\right)}{\underline{P}_{W,max}} = \frac{4 \cdot Z'_s Z'_p}{\left[\left(Z'_p + Z'_s\right)^2 + \left(Z''_p + Z''_s\right)^2\right]} \leq 1 \tag{8.7}$$

with

$$\underline{P}_{W,max} = \frac{1}{8} \frac{|F_p|^2}{Z'_p}. \tag{8.8}$$

As expected the perfect power extraction requires the matched impedance. Furthermore the actuator impedance $\underline{Z}_s$ should be matched to the conjugate complex impedance of the structural environment. Otherwise, if $\underline{Z}_s \neq \underline{Z}_p^*$, the actuator must be highly powered in order to compensate of the mismatch $\left(\underline{Z}_p^* - \underline{Z}_s\right)$ in Eq. (8.5)—realized by the actively controlled part in Eq. (8.6b).

An *optimal active vibration suppression* leads to

$$\underline{F} = \underline{F}_P \quad \text{and} \quad \underline{v} = 0. \tag{8.9}$$

Then the junction impedance becomes infinite ($\underline{Z} \to \infty$) with the adjustment

$$\underline{F}_{s,inf} = \underline{F}_p. \tag{8.10}$$

Otherwise, a "zero impedance" ($\underline{Z} = 0$) is required for an *optimal force flow compensation* that needs the optimal adjustments

$$\underline{F}_{s,comp} = -\frac{\underline{Z}_s}{\underline{Z}_p} \underline{F}_p. \tag{8.11}$$

This secondary force cause the relationships at the junction point

$$\underline{F} = 0 \quad \text{and} \quad \underline{v} = \frac{1}{\underline{Z}_p} \underline{F}_p. \tag{8.12}$$

8.2 Honeycomb Actuator Design, Fabrication and Applications

In order to meet the impedance requirements mentioned above an alternative architecture to the monolithic constructions were propose: piezoelectric actuators in honeycomb design (Fig. 8.1) [7, 8].

The honeycomb geometry was chosen because of its lightweight properties: it has a nearly perfect load distribution in the plane perpendicular to the cell tubes. Furthermore there is a dynamic flexibility in this plane even the static stiffness is very high. The static and the structural dynamic design of the honeycomb geometry (e.g. thickness and length of the partitions) and a special activation allow to match the actuator to desired impedances. A typical mechanical impedance of piezoceramic honeycomb actuators, see Fig. 8.4, has the modal description

$$\underline{Z}_{comb}(s) = \left(\Phi_i^{\mathrm{T}}\right)^{-1} diag[Z_i(s)] \cdot \Phi_i^{-1} \tag{8.13}$$

with

$$\underline{Z}_i(s) = \frac{M_i s^2 + D_i s + K_i}{s}. \tag{8.14}$$

The nomenclature (corresponding to the i-th mode) contains the eigenvectors Φ_i and the elements M_i, D_i and K_i of the generalized mass, damping, and stiffness. In contrast to equivalent monolithic piezoceramic patches that have eigenfrequencies typically in the range of 25 kHz piezoceramic honeycomb actuators have drastically reduced eigenfrequencies In the case of Fig. 8.4 the eigenfrequency is nearly 2 kHz.

The impetus of this actuator design was given by honeybees [8, 9]. Bees do not use their wax combs only for their brood and food. They also correspond to each other using the combs as the vibrating media. Furthermore, the bees manufacture of honeycombs was taken as an example in a bionic sense in order to fabricate ceramic combs. The speculations on how the honeybees so precisely construct the hexagonal cells are centuries old. But recent investigations found out that the hexagons are the result of self-organizing processes in which the beewax reaches a liquid equilibrium [10]. During the building process the honeybees heat up the wax to around 40°C. That is the temperature from that the wax becomes amorphic. Under slightly raised temperatures the thermoplastic beewax flows exactly to that geometry that has the lowest enthalpy. The result of this thermodynamic process are hexagonal cells. The final dimensions only depend on the starting conditions of the self-organizing process.

Matching a ceramic comb structure to a mechanical impedance with low eigenfrequencies the comb should have very thin struts and very precise curved cell junctions. Using conventional ceramic forming methods, such as casting, plastic forming or pressing, and even with new developments such as micro-sandblasting, ionotropic capillary gel forming or electrophoretic deposits the

Fig. 8.4 Impedance of a
piezoceramic honeycomb
actuator

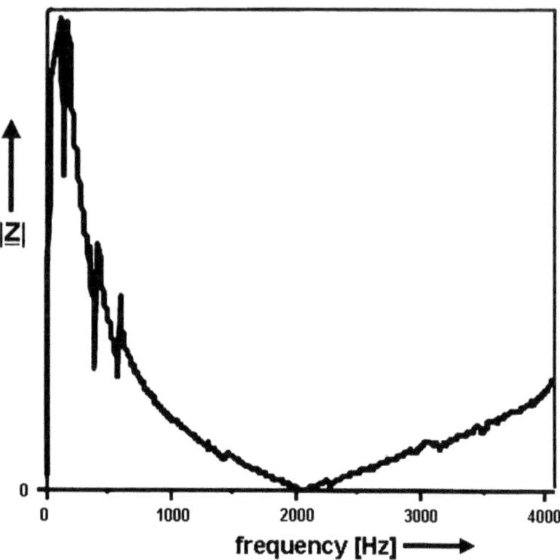

required geometries are not attainable. Therefore, the most promising strategy
seems to be to use physical self-organizing effects similar to the organic comb
manufacture.

In physics self-organization is defined as a forming process in a dissipative
thermodynamic system away from equilibrium. In agreement with the second law
of thermodynamics, self-organizing processes appears spontaneously on all scales
and lead to structural patterns that are distinguished to have a minimum of energy
and free enthalpy and a maximum of entropy. The reason for self-organization is
the presence of mechanical, chemical, electrical, electrochemical, thermodynamic,
acoustical or aerodynamic potentials whose absolute minima are not yet adjusted.

Applied to ceramic forming processes the self-organizing fabrication must rely
at least on four basic attributes: the system is dissipative, non-linear, away from
absolute enthalpy minimum and getting started by clearing or shifting a potential
barrier. A typical experimental result of a ceramic comb is shown in Fig. 8.5.

As long as the object of self-organization is a thin and non-closed membrane,
the result of the self-organizing processes are minimal surfaces immersed in $\Re^3$—a
topology with a disappearing mean curvature. The latter is the average of the
principle curvatures, that are the eigenvalues of the shape operator of given surface
points. The forms of soaps are impressive examples of minimal surfaces. But if the
objects are volume elements with closed surfaces (e.g. spheres), the resulting
topologies differ from typical minimal surfaces. Then the mean curvature is still
constant but non-zero.

Nevertheless the geometry of honeycombs can be analytically estimated as a
periodic composition of trinoids. In potential theory they are knows as symmetric
trinoids of genus 1. Using the Enneper-Weierstrass representation, the analytic
description of trinoids is given by

Fig. 8.5 Self-organized
ceramic cell

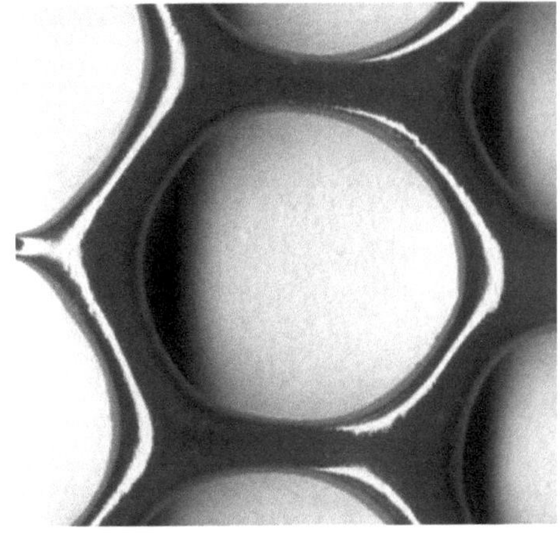

$$\begin{bmatrix} x(r,\varphi) \\ y(r,\varphi) \\ z(r,\varphi) \end{bmatrix} = \Re e\left\{ \int \begin{bmatrix} f(1-g^2) \\ j\cdot f(1+g^2) \\ 2\cdot f\cdot g \end{bmatrix} dz \right\}, \qquad (8.15)$$

whereas the analytic functions f and g are given by

$$f(z) = \left(z^3 - 1\right)^{-2} \quad \text{and} \quad g(z) = z^2. \qquad (8.16)$$

While g is a meromorphic function with an order less than f the function f has a holomorphic form. In order to predict the final comb state of self-organizing processes for given static and dynamic load conditions a fast program library for C++ environments was generated. The implemented algorithms consider thermo-mechanic potentials whose negative gradients are the forces that cause the structure to change its geometry self-organizingly. The recursive iteration processes allow two and three dimensional applications (Figs. 8.6 and 8.7). The self-organizing processes achieve their final states when all stresses are completely compensated indicating a homogeneous blue coloured distribution. A thermal potential is just suitable to control the self-organization because the volume shrinkage can be controlled by temperature [11–13]. Therefore the choice of piezoelectric material is very significant for self-organizing fabrication methods [14, 15].

For the manufacture of honeycomb piezoelectric actuators green ceramic foils are structured with CO_2-lasers. With a focussed laser beam thin foils are structured in a shape of a honeycomb, at first employing an alumosilicate model material for validation purposes, secondly PZT (Fig. 8.8) and $K_{0.5}Na_{0.5}NbO_3$ (KNN) in the same way. In order to obtain thicker honeycomb structures the thin foils are stacked and bonded to a ceramic multilayer element. This procedure allows the fabrication of piezoelectric actuator stacks. A major benefit of this approach is the variability in manufacturing of honeycomb structures with preferred cell size to wall thickness ratios.

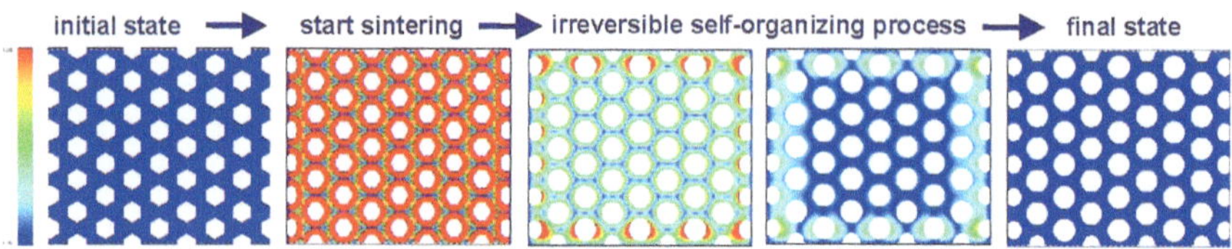

Fig. 8.6 Simulated sequence of self-organized process with colour map: *blue* (no mechanical stress), *red* (maximum of stress)

Fig. 8.7 Simulated self-organized 3D comb: photronic© or diamond lattice structures

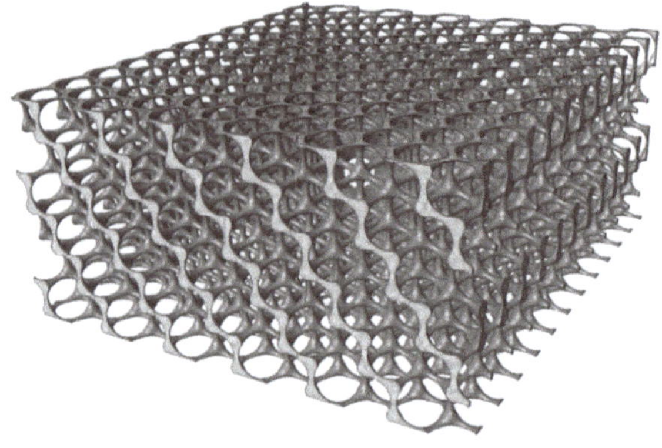

Fig. 8.8 Laser sintered ceramic cell

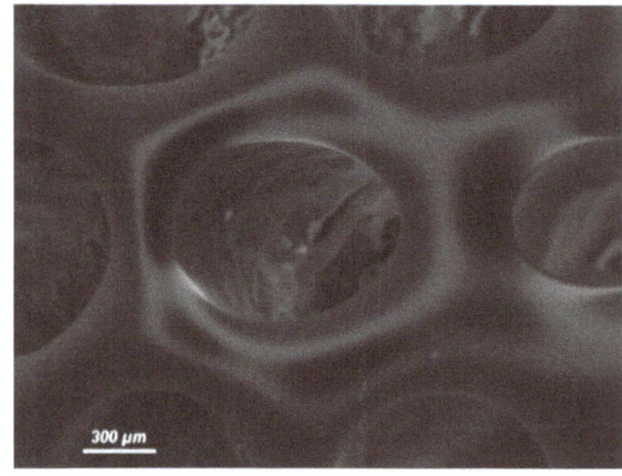

A second approach, in which a ceramic powder compact (green body) is structured in a type honeycomb pattern via a 3D surface milling process and the structured powder compact is locally densified to a dense ceramic body by intense laser radiation, causes multiple phase systems. Therefore it is warmly recommended to prefer this fabrication procedure for lead-free piezoceramic materials.

Application candidates of piezoceramic actuators are in communication technologies (handy vibration alarms that only need 0,02 W), in sport industries (smart skies, snowboards and non-vibrating tennis rackets), in household appliances (vibration heads of electric shavers), in optics (internal anti-vibration system in binoculars), in medicine technologies (vibrating shoe soles increasing the balance or harmonic scalpels) and in robotics (active vibration control for micropositioning).

Acknowledgments The author would like to thank the German Federal Ministry of Education and Research (BMBF) for financial support of the research investigations, that were also performed at the research groups of Prof. Dr. J. G. Heinrich from the Institute of Nonmetallic Materials of the TU Clausthal, Prof. Dr. J. Tautz from the Beegroup of the University of Würzburg, and Dr. W. Braue from the Institute of Materials Research at DLR, Köln. Additionally the author has to thank industrial partners participating the development of piezoceramic honeycomb elements, such as Prof. Dr. J. Günster from CIC Ceramic Institute Clausthal GmbH, R. Wimmel from ERAS GmbH, Göttingen, H. Wichmann and S. Linke from Invent GmbH, Braunschweig and Dr. H.-J. Schreiner and R. Bindig from CeramTec GmbH, Lauf.

References

1. Melcher, J.: Adaptive Impedanzregelung an strukturmechanischen Systemen. Shaker Verlag, ISBN 3-8265-8887-8, Aachen, Germany (2001)
2. Chen, G.S., Lurie, B.J.: Bridge feedback for active damping augmentation. Proceedings of the 31th AIAA/ASME/ASCE/AHS/ASC structures, structural dynamics, and materials conference, April 1990, AIAA Paper No. 90-1243
3. Liang, C., Sun, F.P., Rogers, C.A.: Coupled electro-mechanical analysis of adaptive material systems-determination of the actuator power consumption and system energy transfer. J. Intell. Mater. Syst. Struct. 5(1), 12 Jan (1994)
4. Melcher, J., Junge, M.: Adaptive impedance control using piezoelectric honeycomb actuator concepts. Proceedings of the 13th international conference on adaptive structures and technologies ICASt'02, Potsdam, Germany, pp. 644–655, Oct (2002)
5. Melcher, J., Junge, M., Selle, D.: Als Sensor und/oder Aktuator einsetzbares elastisches Bauteil. German Patent, DE 102 38 932 B3 (2004)
6. Guicking, D., Melcher, J., Wimmel, R.: Active impedance control in mechanical structures. Acustica **69**, 39–52 (1989)
7. Junge, M., Melcher, J.: Von der Natur inspiriert –ein Konzept zur Entwicklung dynamischer Low-Energy-Aktuatoren. In: Wisser, A., Nachtigall, W. (eds.) BIONA-report 16, Akad. Wiss. Lit., Mainz, GTBB, Saarbrücken, pp. 140–151, (2003)
8. Melcher, J., Junge, M., Günster, J., Heinrich, J.G.: New piezoceramic actuators with high displacements for active vibration control. 28th international cocoa beach conference and exp. on advanced ceramics and composites in conjunction with the 8th international symposium on ceramics in energy storage and power conversion systems, Florida, Jan 2004
9. Melcher, J.: Der Trick der Bienen. DLR-Nachrichten Nr. **119**, 36–39 Aug (2008)
10. Pirk, C.W.W., Hepburn, H.R., Radloff, S.E., Tautz, J.: Honeybee combs: construction through a liquid equilibrium process? Naturwissenschaften **91**, 350–353 (2004)
11. Günster, J., Oelgardt, C., Heinrich, J.G., Melcher, J.: The power of light: selforganized formation of macroscopic amounts of silica melts controlled by laser light. Appl. Phys. Lett. **94**(2), 021114, Jan (2009), http://link.aip.org/link/?APL/94/-021114, http://www.physorg.com/news152456596.html
12. Melcher, J., Krämer, M., Heinrich, J., Günster, J., Tautz, J.: Verfahren zum Ausbilden einer Struktur mit optimierter Raumform. German Patent DE 10 2005 025367.9, 2005 (equivalent to US Patent 7,516,639 B2, 2009)

13. Melcher, J., Mund, E., Braue, W., Hildmann, B., Günster, J., Heinrich, J.G.: New piezoceramic actuators using self-organizing fabrication methods. 10th international conference and exhibition of the European ceramic society, Berlin, June (2007)
14. Tian, X., Dittmar, A., Melcher, J., Heinrich, J.G.: Sinterability studies on $K_{0.5}Na_{0.5}NbO_3$ using laser as energy source. J. Appl. Surf. Sci. **256**, 5918–5923 (2010)
15. Tian, X., Günster, J., Melcher, J., Dichen, L., Heinrich, J.G.: Process parameters analysis of direct laser sintering and post treatment of porcelain components using taguchi's method. J. Eur. Ceram. Soc. **29**, 1903–1915 (2009)

Part II
Structural Mechanics

Chapter 9
Validation Approach for Robust Primary Carbon Fiber-Reinforced Plastic Structures

Alexander Kling

Abstract Current industrial demands for fiber composite primary structures in the area of aeronautics require innovative, experimentally validated simulation methods and tools, to support a cost and weight efficient design and to reduce their time-to-market utilizing virtual (simulation based) testing. Reliable application of numerical analysis in upfront design challenges not only verification ("solve the equation right") aspects but also the validation of the numerical methods ("solve the right equations") with trustworthy experimental investigations. This chapter provides an overview on different aspects of the validation process, starting with a concise insight in the terminology in modeling and computational simulation, followed by a description of the selected approach to validate structures at different levels of detail, the comparison of numerical results with experimentally extracted data and finally a brief outlook with respect to transferability prospects and limitations. These aspects will be reflected exemplarily utilizing numerical and experimental investigations on buckling and postbuckling of stiffened CFRP panels.

9.1 Terminology

Figure 9.1 depicts the basic phases of modeling and simulation as presented by the Society of Computer Simulation refer to [1] or by some other fundamental guidelines for Verification and Validation in the area of computational fluid dynamics [2]. The "reality" (e.g. measurements during product use, like accelerations determined via

A. Kling (✉)
Institute of Composite Structures and Adaptive Systems,
German Aerospace Center (DLR e.V.), Lilienthalplatz 7, 38108, Braunschweig, Germany
e-mail: alexander.kling@dlr.de

M. Wiedemann and M. Sinapius (eds.), *Adaptive, Tolerant and Efficient Composite Structures*, Research Topics in Aerospace, DOI: 10.1007/978-3-642-29190-6_9,
© Springer-Verlag Berlin Heidelberg 2013

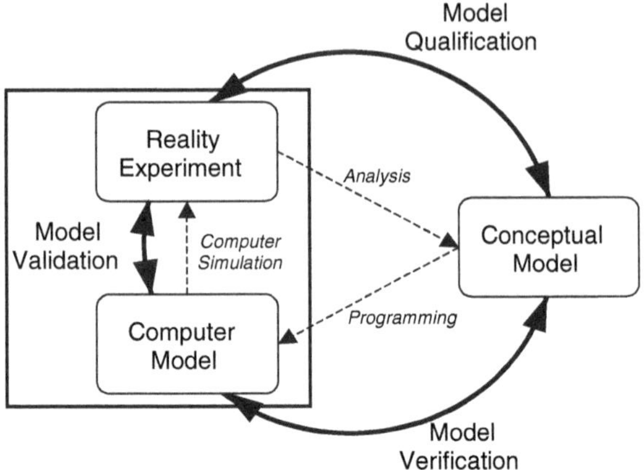

Fig. 9.1 Phases of modeling and simulation (modified from [2])

flight test of an airplane) or experiment represents the problem entity to be modeled, which is in the considered case the physical experiment described in Sect. 9.3 on two thin-walled stiffened CFRP panels. The conceptual model is based on the analysis process of the "reality" or experiment and can be summarized by all necessary mathematical equations to properly describe the physical behavior within the area of interest. Subsequently, these mathematical equations are programmed to obtain the computer model (e.g. by FEM software). Finally, the process of computer simulation connects back the computer model to the "reality" or experiment. The model qualification provides an insight on the adequacy or fidelity of the conceptual model with respect to the real problem. Model verification ("Solve the equations right") deals with the relationship between conceptual model and the computer model to identify and eliminate possible programming errors. The highlighted area of model validation ("Solve the right equations"), the main focus within the subsequent sections, determines the confidence in the simulation capability of the computer model in comparison to reality or experimental findings.

9.2 Validation Process

Figure 9.2 presents a modified and extended version of the numerical modeling process including experimental validation as proposed in the ASME Verification and Validation Guide [3]. The basic ideas and definitions expressed in Fig. 9.1 are still present in this flow-chart, but each process and activity was further subdivided into its next level of detail. In addition, two separate branches were introduced, the right one includes the modeling aspects and the left one the physical testing elements. A brief summary as well as characteristic modifications and extensions of the flow-chart with respect to buckling and postbuckling of stiffened panels are given subsequently. Further information on structural stability analysis may be found in Chaps. 14 as well as 16.

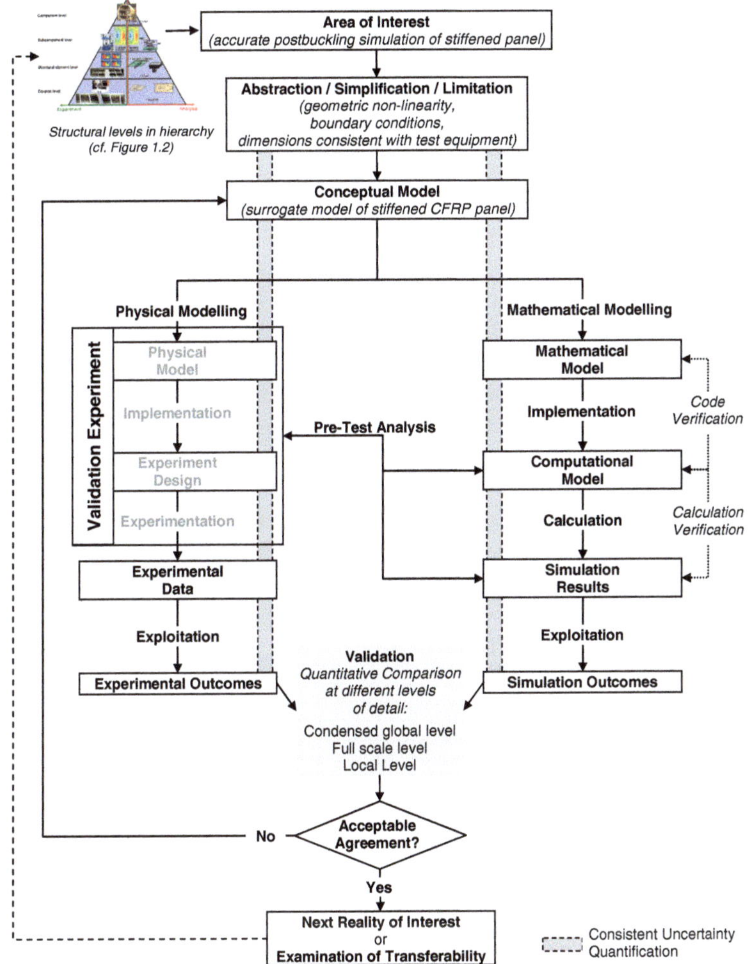

Fig. 9.2 Detailed illustration of the numerical modeling process including experimental validation (modified and extended from [3])

The area of interest in this specific example can be best summarized as to extend the current stability design scenario to exploit the postbuckling regime (weight reduction) of stiffened CFRP panels, which is demonstrated by improved modeling, simulation and experimental validation (cf. [4]). The abstraction/simplification/limitation process is a strongly linked effort of the experimenter and analyst with respect to physical assumptions (e.g. consistency with respect to testing capabilities, discrepancies between real-world boundary conditions and experimentally feasible ones, etc.), which leads to a conceptual model, in the specific case to a surrogate model of stiffened CFRP panels.

The modeling branch's first element mathematical model and its implementation as well as code verification, was not a main issue due to the fact that a commercial FE code was utilized for simulation. The subsequent elements computational model, simulation results and outcomes can be reviewed in [5].

The first two elements in the physical modeling branch are covered by the approach to conduct validation experiments rather than phenomenological investigations with the clear objective to assess the accuracy of the mathematical model. An important link between both branches (or personified between experimenter and analyst) is the so called pre-test analysis, with the main purpose to support the common understanding with respect to the physical and mathematical model and to identify possible sensitivities or critical sources of discrepancies. The subsequent two elements experimental data and outcomes have to be critically assessed to consider deterministic as well as probabilistic deviations.

The validation represents the quantitative comparison, covering different levels of detail, ranging from a condensed global level, via a full scale level to a local level of the examined stiffened panels and in this example the focus on their non-linear buckling and postbuckling behavior. This allows finally a well founded assertion with respect to the applicability of the analysis procedure to predict this complex structure mechanical behavior in the defined domain of interest:

On a condensed global level, the overall load-shortening curves are compared between experiment and analysis. The terminology "condensed" means in this context, that global (panel-level) load as well as shortening are considered to characterize the structural behavior. This is usually the first step to examine characteristics along the load-shortening run and compare them with experimental findings.

On a full scale level, the deformation patterns at different load levels are compared. The applied experimental measurement techniques (digital image correlation [DIC] system, ARAMIS) allow after postprocessing a direct comparison with respect to the numerically determined displacement patterns.

On local level of validation, measurements at certain locations on a relatively punctiform level, like strains from strain gauges or radial displacements from position encoders are compared with numerical findings, with the purpose to evaluate the correspondence locally, on a rather small scale.

Each level has its individual necessity for a validated numerical model and analysis procedure, due to the fact that with this multi-level approach, possible shortages or discrepancies become obvious. Furthermore, first indications can be extracted to identify possible causes of differences as well as promising improvements of the computer model.

The purpose behind the multi-level approach within the problem-oriented (e.g. bucking and postbuckling analysis of a stiffened thin-walled panels) validation process, is to raise the awareness that a simple comparison of the load-shortening curve, on a condensed global level, might be not enough to prove the validity of the model and numerical analysis procedure.

In case an acceptable level of agreement was not obtained within the validation process, first a consistent uncertainty quantification should be done to eliminate any kind of error, second, the solid arrow pointing up to the conceptual model should be considered as one (next to abstraction/simplification or material data base) possible approach of improvement to assess both branches "iteratively". However, based on the experience of the analyst and experimenter or sensitivities extracted from pre-test analysis (e.g. boundary conditions) specific shortcomings or sources of discrepancies might be identified directly. In case an acceptable level of agreement was achieved, the next reality of interest (dashed arrow towards the pyramid of structural levels in hierarchy in Fig. 9.2) might be studied.

9.3 Example: Stiffened CFRP Panel

Within this section, the model with its boundary conditions and the validation approach at three levels of detail is resumed utilizing data from [5] to demonstrate its applicability to evaluate the quality of simulation-based results with respect to non-linear stability analysis of stiffened CFRP panels.

In the numerical model, the axial loading is introduced via a reference node, marked as a black dot in Fig. 9.3, which is rigidly connected, with all nodes (skin and stringer) on the loaded circumferential edge of the panel (marked green). The applied boundary conditions for this node-set is [0/0/-], which represent in the first line the displacement degrees of freedom [R/T/Z], and [0/0/0] in the second line characterizing the rotational degrees of freedom $[\varphi_R/\varphi_T/\varphi_Z]$, both in a cylindrical coordinate system. The red marked nodes model the end blocks (epoxid + filler material applied on both ends for load introduction in the experimental setup), where is assumed that only axial displacements are allowed. The blue accentuated two node lines along the longitudinal edges of the panel stand for the applied boundary conditions [0/-/-], [0/0/-] as a surrogate model of the metallic longitudinal edge supports applied in the test. The node line marked in yellow along the circumferential edge represents the fixed support [0/0/0], [0/0/0] at the end of the panel.

The condensed global level is met by the load-shortening curves depicted in Fig. 9.4. The experimental results of the nominally identical panels B0276 and B0281 are depicted in comparison with the outcomes from the numerical analysis with and without scaled geometric imperfections.

The predicted axial stiffness in the pre- and postbuckling range are very close to the experimentally extracted ones. The onset of global, stringer-based, buckling shows some variation in the experimental outcomes, however, due to its highly non-linear behavior in the postbuckling regime is it still in an acceptable range and corresponds well with the numerical results.

In the transition phase from the unsymmetric global 1/3 (outwards), 2/3 (inwards) buckling deformation (cf. deformed panel in Fig. 9.4) to a single central one, the run of all curves are close up to the point of collapse.

The significant loss in load-carrying capacity at 3.5–3.6 mm shortening from the experimental data are due to structural degradation (e.g. large failure in the stringer-skin interface, major fiber and interfiber failure, extensive delamination),

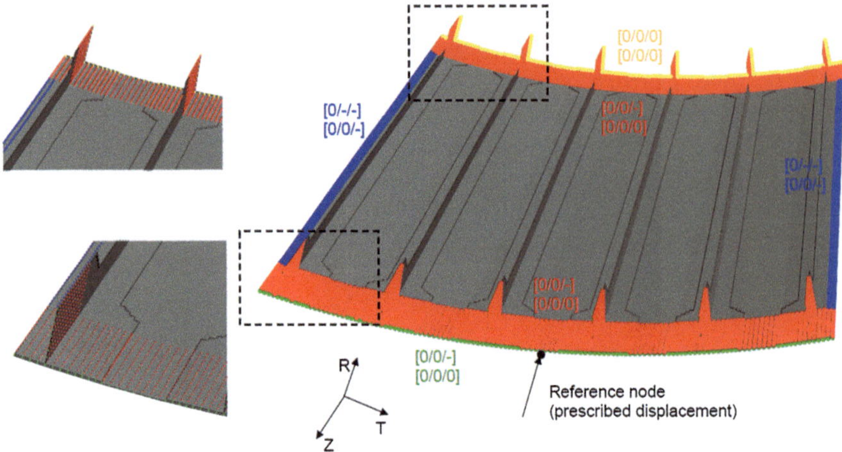

Fig. 9.3 Details of the applied boundary conditions of the FE analysis

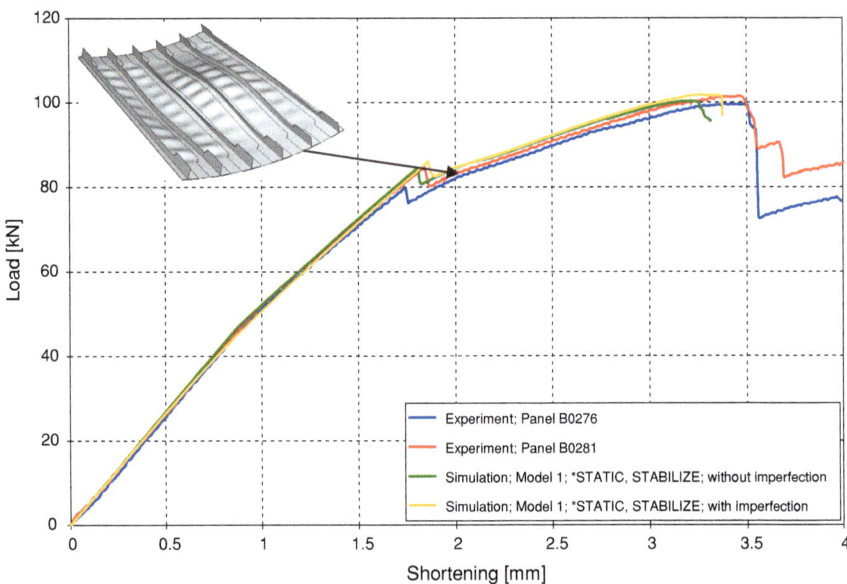

Fig. 9.4 Comparison of experimentally and numerically extracted load-shortening curves (condensed global level of validation), global 1/3 (outwards), 2/3 (inwards) buckling shape

which was located and characterized after the test. The knees (loss in load-carrying capacity) in the numerical results close to the experimental collapse are due to a predicted mode change, followed by the termination of the analysis due to massive convergence problems.

It should be emphasized that the numerical and experimental investigations detailed in this chapter focus mainly on the non-linear buckling and postbuckling behavior, methods to consider structural degradation (e.g. stringer-skin separation) have not been introduced in the analysis procedure.

Overall, the condensed global level, by comparing the load-shortening run, up to the deep postbuckling regime, revealed a high level of agreement between experimentally extracted data and numerical results.

The side-by-side comparison in Fig. 9.5 of deformation patterns at identical load levels was selected to cover the so called full scale level. To simplify the visual comparison, the surface of the panel which was covered by the digital image correlation system—DIC (experimental measurements), excluding clamping boxes on top and bottom and longitudinal edge supports, has been framed in the color rendering of the numerical results. Note, that the white areas in the experimental measurements are due to applied strain gauges and their wiring.

The measured onset of local skin buckling depicts a slightly more inhomogeneous deformation pattern in comparison to the result form the non-linear analysis. This could be caused by the influence of the applied boundary conditions and/or some geometric imperfections introduced during the manufacturing process influencing this sensitive stability phenomenon. However, the total number of half-waves in the axial direction is identical for analysis and experiment. The global, stringer-based, 2/3 to 1/3 buckle depicts a larger expansion in axial as well as circumferential direction for the experimental measurements, however, this is due to different color distribution within the legend of experimental and numerical results. Despite this pure visual discrepancy, further detailed examinations revealed that a high level of agreement was found on the full scale level between the experimental measurements (89 deformation patterns up to collapse) and the numerical results along the whole load-shortening run.

To underpin the findings on the local level, experimentally measured and numerically predicted radial displacements at defined locations on top (free edge) of the T-stringer-blades were extracted and compared in Fig. 9.6. A positive radial displacement corresponds to an outwards deformation from the stringer- to the skin-side. As expected there are almost no deformations prior the onset of global buckling, only small positive radial displacements due to the applied boundary conditions and the axial loading are detected. However, at an axial shortening of approximately 1.8 mm, which comply with the onset of global buckling, the radial displacements reproduce the unsymmetric global 2/3 to 1/3 buckling deformation (refer to Fig. 9.5). The displacement transducer W88 indicates a sudden rise in positive radial direction (1/3). W90 and W91 show an even larger transition in negative radial direction (2/3). The displacement transducer at W89 reflects a linear slope, which can be explained by its initial inflection point location and the subsequent transition to a single central global buckle. This alteration, towards a single global buckle in the centre, explains the run of W88 towards a negative radial displacement and the further negative gain of W89, W90 and W91. The corresponding numerically determined displacements revealed a high degree of agreement with respect to the experimentally extracted data.

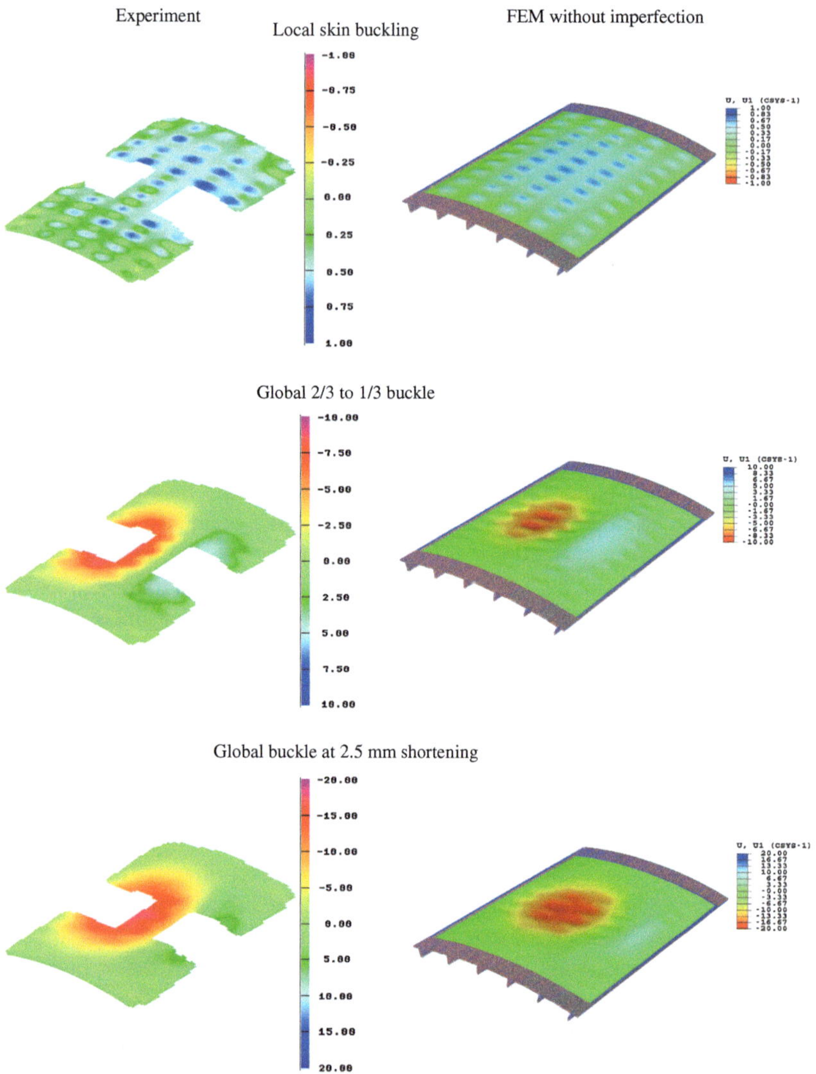

Fig. 9.5 Deformation patterns at characteristic load levels (full scale level of validation)

All three levels (condensed global level, full scale level, local level) of the selected validation approach reflected a high level of correlation between the experimentally extracted and numerically determined results, which allows to consider the modeling and numerical analysis process (cf. Fig. 9.2) as an appropriate methodology—well transferable beyond the exemplarily detailed study on buckling of stiffened panels.

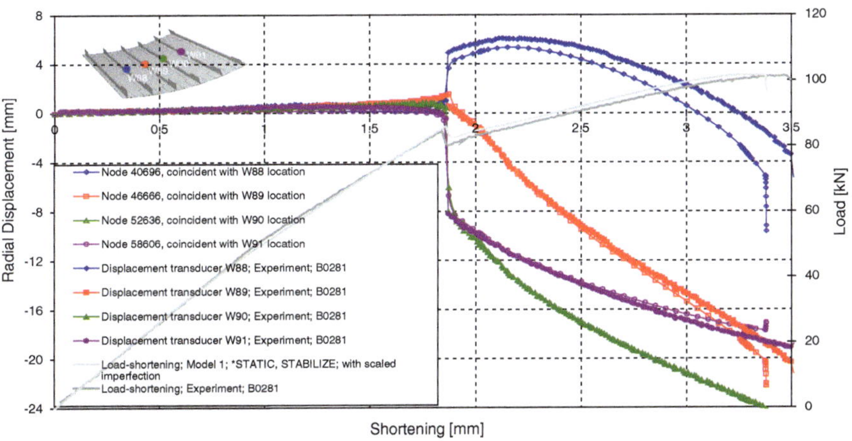

Fig. 9.6 Radial displacement versus load-shortening curve (local level of validation)

9.4 Validation Assessment

After the successful validation of the modeling and simulation process, utilizing a limited number of tests, an important question about its extended validity arises: How well is this analysis procedure applicable within an extended designs space? This defines the border or transition between numerically predicted results and validated ones. The process of validation assessment is characterized by determining the degree to which a model and analysis process is an accurate representation of the so called real world application with respect to the intended use of simulation. It will be briefly covered within this section.

Figure 9.7 depicts exemplarily in two dimensions the relationship between model validation and model prediction for two design variables X_1 and X_2 (e.g. stringer-pitch or radius of curvature for the considered stiffened panels). Validated domains are centered by a limited number of physical experiments and the shaded decreasing confidence-level towards model prediction with increasing distance. Two different scenarios are depicted, on the left hand side, in which the domains of interest are rather limited to acceptable confidence-levels supported by experiments. On the right hand side a significantly larger domain of interest is defined, covering also white areas representing critical regions of low confidence. The latter represents common engineering practice (shear-axial compression interaction curve), simply due to the fact that physical experiments covering the full domain of interest with a high level of confidence is not always practicable due to time and cost constraints. At least the knowledge to be in a critical domain with a low level of confidence is of high relevance. This can be substantiated with expert knowledge as well as numerically predicted sensitivities (e.g. parametric studies to examine the "robustness" of validity within a probabilistic context), to understand

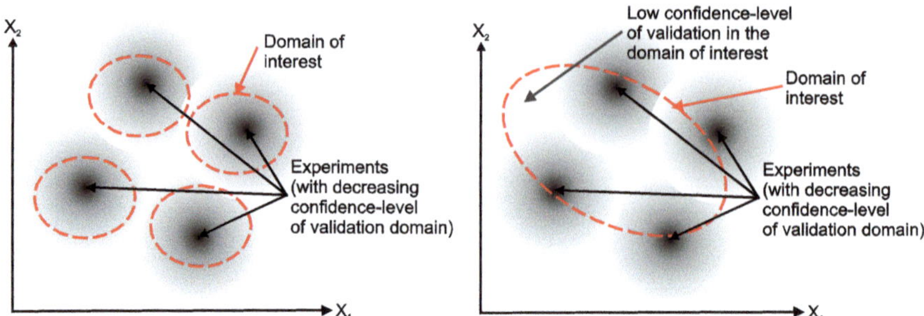

Fig. 9.7 2-D visualization of validation confidence (*left*: domains with *high* level of confidence, *right*: larger domain of interest including areas of *low* confidence)

specific modeling assumptions as well as the structural response to assess the predicted results with a given level of cautiousness.

Furthermore, transfer aspects within the context of validated simulation procedures are often linked to the concept of validation hierarchy (cf. [6]), because it is not always feasible to conduct real validation experiments of the final complex system. Therefore alternatively a building block approach is commonly utilized as depicted in Fig. 9.8. In this case characteristic structural levels are shown for an aircraft fuselage in form of a pyramid (inspired by the pyramid of tests introduced by Rouchon [7]). The solid horizontal arrows represent the validation process on different levels building up on each other.

In addition, the vertical arrows connecting different levels e.g. structural-element level and subcomponent level can be thought as some kind of "elevator" function on the right hand side (analysis) of the pyramid, which can be accomplished utilizing for example a multiscale approach (cf. Chap. 11) e.g. submodelling technique available in commercial FE codes, which allow to build up a barrel by (validated) panel sections (going up) or to identify and analyze "hot-spots" (going down), critical areas of possible failure and subsequent degradation via homogenization, by going up again.

9.5 Conclusion

Within this chapter relevant aspects with respect to a novel multi-level validation approach were exemplarily detailed by the buckling and postbuckling analysis as well as physical testing of stiffened composite panels. A high level of correlation was found up to the deep postbuckling regime between numerical analysis and experiment not only for the load-shortening run (global condensed level of validation) but also on local level e.g. for radial displacement or full scale level for the full field deformation patterns. Finally, transfer aspects were pointed out as a part of the validation assessment, which allow to examine the applicability of a

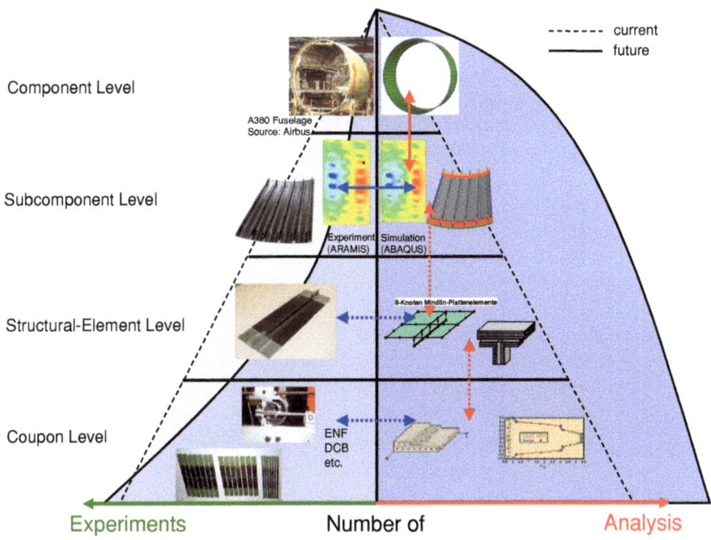

Fig. 9.8 Pyramid of tests with future and current scenario of physical and virtual testing

validated modeling and simulation process to an extended design space, as well as some contributions with respect to the concept of validation hierarchy.

The proposed and applied multi-level validation approach demonstrated its applicability to evaluate the quality of the numerically extracted results. Furthermore, the comparison at different level allows to extract possible discrepancies of the model, which would not be detectable in case of a simple comparison of load-shortening runs. The validation process revealed that the selected numerical analysis procedure (model building and solution) is an appropriate approach to simulate the buckling and postbuckling behavior of the considered structure at a high level of confidence, which allows e.g. an extended design philosophy to use the remaining load carrying capacity in the postbuckling regime as safety region.

Within Fig. 9.8 the current share between experiments and analysis in design of complex primary aircraft structures is marked as dashed lines, assuming a symmetric pyramid. Further improvements with respect to fast and reliable (validated) virtual testing methods (horizontal arrows) linked with the efficient and automated multiscale analysis (vertical arrows) will allow to shift the share towards an extensive use of analysis as depicted with solid lines (distorted pyramid). This will not only allow to reduce to total number of time and cost consuming experiments up to the certification of an aircraft, but also to reduce risks within the development process, due to high quality simulation within an early stage of the design process.

References

1. Schlesinger, S.: Terminology for model credibility. Simulation **32**(3), 103–104 (1979)
2. Computational Fluid Dynamics Committee on Standards: Guide for Verification and Validation of Computational Fluid Dynamics Simulation, AIAA G-077-1998. American Institute of Aeronautics and Astronautics, Reston (1998)
3. The American Society of Mechanical Engineers: Guide for Verification and Validation in Computational Solid Mechanics, ASME V&V 10-2006 (2006)
4. www.cocomat.de
5. Kling, A.: Contributions to improved stability analysis for design of thin-walled composite structures. Dissertation. DLR-Forschungsbericht. DLR-FB 2010-01 (2010)
6. Oberkampf, W.L.: What are validation experiments? Exp. Tech. **25**, 35–40 (2001)
7. Rouchon, J.: Certification of large airplane composite structures, in: recent progress and new trends in compliance philosophy. ICAS 1990 **2**, 1439–1447 (1990)

Chapter 10
Simulation of Fiber Composites: An Assessment

Klaus Rohwer

Abstract Two facts are the main drivers for a steady rise of models simulating fiber composites: an increasing demand on optimal utilization of the material and a drastic improvement in computational power. Though this process is still in full swing a review is considered reasonable since it facilitates guidelines for future research. Topics which comprise major development lines are micromechanics, laminate theories, design and optimization, damage and failure, and manufacturing, though a strict separation is not in all cases useful. In reviewing these topics it will turn out that their model status is of rather different maturity. Areas are identified where there are plenty of models available which are really not needed, whereas other problems cannot be modeled adequately as yet.

10.1 Modeling Aspects

Numerical simulation requires a model, which on one hand should reflect the desired aspects of reality accurately enough but on the other hand be as simple as possible. Due to the complexity of fiber composites the number of models and tools for simulating the structural behavior is horrendous. It is by far not possible to value them all; rather, it is the intention to identify areas where, to the author's opinion, there are still deficits or where there is a surplus of models but a lack of knowledge regarding suitable application. Furthermore, it is of utmost importance to validate the model. To that end it is not sufficient to prove an adequate prediction in two or three test cases. Rather, extensive test series with relevant structural components must be run.

K. Rohwer (✉)
Institute of Composite Structures and Adaptive Systems,
German Aerospace Center (DLR e.V.), Lilienthalplatz 7, 38108, Braunschweig, Germany
e-mail: Klaus.Rohwer@dlr.de

M. Wiedemann and M. Sinapius (eds.), *Adaptive, Tolerant and Efficient Composite Structures*, Research Topics in Aerospace, DOI: 10.1007/978-3-642-29190-6_10,

10.2 Micromechanics

10.2.1 Material Properties

Micromechanics account fibers and the matrix for separate constituents; in some cases an interface layer is assumed in addition. There are chiefly two micromechanical tasks: Determination of stress between fibers and the matrix as well as homogenization of constituents. A major difficulty is the determination of material properties. That holds especially for fiber properties in transverse direction as well as for a possible interface layer. Since proper tests are difficult to perform these values are often determined from an inverse application of homogenization formulas.

10.2.2 Micromechanical Stress

Calculating micromechanical stress is generally based on a representative volume element (RVE) which contains the constituents in sufficiently large numbers. With suitable boundary conditions applied the RVE is then analyzed by means of finite elements. Information about size and boundary conditions are provided by Larsson et al. [1]. Examples of stress distributions are given by Rohwer and Xie [2], Nassehi et al. [3], Zhao et al. [4], Jin et al. [5] and Huang et al. [6]. This well-established procedure is quite useful for understanding the effects of inhomogeneity.

10.2.3 Stiffness Homogenization

Analyzing a complete structure from fiber composite material needs a homogenized approach. Adequate stiffness properties can be determined from an RVE analysis, as for instance proposed by Kari [7]. Garnich and Karami [8] included the effect of fiber waviness. But such a laborious procedure is normally avoided. Instead, simple homogenization formulas are used. For the Young's modulus and Poisson's ration in fiber direction it is the rule of mixture the results of which deviate only slightly from more complicated formulas by Hill [9]. In transverse direction and for shear the formulas are based on the rule of mixture for the compliances. But the results must be reduced by a certain amount due to 3D-effects, as can be visualized from Fig. 10.1. Many different proposals for such a reduction are available, e.g. Tsai and Hahn [10], Hashin [11] or Chamis [12]. With reasonable fiber and matrix properties a deviation of up to $\pm 10\%$ between the different formulas has been determined. Besides the stiffness a number of other properties have been homogenized. Among them are the coefficients of thermal

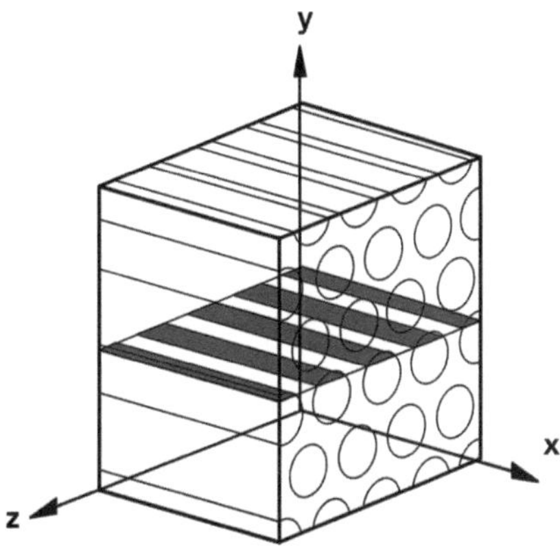

Fig. 10.1 Rule of mixture
for stiffness homogenization

expansion, e.g. Kulkarni and Ozden [13], the coefficients of moisture expansion, and the coefficients of thermal conductivity. Especially the thermal conductivity has been intensively studied; recently Al-Sulaiman et al. [14] even included the effect of voids within the matrix.

10.2.4 Strength Homogenization

The status of strength homogenization is far less advanced. Some effort is put on the determination of tensile strength in fiber direction. Considering the standard composite design with an extension to failure of the matrix much higher than that of the fiber, the composite failure stress can be roughly estimated by the rule of mixture from the failure stress of the fiber and the matrix stress at fiber rupture. However, that does neither account for varying fiber strength along each single fiber nor for different strength between fibers. A number of hypotheses accounting for these variations have been proposed, e.g. Zweben [15], Rosen [16], Rosen and Zweben [17], but the application is not very convincing. More recent developments along this line are the global load sharing scheme by Curtin [18] and the simultaneous fiber-failure model by Koyanagi et al. [19]. Micromechanical simulations on the transverse failure behavior of fiber-epoxy systems are presented by Cid Alfaro et al. [20]. They pointed at a strong influence of the relative strength of the fiber-epoxy interface and the matrix.

Compressive strength models were first set up by studying the buckling of fibers with an elastic support. Depending on the fiber volume fraction Rosen and Hashin [21] found different failure modes. Dharan and Lin [22] extended this approach by accounting for an interface layer around the fibers. Further models were proposed by Xu and Reifsnider [23] and by Budiansky and Fleck [24]. A micromechanical analysis of the kink band formation after fiber buckling including the effect of fiber misalignment was performed by Jensen and Christoffersen [25]. Lately, Gutkin et al. [26, 27] determined two different failure mechanisms, (1) shear-driven fiber compressive failure and (2) kinking/splitting. Mishra and Naik [28] used the inverse micromechanical method to calculate fiber properties and applied them to determine the compressive strength for a composite with a different fiber volume fraction. A formulation capable of obtaining the maximum compression stress, and the post-critical performance of the material once fiber buckling has taken place was proposed by Martinez and Oller [29]. Only recently, micromechanical shear strength analyses were performed by Ng et al. [30] and Totry et al. [31].

In summary, micromechanics of composites have some benefits with respect to understanding the effects of inhomogeneity. All these methods, however, suffer from the difficulty to measure the micromechanical properties, in particular those of the fiber and the interface. Homogenization models in general are well advanced, but determination of transverse and shear strength needs further attention.

10.3 Laminate Theories

10.3.1 Laminate-Wise Approximations

After homogenization the behavior of a single layer of composite material can be described by an anisotropic material law. Structures predominantly consist of several layers with different fiber orientation. Attempts to homogenize these in thickness direction did not prove reasonable. Instead, certain assumptions are made for the laminate behavior in thickness direction. The straightforward one is the Bernoulli hypothesis which, together with a plane state of stress, leads to the classical lamination theory (CLT). The structural behavior is then described by the so-called ABD-matrix. CLT is still first choice for thin-walled composite laminates.

Since the relation between shear modulus and Young's modulus in fiber direction is relatively small Whitney and Pagano [32] suggested using the First-Order Shear Deformation Theory (FSDT) instead. In addition to the ABD-matrix that needs a 2×2-matrix of transverse shear stiffness which cannot be determined without further assumptions. Rohwer [33] proposed an explicit computation, Altenbach [34] has presented a method, which works for sandwiches as well, and Schürg et al. [35] have published an energy minimization approach to reach that aim. FSDT is rather popular also because many plate and shell finite elements are based on this theory, so that only the anisotropic material law must be taken care of.

From the early 1980s onwards an increasing number of higher order theories have been developed. It started with the cubic distribution in thickness direction for membrane displacements by Murthy [36] and Reddy [37]. At the expense of additional functions that allows the determination of more reasonable transverse shear stresses from the material law. Following an idea of Reissner [38], Lo et al. [39] added a quadratic distribution of the transverse displacement which yields even better results but needs to specify eight functions. Using the method of multiple scales, Wu et al. [40] presented an asymptotic expansion of the 3D elasticity equations with material properties piecewise continuous in thickness direction. Shell thickness over radius is used as the perturbation parameter. Successive integration leads to a process embracing the CLT, the FSDT, and the cubic model by Reddy, respectively, as first-order approximations to the three-dimensional theory. The procedure allows improving the solutions obtained by the respective theories in an adaptive and hierarchical manner without increasing the number of functions.

Laminate-wise models have been evaluated by Noor and Burton [41], Rohwer [42], Yang et al. [43], Timarci and Aydogdu [44] and Carrera et al. [45]. As a result it can be stated that for standard applications the FSDT with suitable shear correction delivers sufficiently accurate results at acceptable expenses. But in special cases it may be necessary to increase the order of polynomials in thickness direction. Rohwer et al. [46] have presented such a case, where local thermal loads require polynomials of up to 5th order for modeling in-plane as well as transverse displacements in order to determine reliable stresses. Because of the considerably large effort involved with an application of such high-order models they should be restricted to cases where they are really needed.

10.3.2 Layer-Wise Approximations

In a technical note Pagano and Hatfield [47] have shown a pronounced layer-wise zigzagging distribution of the in-plane displacements at slenderness ratios a/h <10. This observation gave rise to develop a substantial number of analysis models, which consider each layer separately. For several of these models the number of functions depends on the number of layers. Since in real structures the layer number sometimes is in excess of 100 the large number of functions is an evident handicap. Computation time increases drastically since not only the number of degrees of freedom but also the bandwidth of the stiffness matrix is proportionally expanded. Therefore, it can be stated that in general layer-wise models with the number of functions depending on the number of layers are not suitable for practical application.

By means of compatibility and equilibrium conditions applied at layer interfaces one can achieve layer-wise models for which the number of functions does not depend on the number of layers. A considerable research effort in this field has led to many models of that type of which only a limited selection can be cited here. Sun and Whitney [48] assumed the in-plane displacements to vary linearly over each layer, whereas the transverse displacements remain constant. By enforcing not only the compatibility but also the transverse shear equilibrium at layer interfaces they could reduce the number of functions to five. Di Sciuva [49] later

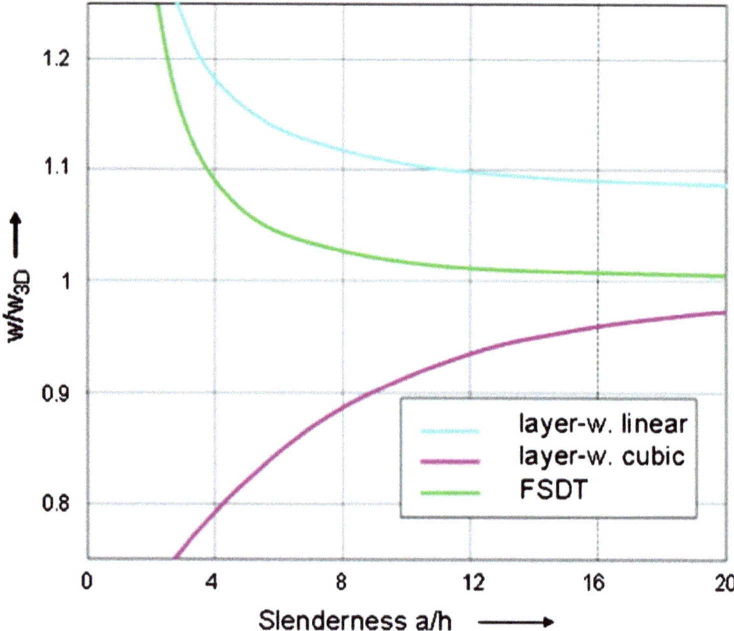

Fig. 10.2 Center deflection of a rectangular [0, 90]s plate with layer thicknesses [0.4, 0.1]s compared to 3D elasticity solution

used a different formulation but ended up in an identical model. Lee et al. [50] assumed the in-plane displacements layer-wise cubically distributed and the transverse displacement remaining constant. Displacement compatibility and equilibrium against transverse shear at layer interfaces are utilized for eliminating unknowns. The model is thus left with five functions. The same approach, somewhat more elegantly presented using the Heaviside unit step functions, was provided by Savithri and Varadan [51]. A study by Rohwer et al. [52] has shown that especially the layer-wise linear approach can deliver rather accurate results, but for certain configurations they are worse than those obtained by the FSDT. As an example Fig. 10.2 shows the center deflection determined by means of different models with five functional degrees of freedom each. Lo et al. [53] have superimposed cubic laminate-wise in-plane displacements with layer-wise cubic ones; the transverse displacements are assumed linear. Using compatibility and equilibrium conditions for elimination the number of functions could be reduced to 16; the results obtained with this approach are remarkably good, but the expenses are high as compared to FSDT.

Another possibility to account for the effect of layer differences is superimposing specific zigzag functions onto a continuous displacement approximation in thickness direction. Examples for such an approach are proposed by Murakami [54], Di and Ramm [55], Brank and Carrera [56] as well as Icardi and Ferrero [57].

Unfortunately there is little information available about the efficiency as compared to more standard models. A unified formulation accounting for higher-order, zig-zag, layer-wise and mixed theories has been provided by Carrera [58].

10.3.3 Transverse Stresses

Multidirectional laminates are susceptible to delaminations. Information about the transverse shear and normal stresses are needed to cover this threat. With simple laminate models like CLT or FSDT these stresses cannot be determined from the material law. But if the membrane stresses are determined accurately enough equilibrium conditions can be applied locally for the transverse shear stresses, and their derivatives in turn can be integrated to determine the transverse normal stress. Corresponding procedures as proposed by Pryor and Barker [59], Lo et al. [60] and Engblom and Ochoa [61] suffer from the need of higher order shape functions, a constraint which has been reduced by the Extended 2D method of Rolfes and Rohwer [62]. An alternative approach is lately proposed by Schürg et al. [35]. Stresses determined by means of the equilibrium conditions can be iteratively improved. For this purpose predictor-corrector processes are invented by Noor et al. [63] and Noor and Malik [64]. Further, the re-analysis method described by Park and Kim [65] and Park et al. [66] can be applied for that purpose. A totally different approach by Guiamatsia [67] bases the homogenization in thickness direction not on a priori displacement assumptions but on the superposition of certain fundamental states. Published results look promising, but limitations due to the superposition process are not clear.

An up-to-date and elaborate overview over the different approaches is provided by Kreja [68]. In conclusion it can be stated that for structures with common slenderness rates and smooth loading the FSDT with improved transverse shear stiffness can be regarded as a good choice. Transverse stresses should be determined by application of the equilibrium conditions. So far, only limited knowledge is available with respect to the accuracy of layer-wise models. Engineering judgment is still needed when deciding upon the model for stress analysis of thicker layered structures. Depending on the conditions at hand it may rather be suitable to use 3D finite elements right away.

10.4 Design and Optimization

10.4.1 Initial Design

Design is usually an ill-posed problem rather than a problem of optimization as Kolpakov and Kolpakov [69] have pointed out. Anyway, structural design aims at a solution well suited to the requirements and thus needs some type of

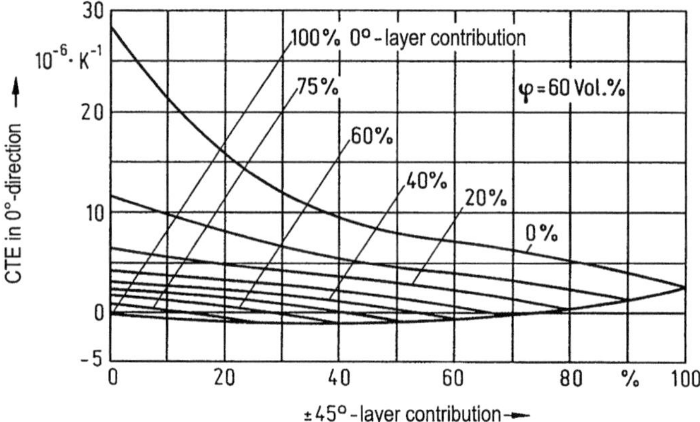

Fig. 10.3 Design diagram for a symmetric laminate with predetermined CTE in 0°-direction (layer properties: $\text{CTE}_{\text{parallel}} = -0,8 \times 10^{-6}\,\text{K}^{-1}$ $\text{CTE}_{\text{transverse}} = 29 \times 10^{-6}\,\text{K}^{-1}$)

optimization. With fiber composites this includes not only the choice of material and thickness but also fiber directions and the stacking sequence. In the early stage of the design process that requires simple and fast procedures which not necessarily yield the optimal solution but lead into the right direction. One such procedure places fibers in the direction of principle stress; another one assumes three fiber directions and applies the netting theory, which ignores the effect of the matrix material, to determine the corresponding layer thickness. Besides other deficiencies as for instance outlined by Evans and Gibson [70] both are applicable for membrane problems only. For special cases like setting up a laminate with predetermined thermal extension or certain stiffness in one direction there are diagrams available. Figure 10.3 provides such a diagram. Moreover, ranking can be applied as a design procedure, where the objective function is calculated for all suitable stacking sequences. Such an approach is used by the laminate design programs RANKHO by Tan [71], LamRank by Tsai [72] or ESAComp [73].

Vannucci et al. [74] lately proposed another procedure, which works for so-called quasi-homogeneous laminates showing identical behavior in bending and in extension. That allows reducing the optimization to the in-plane properties. For these special cases stacking sequences can be determined which provide maximum plate bending stiffness, buckling load or natural frequency. Lopes et al. [75, 76] optimized the plate stacking sequence with respect to their impact resistance and damage tolerance characteristics.

10.4.2 Structural Optimization

According to Eschenauer [77] structural optimization in general is based on three columns: modeling the structural behavior, defining the objective function, and selecting a suitable optimization strategy. The structural model should be as simple as possible since it must be analyzed rather often, but the important features must certainly be well described. The objective function specifies which quantity should be taken to the optimum, e.g. the structural weight or the manufacturing cost; in addition the variables as well as the constraints are defined. There are many optimization strategies on the market. As Zimmermann [78] has pointed out, there is no general optimal strategy, but it depends on the type of problem. Thus choosing the best strategy needs experience. Ghiasi et al. [79] have reviewed a large number of methods suitable for optimizing the stacking sequence of laminated composites. They differentiate between gradient-based methods, direct search methods including the popular simulated annealing and genetic algorithms, as well as specialized algorithms like layer-wise optimization. A modified simulated annealing method was proposed by Javidrad and Nouri [80] where more freedom during the cooling process leads to an improved convergence. With a similar method Akbulut and Sonmez [81] have found out that the design domain is greatly enlarged by increasing the number of distinct lamina angles and the range of values they may take. Gou et al. [82] analyzed the problem of aeroelastic tailoring for aircraft wings using the bending-twisting coupling effect of laminates. They found out that the genetic algorithm is preferable in case the flutter speed is the objective function, whereas for optimal torsion rigidity a gradient method is much faster. Lopez et al. [83] have optimized laminated plates under in-plane load using maximum stress, Tsai-Wu, and Puck failure criteria. They have demonstrated that an optimal design strongly depends on the chosen failure criterion, and that none of the failure criteria is always the most or the least conservative when different load conditions are applied. Azarafza et al. [84] have treated laminated composite circular cylindrical shells subjected to compressive axial and transverse transient dynamic loads. They performed multi-objective optimization of weight and dynamic response using a genetic algorithm. It turned out that the dominant constraint that affects the optimization process most is buckling. A cylindrical shell under pure bending was optimized by Blom et al. [85] considering the restrictions in manufacturing possibilities by means of tailored fiber placement. Increasing the stiffness at the tension side and correspondingly reducing it at the compression side increased the maximum buckling load. With the aid of a genetic algorithm Almeida and Awruch [86] presented a technique for design optimization of composite laminated structures in the geometric nonlinear range. A multiobjective optimization is performed, and a pareto-optimal set is obtained by shifting the optimization emphasis using a weighting factor. Johansen and Lund [87] applied a sensitivity analysis for maximizing the safety against failure of a fully three dimensional laminated composite structure. Gillet et al. [88] used a genetic algorithm to determine the relative importance of several design parameters. They found out that material parameters are of great influence whereas the fiber orientation is of minor importance. Using particle swarm optimization Peng et al. [89] designed fiber–metal laminates for maximum strength. After a thorough investigation Ghiasi et al. [90] rated the optimality criterion methods to be first choice if available. If not, multi-level optimization methods would be the best candidates. If neither of these can be used they recommended a genetic algorithm or a gradient-based method or a combination of these two.

Obviously there are sufficient optimization strategies and computational tools available. Lack of knowledge, however, can be stated for the situation in the early design phase. One of the few exceptions is the dissertation of Zimmermann [78], which provides general information about the optimum layer sequences of axially compressed cylindrical shells depending on the number of layers. Further, a word of caution with respect to optimization seems to be adequate. Ottino [91] has pointed to the fact that the optimal state can be a high-risk state and one should rather strive for a robust solution. That also is one of the aspects treated by Marzcyk [92]. Other than damage tolerant design, robust design adds probabilistic uncertainties to the optimization process as has been discussed by Lee et al. [93] with respect to composite panels.

10.5 Damage and Failure

10.5.1 Failure Criteria

Not even fiber composites can withstand unlimited loads. Because of the anisotropy and inhomogeneity of composites the failure is a rather complex process. Thus the difficulties to decide even for a suitable failure criterion under quasi-static load are understandable to a certain extent. Most of the proposals are formulated in stresses, but there are still some scientists who prefer a formulation in strains. Further there is an alternative between interactive and non-interactive criteria. Though the tendency goes to interactive criteria a few non-interactive ones are still in use, sometimes with astonishingly good results. Regarding the interactive criteria some simply use a tensor polynomial as an extension of the von Mieses criterion for ductile materials, and others apply more physically based considerations. Still under discussion is the question whether or not the failure envelop should be open or closed. Already in the mid 1980s Nahas [94] has reviewed a large number of failure criteria. More recent is the excellent overview provided in the compendium by Hinton et al. [95]. Further failure criteria are developed for instance by Luccioni [96] and by Pinho et al. [97]. Cuntze [98] has improved his own criterion through a single but effective modification of one inter-fiber-failure condition. Stamblewski et al. [99] have used micromechanical analysis results to feed into a quadratic criterion for which the enclosed volume is maximized. Lee and Soutis [100] found out that the compressive strength decreases with increasing size of the test specimen, an effect which is not covered by the available failure criteria.

In general one gets the impression that there are more than enough criteria; urgently needed is a validation against reliable test data. In this respect the world-wide failure exercise, the first phase of which is published by Hinton et al. [95] is of great value.

10.5.2 Damage Progression

When the material strength limit is reached at a certain point in one single ply of the laminated structure, as specified by a failure criterion, it usually does not mean total failure. Damage progression under a further increasing load must be modeled in order to determine the actual behavior. In principle the gradual failure process can be modeled by fracture mechanical, damage mechanical and phenomenological approaches. To that end Knops and Bögle [101] have set up a computer program using degradation models which account for the gradual loss of stiffness. Maimi et al. [102] proposed a thermodynamically consistent damage model based on the LaRC04 failure criteria. A 3D progressive damage theory was used by Cui et al. [103] to analyze the whole process of damage initiation and development for composite laminates under impact loading as well as tensile load after impact. Basu et al. [104] have developed a progressive damage growth model for composite laminates under compression, and implemented it into the finite element package ABAQUS. Based on a micromechanical approach Ha et al. [105] proposed a progressive damage model to predict failure of composite laminates under multi-axial mechanical as well as under thermal loads. Similarly, Zhang and Zhang [106] accounted for the nonlinear material behavior as well as damage on the micro-scale in the development of a macro-scale constitutive model which describes the progressive failure process of composite laminates. A review of degradation models for progressive failure of composites is provided by Garnich and Akula [107].

10.5.3 Delamination

Delamination is a damage mode which needs special attention. It may be initiated from manufacturing flaws, low velocity impact or the free edge effect. The usual stress-based failure criteria are not suitable for calculating the delamination progression; instead fracture mechanics methods are preferred. Rather popular is the Vitual Crack Closure Technique as described by Rybicki and Kanninen [108], which is implemented into several finite element programs. A modified version was proposed by Riccio and Gigliotti [109]. As an alternative, Alfano and Crisfield

[110] have proposed an analysis method using a cohesive-zone model combined with interface elements. A theory presented by Xiao and Gillespie [111] improves the prediction of the shear strength enhancement in the presence of friction and compression. Davidson and Zhao [112] have set up a 'limited input bilinear criterion' for mixed-mode delamination failure which needs data only from DCB, SLB, and ENF tests for its characterization. Thus the models for simulating the delamination progression seem to be well developed.

More involved is the simulation of buckling in the presence of delaminations. The effects of rectangular and triangular local delaminations as well as asymmetric sub-layers were treated by Wang and Lu [113]. The influence of stacking sequence on the buckling load of plates with strip delaminations was analysed by Pekbey and Sayman [114]. Lee and Park [115] applied an enhanced assumed strain solid element for better simulate the local buckling mode at the delamination zone. Two enveloped delaminations, one fully covering the other, were treated by Parlapalli et al. [116]. They determined upper and lower bounds for the buckling load which strongly depend on the locations and sizes of the two delaminations. Correspondingly, Tafreshi [117] studied the effect of the size and location of a delamination. He found out that for delaminations closer to the free surface of the laminate failure will be due to delamination growth rather than buckling. On the other hand, Aslan and Sahin [118] found out that for multiple delaminations the largest and near-surface delamination affects the buckling load most, the size of a beneath delamination has no significant effect. Kremer and Schürmann [119] treated the problem of tension-loaded plates with cut-outs, which can buckle because of local compression or shear. They showed that a shape optimized cut-out runs risk to initiate buckling before reaching the fracture load.

If the composite structure undergoes impact or crash load the dynamic aspects must be considered. Explicit finite element codes like PAM CRASH or LS-DYNA contain suitable tools. Special procedures for treating delamination under impact are proposed by Fleming [120] and Iannucci [121]. Williams et al. [122] used the strain equivalence approach for setting up a model to predict impact damage growth and its effects on the impact force histories in carbon fiber reinforced plastic laminates. A damage model proposed by Iannucci and Ankersen [123] uses damage variables assigned to tensile, compressive and shear damage at a lamina level, thus allowing the total energy dissipated for each damage mode to be controlled during a dynamic or impact event.

10.5.4 Fatigue

For structures from fiber composite material strength reduction due to fatigue loading is well known to be comparatively small, but not negligible in the first place. In his PhD thesis Talreja [124] has comprehensively described the failure mechanisms which happen during fatigue loading. Since then the subject has been intensively worked on leading to a large number of relevant publications.

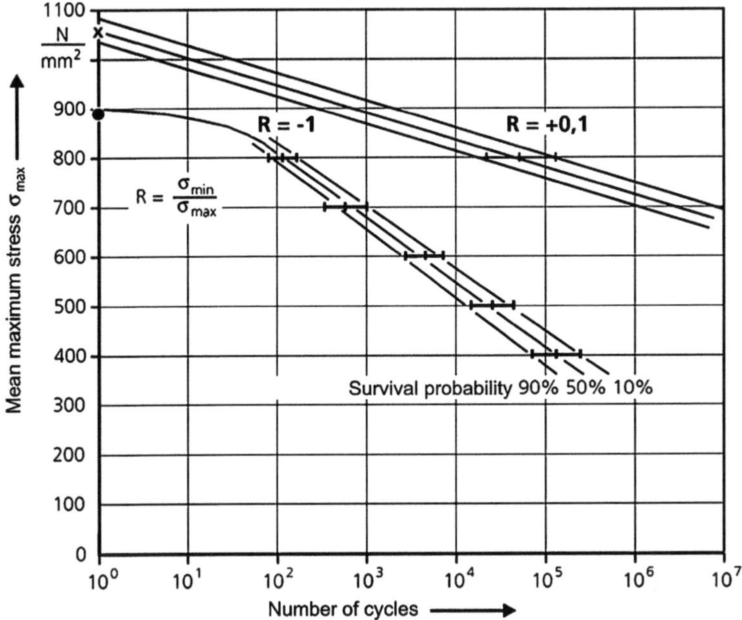

Fig. 10.4 Effect of stress ratio on the fatigue life for a $[0_2,-45,0_2,+45,0,90]$s laminate from T300/914C

The book recently edited by Harris [125] gives an insight into the research state of the art. Tests have shown that besides other parameters fatigue life heavily depends on the stress ratio as depicted in Fig. 10.4. Simulation models which accurately describe the damage progression during fatigue, however, are rare.

A linear damage accumulation hypothesis after Palmgren and Miner does not account for the load sequence which is important for composites. The Marco-Starkey model [126] is a nonlinear Miner rule the exponents of which must be determined by tests. The Strength-Degradation model by Yang and Liu [127] needs to measure the cycles to failure and the residual strength for every load level, and in addition it is not applicable for notched structures. The Percent-Failure rule requires fatigue tests for every load level. It can account for specific load sequences by switching between corresponding S-N curves. Moisture reduces the residual strength of GFRP equivalently to the reduction in static strength as has been observed by Cerny and Mayer [128], an effect which is not accounted for in the available models. For cross-ply laminates under constant amplitude tension-tension or compression-compression loading a progressive fatigue damage model was proposed by Shokrie and Taheri-Behrooz [129] and Taheri-Behrooz et al. [130]. As input, however, static and fatigue properties of unidirectional composites in longitudinal and transverse direction are needed. They must be measured in corresponding tests. Damage initiation and damage propagation was studied by May and Hallett [131] and May et al. [132] using cohesive interface elements.

The model requires anticipating a potential matrix crack where the interface elements are to be placed. After extensive discussions on the pros and cons of existing models Quaresimin et al. [133] conclude that they are not fully satisfying, some models even lead to non-conservative results. Especially for very high cycle fatigue and the effect of temperature increase in epoxy resin reliable models do not yet exist.

10.6 Manufacturing

10.6.1 Draping

In case of structures with a plane or developable surface placing the fibers in the required direction is not a problem which needs simulation. However, if the structures have a non-developable surface accurate draping is not simple and simulation software can help finding an optimal solution. Accuracy and reliability of simulation models, however, are still under discussion. Hancock and Potter [134], for instance, investigated the simple pin jointed net model applied in reverse to generate formable geometries. Later, the same authors [135] stated that the conventional outputs of kinematic modeling are of limited applicability in informing the hand lay-up process for complex surfaces. They proposed a novel strategy for generating detailed unambiguous manufacturing instructions as a method for enhancing the practicality of kinematic simulation tools. Vanclooster et al. [136] compared the kinematic mapping against the explicit finite element method and found out that the kinematic mapping approach severely fails in predicting the fiber reorientation that occurs during stamp forming for non-symmetrical forming configurations. The FEM-simulation gave a reasonably good prediction of the fiber reorientation and seemed the most promising technique for draping simulations. In spite of the ongoing discussion several software packages provide tools for draping. In this context the program systems Catia V5 CPD [137], Simulayt Advanced Fiber Modeler [138], Vistagy FiberSIM [139] and Anaglyph Laminate Tools [140] are to be mentioned.

10.6.2 Resin Flow and Curing

Resin transfer molding has proven to be an efficient manufacturing technology. Simulating the resin flow and the curing can help avoiding pitfalls. To prevent dry spots Minaie and Chen [141] simplified the multiport resin flow as many individual one-dimensional flows each of which extend from the gate to the vent and are controlled independently by adjusting the pressure of the related inlet. In a vacuum assisted resin transfer molding process Johnson and Pitchumani [142]

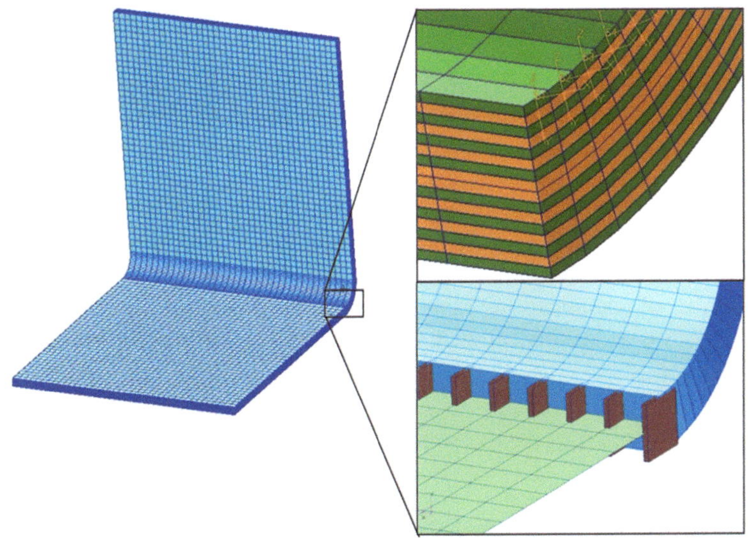

Fig. 10.5 Modeling L-profile for spring-in simulation

used induction heating as a means of locally reducing viscosity to counteract the effects of nonhomogeneity in the permeability of preform lay-ups. García et al. [143] simulated the mould filling process within an updated Lagrangian framework. The use of a meshless technique allowed avoiding the numerical difficulties associated to the fluid properties transport through the whole domain in fixed mesh simulations. For saving computer time Dong [144] proposed a two-step method, where first, a design of experiment (DOE) approach is coupled with a 2-D control volume finite element method simulation to calculate the equivalent permeability and porosity for various process variable combinations. Second, the algorithm for the through-thickness flow front construction is developed. A commercially available software packages for flow simulation is RTM-Worx [145], which is based on the finite element and control volume methods to solve the physical equations that govern flow of a resin through a porous medium. Further, there is PAM-RTM [146], which simulates preforming, considers different types of resin, preheating of the mold, filling and curing using the Kamal Sourour model.

Manufacturing fiber composites accurate to measurement needs to account for the thermal and chemical reactions at curing. They may lead to the so-called spring-in which must be determined in order to counteract by forming a suitable mould. Processing induced warping has been treated by Darrow and Smith [147], who considered thickness cure shrinkage, mould expansion, and fiber volume fraction gradients. More involved, but also more expensive is the simulation presented by Cheung et al. [148]. They obtained the thermal equilibrium and chemical kinetics during the curing phase of the resin transfer molding process subject to mould temperature history and corresponding manufacturing process plans. Sweeting et al. [149] have studied numerically the distortion of circularly curved flanged laminates. Different parameters influencing the distortion of straight L and curved C profiles have been treated by Svanberg [150]. Straight L- and Z-profiles have been analyzed by Spröwitz et al. [151]. They found out that it is sufficient to model only the curved region with volume elements whereas the flanges can do with shell elements, cf. Fig. 10.5. Based on a micromechanical analysis Brauner et al. [152] concluded that thermal and chemical shrinkage must be supplemented by matrix yielding and degradation in order to more accurately simulate the curing stresses.

A number of manufacturing aspects can be adequately modeled as yet. It seems desirable, however, to connect them to simulate the complete manufacturing chain.

References

1. Larsson, F., Runesson, K., Saroukhani, S., Vafadari, R.: Computational homogenization based on a weak format of micro-periodicity for RVE-problems. Comput. Methods Appl. Mech. Eng. **200**, 11–26 (2011). doi:10.1016/j.cma.2010.06.023
2. Rohwer, K., Xie, M.J.: Micromechanical curing stresses in CFRP. Compos. Sci. Technol. **25**, 169–186 (1986)
3. Nassehi, V., Dhillon, J., Masciaet, L.: Finite element simulation of the micromechanics of interlayered polymer/fibre composites: a study of the interactions between the reinforcing phases. Compos. Sci. Technol. **47**, 349–358 (1993)
4. Zhao, L.G., Warrior, N.A., Long, A.C.: A micromechanical study of residual stress and its effect on transverse failure in polymer–matrix composites. Int. J. Solids Struct. **43**, 5449–5467 (2006)
5. Jin, K.-K., Huang, Y., Lee, Y.-H., Ha, S.K.: Distribution of micro stresses and interfacial tractions in unidirectional composites. J. Compos. Mater. **42**, 1825–1849 (2008). doi:10.1177/0021998308093909
6. Huang, Y., Jin, K.-K., Ha, S.K.: Effects of fiber arrangement on mechanical behavior of unidirectional composites. J. Compos. Mater. **42**, 1851–1871 (2008). doi:10.1177/0021998308093910
7. Kari, S.: Micromechanical modelling and numerical homogenization of fibre and particle reinforced composites. Dissertation, University of Magdeburg (2006)
8. Garnich, M.R., Karami, G.: Finite element micromechanics of stiffness and strength of wavy fiber composites. J. Compos. Mater. **38**, 273–292 (2004). doi:10.1177/0021998304039270
9. Hill, R.: Theory of mechanical properties of fibre-strengthened materials. J. Appl. Mech. **31**, 223–232. Errata (1965) J. Appl. Mech. **32**, 219 (1964)
10. Tsai, S.W., Hahn, H.T.: Introduction to Composite Materials. Technomic Publishing Company, Westport (1980)
11. Hashin, Z.: Analysis of composite materials—A survey. J. Appl. Mech. **50**, 481–505 (1983)
12. Chamis, C.C.: Simplified composite micromechanics equations for hygral, thermal, and mechanical properties. SAMPE Q. **15**, 14–23 (1984)
13. Kulkarni, R., Ozden, O.: Transverse and longitudinal CTE measurements of carbon fibers and their impact on interfacial residual stresses in composites. J. Compos. Mater. **40**, 733–754 (2006). doi:10.1177/0021998305055545

14. Al-Sulaiman, F.A., Al-Nassar, Y.N., Mokheimer, E.M.A.: Prediction of the thermal conductivity of the constituents of fiber-reinforced composite laminates: Voids effect. J. Compos. Mater. **40**, 797–814 (2006). doi:10.1177/0021998305055548
15. Zweben, C.: A bounding approach to the strength of composite materials. Eng. Fract. Mech. **4**, 1–8 (1972)
16. Rosen, B.W.: Tensile failure of fibrous composites. AIAA-J. **2**(11), 1985–1991 (1964)
17. Rosen, B.W., Zweben, C.H.: Tensile failure criteria for fiber composite materials. NASA CR-2057, Washington DC (1972)
18. Curtin, W.A.: Theory of mechanical properties of ceramic-matrix composites. J. Am. Ceram. Soc. **74**, 2837–2845 (1991)
19. Koyanagi, J., Hatta, H., Kotani, M., Kawadael, H.: A comprehensive model for determining tensile strengths of various unidirectional composites. J. Compos. Mater. **43**, 1901–1914 (2009). doi:10.1177/0021998309341847
20. Cid Alfaro, M.V., Suiker, A.S.J., de Borst, R.: Transverse failure behavior of fiber-epoxy systems. J. Compos. Mater. **44**, 1493–1516 (2010). doi:10.1177/0021998309360941
21. Rosen, B.W., Hashin, Z.: Analysis of material properties. In: Reinhart, T.J., et al. (eds.) Engineered Materials Handbook, Composites. Metal Park, Ohio (1989)
22. Dharan, C.K.H., Lin, Ch.-L.: Longitudinal compressive strength of continuous fiber composites. J. Compos. Mater. **41**, 1389–1405 (2007). doi:10.1177/0021998306068078
23. Xu, Y.L., Reifsnider, K.L.: Micromechanical modeling of composite compressive strength. J. Compos. Mater. **27**, 572–588 (1993). doi:10.1177/002199839302700602
24. Budiansky, B., Fleck, N.: Compressive failure of fiber composites. J. Mech. Phys. Solids **41**, 183–211 (1993)
25. Jensen, H.M., Christoffersen, J.: Kink band formation in fiber reinforced materials. J. Mech. Phys. Solids **45**, 1121–1136 (1997)
26. Gutkin, R., Pinho, S.T., Robinson, P., Curtis, P.T.: Micro-mechanical modelling of shear-driven fibre compressive failure and of fibre kinking for failure envelope generation in CFRP laminates. Compos. Sci. Technol. **70**, 1214–1222 (2010). doi:10.1016/j.compscitech.2010.03.009
27. Gutkin, R., Pinho, S.T., Robinson, P., Curtis, P.T.: On the transition from shear-driven fibre compressive failure to fibre kinking in notched CFRP laminates under longitudinal compression. Compos. Sci. Technol. **70**, 1223–1231 (2010). doi:10.1016/j.compscitech.2010.03.010
28. Mishra, A., Naik, N.K.: Inverse micromechanical models for compressive strength of unidirectional composites. J. Compos. Mater. **43**, 1199–1211 (2009). doi:10.1177/0021998308104133
29. Martinez, X., Oller, S.: Numerical simulation of matrix reinforced composite materials subjected to compression loads. Arch. Comput. Methods Eng. (2009). doi:10.1007/s11831-009-9036-3
30. Ng, W.H., Salvi, A.G., Waas, A.M.: Characterization of the in-situ non-linear shear response of laminated fiber-reinforced composites. Compos. Sci. Technol. **70**, 1126–1134 (2010). doi:10.1016/j.compscitech.2010.02.024
31. Totry, E., Molina-Aldareguía, J.M., González, C., LLorca, J.: Effect of fiber, matrix and interface properties on the in-plane shear deformation of carbon-fiber reinforced composites. Compos. Sci. Technol. **70**, 970–980 (2010). doi:10.1016/j.compscitech.2010.02.014
32. Whitney, J.M., Pagano, N.J.: Shear deformation in heterogeneous anisotropic plates. J. Appl. Mech. **37**, 1031–1036 (1970). doi:10.1115/1.3408654
33. Rohwer, K.: Improved transverse shear stiffness for layered finite elements. DFVLR-Forschungsbericht 88-32, Braunschweig, (1988)
34. Altenbach, H.: An alternative determination of transverse shear stiffnesses for sandwich and laminated plates. Int. J. Solids Struct. **37**, 3503–3520 (2000)

35. Schürg, M., Wagner, W., Gruttmann, F.: An enhanced FSDT model for the calculation of interlaminar shear stresses in composite plate structures. Comput. Mech. **44**, 765–776 (2009). doi:10.1007/s00466-009-0410-7

36. Murthy, M.V.V.: An improved transverse shear deformation theory for laminated anisotropic plates. NASA Technical Paper 1903, Hampton VA (1981)

37. Reddy, J.N.: A simple higher-order theory for laminated composite plates. J. Appl. Mech. **51**, 745–752 (1984). doi:10.1115/1.3167719

38. Reissner, E.: On transverse bending of plates, including the effect of transverse shear deformation. Int. J. Solids Struct. **11**, 569–573 (1975). doi:10.1016/0020-7683(75)90030-X

39. Lo, K.H., Christensen, R.M., Wu, E.M.: A high-order theory of plate deformation, parts 1 and 2. J. Appl. Mech. **44**, 663676 (1977). doi:10.1115/1.3424154 and 10.1115/1.3424155

40. Wu, Ch.-P., Tarn, J.-Q., Chen, P.-Y.: Refined asymptotic theory of doubly curved laminated shells. J. Eng. Mech. **123**, 1238–1246 (1997). doi:10.1061/(ASCE)0733-9399(1997)123:12(1238)

41. Noor, A.K., Burton, W.S.: Assessment of computational models for multilayered composite shells. Appl. Mech. Rev. **43**, 67–96 (1990). doi:10.1115/1.3119162

42. Rohwer, K.: Application of higher order theories to the bending analysis of layered composite plates. Int. J. Solids Struct. **29**, 105–119 (1992). doi:10.1016/0020-7683(92)90099-F

43. Yang, H.T.Y., Saigal, S., Masud, A., Kapania, R.K.: A survey of recent shell finite elements. Int. J. Numer. Methods Eng. **47**, 101–127 (2000). doi:10.1002/(SICI)1097-0207(20000110/30)47:1/3<101::AID-NME763>3.0.CO;2-C

44. Timarci, T., Aydogdu, M.: Buckling of symmetric cross-ply square plates with various boundary conditions. Compos. Struct. **68**, 381–389 (2005). doi:10.1016/j.compstruct.2004.04.003

45. Carrera, E., Miglioretti, F., Petrolo, M.: Accuracy of refined finite elements for laminated plate analysis. Compos. Struct. **93**, 1311–1327 (2011). doi:10.1016/j.compstruct.2010.11.007

46. Rohwer, K., Rolfes, R., Sparr, H.: Higher-order theories for thermal stresses in layered plates. Int. J. Solids Struct. **28**, 3673–3687 (2001)

47. Pagano, N.J., Hatfield, S.J.: Elastic behavior of multilayered bidirectional composites. AIAA-J. **10**, 931–933 (1972)

48. Sun, C.-T., Whitney, J.M.: Theories for the dynamic response of laminated plates. AIAA-J. **11**, 178–183 (1973)

49. Di Scuiva, M.: A refinement of the transverse shear deformation theory for multilayered anisotropic plates. Atti del Dipartimento di Ingegneria Aeronautica e Spaziale del Politecnico di Torino, Publication, No. 5 (1983)

50. Lee, K.H., Senthilnathan, N.R., Lim, S.P., Chow, S.T.: An improved zig-zag model for the bending of laminated composite plates. Compos. Struct. **15**, 137–148 (1990). doi:10.1016/0263-8223(90)90003-W

51. Savithri, S., Varadan, T.K.: Accurate bending analysis of laminated orthotropic plates. AIAA-J. **28**, 1842–1844 (1990)

52. Rohwer, K., Friedrichs, S., Wehmeyer, C.: Analyzing laminated structures from fibre-reinforced composite material—An assessment. Tech. Mechanik **25**, 59–79 (2005)

53. Lo, S.H., Zhen, W., Sze, K.Y., Wanji, C.: C^0-type global–local theory with non-zero normal strain for the analysis of thick multilayer composite plates. Comput. Mech. **47**, 479–491 (2011). doi:10.1007/s00466-010-0554-5

54. Murakami, H.: Laminated composite plate theory with improved in-plane responses. J. Appl. Mech. **53**, 661–666 (1986). doi:10.1115/1.3171828

55. Di, S., Ramm, E.: Hybrid stress formulation for higher-order theory of laminated shell analysis. Comput. Methods Appl. Mech. Eng. **109**, 359–376 (1993). doi:10.1016/0045-7825(93)90087-E

56. Brank, B., Carrera, E.: Multilayered shell finite element with interlaminar continuous shear stresses: A refinement of the Reissner-Mindlin formulation. Int. J. Numer. Methods Eng. **48**, 843–874 (2000). doi:10.1002/(SICI)1097-0207(20000630)48:6<843::AID-NME903>3.0.CO;2-E

57. Icardi, U., Ferrero, L.: Multilayered shell model with variable representation of displacements across the thickness. Composites: Part B **42**, 18–26 (2011). doi:10.1016/j.compositesb.2010.09.022

58. Carrera, E.: Theories and finite elements for multilayered plates and shells: a unified compact formulation with numerical assessment and benchmarking. Arch. Comput. Methods Eng. **10**(3), 215–296 (2003)

59. Pryor, C.W., Barker, R.M.: A finite element analysis including transverse shear effects for application to laminated plates. AIAA-J. **9**, 912–917 (1971)

60. Lo, K.H., Christensen, R.M., Wu, E.M.: Stress solution determination for high order plate theory. Int. J. Solids Struct. **14**, 655–662 (1978). doi:10.1016/0020-7683(78)90004-5

61. Engblom, J.J., Ochoa, O.O.: Through-the-thickness stress predictions for laminated plates of advanced composite materials. Int. J. Numer. Methods Eng. **21**, 1759–1776 (1985). doi:10.1002/nme.1620211003

62. Rolfes, R., Rohwer, K.: Improved transverse shear stresses in composite finite elements based on first order shear deformation theory. Int. J. Numer. Methods Eng. **40,** 51–60 (1997). doi:10.1002/(SICI)1097-0207(19970115)40:1<51::AID-NME49>3.0.CO;2-3

63. Noor, A.K., Burton, W.S., Peters, J.M.: Predictor-corrector procedures for stress and free vibration analyses of multilayered composite plates and shells. Comput. Methods Appl. Mech. Eng. **82**, 341–363 (1990). doi:10.1016/0045-7825(90)90171-H

64. Noor, A.K., Malik, M.: An assessment of five modelling approaches for thermo-mechanical stress analysis of laminated composite panels. Comput. Mech. **25**, 43–58 (2000). doi:10.1007/s004660050014

65. Park, J.W., Kim, Y.H.: Re-analysis procedure for laminated plates using FSDT finite element model. Comput. Mech. **29**, 226–242 (2002). doi:10.1007/s00466-002-0336-9

66. Park, J.W., Lee, K.C., Kim, Y.H.: Comparative study of finite element based response evaluation methods for laminated plates. Comput. Mech. **32**, 115–133 (2003). doi:10.1007/s00466-003-0466-8

67. Guiamatsia, I.: A new approach to plate theory based on through-thickness homogenization. Int. J. Numer. Meth. Eng. **84**, 1139–1165 (2010). doi:10.1002/nme.2934

68. Kreja, I.: A literature review on computational models for laminated composite and sandwich panels. Cent. Eur. J. Eng. **1**(1), 59–80 (2011). doi:10.2478/s13531-011-0005-x

69. Kolpakov, A.A., Kolpakov, A.G.: Solution of the laminated plate design problem: new problems and algorithms. Comput. Struct. **83**, 964–975 (2005). doi:10.1016/j.compstruc.2004.08.012

70. Evans, J.T., Gibson, A.G.: Composite angle ply laminates and netting analysis. Proc. R Soc. London A **458**, 3079–3088 (2002)

71. Tan, S.C.: Stress Concentrations in Laminated Composites. Technomic Publishing Co, Lancaster (1994)

72. Tsai, S.W.: Composites design. Think Composites, Dayton, (1988)

73. Anonymous: http://www.esacomp.com (2011). Accessed 20 July 2011

74. Vannucci, P., Barsotti, R., Bennati, S.: Exact optimal flexural design of laminates. Compos. Struct. **90**, 337–345 (2009). doi:10.1016/j.compstruct.2009.03.017

75. Lopes, C.S., Seresta, O., Coquet, Y., Gürdal, Z., Camanho, P.P., Thuis, B.: Low-velocity impact damage on dispersed stacking sequence laminates. Part I: Experiments. Compos. Sci. Technol. **69**, 926–936 (2009). doi:10.1016/j.compscitech.2009.02.009

76. Lopes, C.S., Camanho, P.P., Gürdal, Z., Maimí, P., Gonzálezet, E.V.: Low-velocity impact damage on dispersed stacking sequence laminates. Part II: Numerical simulations. Compos. Sci. Technol. **69**, 937–947 (2009). doi:10.1016/j.compscitech.2009.02.015

77. Eschenauer, H.A.: The 'three columns' for treating problems in optimum structural design. In: Bergmann, H.W. (ed.) Optimization: Methods and Applications, Possibilities and Limitations. Lecture Notes in Engineering 47, Springer-Verlag Berlin, Heidelberg, (1989)
78. Zimmermann, R.: Optimierung axial gedrückter CFK-Zylinderschalen. Dissertation, Universität-Gesamthochschule Siegen (1991)
79. Ghiasi, H., Pasini, D., Lessard, L.: Optimum stacking sequence design of composite materials Part I: Constant stiffness design. Compos. Struct. **90**, 1–11 (2009). doi:10.1016/j.compstruct.2009.01.006
80. Javidrad, F., Nouri, R.: A simulated annealing method for design of laminates with required stiffness properties. Compos. Struct. **93**, 1127–1135 (2011). doi:10.1016/j.compstruct.2010.10.011
81. Akbulut, M., Sonmez, F.O.: Design optimization of laminated composites using a new variant of simulated annealing. Comput. Struct. **89**, 1712–1724 (2011). doi:10.1016/j.compstruc.2011.04.007
82. Gou, S., Cheng, W., Cui, D.: Aeroelastic tailoring of composite wing structures by laminate layup optimization. AIAA-J. **44**, 3146–3149 (2006)
83. Lopez, R.H., Luersen, M.A., Cursiet, E.S.: Optimization of laminated composites considering different failure criteria. Composites: Part B **40,** 731–740 (2009). doi:10.1016/j.compositesb.2009.05.007
84. Azarafza, R., Khalili, S.M.R., Jafari, A.A., Davar, A.: Analysis and optimization of laminated composite circular cylindrical shell subjected to compressive axial and transverse transient dynamic loads. Thin-Walled Struct. **47**, 970–983 (2009). doi:10.1016/j.tws.2009.01.004
85. Blom, A.W., Stickler, P.B., Gürdalet, Z.: Optimization of a composite cylinder under bending by tailoring stiffness properties in circumferential direction. Composites: Part B **41**, 157–165 (2010). doi:10.1016/j.compositesb.2009.10.004
86. Almeida, F.S., Awruch, A.M.: Design optimization of composite laminated structures using genetic algorithms and finite element analysis. Compos. Struct. **88**, 443–454 (2009). doi:10.1016/j.compstruct.2008.05.004
87. Johansen, L., Lund, E.: Optimization of laminated composite structures using delamination criteria and hierarchical models. Struct. Multi. Optim. **38**, 357–375 (2009). doi:10.1007/s00158-008-0280-1
88. Gillet, A., Francescato, P., Saffre, P.: Single- and multi-objective optimization of composite structures: the influence of design variables. J. Compos. Mater. **44**, 457–480 (2010). doi:10.1177/0021998309344931
89. Peng, W., Chen, J., Wei, J., Tu, W.: Optimal strength design for fiber-metal laminates and fiber-reinforced plastic laminates. J. Compos. Mater. **45**, 237–254 (2011). doi:10.1177/0021998310373521
90. Ghiasi, H., Fayazbakhsh, K., Pasini, D., Lessard, L.: Optimum stacking sequence design of composite materials Part II: Variable stiffness design. Compos. Struct. **93**, 1–13 (2010). doi:10.1016/j.compstruct.2010.06.001
91. Ottino, J.M.: Engineering complex systems. Nature **427**, 399 (2004)
92. Marczyk, J. Future trends in computer-aided engineering. In: Proceedings NAFEMS World Congress, Crete (2009)
93. Lee, M.C.W., Mikulik, Z., Kelly, D.W., Thomson, R.S., Degenhardt, R.: Robust design—A concept for imperfection insensitive composite structures. Compos. Struct. **92**, 1469–1477 (2010). doi:10.1016/j.compstruct.2009.09.054
94. Nahas, M.N.: Survey of failure and post-failure theories of laminated fiber-reinforced composites. J. Compos. Technol. Res. **8**, 138–153 (1986)
95. Hinton, M.J., Kaddour, A.S., Soden, P.D. (eds.): Failure Criteria in Fibre Reinforced Polymer Composites: The World-Wide Failure Exercise. Elsevier, Amsterdam (2004)
96. Luccioni, B.M.: Constitutive model for fiber-reinforced composite laminates. J. Appl. Mech. **73**, 901–910 (2006). doi:10.1115/1.2200654

97. Pinho, S.T., Davila, C.G., Camanho, P.P., Iannucci, L., Robinson, P.: Failure models and criteria for FRP under in-plane or three-dimensional stress states including shear non-linearity. NASA/TM-2005-213530, Hampton (2005)
98. Cuntze, R.: Efficient 3D and 2D failure conditions for UD laminae and their application within the verification of the laminate design. Compos. Sci. Technol. **66**, 1081–1096 (2006). doi:10.1016/j.compscitech.2004.12.046
99. Stamblewski, C., Sankar, B.V., Zenkert, D.: Analysis of three-dimensional quadratic failure criteria for thick composites using the direct micromechanics method. J. Compos. Mater. **42**, 635–654 (2008). doi:10.1177/0021998307088609
100. Lee, J., Soutis, C.: A study on the compressive strength of thick carbon fibre–epoxy laminates. Compos. Sci. Technol. **67**, 2015–2026 (2007). doi:10.1016/j.compscitech.2006.12.001
101. Knops, M., Bögle, C.: Gradual failure in fibre/polymer laminates. Compos. Sci. Technol. **66**, 616–625 (2006). doi:10.1016/j.compscitech.2005.07.044
102. Maimi, P., Camanho, P.P., Mayugo, J.-A., Davila, C.G.: A thermodynamically consistent damage model for advanced composites. NASA/TM -2006-214282, Hampton, VA (2006)
103. Cui, H.-P., Wenb, W.-D., Cui, H.-T.: An integrated method for predicting damage and residual tensile strength of composite laminates under low velocity impact. Comput. Struct. **87**, 456–466 (2009). doi:10.1016/j.compstruc.2009.01.006
104. Basu, S., Waas, A.M., Ambur, D.R.: Computational modeling of damage growth in composite laminates. AIAA-J. **41**, 1158–1166 (2003)
105. Ha, S.K., Huang, Y., Han, H.H., Jinet, K.K.: Micromechanics of failure for ultimate strength predictions of composite laminates. J. Compos. Mater. **44**(20), 2347–2361 (2010). doi:10.1177/0021998310372464
106. Zhang, Y.X., Zhang, H.S.: Multiscale finite element modeling of failure process of composite laminates. Compos. Struct. **92**, 2159–2165 (2010). doi:10.1016/j.compstruct.2009.09.031
107. Garnich, M.R., Akula, V.M.K.: Review of degradation models for progressive failure analysis for fiber reinforced polymer composites. Appl. Mech. Rev. **62**, 1–33 (2009). doi:10.1115/1.3013822
108. Rybicki, E.F., Kanninen, M.F.: A finite element calculation of stress intensity factor by a modified crack closure integral. Eng. Fract. Mech. **9**, 931–938 (1977). doi:10.1016/0013-7944(77)90013-3
109. Riccio, A., Gigliotti, M.: A novel numerical delamination growth initiation approach for the preliminary design of damage tolerant composite structures. J. Compos. Mater. **41**, 1939–1960 (2007). doi:10.1177/0021998307069908
110. Alfano, G., Crisfield, M.A.: Solution strategies for the delamination analysis based on a combination of local-control arc-length and line searches. Int. J. Numer. Methods Eng. **58**, 999–1048 (2003). doi:10.1002/nme.806
111. Xiao, J.R., Gillespie Jr, J.W.: A phenomenological Mohr–Coulomb failure criterion for composite laminates under interlaminar shear and compression. J. Compos. Mater. **41**, 1295–1309 (2007). doi:10.1177/0021998306067318
112. Davidson, B.D., Zhao, W.: An accurate mixed-mode delamination failure criterion for laminated fibrous composites requiring limited experimental input. J. Compos Mater **41**, 679–702 (2007). doi:10.1177/0021998306071031
113. Wang, X., Lu, G.: Local buckling of composite laminar plates with various delaminated shapes. Thin-Walled Struct. **41**, 493–506 (2003). doi:10.1016/S0263-8231(03)00020-X
114. Pekbey, Y., Sayman, O.: A numerical and experimental investigation of critical buckling load of rectangular laminated composite plates with strip delamination. J. Reinf. Plast. Compos. **25**, 685–697 (2006). doi:10.1177/0731684406060566
115. Lee, S.-Y., Park, D.-Y.: Buckling analysis of laminated composite plates containing delaminations using the enhanced assumed strain solid element. Int. J. Solids Struct. **44**, 8006–8027 (2007). doi:10.1016/j.ijsolstr.2007.05.023

116. Parlapalli, M.S.R., Shu, D., Chai, G.B.: Buckling of composite beams with two enveloped delaminations: lower and upper bounds. Comput. Struct. **86**, 2155–2165 (2008). doi:10.1016/j.compstruc.2008.06.008

117. Tafreshi, A.: Instability of delaminated composite cylindrical shells under combined axial compression and bending. Compos. Struct. **82**, 422–433 (2008). doi:10.1016/j.compstruct.2007.01.021

118. Aslan, Z., Sahin, M.: Buckling behavior and compressive failure of composite laminates containing multiple large delaminations. Compos. Struct. **89**, 382–390 (2008). doi:10.1016/j.compstruct.2008.08.011

119. Kremer, T., Schürmann, H.: Buckling of tension-loaded thin-walled composite plates with cut-outs. Compos. Sci. Technol. **68**, 90–97 (2008). doi:10.1016/j.compscitech.2007.05.035

120. Fleming, D.C.: Delamination modeling of composites for improved crash analysis. J. Compos. Mater. **35**, 1777–1792 (2001). doi:10.1106/3V9W-9099-HYQ3-08GR

121. Iannucci, L.: Dynamic delamination modelling using interface elements. Comput. Struct. **84**, 1029–1048 (2006). doi:10.1016/j.compstruc.2006.02.002

122. Williams, K.V., Vaziri, R., Poursartip, A.: A physically based continuum damage mechanics model for thin laminated composite structures. Int. J. Solids Struct. **40**, 2267–2300 (2003). doi:10.1016/S0020-7683(03)00016-7

123. Iannucci, L., Ankersen, J.: An energy based damage model for thin laminated composites. Compos. Sci. Technol. **66**, 934–951 (2006). doi:10.1016/j.compscitech.2005.07.033

124. Talreja, R.: Fatigue of composite materials. Dissertation, Technical University of Denmark, Lyngby, (1985)

125. Harris, B. (ed.): Fatigue in Composite. Woodhead Publishing Ltd, Abington Cambridge (2003)

126. Marco, S., Starkey, W.L.: A concept of fatigue damage. Trans. ASME **76**, 627–632 (1954)

127. Yang, J.N., Liu, M.D.: Residual strength degradation model and theory of periodic proof tests for graphite/epoxy laminates. J. Compos. Mater. **11**, 176–204 (1977). doi:10.1177/002199837701100205

128. Cerny, I., Mayer, R.M.: Evaluation of static and fatigue strength of long fiber GRP composite material considering moisture effects. Compos. Struct. **92**, 2035–2038 (2010). doi:10.1016/j.compstruct.2009.11.024

129. Shokrieh, M.M., Taheri-Behrooz, F.: Progressive fatigue damage modeling of cross-ply laminates, I: Modeling Strategy. J. Compos. Mater. **44**, 1217–1231 (2010). doi:10.1177/0021998309351604

130. Taheri-Behrooz, F., Shokrieh, M.M., Lessard, L.B.: Progressive fatigue damage modeling of cross-ply laminates, II: Experimental evaluation. J. Compos. Mater. **44**, 1261–1277 (2010). doi:10.1177/0021998309351605

131. May, M., Hallett, S.R.: An advanced model for initiation and propagation of damage under fatigue loading—Part I: Model formulation. Compos. Struct. **93**, 2340–2349 (2011). doi:10.1016/j.compstruct.2011.03.022

132. May, M., Pullin, R., Eaton, M., Featherston, C., Hallettet, S.R.: An advanced model for initiation and propagation of damage under fatigue loading—Part II: Matrix cracking validation cases. Compos. Struct. **93**, 2350–2357 (2011). doi:10.1016/j.compstruct.2011.03.023

133. Quaresimin, M., Susmel, L., Talreja, R.: Fatigue behaviour and life assessment of composite laminates under multiaxial loadings. Int. J. Fatigue **32**, 2–16 (2010). doi:10.1016/j.ijfatigue.2009.02.012

134. Hancock, S.G., Potter, K.D.: Inverse drape modelling—an investigation of the set of shapes that can be formed from continuous aligned woven fibre reinforcements. Compos. Part A **36**, 947–953 (2005). doi:10.1016/j.compositesa.2004.12.001

135. Hancock, S.G., Potter, K.D.: The use of kinematic drape modelling to inform the hand lay-up of complex composite components using woven reinforcements. Compos. Part A **37**, 413–422 (2006). doi:10.1016/j.compositesa.2005.05.044

136. Vanclooster, K., Lomov, S.V., Verpoest, I.: Experimental validation of forming simulations of fabric reinforced polymers using an unsymmetrical mould configuration. Composites: Part A **40**, 530–539 (2009). doi:10.1016/j.compositesa.2009.02.005
137. Dassault Systems: http://www.3ds.com (2011). Accessed 7 Aug 2011
138. Simulayt: http://www.simulayt.com (2011). Accessed 7 Aug 2011
139. Vistagy http://www.vistagy.com (2011). Accessed 7 Aug 2011
140. Anaglyph: http://www.anaglyph.co.uk (2011). Accessed 7 Aug 2011
141. Minaie, B., Chen, Y.F.: Adaptive control of filling pattern in resin transfer molding process. J. Compos. Mater. **39**, 1497–1513 (2005). doi:10.1177/0021998305051082
142. Johnson, R.J., Pitchumani, R.: Flow control using localized induction heating in a VARTM process. Compos. Sci. Technol. **67**, 669–684 (2007). doi:10.1016/j.compscitech.2006.04.012
143. García, J.A., Gascón, L., Cueto, E., Ordeig, I., Chinesta, F.: Meshless methods with application to liquid composite molding simulation. Comput. Methods Appl. Mech. Eng. **198**, 2700–2709 (2009). doi:10.1016/j.cma.2009.03.010
144. Dong, C.: A modified rule of mixture for the vacuum-assisted resin transfer moulding process simulation. Compos. Sci. Technol. **68**, 2125–2133 (2008). doi:10.1016/j.compscitech.2008.03.019
145. Polyworx: http://www.polyworx.com (2011). Accessed 7 Aug 2011
146. ESI: http://www.esi-group.com (2011). Accessed 7 Aug 2011
147. Darrow Jr, A.D., Smith, L.V.: Isolating components of processing induced warpage in laminated composites. J. Compos. Mater. **36**, 2407–2419 (2002). doi:10.1177/0021998302036021784
148. Cheung, A., Yu, Y., Pochiraju, K.: Three-dimensional finite element simulation of curing of polymer composites. Finite Elem. Anal. Des. **40**, 895–912 (2004). doi:10.1016/S0168-874X(03)00119-7
149. Sweeting, R., Liu, X.L., Paton, R.: Prediction of processing-induced distortion of curved flanged composite laminates. J. Compos. Struct. **57**, 79–84 (2002)
150. Svanberg, J.M.: Predictions of manufacturing induced shape distortions—high performance thermoset composites. Dissertation, Lulea University of Technology, Sweden, (2002)
151. Spröwitz, T., Tessmer, J., Wille, T.: Process simulation in fiber-composite manufacturing – spring-in. In: NAFEMS Seminar: Simulating Composite Materials and Structures, Bad Kissingen, Germany, (2007)
152. Brauner, C., Block, T.B., Hoffmeister, C., Herrmann, A.S.: Process simulation of carbon fibre/epoxy composites on the micro level to analyse chemical and thermal induced residual stresses. In: NAFEMS Seminar: Progress in Simulating Composites, Wiesbaden, Germany 2011

Chapter 11
Modeling of Manufacturing Uncertainties by Multiscale Approaches

Janko Kreikemeier and David Chrupalla

Abstract In this chapter, a numerical multiscale modeling approach is presented and discussed. It bases on the FE^2 approach in which a simultaneous finite element computation of the mechanical response at two different length scales is carried out at each macroscopic integration point. The approach is suitable to obtain the global load response of composite structures without omitting the effect of physical phenomena at the local scale, as for example process-induced defects like voids or fiber waviness.

11.1 Classification of Manufacturing Uncertainties

The manufacturing of composite structures includes the risk of induced defects. Voids and fiber waviness are critical and can be frequently found in composite structures. These defect types reduce the material stiffness and strength and deteriorate the performance of the structure. The predominant reasons for voids are entrapped air that develops during the resin infiltration process, volatile gases that occur during the curing process, or bad wettability behavior of the fibers. Voids develop either within a ply (intralaminar) or between adjacent plies (interlaminar). Fiber waviness defects on the other hand are often resulting during the manufacturing of fiber textiles or the draping of dry textiles and prepregs.

In addition to research on how the manufacturing process can be improved to avoid the occurrence of defects in composite structures, it is important to account

J. Kreikemeier (✉) · D. Chrupalla
Institute of Composite Structures and Adaptive Systems,
German Aerospace Center (DLR e.V.), Lilienthalplatz 7, 38108, Braunschweig, Germany
e-mail: Janko.kreikemeier@dlr.de

D. Chrupalla
e-mail: david.chrupalla@dlr.de

M. Wiedemann and M. Sinapius (eds.), *Adaptive, Tolerant and Efficient Composite Structures*, Research Topics in Aerospace, DOI: 10.1007/978-3-642-29190-6_11,

for the influence of the defects on the load bearing behavior. Experimental, numerical and analytical studies can be found in the literature [1–5].

Generally, the challenge is to obtain accurate results at the large structural (global) scale at reasonable computational cost without omitting the effects of relevant physical phenomena at a smaller (local) scale. Multiscale modeling techniques are suitable methods to meet this challenge.

11.2 Brief Review on Multiscale Modeling

The most common method for the definition of complex microstructures is the representative volume element (RVE) approach in conjunction with a numerical homogenization scheme [6–8]. Dealing with multiscale modeling the dimensions on the micro scale must be orders of magnitudes smaller compared to the structural dimensions, i.e. $L_{Micro} \ll L_{Macro}$. Then the prerequisite of scale separation is fulfilled. The scale separation as well as the homogenization scheme are point wise procedures, i.e., the effective material behavior of one macroscopical material point is obtained by means of a RVE of finite size [9].

In general the distinction between uncoupled as well as coupled homogenization schemes is made. The uncoupled approach is used in case of linear elastic materials and small deformations to obtain the effective response just once [10], whereas the coupled approach should be used in case of finite deformations and inelastic material behavior, e.g. to describe plasticity or damage phenomena [11–14]. In contrast to the uncoupled homogenization scheme, both, the microscopic as well as the macroscopic boundary value problem must be solved simultaneously for the coupled approach at any time, because the effective material response is highly dependent on the actual microscopic behavior. Hence, a smart solution strategy should be used for minimizing computational effort.

If the assumption of scale separation does not hold true anymore, the scale transition can be defined by using hierarchical multi scale approaches. A fine discretization of the whole underlying microstructure is made in particular interesting macroscopic regions [11, 15–17].

A FE^2 approach is developed and implemented in [18, 19] into a general purpose finite element program. A two scale approach is introduced to obtain the effective constitutive response of the macroscopic region. This approach is named FE^2 since a simultaneous finite element computation of the mechanical response at two different geometric scales is performed at each macroscopic integration point. The whole approach is handled as a classical internal variable model where the internal variables are taken from the microscopic data which are required by the microscopic RVE calculation.

A multilevel finite element method is developed in [12] accounting for the visco-plastic behavior of an aluminum material with a certain amount of voids. The focus lies on the higher order homogenization scheme using the gradients of the deformation measures within the constitutive relation.

Continuous as well as discrete variational formulations are developed in [13] for the homogenization of inelastic materials at finite strains by using an energy storage function as well as a dissipation function to account for the effective response of a general standard medium.

A two scale approach is introduced in [20] within the framework of small deformations by using an analytical homogenization scheme, which is based on the generalized method of cells (GMC) [21].

A micromechanical elasto-plastic damage model is presented in [22] in conjunction with a RVE formulation to get the overall damage behavior of fiber reinforced matrix composites. The approach is based on an exterior point Eshelby tensor for circular inclusion problems [23] and the ensemble-averaged effective yield criterion.

In conclusion the development of multiscale modeling strategies is an ongoing process. The great advantage must be seen within a detailed description of the underlying microstructure. The macroscopic constitutive relations are completely obtained by applying an appropriate homogenization scheme onto the microscopic boundary value problem. The common feature of all of the approaches is a simultaneous computation on the macro scale as well as on the micro scale. Thus, the computational effort is very large, but the detailed description of all of the micromechanical features as well as the ongoing improvements of computational power and performance the research within this field is justified.

11.3 A Novel Multiscale Modeling Approach

A newly developed automated iterative loose coupling approach can be used to describe the global behavior of heterogeneous structures [24]. The term "loose coupling" refers to the indirect connection of the global and local finite element (FE) models that are described by different systems of equations solved separately. It works fully automated and takes into account the influence of the local FE results at each iteration of the nonlinear global FE-analysis.

The global-to-local transition is realized by prescribed displacements at the local boundary nodes. The displacements are derived from the strains obtained at the Gaussian points of the global elements. The local-to-global transition is realized via assigning homogenized stresses obtained from the solution of the local boundary value problems to the global Gaussian points and updating the global tangent stiffness operator for the Newton–Raphson scheme. Hotspot criteria, e.g. failure criteria, are used at the global level to decide when it becomes necessary to account for local effects. The steps for this approach can be summarized as follows:

1. Definition of local models via hotspot criterion (Hashin failure criterion)
2. Global-to-local transition
3. Local-to-global transition
4. Numerical determination of the global tangent stiffness operator C_M

11.3.1 Definition of the Local Models

The global areas have to be chosen in which a local analysis becomes necessary. This is done by evaluating Hashin's failure criterion at the global Gaussian points. Adjacent critical global areas are merged into one local model. The local models represent a global area and not only one global Gaussian point. The local models are much more detailed in terms of material modeling and mesh density.

11.3.2 Global-to-Local Transition

For the transition from global to local level, the displacements of the global nodes at the surface of the local model are used as boundary conditions for the local model. Thereby the global nodal displacements are only applied on the boundary nodes of the local models, but not on its inner nodes. Thus, all inner local nodes are unrestrained. On the other hand, the local boundary nodes that coincide with a global node will get the exact value of the global nodal displacement. For all other local boundary nodes in between the global nodes the respective nodal displacements are interpolated via the global shape functions N whose order depends on the used global elements. The displacements at the global nodes are calculated from the strain field at the global Gaussian points. In case of small deformations (geometrical linearity), higher order terms in the strain tensor are disregarded and the displacement field is related to the strains via:

$$\mathbf{E}_{M}^{L} = \frac{1}{2}\left(\mathbf{H}_{M} + \mathbf{H}_{M}^{T}\right) \tag{11.1}$$

$\mathbf{E}_{M}^{L}$ Linearized Green–Lagrange strain tensor at the global level
$\mathbf{H}_{M}$ Displacement gradient at the global level

The displacements at the global Gaussian points are calculated via:

$$\mathbf{u}_{\text{Gauss}} = \mathbf{E}_{M}^{L} \cdot \mathbf{x_0} \tag{11.2}$$

$\mathbf{u}_{\text{Gauss}}$ Displacement at one global Gaussian point
$\mathbf{x}_0$ Position vector of the global Gaussian point

The displacements at the global nodes are calculated via:

$$\mathbf{u}_{\text{node}} = \mathbf{N}^{-1} \cdot \mathbf{u}_{\text{Gauss}} \tag{11.3}$$

$\mathbf{u}_{\text{node}}$ Vector of displacements at the global nodes
$\mathbf{N}$ Matrix of element shape functions

In general, the interpolation of displacements at the local edge nodes in between global nodes is calculated as:

$$\mathbf{u}_{\text{localnode}} = \mathbf{N}_{\text{localnode}} \cdot \mathbf{u}_{\text{node}} \tag{11.4}$$

$\mathbf{u}_{\text{localnode}}$ Vector of displacements at the local boundary nodes
$\mathbf{N}_{\text{localnode}}$ Matrix of global element shape functions at locations of local nodes

11.3.3 Local-to-Global Transition

The transition from local to global level is realized via averaged local stresses. The local elements of one sub-domain which refers to one global integration point are used to calculate the averaged quantities of the corresponding global integration point. Thus, for each global integration point a separate averaged stress tensor is calculated. In case of small deformations, the following simplifying assumption is justified:

$$\mathbf{T}_{\text{m}} \approx \mathbf{P}_{\text{m}} \approx \mathbf{S}_{\text{m}} \tag{11.5}$$

$\mathbf{T}_{\text{m}}$ Cauchy stress tensor at the local level
$\mathbf{P}_{\text{m}}$ First Piola–Kirchhoff stress tensor at the local level
$\mathbf{S}_{\text{m}}$ Second Piola–Kirchhoff stress tensor at the local level

Based on Eq. (11.5) the second Piola–Kirchhoff stress tensor at the global level is calculated by averaging the stresses at the local level:

$$\mathbf{S}_{\text{M}} = \frac{1}{V_{\text{Sub}}} \int_{V_{\text{Sub}}} \mathbf{S}_{\text{m}} dV_{\text{Sub}} \tag{11.6}$$

$$\mathbf{S}_{\text{M}} = \frac{1}{V_{\text{Sub}}} \sum_{i=1}^{N_p} \left(\mathbf{S}_{\text{mi}} V_{\text{mi}} \right) \tag{11.7}$$

$\mathbf{S}_{\text{M}}$ Second Piola–Kirchhoff stress tensor of a single global integration point
$\mathbf{S}_{\text{m}i}$ Second Piola–Kirchhoff stress tensor of the local integration points i
V_{sub} Reference volume of the local sub-domain
$V_{\text{m}i}$ Volume associated with local integration point i
N_{p} Number of integration points of the local sub-domain

11.3.4 Numerical Determination of the Global Tangent Stiffness Operator C_M

Since the global analysis is nonlinear due to material nonlinearities at the local scale, an updated global tangent stiffness operator for the Newton–Raphson scheme has to be determined. This has to be done numerically since no analytical solution is available. The differential quotient has to be formed for the numerical calculation. This is done by solving six additional local boundary value problems, where a small strain perturbation δE_j is added to the j-th component of the global strain tensor. The local boundary value problem is solved 6 more times within each global iteration. The components of the global tangent stiffness operator are defined in matrix notation by:

$$C_{M,ij}^{S} = \frac{S_{\delta Ej,i} - S_{\text{homogenized},i}}{\delta E_j}, \quad i = 1\ldots6,\ j = 1\ldots6 \tag{11.8}$$

$C_{M,ij}^{S}$ Components of the global tangent stiffness operator

$S_{\text{homogenized},i}$ i-th component of the stress tensor at one local integration point, obtained from averaging the local stress tensor

δE_j Small perturbation to the j-th component of the global strain tensor

$S_{\delta Ej,i}$ i-th component of the perturbed stress tensor

11.4 Numerical Example

A simple test case is investigated for a solid-to-solid coupling in order to illustrate the feasibility of the approach in the commercial FE code ABAQUS^{TM} [25] via manual implementation of subroutines and scripts. The example consists of a bar subjected to a tension load in its axial direction. One reference calculation without the coupling of different modeling levels is compared to a calculation with the proposed multiscale approach to verify the implementation of the multiscale approach.

11.4.1 Reference Calculations

The reference model has a mesh that consists of linear solid elements with a cubical shape and an edge length of 0.5 mm. The bar has a length of 10 mm and height and width are 2 mm. It is clamped on one side and has a displacement of 0.07 mm in the axial direction applied on the other side, as is shown in Fig. 11.1. Geometric linearity is assumed due to small deformations. Orthotropic material properties are defined for all elements. For the critical area in the center of the bar

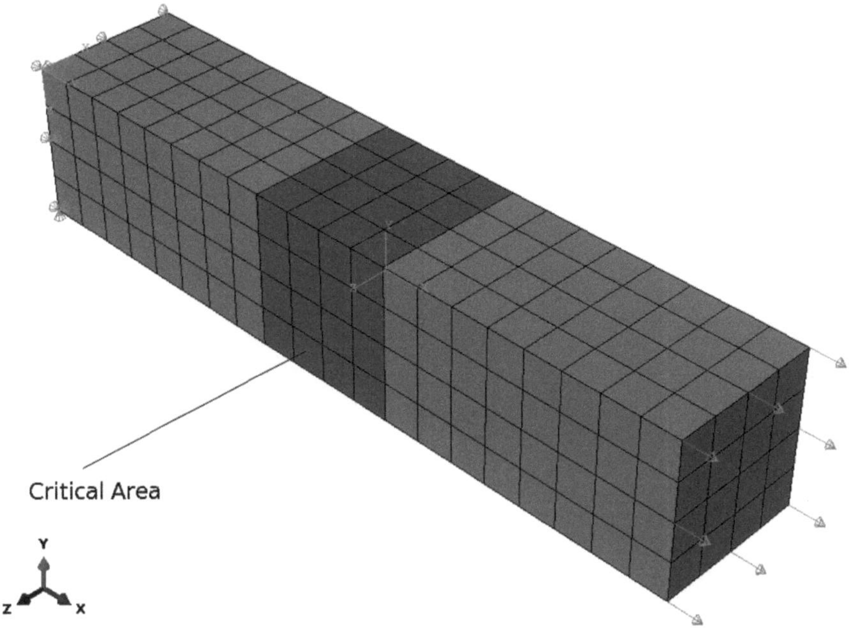

Fig. 11.1 Reference model: critical area and mesh

a nonlinear material model is used which is based on the model developed by Ladevèze [26, 27]. Internal damage variables degrade the material stiffness with increasing load in this model. While material nonlinearity is assumed in the elements that cover the critical area of the bar, all other elements have linear elastic material behavior by definition. Fig. 11.1 shows the critical area in dark color and the mesh of the reference model.

11.4.2 Calculation with Homogenization-Based Two-Way Multiscale Approach

The global model has the same geometry, initial material properties, and boundary conditions for the calculation with the homogenization-based two-way multiscale approach as for the reference model. The difference is that the mesh consists only of five elements. Moreover, the material nonlinearity is modeled by the multiscale approach in the critical area of the bar, which is covered by one element at the global level. The local model has the same size as the critical global element and the same mesh density as the reference model. The material model is identical at the local level and in the reference calculation. An overlay plot of the global and the local model is shown in Fig. 11.2.

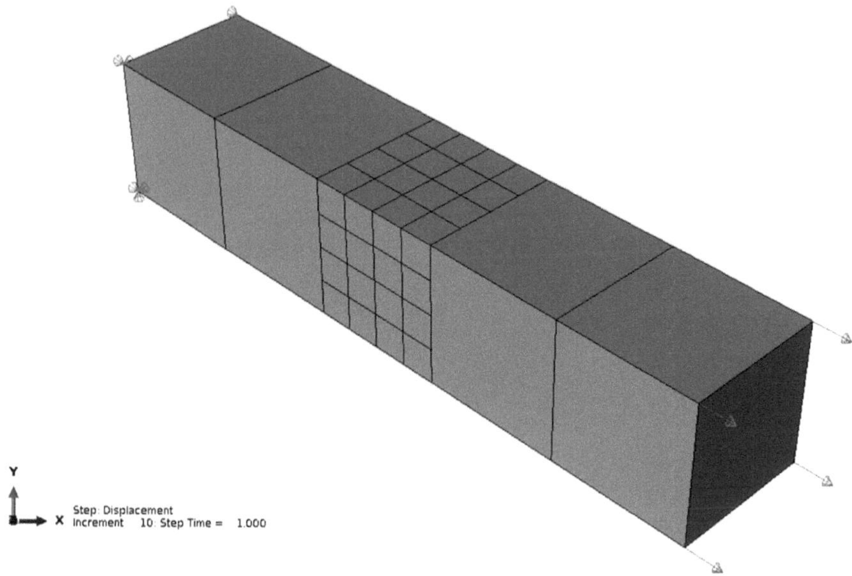

Fig. 11.2 Overlay plot of the global and the local FE-model

In order to verify the feasibility of the coupling procedure in ABAQUS^{TM}, no hotspot criterion has been implemented at the global level yet to determine critical areas. The global–local coupling procedure is performed in each global equilibrium iteration step of the solution process. The algorithm has the following form:

1. Initialize the global model
2. Begin next global load increment

   (a) Perform prediction for next load increment
   (b) Compute initial residuum

3. Begin next global iteration

   (a) Compute initial tangent stiffness
   (b) Perform correction: Compute displacement field from strains at global Gaussian points

4. Create local boundary value problem
5. Solve boundary value problem
6. Homogenize stresses and assign them to the respective global iteration points
7. Calculate the global tangent stiffness operator numerically by solving six local boundary value problems with a small perturbation applied successively to the six components of the global strain tensor.
8. Compute updated global residuum.

   (a) If residuum > eps: GO TO 3b.

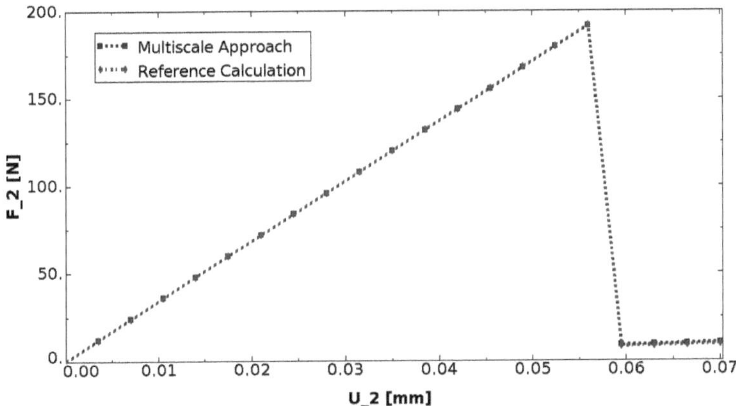

Fig. 11.3 Load-displacement curves of the multiscale approach and the reference calculation

(b) If residuum < eps: GO TO 2.
(c) If residuum < eps and final global load increment is reached: analysis complete.

Figure 11.3 shows the load–displacement behavior of the two models in the direction of the applied displacement. It is obvious that the results obtained from the two calculations are almost identical which verifies the implementation of the presented multiscale approach in ABAQUS. The deviations are less than 1% with respect to the reference calculation and can be explained by numerical inaccuracies resulting from small rounding errors during the calculation.

11.5 Conclusion and Outlook

A novel automated iterative loose coupling approach is presented. It is possible with this approach to describe the global behavior of a composite structure without omitting the effect of physical phenomena like process-induced manufacturing defects at a lower scale. The implementation in the commercial FE-code ABAQUS™ is verified for the coupling of a solid element at the global scale with several solid elements at the local scale. Only small deviations are observed comparing the reference calculation and the multiscale approach.

The implementation of the approach will be enhanced to geometric nonlinearity in future research. Furthermore, two-dimensional shell models and three-dimensional solid models shall be coupled to compute large structures like panels or barrels with manufacturing defects.

References

1. Garnich, M.R., Karami, G.: Finite element micromechanics for stiffness and strength of wavy fiber composites. J. Compos. Mater. **38**(4), 273–292 (2004)
2. Hsiao, H.M., Daniel, I.M.: Effect of fiber waviness on stiffness and strength reduction of unidirectional composites under compressive loading. Compos. Sci. Technol. **56**(5), 581–593 (1996)
3. Huang, H., Talreja, R.: Effects of void geometry on elastic properties of unidirectional fiber reinforced composites. Compos. Sci. Technol. **65**(13), 1964–1981 (2005)
4. Jeong, H.: Effects of voids on the mechanical strength and ultrasonic attenuation of laminated composites. J. Compos. Mater. **31**(3), 276–292 (1997)
5. Potter, K., Khan, B., Wisnom, M., Bell, T., Stevens, J.: Variability, fibre waviness and misalignment in the determination of the properties of composite materials and structures. Compos. Part A, **39**(9), 1343–1354 (2008)
6. Böhlke, T.: Crystallographic Texture Evolution and Elastic Anisotropy—Simulation, Modeling and Application. Shaker-Verlag, Germany (2001). ISBN 3-8265-8758-8
7. Lubarda, V.: Elastoplasticity Theory. CRC Press, Boca Raton, ISBN 978-0-8493-1138-3 (2002)
8. Nemat-Nasser, S.: Averaging theorems in finite deformation plasticity. Mech. Mat. **31**, 493–523 (1999)
9. Kreikemeier, J.: A Two Scale Finite Element Approach to Analyse the Damage State of Composite Structures. Forschungsbericht DLR 2011-05 (2011)
10. Suquet, P.: Elements of Homogenization for Inelastic Solid Mechanics. Homogenization Techniques for Composite Media, pp. 193–278 (1987)
11. Gitman, I.: Representative Volumes and Multi-Scale Modelling of Quasi-Brittle Materials. Technische Universität Delft (2006)
12. Kouznetsova, V.: Computational Homogenization for the Multi-Scale Analysis of Multiphase Materials. Technische Universität Eindhoven (2002)
13. Miehe, C., Schotte, J., Lambrecht, M.: Homogenization of inelastic solid materials at finite strains based in incremental minimization principles. J. Mech. Phys. Solids **50**, 2123–2167 (2002)
14. Schröder, J.: Theoretische und algorithmische Konzepte zur phänomenologischen Beschreibung anisotropen Materialverhaltens. Institut für Mechanik (Bauwesen), Universität Stuttgart (1996)
15. Fish, J.: The s-version of the finite element method. Comput. Struct. **43**, 539–547 (1992)
16. Hughes, T., Feijóo, G., Mazzei, L., Quincy, J.: The variational multiscale method—a paradigm for computational mechanics. Comp. Meth. Appl. Mech. Eng. **166**, 3–24 (1998)
17. Kadowaki, H., Liu, W.: Bridging multiscale method for localization problems. Comp. Meth. Appl. Mech. Eng. **193**, 3267–3302 (2004)
18. Feyel, F.: A multilevel finite element method (FE^2) to describe the response of highly non-linear structures using generalized continua. Comp. Meth. Appl. Mech. Eng. **192**, 3233–3244 (2003)
19. Feyel, F.: Multiscale FE^2 elastoviscoplastic analysis of composite structures. Comput. Mater. Sci. **16**, 344–354 (1999)
20. Matzenmiller, A., Köster, B.: Consistently linearized constitutive equations of micromechanical models for fibre composites with evolving damage. Int. J. Sol. Struct. **44**, 2244–2268 (2007)
21. Aboudi, J.: Mechanics of Composite Materials—A Unified Micromechanical Approach. Elsevier, Amsterdam, ISBN 0-4448-8452-1 (1991)
22. Kim, B., Lee, H.: Elastoplastic modelling of circular fiber-reinforced ductile matrix composites considering a finite rve. Int. J. Sol. Struct. **47**, 827–836 (2010)
23. Li, S., Sauer, R., Wang, G.: A circular inclusion in a finite domain I. The Dirichlet-Eshelby problem. Acta. Mechanica. **179**, 67–90 (2005)

24. Chrupalla, D., Berg, S., Kärger, L., Doreille, M., Ludwig, T., Jansen, E., Rolfes, R., Kling, A.: A homogenization-based two-way multiscale approach for composite materials. In: Rolfes, R., Jansen, E.L. (eds.) Proceedings of the 3rd ECCOMAS Thematic Conference on The Mechanical Response of Composite Materials, Institute of Structural Analysis, Leibniz University of Hannover (2011)
25. ABAQUS^{TM} Version 6.8 Documentation, ABAQUS 6.8.1 (2008)
26. Hartung, D.: Materrialverhalten von Faserverbundwerkstoffen unter dreidimensionalen Belastungen. Forschungsbericht DLR 2009-12 (2009)
27. Ladevèze, P., Allix, O., Deü, J.F., Lévêque, D.: A mesomodel for localisation and damage computation in laminates. Comp. Meth. Appl. Mech. Eng. **183**(1–2), 105–122 (2000)\

Chapter 12
Experimental Determination of Interlaminar Material Properties of Carbon Fiber Composites

Daniel Hartung and Martin Wiedemann

Abstract Non Crimp Fabric (NCF) provides a low-cost potential and competitive advantages for thick composite structures. In this chapter, a method will be presented to determine the interlaminar failure under combined through-thickness load conditions. Additionally, the in-plane failure behaviour of NCF composite is discussed and analysed. A new test setup, based on the idea of Arcan, determines the material properties. Test results of combined through-thickness loading are presented by in the form of a shear-compression failure curve. The tests are reproducible and reliable. The failure envelope is finally used to verify known failure criteria.

12.1 Introduction

Carbon fiber Non Crimp Fabric (NCF) provides advantages for industrial manufacturing i.e. automated preforming capabilities, efficient infusion technologies and thus reduced processing time. NCF materials are made by stitching together adjacent plies with a thin polyester yarn. The use of NCF composites for thick structures is growing; therefore through-thickness stresses cannot be neglected. Consequently, the failure and stress analysis requires the consideration of the three dimensional stress states. Failure of thick composite material is induced by interlaminar stresses between adjacent plies, which cause delaminations. Generally, the delamination strength in the thickness direction is small

D. Hartung
Institute of Composite Structures and Adaptive Systems, German Aerospace
Center (DLR e.V.), Lilienthalplatz 7, 38108, Braunschweig, Germany

M. Wiedemann (✉)
Institute of Composite Structures and Adaptive Systems, German Aerospace
Center (DLR e.V.), Lilienthalplatz 7, 38108, Braunschweig, Germany
e-mail: martin.wiedemann@dlr.de

M. Wiedemann and M. Sinapius (eds.), *Adaptive, Tolerant and Efficient Composite Structures*, Research Topics in Aerospace, DOI: 10.1007/978-3-642-29190-6_12,
© Springer-Verlag Berlin Heidelberg 2013

Table 12.1 Maximum strength of unidirectional single ply (fiber: Tenax UTS 5632 12 K, resin: RTM6)

Fiber strength tension	Fiber strength compression	Intralaminar strength tension	Intralaminar strength compression	Intralaminar strength shear
R_{11}^t	R_{11}^c	R_{22}^t	R_{22}^c	R_{12}
[MPa]	[MPa]	[MPa]	[MPa]	[MPa]
2,208	1,095	43	189	72
±84	±0.1	±1.0	±1.0	±5.2

Fig. 12.1 Section of the NCF with characteristic damages

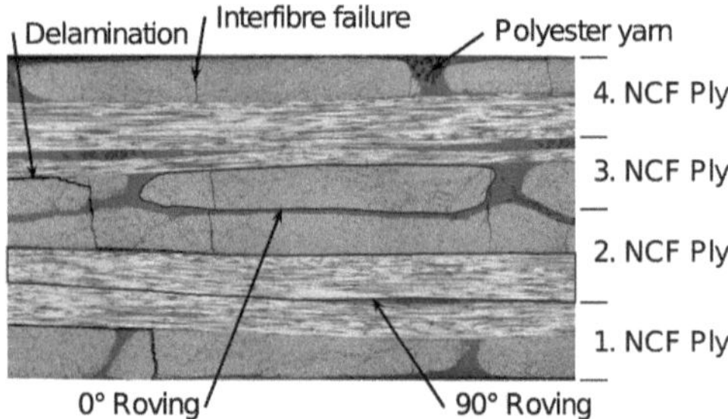

compared to the fiber strength and restricts the applicability of composites structures. To test composite materials in their thickness direction under combined load conditions requires an adequate test device.

12.2 Intra- and Inter-Laminar Failure Behaviour of NCF

It is often assumed that the failure behaviour of fiber composites can be separated into fiber and inter-fiber failure modes: intralaminar (in the laminate plane) and interlaminar (normal to the laminate plane). The allowable material strength along the fiber is significantly higher than the inter-fiber strength. Table 12.1 shows the difference between fiber strength and the intralaminar strength for a carbon Ultra High Tension Strength (UTS) unidirectional single ply.

Due to the multi-axial fiber orientation of NCFs, the failure behaviour of the laminate is in most cases driven by inter-fiber failure modes. Several authors investigated the damage growth of composites under quasi static load conditions. Many of them concluded that micro cracks accumulate continuously during loading and drive the final failure behaviour of composites.

The micrograph image analysis supports this explanation. Figure 12.1 shows different inter-fiber failures, i.e. from micro cracks up to delaminations of a composite that consists of four NCF plies, each with two single plies in an orthogonal orientation. The specimen has been loaded along the 90° direction.

Figure 12.2 shows micrograph images at four strain levels: 0.3, 0.6, 0.75 and 1.5%. No damages can be found at a strain level of 0.3%. At a strain level of 0.6%, first inter-fiber damages in the form of micro cracks inside of a roving can be found.

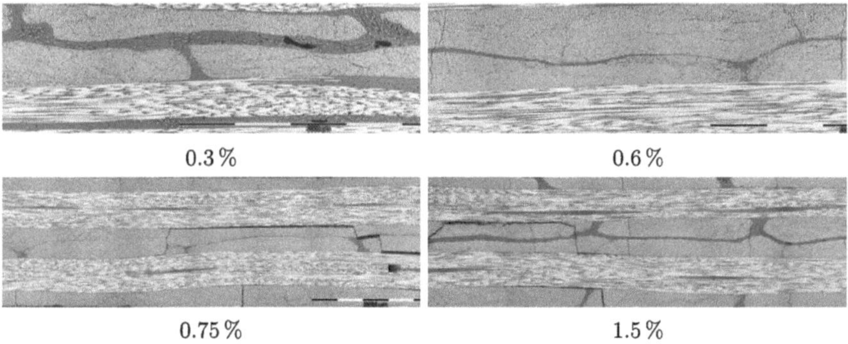

0.3 %   0.6 %

0.75 %   1.5 %

Fig. 12.2 Micrograph images at different strain levels

Fig. 12.3 Stress strain curve of NCF under quasi static load

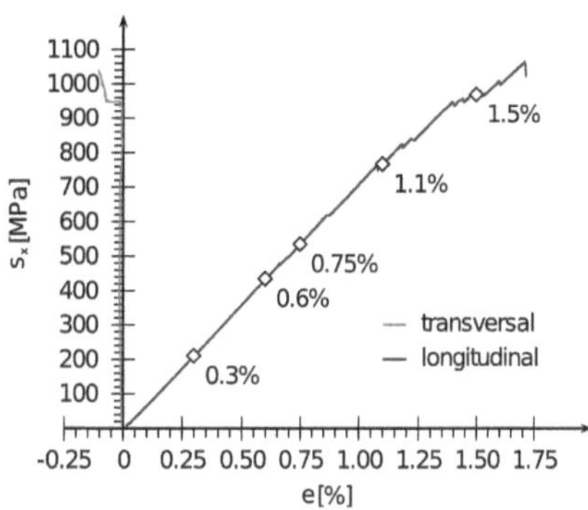

At higher strain levels, these micro cracks starts to accumulate. At the strain level of 0.75%, first delamination damages are visible. In principle, half the ultimate strain level is sufficient to produce all damage phenomena, which finally results in the composite failure. At the strain level of 1.5% a high damage density can be observed.

It is important to understand that micro cracks often are delamination onsets and enable a delamination growth through the plies. Furthermore they can bridge delaminations across single plies. In general, it is not possible to test a single intralaminar failure mode separately.

The corresponding stress strain curve in Fig. 12.3 shows nearly linear material behaviour. The ultimate material strength is identified by the load drop-off.

12.3 Interlaminar Test Methods

The complex failure behaviour of composites motivated many authors to develop different interlaminar test methods. Several test concepts, specimen designs and test-setups exist, but only a few of them produce test results for combined through-thickness loading conditions and none of them with acceptable accuracy and reproducibility.

Fig. 12.4 Specimen
geometry used by Arcan [6]

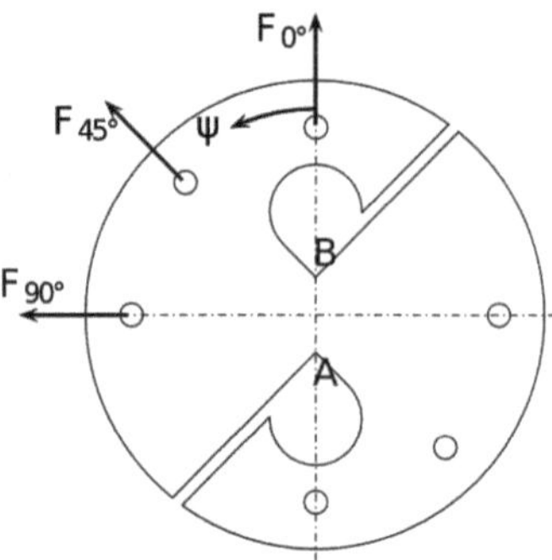

Some test methods determine appropriate single interlaminar properties. The Iosipescu test used by Adams et al. [1] determines representative interlaminar shear properties of composites. In principle, this test can also be used to test combined load conditions. Balakrishnan et al. [2] and Bansal et al. [3] determined combined interlaminar shear and compressive properties through a modified bi-axial Iosipescu test. The double notched and rail shear tests are further possibilities to determine interlaminar shear properties, but with these tests no combined load conditions can be applied. Furthermore, Uenal et al. [4] and Hussain et al. [5] noticed that bending stresses are superposed on the shear stress state, which causes inaccurate interlaminar properties. Another popular method is to bend flat or curved specimens. Avva et al. [6] and Greszczuk et al. [7] developed curved specimens and determined the interlaminar properties. In general, bending causes normal and shear stress components leading to relative complex stress states. The determination of one single strength property is often difficult or not possible.

A method to analyse combined tensile and shear loads was originally published by Arcan et al. [8]. Arcan has developed the specimen shown in Fig. 12.4 with two asymmetric cut-outs. The specimen can be installed at different angles into a uni-axial standard test device. Depending on the direction, single and combined shear or tensile stress states are generated within a defined cross section (points A and B). The sophisticated specimen geometry is a disadvantage, as it is generally difficult to produce.

El-Hajjar et al. [9] published a modified concept with a smaller specimen geometry, which is clamped to a steel fixture. Both test setups determine intralaminar in-plane material properties and focus on determining properties for shear loading.

12.4 A Test Setup for Interlaminar Properties Under Combined Loading

The concept acc. Fig. 12.5 has been developed to test combined and single interlaminar properties. The test device allows measurement of properties for single shear, tension and compression loads, as well as for combined shear loads.

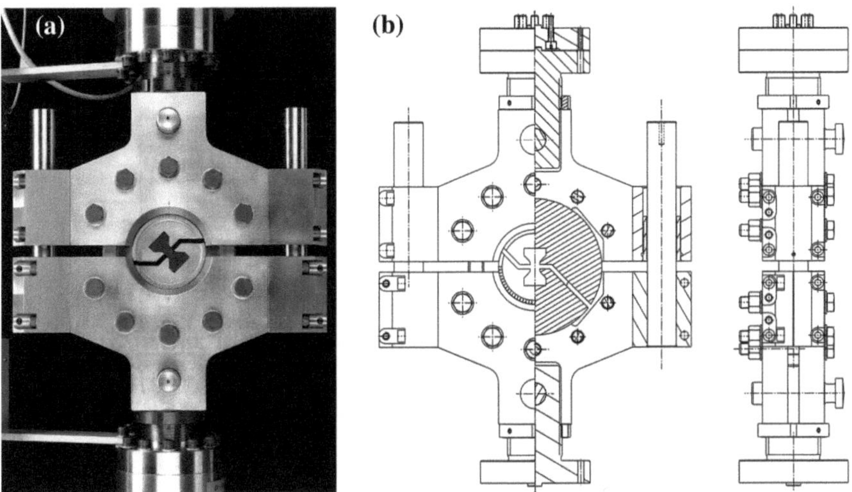

Fig. 12.5 Bi-axial test device to determine interlaminar properties

The concept differs from other Arcan test concepts by the pair of lubricated linear ball bearing systems on both sides of the test device. They provide a defined movement of the test rig and therewith a precise deformation of the specimen. Different types of specimen geometries can be tested by changing a rotatable insert inside the test rig. This insert can be used in different orientations between 0 and 90° angles. The specimen is tested without any clamping or adhesive bonding. This enables high test rates, but requires accurate production of specimen geometries.

12.4.1 Material Preparation and Specimen Production

The specimens are produced from a flat NCF plate with a moderate thickness of approximately 30 mm. The specimen plate used for the test is made of 60 NCF plies in a symmetrical layup. Each NCF ply consists of two orthogonal single unidirectional plies. The specimen plate is produced with the resin infusion technology through an autoclave process.

A representative test requires homogeneous stress distributions within the test sections of the specimens. It was found that tensile and shear loads cannot be tested with identical geometries. Therefore, two types of specimen geometries are designed. The tension specimen in Fig. 12.6 was primarily developed to determine interlaminar properties under tensile loads. The tensile specimen is manufactured with additional E-glass/epoxy composite tabs on the upper and the lower side of the specimen. FE simulations revealed that the best homogenous tensile stress distribution is generated with a specimen radius of 12 mm.

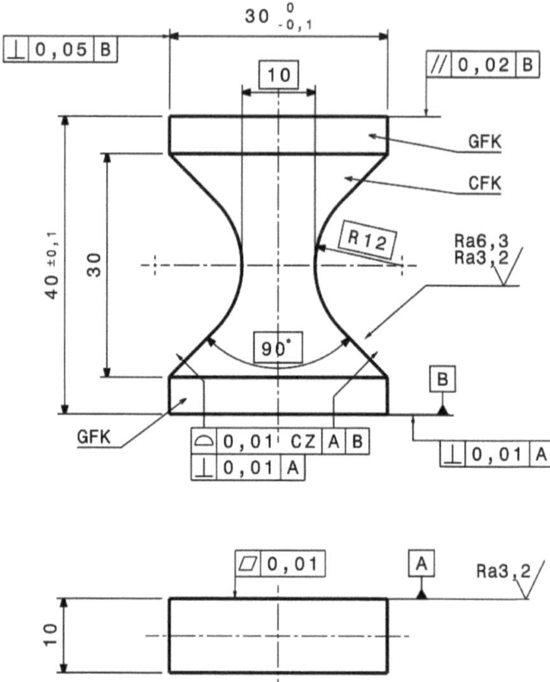

Fig. 12.6 Interlaminar tensile specimen

The specimen in Fig. 12.7 is used to determine interlaminar properties under shear loads as well as for combined loading. It was found that a notch radius of 1 mm provides the best compromise between a narrow notch to generate homogeneous shear stress distributions and a representative part of the cross-ply NCF for the shear specimen.

The tested cross section of both specimens has a width of 10 mm and a length of 10 mm. The specimens are tested without clamping or bonding. Therefore the test load introduction into the specimen is critical. The rectangular contour of the shear specimen provides a reliable load transfer. The load is introduced into the tension specimen via wedges on the upper and the lower side of the specimen.

12.4.2 Testing and Analysis of the Results

It is assumed that the applied test force F_p is always a superposition of a normal

$$F_{p,N} = F_p \cos \psi \qquad (12.1)$$

and a transversal

$$F_{p,T} = F_p \sin \psi \qquad (12.2)$$

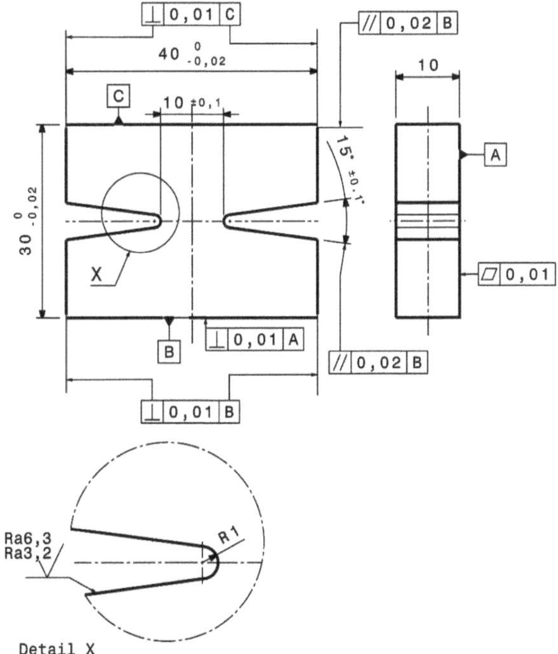

Fig. 12.7 Interlaminar shear specimen

component. It can therefore be split into two terms according to the orientation angle ψ of the specimen in relation to the machine coordinate system of the test device. The load ratio

$$f_{biax} = \frac{F_{p,N}}{F_{p,T}} \tag{12.3}$$

determines the ratio between longitudinal and transversal loading. Table 12.2 uses this load ratio to determine the failure load under combined loading.

Furthermore, it is assumed that the specimens will fail at their narrowest cross section. Accordingly, this section determines the material strength properties.

Figure 12.8 shows the stress strain curves of the interlaminar shear and interlaminar tensile tests. The results for the various similar test specimens show that the test determines consistent and reliable material properties.

The specimen failure and the maximum interlaminar strength are clearly identified by the load drop of the stress strain curve. The determination of the material strength under compression is more difficult, because the measured test result could not be conclusively assigned to the compression strength. Figure 12.9 shows two test results for combined compression-shear loads. The 45° loading angle corresponds to a balance of shear and compression. It shows a load drop at approximately 13 kN test force. At a load ratio corresponding to a 60° angle in the test device, i.e. leading to a slightly higher compression, no load drop is observed anymore. Although compressive loads cause micro-cracks, the material can be

Table 12.2 Interlaminar strength under combined and single loads

Specimen	Load ratio (%) f_{biax}	Tension/compression max. strength [MPa]	Shear max. strength [MPa]
Tension	0	41,123.83	0
Shear	0	0	47,725.51
Shear	15	−15,861.01	59,203.73
Shear	30	−42,464.39	73,544.39
Shear	45	−91,628.24	92,628.24
Shear	60	−137,6679.48	11,366.56

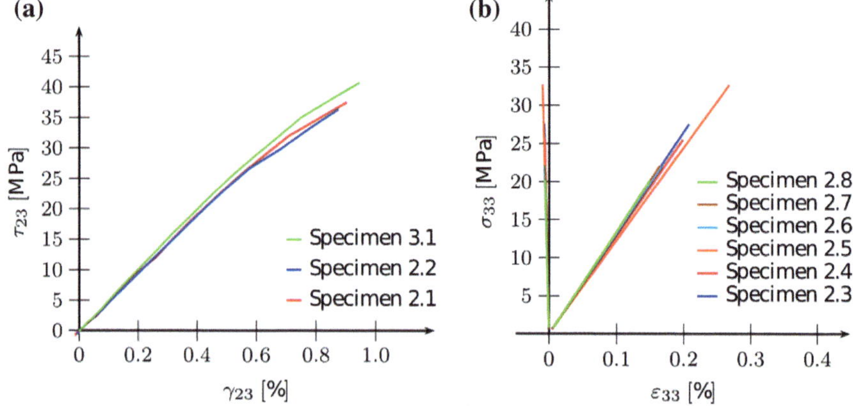

Fig. 12.8 Load–displacement curve for the shear and tensile test. **a** Interlaminar shear load. **b** Interlaminar tension load

continuously loaded. Under high compression load the specimen does not fail by separation, i.e. breaking into two parts, as observed under tensile or shear loads. Therefore, combined shear-compression properties at higher compression or pure compression strength properties could not be determined by the analysis of the load displacement curve. But combined shear-compression strength can be determined up to the limit of a balanced compression-shear load ratio corresponding to a 45° angle on the test device.

To determine compression strength at higher compression-shear load ratios, it is assumed that the strength can be specified at the first discontinuity of the load displacement curve.

12.4.3 Results for Combined Interlaminar Loads

Figure 12.10 shows test results for tensile as well as for combined shear and compression loads. It was found that the overall strength increases with a growing compression to shear ratio up to a certain limit. It was observed that the scatter of

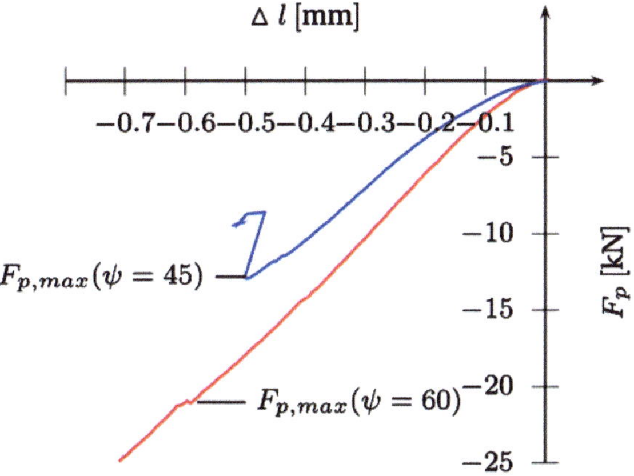

Fig. 12.9 Combined compression-shear load-displacement curve

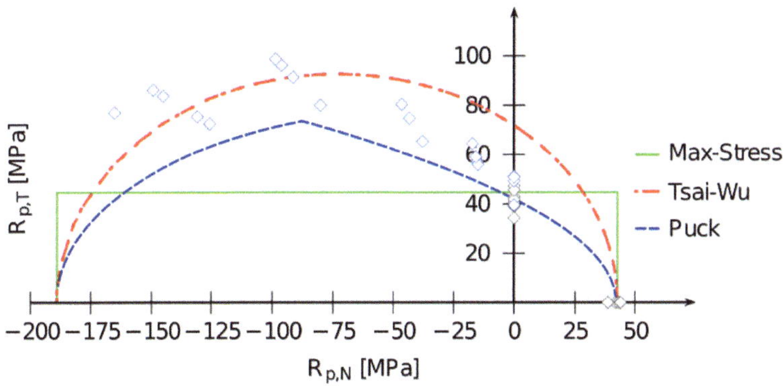

Fig. 12.10 Interlaminar material strength and failure criteria envelops

test results becomes larger for increasing load ratios. The interlaminar tensile strength is nearly equal to the intralaminar strength in Table 12.1. This is in good agreement with the assumption that a thick composite material provides the same intralaminar and interlaminar tensile strength transversal to the fiber direction.

The measured strength values are summarised in Table 12.2. The interlaminar tension strength has been determined with tension specimens, the combined interlaminar shear and compression strength with shear specimens. The biaxial load ratio f_{biax} defines a relation of the compression or tensile strength normal and transversal to the material orientation according to equation [6].

Figure 12.10 also shows the interlaminar strength prediction according to the maximum stress criterion, to the polynomial criterion of Tsai and Wu [10] and to the fracture hypothesis of Puck. According to Puck and Tsai-Wu the failure envelopes increase under combined shear and compression loads. The maximum stress criterion provides a rough approach with a constant behaviour under combined interlaminar loads. Especially the hypothesis of Puck provides an adequate approximation of the failure envelope. Although the increase in strength under increasing compression-shear ratio is also predicted by Tsai and Wu, their criterion uses the intralaminar shear strength to also predict the interlaminar shear strength and overestimates the maximum strength under interlaminar shear loads.

It shall be noticed that all failure approaches are based on linear elastic material behaviour, whereas the measurements reveal a nonlinear material behaviour, even before the material strength limit is reached.

12.5 Conclusion

Research so far has been mainly focused to the composite failure mechanism due to intralaminar loading, whereas the interlaminar failure behaviour under combined loads in the thickness direction has been treated relatively seldom in the literature. Micrograph images of NCF laminates tested under in-plane loads show damages in different interacting inter-fiber modes. Micro cracks, which are an intralaminar failure mode, interact with interlaminar damage modes. A new test setup enables the determination of interlaminar properties under combined load conditions. The advantage of the presented test method is the determination of interlaminar strength properties of composite under well defined combined load cases. The test results show an increasing strength towards increasing compression-shear load ratios. This behaviour corresponds to the failure envelopes of the hypothesis from Puck and Tsai-Wu, but the onset of damage starts at lower load levels. Although the known failure hypothesis corresponds to the measurements, the nonlinear damaging behaviour is not covered.

References

1. Adams, D.F., Lewis, E.Q.: Experimental strain analysis of the Iosipescu shear test specimen. Exp. Mech. **6**, 352–360 (1995)
2. Balakrishnan, M.V., Bansal, B., Kumosa, M.: Biaxial testing of unidirectional carbon-epoxy composite using biaxial Iosipescu test fixture. J. Compos. Mater. **31**, 486–508 (1997)
3. Bansal, A., Kumosa, V.: Application of the biaxial Iosipescu method to mixed-mode fracture of unidirectional composites. Int. J. Fract. **71**, 131–150 (1995)
4. Ünal, O., Bansal, N.P.: In-plane and interlaminar shear strength of a unidirectional hi-nicalon fiber-reinforced celsian matrix composite. Nasa Technical Report NASA/TM 2000-210608, National Aeronautics and Space Administration, Glenn Research Center, NASA Center for Aerospace Information 7121 Standard Drive Hanover, MD 21076, Dezember 2000.
5. Hussain, A.K., Adams, D.F.: Analytical evaluation of the two-rail shear test method for composite materials. Compos. Sci. Tech. **64**(2), 221–238 (2004)
6. Avva, H.S., Allen, H.G., Shivakumar, K.N.: Through the thickness tension strength of 3-d braided composites. J. Compos. Mater. **30**(1), 51–69 (1996)

7. Hartung, D.: Materialverhalten von Faserverbundwerkstoffen unter dreidimensionalen Belastungen. DLR Forschungsbericht 2009–2012, Fakultät für Maschinenbau, Technische Universität Braunschweig (2009)
8. Arcan, M., Hashin, Z., Voloshin, A.: A method to produce uniform plane-stress states with applications to fibre-reinforced materials. Exp. Mech. **18**, 141–146 (1977)
9. Gning, P.B., Delsart, D., Mortier, J.M., Coutellier, D.: Through-thickness strength measurements using Aran's method. Compos. B **41**, 308–316 (2010)
10. Tsai, S.W.: Theory of composites design. ISBN 0-9618090-3-5. Think composites, think composites, Dayton, USA, 1 edn, 1992

Chapter 13
Impact and Residual Strength Assessment Methodologies

Luise Kärger, Jens Baaran and Anja Wetzel

Abstract In this chapter, efficient methodologies to evaluate impact resistance and damage tolerance of composite structures are introduced. Internal non-visible or barely visible impact damage (NVID, BVID) can provoke a significant strength and stability reduction in monolithic composite structures as well as in composite sandwich structures. Therefore, methodologies have been developed to reliably simulate the dynamic response and to predict the impact damage size that develops during low-velocity impact (LVI) events. Additionally, methods for the prediction of the compression-after-impact (CAI) strength are presented. Special attention is given to the impact assessment methodologies, which have been implemented in the DLR in-house tool CODAC. Simulation results of CODAC are presented and compared to experimental results.

13.1 Failure Analysis with Damage Initiation and Degradation

Adequately modelling the failure behaviour of composite materials is essential for an effective impact assessment. Firstly, the amount of damage indicates the residual strength of an impacted structure and is, therefore, the most important result of an impact analysis. Secondly, failure progression during an impact event can considerably influence the impact behaviour of the specimen. Consequently, material degradation needs to be taken into account. A comprehensive review of damage and failure models is given in Sect. 10.5. Specific failure assessment methodologies on a

L. Kärger (✉) · J. Baaran · A. Wetzel
Institute of Composite Structures and Adaptive Systems, German Aerospace
Center (DLR e.V.), Lilienthalplatz 7, 38108, Braunschweig, Germany
e-mail: alexander.kling@dlr.de

M. Wiedemann and M. Sinapius (eds.), *Adaptive, Tolerant and Efficient Composite Structures*, Research Topics in Aerospace, DOI: 10.1007/978-3-642-29190-6_13,

local scale are presented in Chaps. 11 and 12. The present section focuses on damage models, which are suitable for impact and CAI analysis.

13.1.1 Damage Models for Monolithic Composites

The three failure modes that typically occur during impact are fiber breakage, matrix cracking and delamination. They greatly differ in their behaviour and need to be treated separately in simulation. Thus, mode specific stress based failure criteria are usually applied. Damage evolution is accounted for by reducing specific stiffness components to specific values, depending on the failure mode.

An insight into the diversity of available composite failure criteria is given in Sect. 10.5.1. Many such criteria are available for application to static problems. However, their application for impact analysis is contentious. For instance Hashin [1] as well as Puck and Schürmann [2] recommend not to use their physically-based failure criteria for impact analysis. On the other hand, Choi and Chang [3] as well as Chai [4] developed delamination criteria particularly for the modelling of low-velocity impacts. Regarding fiber breakage, the maximum stress criterion is sufficient in most practical cases. However, criteria for impact induced matrix failure are difficult to validate due to a lack of experimental results.

An overview of modelling strategies of degradation and damage progression can be found in Sect. 10.5.2. The softening effect due to failure can be critical for both, impact and CAI. The reduction of the stiffness Q_{ij} of the damaged region via degradation factors D_{ij} as in

$$\begin{bmatrix} \sigma_{xx} \\ \sigma_{yy} \\ \tau_{xy} \end{bmatrix} = \begin{bmatrix} D_{11}Q_{11} & D_{12}Q_{12} & \\ D_{12}Q_{12} & D_{22}Q_{22} & \\ & & D_{66}Q_{66} \end{bmatrix} \begin{bmatrix} \varepsilon_{xx} \\ \varepsilon_{yy} \\ \gamma_{xy} \end{bmatrix}, \qquad (13.1)$$

is a straight forward damage mechanics approach. The reduction depends on the combination of damage modes, cf. Kärger et al. [5]. Fiber failure is important for any type of loading and has a strong influence on in-plane stiffness and strength. Delamination strongly affects the CAI strength, cf Sect. 13.3.1, but does not have much influence on the impact behaviour. Matrix cracks have little direct influence on the stiffness, but can extend to interfaces between layers and, thus, lead to delaminations or even fiber cracks.

13.1.2 Core and Skin Damage in Sandwich Structures

Composite sandwich structures consist of two thin, stiff composite skins and an intermediate lightweight core. Nevertheless, impact damage in sandwich structures (Fig. 13.1) can provoke a significant strength and stability reduction.

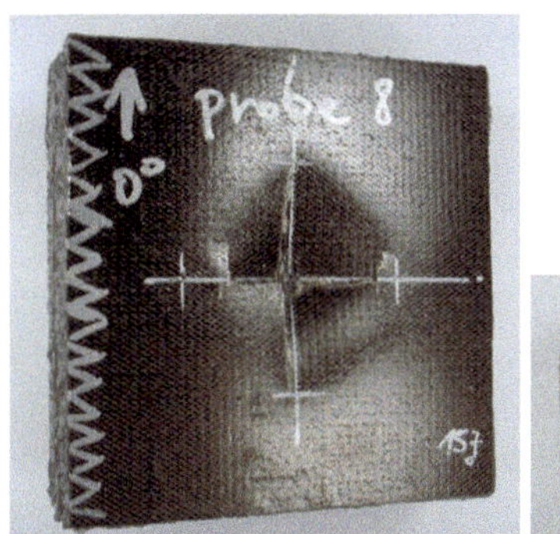

Fig. 13.1 Impact damage in a sandwich with composite skin and folded core

While skin damage, such as large cracks and indentations, may be visible, the amount of core damage can only be detected by expensive Non Destructive Testing (NDT) methods. The core fails due to combined shear and compression. Consequently, a failure criterion, which includes both, transverse normal and transverse shear stresses, such as Besant et al. [6], is recommended. To model the damage progression of the core, the stress–strain-path of the core material was investigated exemplarily [7]. After reaching ultimate strength, the core crushes at a certain stress level. The remaining crush strength plus a residual crush stiffness are taken into account by a step-wise linear failure function [8].

Damage models of monolithic laminates (Sect. 13.1.1) are available for the prediction of skin failure. Impacted, thin sandwich skins are dominated by membrane stresses. Since fiber breakage is the dominant failure mode, the maximum stress criterion is sufficient to predict impact failure. To describe skin failure progression during impact, a macro-mechanical, step-wise linear function was proposed by Kärger et al. [9]. For an efficient CAI analysis, Wetzel [10] considered initial damage (indentation, stiffness degradation) and core damage progression, but neglected skin damage progression.

13.2 Impact Analysis

For low-velocity impacts, the amount of damage can be analysed using iterative static analysis combined with the energy theorem of the contact problem [11]. Alternatively, a dynamic analysis can be chosen. The advantage of the latter is that the numerically obtained transient impact response can be validated by comparing it to respective experimental results. The impactor is usually coupled to the composite plate using the Hertzian contact law.

The DLR in-house tool CODAC, which aims for an efficient impact analysis, applies the implicit Newmark time integration scheme in order to reproduce the transient impact event. Further information on the impact analysis in CODAC can be found in [8, 9, 12].

13.2.1 Impact on a Monolithic Composite Panel

In the past, a great amount of research has be done to enable the prediction of impact damage in composite structures. Abrate [11] gives a good overview of impact studies on composite materials. The effects of local damage on the impact behaviour and on the load bearing capacity have been studied by several authors with differing focus. A short literature review can be found in [5].

Impacted composite panels are usually modelled with shell elements based on the first order shear deformation theory. However, transverse stiffness and stresses are important for the damage progression during impact. Thus, the Extended 2D Method by Rolfes and Rohwer [13] has been implemented in CODAC to improve the transverse properties.

Impact simulation results in the shape of damage areas are illustrated in Fig. 13.2. To model the complex damage state that develops during impact, a damage mechanics methodology according to Sect. 13.1.1 is applied in CODAC. For each Gauss point and each ply, the damage states are stored in a database and updated in real-time during the analysis [12]. In the presented example [14] a 5 mm thick composite plate was impacted with 30 J. Two different failure criteria have been applied and compared to test results. Due to a lack of experimental data, unfortunately, fiber and matrix cracking often cannot be properly assessed. In terms of delamination areas, the Chai failure criterion [4] provided better results than the Choi/Chang criterion [3].

In addition to the computed damage state, the force–time history is another important result of an impact analysis. Through maximum contact force and sudden curve drops or kinks, it provides information on the overall impact behaviour and on the damage progression. For the above presented impact test program, unfortunately, force–time histories were not available. A further impact test program is discussed in [12], where force–time histories are compared to experimental results, cf. Fig. 13.3.

13.2.2 Impact on a Composite Sandwich Panel

Since impact damage in sandwich structures can provoke a significant reduction in strength and stability, the number of publications has greatly increased in the last 10 years. Numerous experimental studies were conducted to observe the impact damage progression. Several methodologies have been developed to predict the impact response and the amount of impact damage. While simple analytical approaches require less computational effort, FE-based approaches generally attain a more accurate description of the impact response. Literature reviews can be found in [8, 9].

An important requirement for suitably simulating impact damage is the accurate approximation of in-plane and transverse stresses. For that reason, two new three-layered shell elements have been developed and implemented into CODAC.

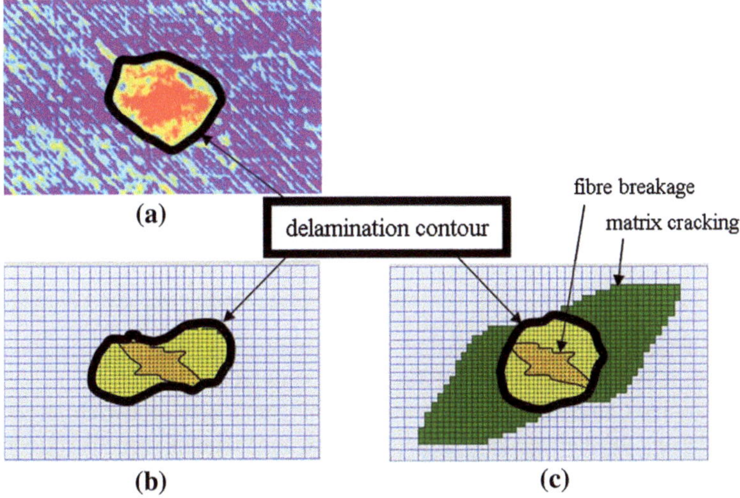

Fig. 13.2 Test result versus damage areas of two damage models [14]. **a** Test result. **b** Choi/Chang with degradation. **c** Chai with degradation

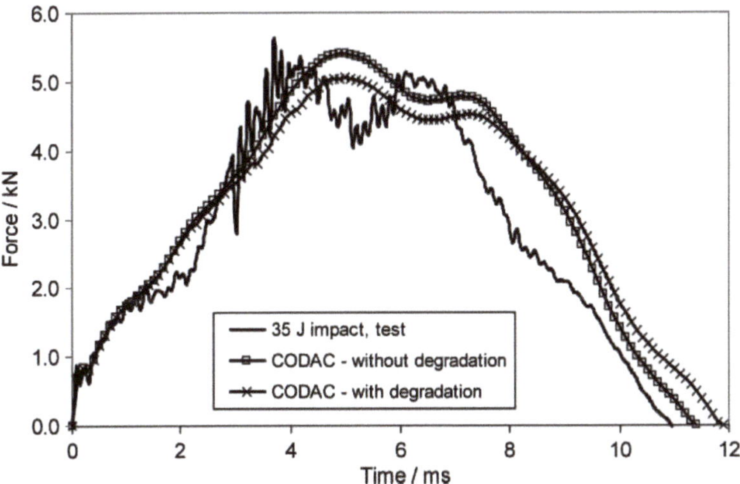

Fig. 13.3 Force–time history for a mid-bay impact on a two-stringer panel [12]

The first element, proposed by Kärger et al. [15], is based on a plane stress assumption and applies improved transverse stiffness and stresses by means of an enhanced Extended 2D Method. The second sandwich element, proposed by Wetzel et al. [16], accounts for the full 3D stress state and, thus, provides a higher accuracy for concentrated out-of-plane loads. Moreover, the strain–displacement relation is enhanced by nonlinear terms to take into account the membrane effect

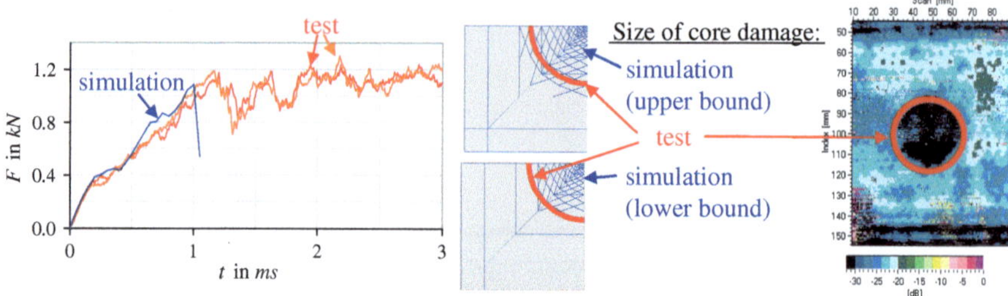

Fig. 13.4 Simulation results: Force-time history and core damage compared with test results

of large deflections in the impacted skin. Although it is computationally more expensive, it is recommended for the impact simulation in CODAC.

The complex impact damage behaviour of composite sandwich structures is modelled using the damage mechanics approach described in Sect. 13.1.2. Similar to the impact analysis of monolithic structures, the damage states of each Gauss point and each ply are stored in a database and updated in real-time during the analysis.

Impact simulation results, such as force–time histories and core damage areas, are illustrated in Fig. 13.4. In this study [9], a sandwich plate, with a 28 mm thick honeycomb core and a 0.63 mm thin CFRP skin, was impacted by energies between 1 and 15 J. Up to the onset of skin tearing at contact forces larger than 1 kN, it could be experimentally validated that force–time histories as well as core damage areas are well predicted by simulation. For higher energies, the core damage is properly limited by a lower and an upper bound. Furthermore, the presence of clearly visible skin damage is properly predicted. The lower bound represents the core damage at the moment when the skin starts to tear. The upper bound of the core damage is computed assuming an intact skin during the whole impact simulation.

13.3 Residual Strength Analysis

NVID/BVID can have a substantial effect on the load bearing capability of the composite laminate. This is particularly pronounced when delaminations locally split a monolithic laminate into several sublaminates, or if the core material of a sandwich structure loses its out-of-plane stiffness. Such damage not only reduces bending stiffness and strength, but also causes local instability. Compressive loading may cause these local instability effects, requiring an assessment of the residual, compression-after-impact (CAI) strength.

13.3.1 CAI Analysis for Monolithic Composites

Impact induced delaminations split a monolithic laminate into several sublaminates and, thus, cause a considerable local stiffness reduction. During postbuckling of the

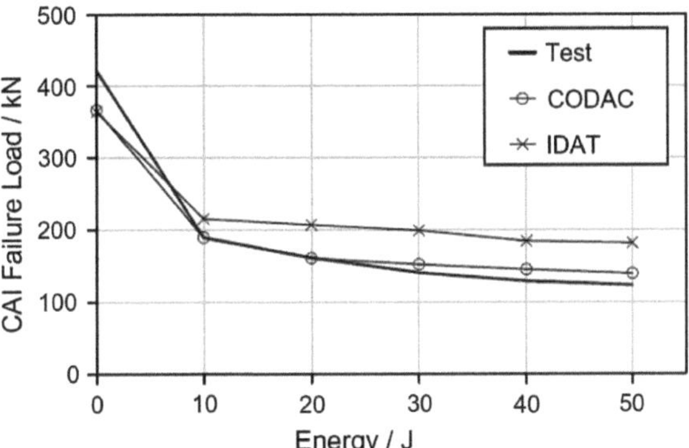

Fig. 13.5 Reduction of residual strength for increasing impact energy. Present methodology (CODAC) and alternative approach (IDAT) compared to test results [14]

sublaminates, the delamination region will behave like a soft inclusion. This leads to stress concentrations at the crack tips of the delamination. These stress concentrations may cause further delamination and damage growth and, in the worst case, lead to failure of a complete structure before the compressive strength of the undamaged laminate is reached.

Assuming a composite laminate of negligible curvature under plane stress, the stiffness loss of the delaminated region can be modeled using a Ritz approach to analyze sublaminate buckling [17]. This Ritz approach can be repeatedly applied to damaged regions with multiple stacked delaminations. It accounts for the fact that buckling of inner sublaminates is prevented by yet unbuckled adjacent sublaminates. Subsequently, stiffness reduction factors can be derived from these buckling loads [18]. Additional stiffness reduction caused by matrix cracking and fiber breakage should also be considered (Sect. 13.1.1), since the presence of these damage modes can further reduce the CAI strength.

A standard FE evaluation of the composite structure that accounts for reduction of stiffness can predict the stress concentration. A point-stress or average-stress fracture criterion can be used to predict damage growth. Tang et al. [18] proposed a semi-empirical point-stress fracture criterion, the so-called Damage Influence Criterion (DIC), for application to NVID/BVID. The DIC assumes failure in the cross section through the center of the damage, perpendicular to the uniaxial loading of the laminate.

Figure 13.5 shows a typical development of the uniaxial residual compressive strength for 4.5 mm thick specimens that were impacted with increasing impact energies. In this case, an impact energy of 10 J reduces the compressive strength already by about 50%. For higher energies, the residual strength is further reduced, but to a much smaller extent.

The present simulation approach, which applies the DIC (CODAC), reproduces this behavior quite well. The CODAC approach is compared to another approach (IDAT), where the CAI analysis is performed by an implicit, non-linear static FE solution using ABAQUS/Standard. Further details about the CAI analysis and validation results can be found in [5, 14].

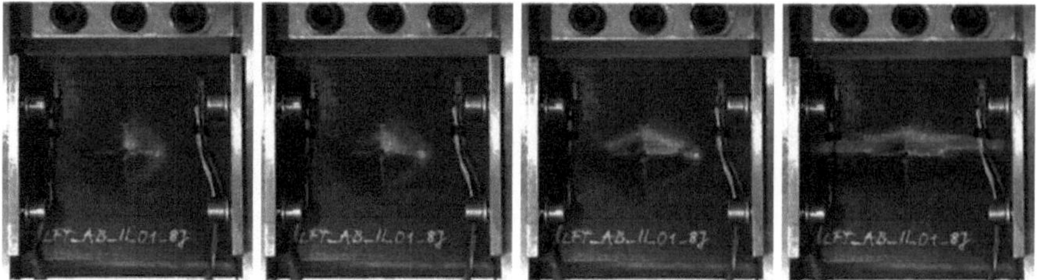

Fig. 13.6 Growth of impact indentation for compressive loading in vertical direction (increasing from left to right) [10]. *Source* ILR, TU Dresden

13.3.2 CAI Analysis for Composite Sandwich Structures

Low-velocity impact on composite sandwich structures typically causes a barely visible permanent indentation accompanied by invisible core damage below the indented face sheet. Compressive failure of impact damaged composite sandwich structures is often initiated by the growth of the initial permanent indentation in the direction perpendicular to the applied loading, cf. Fig. 13.6. This is accompanied by further core damage [19].

For the analysis of unsymmetrical honeycomb sandwich structures, a 3D FE model was used. The relatively thick core was modelled with volume elements and the thin face sheets with composite shell elements. In order to obtain accurate results, a careful evaluation of transverse stiffness and strength properties is essential. According to experiments (cf. Sect. 13.1.2), the post-failure reduction of transverse normal and transverse shear stiffness of the sandwich core was modelled using degradation factors of 0.2 and 0.01, respectively, cf. Eq. (13.1). This simple approach has been found to model the crushing behaviour of the core in an acceptable manner [10, 20]. The example in Fig. 13.7 shows that strain gauge measurements outside of the damage area (strain gauges 10 and 11) correspond well with the simulation results. Also, inside the damage area (strain gauge 13), the strains are predicted quite accurately up to a compressive load of about 40 kN. Close to the border of the damage the stiffness reduction is less severe than in the centre of the damage area, where a larger amount of fiber and matrix fracture is expected. This might be the reason, why the simulation overestimates compressive strains at strain gauge 12, which is located on the border of the face sheet and the core damage resulting from the 4 J impact.

In order to improve computational efficiency, a non-linear semi-analytical Ritz method has been developed, which uses a 1-parameter stiffness model of the Winkler type for the core material. Instead of the core degradation model used for the 3D approach, a simplified model by Olsson [21] is applied here.

Unless face sheet damage substantially influences the compressive failure of the sandwich structure, the semi-analytical Ritz approach provides a very efficient

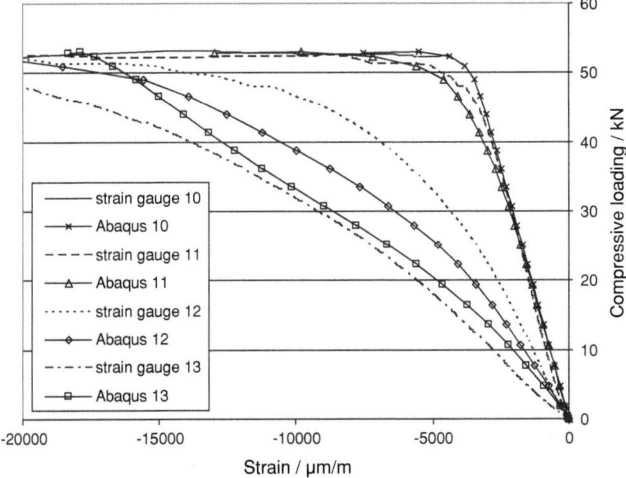

Fig. 13.7 Comparison between simulation results and test results: strains in loading direction in the vicinity of the 4 J impact versus compressive loading of the panel. Strain gauges are located abreast the impact damage with a distance between 7.5 mm (strain gauge 13) and 31.5 mm (strain gauge 10) from the damage centre [20]

means for the assessment of the CAI strength of sandwich structures. While accounting for face sheet damage is possible in principle, the Ritz approach loses its advantage in computational efficiency over a FE approach [10, 20].

References

1. Hashin, Z.: Failure criteria for unidirectional fibre composites. J. Appl. Mech. **47**, 329–334 (1980)
2. Puck, A., Schürmann, H.: Failure analysis of FRP laminates by means of physically based phenomenological models. Compos. Sci. Technol. **58**, 1045–1067 (1998)
3. Choi, H.Y., Chang, F.K.: A model for predicting damage in graphite/epoxy laminated composites resulting from low-velocity impact. J. Compos. Mater. **14**, 2134–2168 (1992)
4. Chai, Y.: A finite element method for predicting impact damage in composite stiffened-panels. Master degree dissertation, North-Western Polytechnical University (NPU) (1996)
5. Kärger, L., Baaran, J., Gunnion, A., Thomson, R.: Evaluation of impact assessment methodologies. Part I: applied methods. Compos. B **40**, 65–70 (2009)
6. Besant, T., Davies, G.A.O., Hitchings, D.: Finite element modelling of low velocity impact of composite sandwich panels. Compos. A **32**, 1189–1196 (2001)
7. Kintscher, M., Kärger, L., Wetzel, A., Hartung, D.: Stiffness and failure behaviour of folded sandwich cores under combined transverse shear and compression. Compos. A **38**, 1288–1295 (2007)
8. Kärger, L.: Effiziente Simulation von Schlagschädigungen in Faserverbund-Sandwichstrukturen. Dissertation, TU Braunschweig (2007)
9. Kärger, L., Baaran, J., Teßmer, J.: Efficient simulation of low-velocity impacts on composite sandwich panels. Comput. Struct. **86**, 988–996 (2008)

10. Wetzel, A.: Effiziente Ermittlung der Restfestigkeit schlaggeschädigter Sandwichstrukturen aus Faserverbundwerkstoff. Dissertation, TU Braunschweig (2008)
11. Abrate, S.: Impact on Composite Structures. Cambridge University Press, Cambridge (1998)
12. Baaran, J., Kärger, L., Wetzel, A.: Efficient prediction of composite damage resistance and damage tolerance. J. Aerosp. Eng. **222**(1), 179–188 (2008)
13. Rolfes, R., Rohwer, K.: Improved transverse shear stresses in composite finite elements based on first order shear deformation theory. Int. J. Numer. Methods Eng. **40**, 51–60 (1997)
14. Kärger, L., Baaran, J., Gunnion, A., Thomson, R.: Evaluation of impact assessment methodologies. part ii: experimental validation. Compos. B **40**, 71–76 (2009)
15. Kärger, L., Wetzel, A., Rolfes, R., Rohwer, K.: A three-layered sandwich element with improved transverse shear stiffness and stresses based on FSDT. Comput. Struct. **84**, 843–854 (2006)
16. Wetzel, A., Kärger, L., Rolfes, R., Rohwer, K.: Evaluation of two finite element formulations for a rapid 3D stress analysis of sandwich structures. Comput. Struct. **83**, 1537–1545 (2005)
17. Chai, H., Babcock, C.D.: Two-dimensional modeling of compressive failure in delaminated laminates. J. Compos. Mater. **19**, 67–99 (1984)
18. Tang, X., Shen, Z., Chen, P., Gädke, M.: Methodology for residual strength of damaged laminated composites. AIAA-97-1220 (1997)
19. Tsang, P.H.W., Lagace, P.A.: Failure mechanisms of impact-damaged sandwich panels under uniaxial compression. In: Proceedings of the AIAA/ASME/ASCE/AHS/ASC 35nd structures, structural dynamics, and materials conference, hilton head, pp. 745–754 (1994)
20. Wetzel, A., Baaran, J.: A semi-analytical methodology for the residual strength prediction of damaged sandwich structures. In: Proceedings of the ECCM 13, Stockholm, Sweden. 2–5 June 2008
21. Olsson, R.: Methodology for predicting the residual strength of impacted sandwich panels. Report FFA TN 1997-09, The Aeronautical Research Institute of Sweden (1997)

Chapter 14
Improved Stability Analysis of Thin Walled Stiffened and Unstiffened Composite Structures: Experiment and Simulation

Dirk Wilckens, Alexander Kling and Richard Degenhardt

Abstract This section deals with the experimental and numerical stability analysis of thin walled composite structures. Two different types of composite structures are considered. An unstiffened cylinder which is very sensitive to imperfections and used for basic research, and a stringer stiffened panel that is more related to aircraft fuselage structures. Both types of structures are loaded in axial compression. A set of different measurement systems has been applied before and during the tests in order to gather as much information as possible about the structure and its imperfections, as well as about the behaviour under load. Advanced optical measurement systems, for instance, have been utilized to monitor the failure in the skin-stringer interface of the stiffened panels. Exemplary test results are presented and compared to numerical simulations.

14.1 Stability Analysis of Stringer Stiffened Curved CFRP Panels

The present paragraph deals with stiffened structures that are regarded as representative parts of an aircraft fuselage which to date in most cases consists of orthotropic stiffened shells. This research work aimed to contribute to establish an advanced design scenario (Fig. 14.1) in which local skin buckling is permitted clearly below limit load (LL). This can for instance, depending on the design scenario, be 0.7 LL for axial compression loading. For different loading, i.e. shear or combined loading, this factor can be different. Such a scenario contributes to

D. Wilckens (✉) · A. Kling · R. Degenhardt
Institute of Composite Structures and Adaptive Systems, Structural Mechanics, German Aerospace Center (DLR e.V.), Lilienthalplatz 7, 38108, Braunschweig, Germany
e-mail: dirk.wilckens.dlr.de

M. Wiedemann and M. Sinapius (eds.), *Adaptive, Tolerant and Efficient Composite Structures*, Research Topics in Aerospace, DOI: 10.1007/978-3-642-29190-6_14,
© Springer-Verlag Berlin Heidelberg 2013

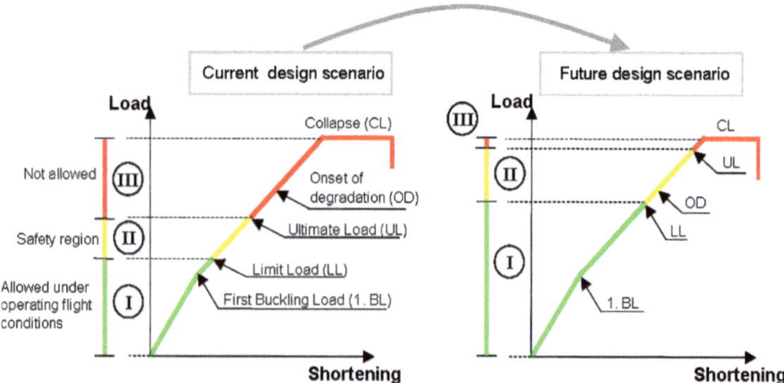

Fig. 14.1 Current and future design scenario

exploit the load carrying capacities in the post buckling region and finally leads to a remarkable weight reduction. However, it has to be secured that global stringer based buckling, which can be regarded as the total loss of stability and the load carrying capacity of the panel, occurs clearly above ultimate load (UL). Current design approaches do not take this potential into account since in the post critical area damage with the corresponding degradation in the laminate and the skin stringer interface may occur which finally leads to the collapse of the structure; this presently is not simulated accurately enough numerically in the design process. Including these effects in the nonlinear post buckling simulation of stiffened CFRP structures requires numerical models that include the geometric as well as the material nonlinear nature of the structural response. These models are essential in the design and analysis process to investigate and to predict how deep into the post buckling regime can be loaded without onset of degradation. The test structures described here are investigated under axial compression to study their post buckling and collapse behaviour und finally to build a sound basis for the validation of computational methods, either numerical approaches or analytical ones.

This contribution summarizes activities on stiffened panels at DLR and describes the capabilities of the buckling test facility in combination with the applied measurement systems. The current extension of the DLR buckling test facility to apply axial compression and shear loading at the same time is described in Chap. 16 of this book. Results of stability tests of stringer stiffened structures are also published by Cordisco and Bisagni [1] as well as Abramovich and Weller [2]. Additional results from previously performed buckling tests of stiffened structures were published by Zimmermann [3], Bisagni and Cordisco [4] and Abramovich [5].

14.1.1 Test Structures

In the context described above, several stringer stiffened panels were tested under axial compression. Two different panel designs are exemplarily described that

Fig. 14.2 Test panel representing a specific design for a large post buckling regime (*left*) and an industrial design (*right*)

were aimed to build up such a data base for validation purposes. The first one (Fig. 14.2 left) was designed with the objective to have a large post buckling region even after the transition to global buckling where skin stringer separation is expected to occur and hence can be monitored over a reasonable shortening regime. The second design (Fig. 14.2 right) reflects a more industrial oriented design without a distinct global buckling regime. The design process is described in more detail in [6]. In order to ensure repeatability and confidence in the test results a reasonable number of test panels per design is required within a validation context, cf. Chap. 9 of this book. The panels were made of the CFRP prepreg IM7/8552, the properties of which are determined in coupon tests for each study. The detailed geometry and the material properties are included in the detailed discussion of the buckling tests published by Wilckens et al. [7] and Degenhardt et al. [8, 9]. Undamaged as well as pre-damaged test panels have been considered whereas the kind of pre-damage was introduced either by a Teflon strip beneath the stringer foot across the whole width of the stiffener or by an impact from the skin side of the panel in the area of the stringer foot. Moreover, in order to trigger and to examine the progressing separation in the skin-stringer interface, cyclic loading by means of repeated quasi static loading was applied on both, the undamaged and also the pre-damaged panels. The load was applied displacement controlled and reduced to zero in between the cycles.

14.1.2 Test Preparation

In order to examine the quality of the structures and therefore to be able to evaluate the test results properly, they were examined with NDI by means of ultra sonic inspection and photogrammetric imperfection measurement. To ensure a uniform load introduction during the buckling test the end blocks are milled before casting into the final end blocks. The boundary conditions can be approximated as clamped

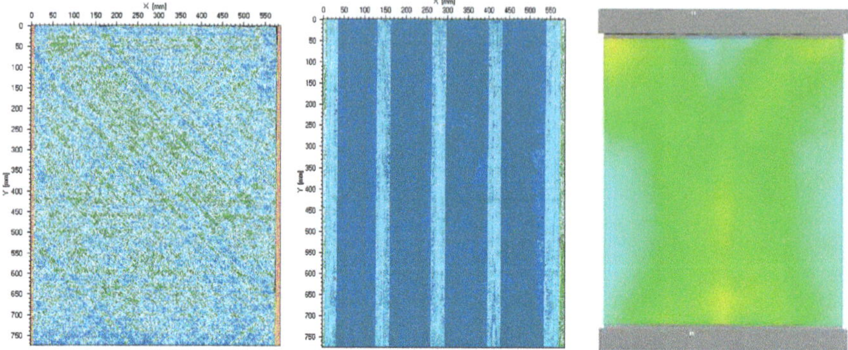

Fig. 14.3 US inspection (*left* and *middle*) and imperfection measurement (*right*)

with a fixed axial displacement at the upper panel edge and the displacement controlled load introduction at the lower edge. In Fig. 14.3, US inspection results for the panel exemplarily discussed in the next paragraph are depicted in terms of the flaw echo of the skin (left) and the rear side echo of the skin stringer connection (middle). They reveal only minor inhomogeneities. The plot at the right hand side shows the deviations from a best fit cylinder generated from the measured points of the real panel. The initial imperfection is smaller than the skin thickness and thus the panel is well suited for the investigation of the stability behaviour. During the test, depending on the objective of the test, different requirements arise on the experiments in terms of the monitoring of what has to be measured. This can be the buckling load, the load shortening path, i.e. the structural stiffness in the pre- and post buckling regime but also the onset and propagation of damage and finally the collapse of the structure. For this reason, appropriate measurement systems need to be utilized. In the investigations summarized here, a major objective was to study the collapse behaviour of the panel that in many cases is initiated or driven by the separation of the stiffeners from the skin. Therefore, load shortening and strain gauge data as well as DIC (photogrammetric deformation measurement) measurements were utilized to gather information about the structural response. Optical Lock-in Thermography (OLT) was applied in order to track the initiation process and the propagation of the skin stringer separation. The DIC pictures were taken from the stringer side of the panel. The front side was kept clean for the thermography measurements. A detailed discussion of the test setup and the application of the measurement systems was published by Degenhardt et al. [9].

14.1.3 Test Results of a Cyclic and Collapse Test

The test result for an undamaged panel is described in this paragraph. The panel was designed for a large post buckling regime (radius = 1000 mm, free buckling

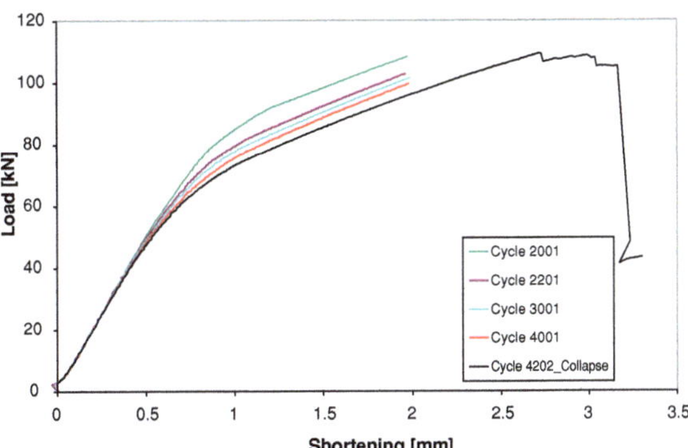

Fig. 14.4 Load shortening curves of the panel for selected load cycles

length = 660 mm, skin thickness = 1 mm, stringer pitch = 132 mm). It was subjected to cyclic axial compression. Before the collapse test, the panel was cycled 4,200 times with different amplitudes. The sequence and the load amplitude for the cycling was determined based on results from previous tests of a nominally identical panel and was related to the buckling load, the onset of global buckling or the collapse load of these panels.

The load shortening curves in Fig. 14.4 show the decreasing load carrying capacity of the panel during cycling in the regime beyond global stringer based buckling. The buckling load as well as the local post buckling stiffness and the point of transition to global bucking decreases with progressing failure in the skin-stringer connection leading to separation of the stiffener from the skin. This occurs in regions where large deflections resulting from local and global buckling lead to severe stress concentrations. An example of the progressing damage in the skin stringer interface is given in Fig. 14.5 showing the thermography images at cycle 2,400 and 3,600 (middle and right) as well as the corresponding typical out-of plane deformation recorded through DIC.

It is remarkable that for cycling in the local post buckling regime for the first 2,000 cycles up to about 1.0 mm shortening the stiffness has neither changed nor did the thermography images show any indication of failure. This observation was also made during further tests on undamaged and pre-damaged panels [7, 8].

After the test the structure was again inspected by ultrasonic inspection. The comparison of the US image and the OLT measurement after the test (Fig. 14.6) shows a good agreement and thus proves the applicability of the OLT method for the detection of damages in the structure.

14.1.4 Test–Simulation Correlation

As mentioned above, the buckling tests were conducted in order to create a data basis for the validation of the simulation models that are capable of accounting for the effects occurring during the tests. The curves in Fig. 14.7 show the comparison

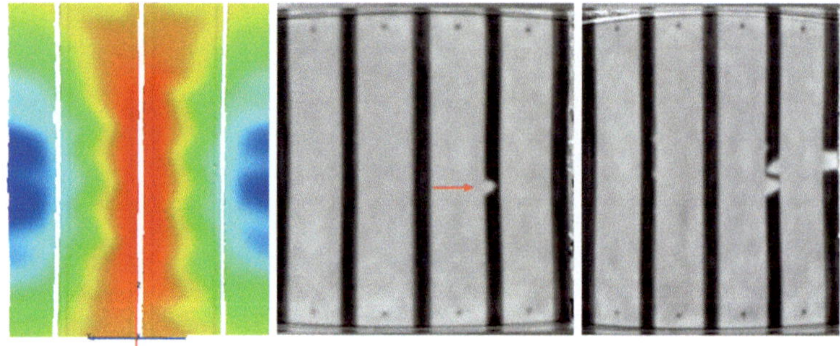

Fig. 14.5 DIC image at u = 2.0 mm (*left*) and OLT images after 2,400 cycles (*middle*) and 3,600 cycles (*right*)

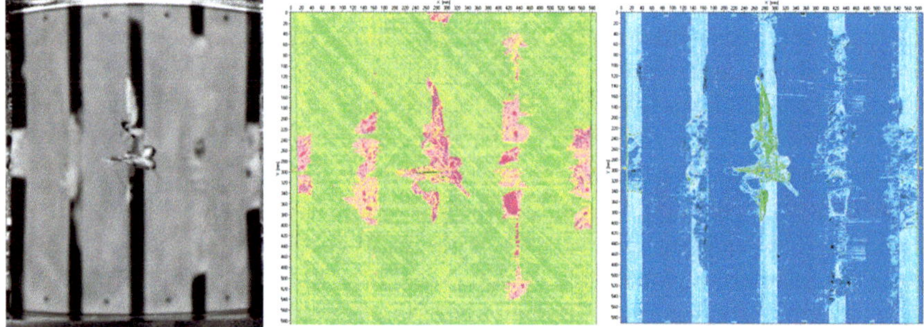

Fig. 14.6 Comparison of OLT image (*left*) and US measurements after collapse

Fig. 14.7 Comparison of test and simulation results

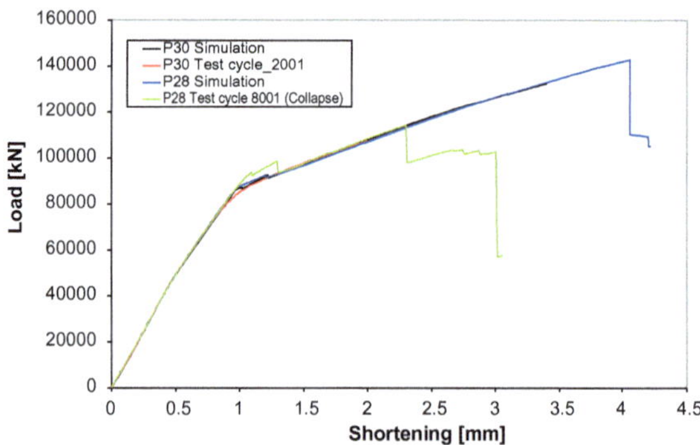

of the simulation results with the respective test curves for a stiffened CFRP panel of the same design as discussed in the previous paragraph. The simulation was conducted using ABAQUS/Standard utilizing the Newton–Raphson Method with artificial damping. The model was build up with four node shell elements of type S4R. It can be concluded that the pre- and post buckling stiffness as well as the

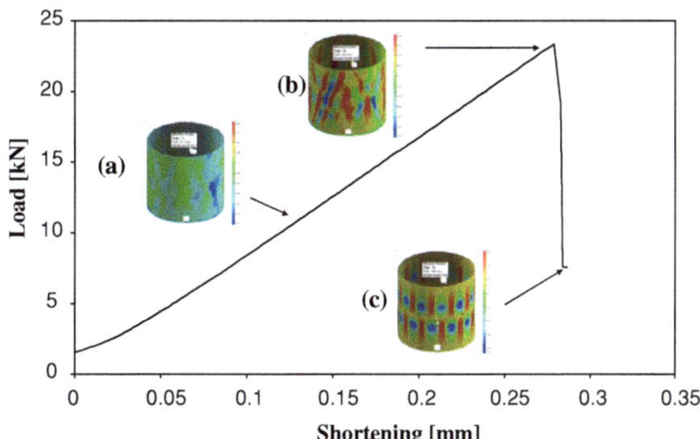

Fig. 14.8 Load shortening curve with selected deformation images

buckling load corresponds very well to the experimental results. Hence, the geometric nonlinear behaviour can be simulated reliably with current non-linear FEM software. Since the FEM simulation of the post buckling behaviour of CFRP structures is still time consuming, approaches have been developed suitable for the efficient simulation of the nonlinear structural behaviour and which can be used in a rather early stage of the development of a fuselage structure. A Ritz based simulation tool was developed by Bürmann for isotropic material [10] and extended by Wilckens for CFRP [11]. An alternative approach based on a reduced basis technique was published by Kling [12]. The failure and collapse mechanism, however, is still challenging. An approach was reported by Degenhardt et al. [6] using a simple degradation model implemented via ABAQUS User Subroutines. Another approach is to model the adhesive layer with interface elements with a cohesive law that are able to simulate the debonding process. This approach was found to be very time consuming and depended on properties for the energy release rates which themselves reveal a large scatter.

14.2 Stability Analysis of Unstiffened CFRP Cylindrical Shells

Unstiffened thin-walled cylindrical shells subjected to axial compression exhibit not only a high load carrying capacity but are also prone to buckling. Unlike for stiffened structures described in the previous paragraph, the stability behaviour of these structures is extremely sensitive to imperfections. These imperfections are defined as deviations from perfect parameters that can be for instance the geometry, the material properties and also the distribution of the applied external loads. To account for these imperfections lowering the real buckling load from the theoretical one, knock down factors are applied as proposed in the NASA SP-8007 guideline. However, these factors are very conservative and they do not reflect the characteristics of composite material adequately. In this context, studies with

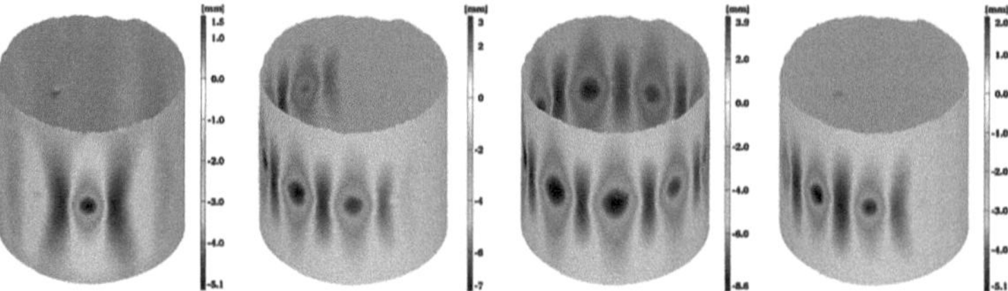

Fig. 14.9 Results of a 360° measurement on a CFRP cylinder (selected deformation patterns of one loading and unloading sequence)

unstiffened cylindrical CFRP shells aim to achieve more realistic knock down factors and also to build a validation basis for the numerical methods.

For the test of CFRP cylindrical shells loaded under axial compression an example is given here. The radius of the cylinder was 250 mm; the free buckling length was 500 mm with a skin thickness of 0.5 mm. The material in this study was the CFRP prepreg system IM7/8552. The detailed geometry and the material properties are given in [13]. The preparation of the test structure comprises among others the digital measurement of the deviation of the skin from the ideal shape as well as ultra sonic inspection of the skin. During the test, the cylinder was loaded by axial compression just beyond buckling load and was then unloaded. In that loading regime, the structure behaves elastically and thus will not be damaged. The full scale deformation and buckling shapes were measured using DIC (AR-AMIS). Figure 14.8 illustrates the measured load shortening curve with three selected DIC images obtained from a 360° measurement. The images a and b depict the pre-buckling deformation while image c is taken in the post buckling region. Another sequence of 360° pictures of the cylinder is given in Fig. 14.9 showing the development of the deformation pattern during of one load and unloading sequence.

Within a numerical simulation, all imperfections need to be included in the model to get the bucking load obtained from a test. However, even if all known deviations are considered in the simulation, the result often diverges from the experimental result. This is because not all imperfections can be specified by a deterministic value like for instance material inhomogeneities or load imperfections so that also probabilistic methodologies are needed. This applies in particular in the context of the design of cylindrical shells, where the imperfections are usually not known in advance. Investigations covering deterministic and probabilistic analysis as well as test results were published by Degenhardt et al. [13]. An alternative approach has been published by Hühne [14] that is based on a single buckle as worst imperfection leading to the realistic load carrying capacity. The probabilistic as well as the deterministic approaches are currently subject to further research at DLR [15].

References

1. Cordisco, P., Bisagni, C.: Design, testing and validation of a composite box under combined loading up to collapse. Int. J. Struct. Stab. Dyn. **10**(4), 853–869 (2010)
2. Abramovich, H., Weller, T.: Repeated buckling and postbucking behavior of laminated stringer stiffened composite panels with and without damage. Int. J. Struct. Stab. Dyn. **10**(4), 807–825 (2010)
3. Zimmermann, R., Klein, H., Kling, A.: Buckling and postbuckling of stringer stiffened fibre composite curved panels—tests and computations. Compos. Struct. **73**, 150–161 (2006)
4. Bisagni, C., Cordisco, P.: Post-buckling and collapse experiments of stiffened composite cylindrical shells subjected to axial loading and torque. Compos. Struct. **73**, 138–149 (2006)
5. Abramovich, H., Weller, T., Bisagni, C.: Buckling behavior of composite laminated stiffened panels under combined shear–axial compression. J. Aircraft **45**(2), 402–413 (2008)
6. Degenhardt, R., Kling, A., Rohwer, K., Orifici, A.C., Thomson, R.S.: Design and analysis of stiffened composite panels including postbuckling and collapse. Comput. Struct. **86**, 919–929 (2008)
7. Wilckens, D., Degenhardt, R., Rohwer, K., Zimmermann, R., Kepke, M., Hildebrand, B., Zipfel, A.: Cyclic buckling tests of pre-damaged CFRP stringer-stiffened panels. Int. J. Struct. Stab. Dyn. **10**(4), 827–851 (2010)
8. Degenhardt, R., Wilckens, D., Klein, H., Kling, A., Hillger, W., Goetting, Ch., Rohwer, K., Gleiter, A.: Experiments to detect the damage progress of axially compressed CFRP panels under cyclic loading. Spec. Volume Key Eng. Mater. **383**, 1–24 (2008)
9. Degenhardt, R., Kling, A., Klein, H., Hillger, W., Goetting, Ch., Zimmermann, R., Rohwer, K., Gleiter, A.: Experiments on buckling and postbuckling of thin-walled CFRP structures using advanced measurement systems. Int. J. Struct. Stab. Dyn. **7**(2), 337–358 (2007)
10. Bürmann, P., Rolfes, R., Tessmer, J., Schagerl, M.: A semi-analytical model for local postbuckling analysis of stringer- and frame-stiffened cylindrical panels. Thin-Walled Struct. **44**, 102–114 (2006)
11. Wilckens, D., Degenhardt, R., Rohwer, K.: IBUCK–A fast simulation tool for the design of CFRP aerospace structures. In: Proceedings of the 2nd International Conference on Buckling and Postbuckling Behaviour of Composite Laminated Shell Structures, Braunschweig, Germany, 3–5 Sept 2008
12. Kling, A., Degenhardt, R., Zimmermann, R.: A hybrid subspace analysis procedure for nonlinear postbuckling calculation. Compos. Struct. **73**, 162–170 (2006)
13. Degenhardt, R., Kling, A., Bethge, A., Orf, J., Kärger, L., Rohwer, K., Zimmermann, R., Calvi, A.: Investigations of imperfection sensitivity of unstiffened CFRP cylindrical shells. Compos. Struct. **92**(8), 1939–1946 (2010)
14. Hühne, C., Rolfes, R., Breitbach, E., Teßmer, J.: Robust design of composite cylindrical shells under axial compression—Simulation and validation. Thin-Walled Struct. **46**, 947–962 (2008)
15. www.desicos.eu

Chapter 15
Composite Process Chain Towards As-Built Design

Tobias Wille, Luise Kärger and Robert Hein

Abstract The relation between design and manufacture is of particular importance within the composite structure development process. Therefore, a continuous composite process chain beyond state-of-the-art is described in this section. Such an all-embracing process chain realizes a concurrent engineering, where iteration loops are enabled and, thus, product improvements and higher process efficiency are achieved. Concurrent engineering comprises the various interdisciplinary working phases and provides the necessary connectivity. In contrast to the traditional one-way relation from design to manufacture, the improved process chain also deals with the feedback from manufacture to design, based on numerical simulations. This is demonstrated by the example of composite parts made by Tailored Fiber Placement (TFP), including effects of the feedback on load bearing capacity.

15.1 Current State of Composite Processes

In spite of extensive effort within several research projects, the current development process of composite structures still exhibits a mostly sequential work flow, beginning with a first feasibility and concept phase, over the more detailed definition

T. Wille (✉) · R. Hein
Institute of Composite Structures and Adaptive Systems, German Aerospace Center
(DLR e.V.), Lilienthalplatz 7, 38108, Braunschweig, Germany
e-mail: tobias.wille@dlr.de

R. Hein
e-mail: robert.hein@dlr.de

L. Kärger
Institute of Composite Structures and Adaptive Systems,
German Aerospace Center (DLR e.V.), Lilienthalplatz 7, 38108, Braunschweig, Germany

M. Wiedemann and M. Sinapius (eds.), *Adaptive, Tolerant and Efficient Composite Structures*, Research Topics in Aerospace, DOI: 10.1007/978-3-642-29190-6_15,

Fig. 15.1 Sequential process chain of composites (state-of-the-art)

and development phase, the production and assembly phase up to the final phase of series production (Fig. 15.1). Several different methods and software tools are applied within these development phases in order to solve particular questions. Accompanying it, a heterogeneous set of tools with generally incomplete tool interfaces hinders a continuous development process with seamless interfaces. This directly affects process efficiency both short-term feasibility questions as well as comprehensive detailed design of composite structures. Therefore great potential could be tapped by a concurrent engineering approach, which may for example arise from an early feedback of manufacturing data to the design phase.

15.2 Continuous Composite Process Chain

Due to uncertainties during the production and assembly processes the structural capability of composites cannot fully be exploited within current composite design. In order to increase the performance of composite parts, the associated process steps were analyzed regarding their respective interdependencies and related exchange data. Aiming for a future "as-built" design process, which already includes all relevant production aspects beforehand, the development and the production phases have to be consolidated. Results of production process simulation and optimization are to be returned to the earlier development phase (Fig. 15.2). This feedback of process simulation data enables to take into account geometrical, material and process-induced deviations, such as local degradation of the composite stiffness and strength as well as manufacturing induced deformations and stresses. Hence, more precise and more robust prediction methods can be achieved and provide the basis for a less conservative dimensioning regarding weight reduction while assuring part quality at minimum process time. Therefore, an efficient and production conform preliminary design, a more reliable and comprehensive analysis as well as a time and cost efficient production shall be realized. These ambitious objectives are especially advantageous in view of high cadence processes, large components as well as complex assembly tasks of composites.

Within the continuous composite process chain many different composite manufacturing technologies, such as vacuum bagging, autoclave, and resin injection technologies as well as a great variety of derivatives and combinations of the afore-mentioned have to be considered [1]. Furthermore, the following main process steps are generally performed with high requested accurateness in order to accomplish the required composite part quality: tooling development (incl. heating and injection concept), draping/lay-up of woven or non-crimp fabrics (dry or pre-impregnated), resin injection (in case of dry fabrics) and curing (at elevated temperatures). With the

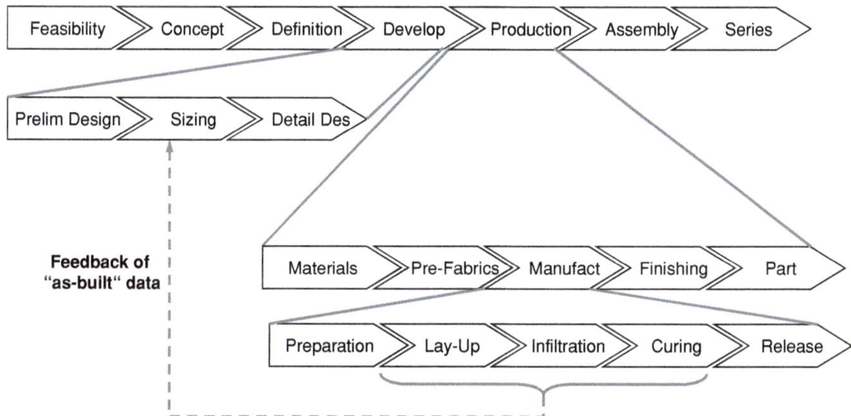

Fig. 15.2 Feedback of "as-built" data from the production phase to the earlier development phase within the continuous composite process chain

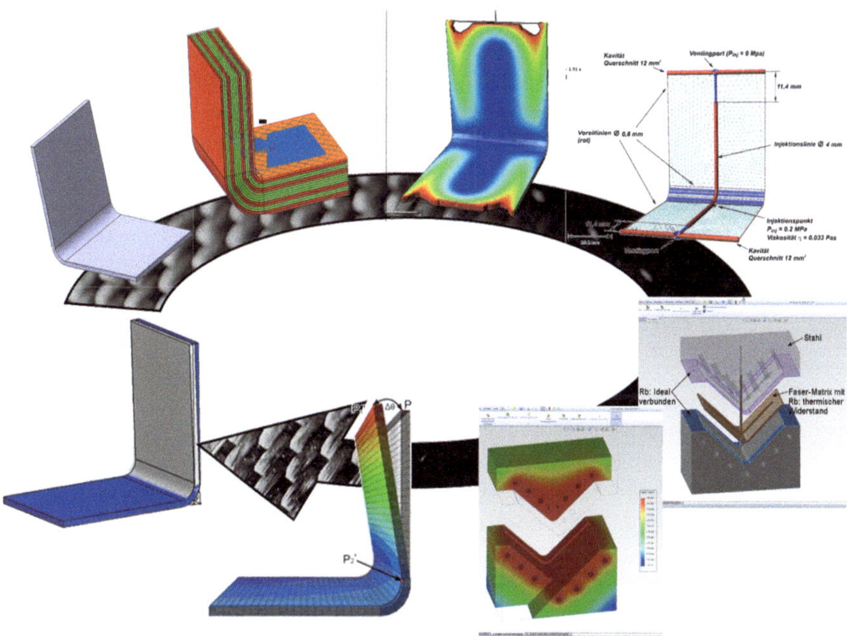

Fig. 15.3 Application of the continuous composite process chain to consider as-built information within the design phase

objective of an "as-built" composite part development, these process steps have to be analyzed regarding their effects on the composite part quality within the early development phase. For this purpose numerical process simulations are to be performed, and resulting data have to be returned into a more precise "as-built" component analysis via appropriate feedback algorithms. The example of a typical process flow depicted in Fig. 15.3 shall be described in more detail.

The starting point represents a mould with the desired dimensions. Considering the fabric lay-up process, draping, fiber placement or forming simulations can be accomplished, in order to receive spatially distributed geometric composite layer properties, such as fiber orientations, thickness distributions or shear angles. With respect to an optimal infusion concept, resin flow simulation can be used to predict flow fronts, dry spots or entrapments of air. Moreover, an optimal tooling and heating concept can be achieved by performing numerical heat transfer analyses coupled with mechanical analyses. Thus, thermal and mechanical boundary conditions can be provided for subsequent composite curing simulations. Since the actual composite properties are depending on the curing process, a prediction of spatially distributed exothermic reactions and degree of cure is of great interest. Furthermore, the process induced residual stresses and distortions, caused by chemical shrinkage and thermal expansion, can be received. These results are highly relevant with respect to a "first-time-right" part development and therefore need to be considered within the development phase. On one hand they can be transferred to Computer Aided Design (CAD) models in order to map/morph composite and tooling surfaces from a first nominal design to an "as-built" design. On the other hand residual stress fields can be integrated as initial loadings into the structural analysis, even superposed by impressed deformations (causing stress distributions) due to assembly, and therefore allow for a more realistic prediction of critical areas under operational loading conditions.

Since all these data are generally spatially distributed within the composite part, they have to be handled as relatively large field values, some of them even as a function of time. Specific aspects can already be handled by applying commercial software. For example, the main fiber direction can be transferred from draping simulation into the detailed finite element analysis. Also, temperature distributions, which were calculated within a tooling heat transfer analysis, can be imported as boundary conditions into a curing simulation. Even residual distortions can be mapped from curing and spring-in analysis onto the original part design. However, with respect to the multiplicity of phenomena, of simulations methods and of software systems, commercial software is currently not fully supporting all aspects of the composite tool chain.

15.3 Application of Manufacturing Feedback for Fiber Placement and Curing

Tailored or Advanced Fiber Placement (TFP or AFP) is a textile process for the production of optimized fiber reinforced structures. TFP follows the examples in nature, where an adapted, perfect design provides an optimal survive, cf. Mattheck et al. [2]. Accordingly, the carbon fiber rovings in TFP structures may be placed in

Fig. 15.4 TFP tension
sample with a central open
hole

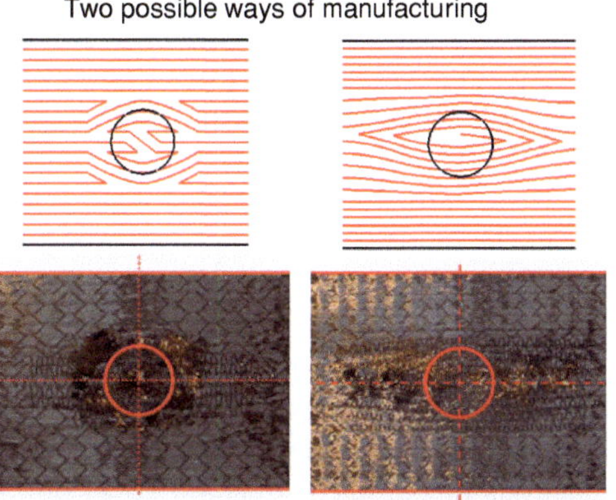

almost any desired orientation, thus deploying calculated optimum fiber quantities and orientations for optimal performance [3]. However, the TFP manufacturing process also affects the material properties. Specific material properties and appropriate material models of the textile fiber composites are of critical importance for a successful application [4, 5].

To reliably predict the material and structural behavior, the TFP As-built Feedback Method has been developed. By means of this method, manufacturing data such as fiber alignments and effective global material properties of local fiber features (e.g. fiber turns) are suitably transformed to as-built finite element models. Consequently, high-fidelity Computer Aided Engineering (CAE) models are generated, which may significantly differ from the original CAE models used for design. Based on the accordingly improved analysis results, recommendations for enhanced design guidelines can be deduced.

Figure 15.4 shows an example, where it is clearly necessary to establish an as-built FE model instead of using the primary as-design FE model for certification: The TFP optimization provides a fiber alignment, which runs smoothly in a curvature around the hole. Since manufacturing constraints have to be considered, different solutions have to be evaluated to realize the curved fiber orientation: The left solution in Fig. 15.4 shows a TFP structure with constant fiber volume fraction, while the right solution shows a higher fiber volume fraction next to the hole. Since the failure behavior of these two examples will be different, various FE models are needed for an appropriate solution.

15.3.1 Feedback of Effective Material Properties

The first part of the TFP As-built Feedback Method embodies a Multiscale Analysis which considers local discontinuities and establishes a suitable

Fig. 15.5 First part of the feedback method: multiscale analysis

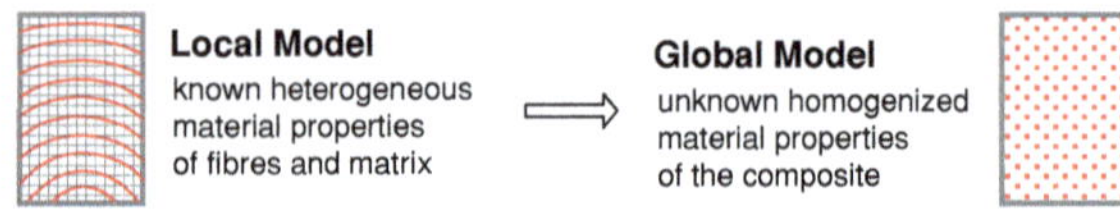

macroscopic resolution of material particularities (e.g. fiber turns, cf. Fig 15.5). Accordingly, suitable mesoscopic FE models (super-elements) are generated to precisely represent the real fiber alignment. Subsequently, the Asymptotic Homogenization Method (AHM) is applied to compute the effective macroscopic elasticity matrix. The AHM is a rigorous mathematical technique to predict the effective global properties of inhomogeneous media [6]. Coupled with the finite element method, the AHM has been supposed to be the most effective tool for the analysis of the microstructural effect on the global behavior of a composite material [7]. Accordingly, it is widely used in the field of composite and multi-scale research, it can be referred to e.g. Böhm [8], Chen et al. [9] and Hassani and Hinton [10, 11].

In the present work, the displacement method, as proposed e.g. by Lukkassen et al. [12], is applied to solve the cell problem of the AHM. By means of the displacement method, unit strain fields are introduced to the mesoscopic FE model (super-element). The corresponding boundary conditions are suchlike that points on opposite sites of the super-element are coupled to each other in all three translational directions except for one point, which is coupled only in two directions and in the third direction it moves by the length of the super-element. Figure 15.6 illustrates the reduction of the resulting effective stiffness component Q11 (stiffness in nominal fiber direction) depending on the radius of a fiber curvature: Q11 drops dramatically for increasing fiber curvatures, particularly if the curvature radius becomes smaller than the tenfold length of the super element.

The computation of effective strength components is build upon the so-called Direct Micromechanics Method (DMM), which was initially proposed by Zhu et al. [13]. Reduction of strength due to curvature is similar to the stiffness reduction shown in Fig. 15.6. The computed effective material properties, stiffness and strength, are transferred and used for the macroscopic as-built FE models.

15.3.2 Feedback of Fiber Alignment

Within the second part of the TFP As-built Feedback Method, the As-Design to As-Built Data Transformation is performed. Based on the preceding multiscale analysis, this transformation utilizes the effective global material properties. Furthermore, the actual fiber alignment of the manufactured structure, stored e.g. in Fiber Placement Manager (FPM) files, is automatically transformed into updated fiber orientations specified in the global FE model in Fig. 15.7.

In Fig. 15.7, the concept of the transformation from "as-design" (resultant model after optimization and conceptual sizing) to "as-built" (resultant model

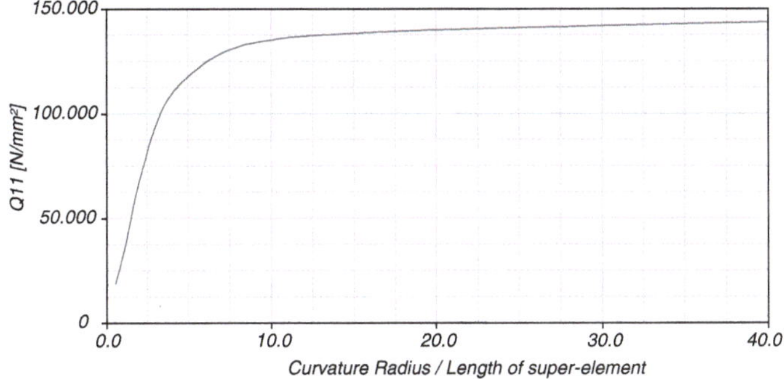

Fig. 15.6 Effective stiffness component *Q11* depending on the radius of fiber turns

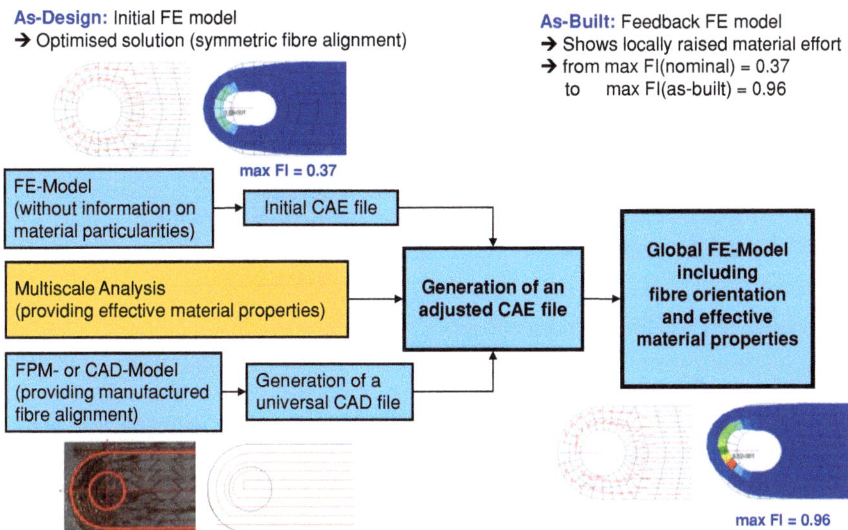

Fig. 15.7 As-design to as-built data transformation as one part of the TFP As-built Feedback Method: Reduction of load bearing capacity of a manufactured TFP tension strap compared to the initial design

from the TFP As-built Feedback Method) is illustrated using the example of a tension strap with two holes. The two contour plots show the computed material effort of the "as-design" (top left) and the "as-built" FE model (bottom right). The material effort is illustrated by the Failure Index FI which has been determined according to the Tsai-Wu failure criterion [14]

$$\mathrm{FI} = \left(\frac{1}{R_{1t}} + \frac{1}{R_{1c}}\right)\sigma_{11} + \left(\frac{1}{R_{2t}} + \frac{1}{R_{2c}}\right)\sigma_{22} - \frac{1}{R_{1t}R_{1c}}\sigma_{11}^2 - \frac{1}{R_{2t}R_{2c}}\sigma_{22}^2 + \left(\frac{1}{R_{12}}\right)^2\sigma_{12}^2,$$

where σ_{ij} are the stress components and $R_{1t}, R_{1c}, R_{2t}, R_{2c}, R_{12}$ are the strength components. The results show the reduction of load-bearing capacity: The optimization process provides a roving alignment, which runs smoothly in a curvature around the hole. For manufacturing, however, several constraints need to be considered, which result in unsymmetrical fiber alignments with local fiber particularities. Thus, the maximum failure index rises from FI (nominal) = 0.37 to FI (as-built) = 0.96 (i.e. close to damage at FI = 1). This signifies an alarming reduction of load-bearing capacity and thus, two important conclusions need to be drawn:

- It is certainly essential to consider the actual manufactured fiber alignment by an "as-built" FE model rather than analyzing the "as-design" FE model of a preliminary optimization process.
- For improved design and manufacturing guidelines, the negative effect of irregular fiber alignments must be considered. Hence, the communication from "as-built" FE analysis to the design and manufacturing departments is essential.

15.3.3 Feedback of Process Induced Residual Stresses and Distortions

Subsequent to the fiber placement process the curing process of composite parts takes place usually at elevated temperatures in heated toolings, ovens, autoclaves or alike. During curing the resin reacts exothermally, which may lead to inhomogeneous temperature distributions depending on the part thickness and on the applied heating concept. For that reason the final degree of cure may be varying spatially within the part, and thus, affecting the mechanical properties of the part. Additionally, chemical shrinkage takes place during curing, and thermal contraction occurs when cooling down the composite part to ambient temperature. Therefore, curing stresses build up, depending on both the thermal and mechanical boundary conditions as well as on the viscoelastic behavior of composites in rubbery and glassy state [1].

Within the last years several methods on curing simulation were published in order to predict both, the residual curing stresses and distortions of composites, e.g. White and Hahn [15], Karkanas and Partridge [16], Johnston [17], Svanberg [18], Ruiz and Trochu [19]. Aiming for a more realistic "as-built" design, these process induced preloads have to be transferred back to the earlier development phase. Accordingly, curing stresses and distortions can be considered as initial loading conditions when evaluating the component behavior under service loading conditions. Applying a coupled transient thermo-mechanical curing analysis, computation time can massively increase, which often limits the simulation of large composite parts. Therefore, it is advisable to consider more efficient

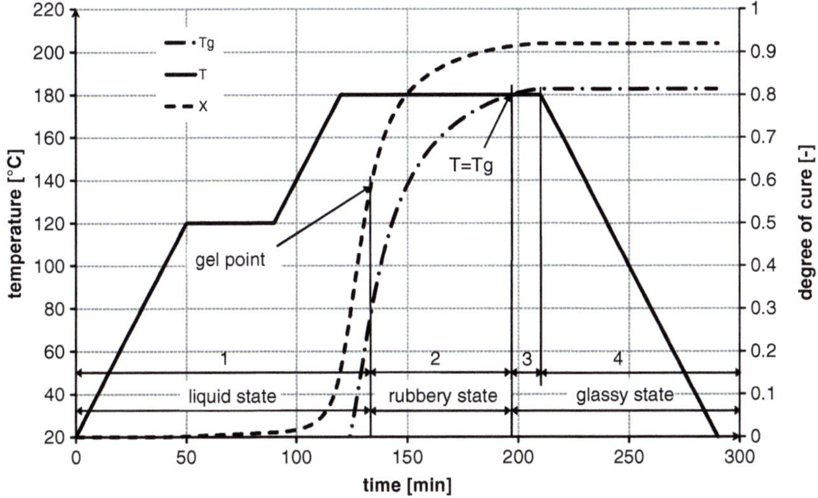

Fig. 15.8 Thermo-chemical simulation of degree of cure X and glass transition temperature Tg

Table 15.1 Steps during curing

Step	State	Effect
1	Before gelation	No effect on stress or distortion
2	Isothermal curing in rubbery state	Shrinkage
3	Isothermal curing in glassy state	Shrinkage
4	Non isothermal cooling in glassy state	Thermal expansion

Fig. 15.9 Resulting residual curing stresses in fiber direction [MPa] on deformed plot

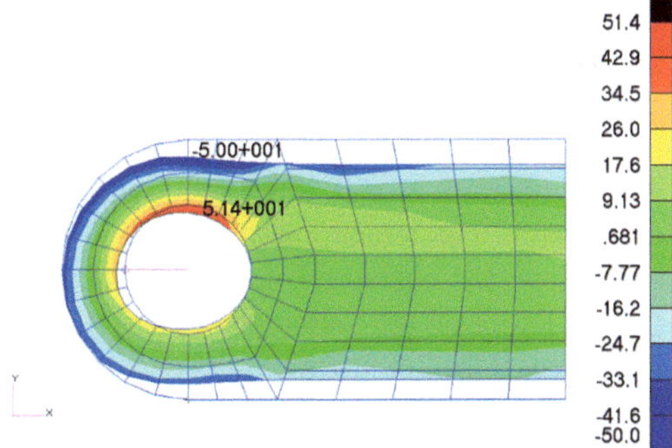

simulation methods for real applications. For instance, the path-dependent constitutive (time-independent) pseudo-viscoelastic material model from Svanberg [18] can be applied to predict process induced distortions and stresses. If the spatial temperature distribution is homogeneous within the composite, e.g. in thin parts, a further simplification by Svanberg of evaluating characteristic process

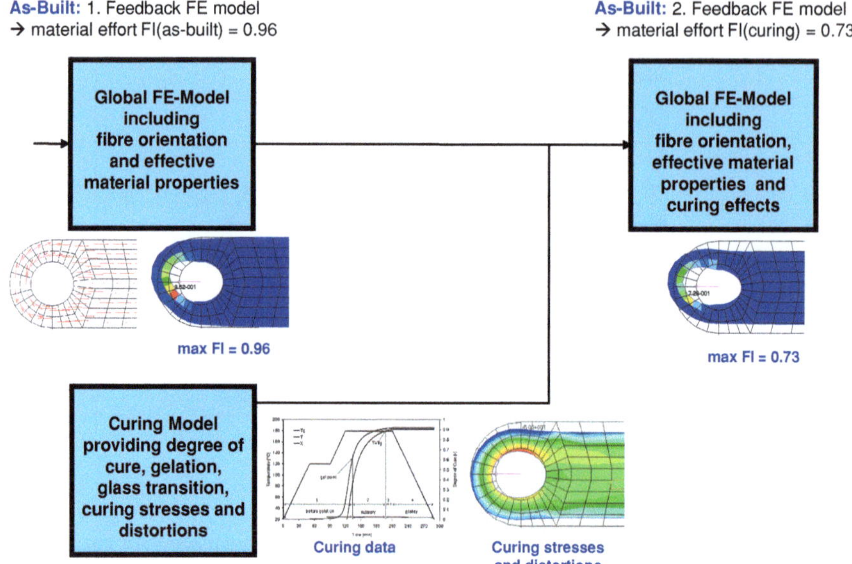

Fig. 15.10 Application of the feedback method to estimate the reduction of load bearing capacity of a manufactured TFP tension strap compared to the first as-built design

steps should be taken into account in order to reduce computation time by achieving the same results. Numerical simulations on simple coupons have already exhibit great accordance between the results from a full transient calculation (applying the path-dependent method) and the results from simplified thermo-elastic calculations (based on characteristic process steps), ref. Wille et al. [20]. For the present TFP strap example, a combination of both simplified simulation methods can be applied in order to predict process induced residual stresses and shape distortions.

Initially, a thermo-chemical simulation is performed in order to predict the degree of cure X and the glass transition temperature Tg at a given process temperature T as a function of time (X = 0 uncured, X = 1 fully cured). From this simulation, the characteristic process steps "gelation" (here X = 0.6), "vitrification" (T = Tg) and "change of curing temperature" (e.g. cool down) can be identified. Based on these results chemical shrinkage effects within the isothermal curing phase can be separated from thermal expansion effects within the heating or cooling phases. As depicted in Fig. 15.8, the curing process is subdivided into four respective steps as listed in Table 15.1.

Hereupon, thermo-elastic calculations are conducted for each step applying the individual material parameters (stiffness, Coefficient of Thermal Expansion (CTE), shrinkage coefficient) of the respective rubbery and glassy states, as described by Svanberg. From these calculations, component stresses and distortions (Ref. Fig. 15.9) can be taken into account as initial conditions for the subsequent structural mechanical analysis and will therefore lead to a change in material effort as well as in the composite failure behavior.

In Fig. 15.10, the "as-built" work flow is extended by the curing simulation step, in order to account for the effect of residual stresses and distortions within a CAE model. By applying the same failure criterion as for the corresponding result plots of the TFP strap example (Tsai-Wu failure criterion), the computed maximum failure index changes from FI(1. as-built) = 0.96 to FI(2. as-built) = 0.73. This reduction of 24% is due to the change of prestressing of the inner fiber tows, mainly arising from resin shrinkage transverse to the fiber direction (radial to the whole), while the length of the tows (in fiber direction, tangential to the whole) are kept nearly constant.

15.4 Outcome of the As-Built Feedback Method

A concurrent engineering, realized by a continuous composite process chain, enables iteration loops for product improvements and higher process efficiency. By means of the developed feedback methods, relevant manufacturing data, e.g. fiber alignments, process induced distortions or residual stresses, can be automatically integrated within the early phase of the future development chain of composite structures. Thus, a less conservative, more robust design is achieved by accounting for different types of constraints and manufacturing uncertainties within the preliminary and detailed design phase.

References

1. Ehrenstein, G.W.: Faserverbundkunststoffe. Carl Hanser Verlag, Munich (2006)
2. Mattheck, C.: Design in nature. Interdisc. Sci. Rev. **19**(4), 298–314 (1994)
3. Tosh, M.W., Kelly, D.W.: On the design, manufacture and testing of trajectorial fibre steering for carbon fibre composite laminates. Compos. A **31**(10), 1047–1060 (2000)
4. Mattheij, P., Gliesche, K., Feltin, D.: Tailored fiber placement—mechanical properties and application. J. Reinf. Plast. Compos. **17**(9), 774–786 (1998)
5. Mattheij, P., Gliesche, K., Feltin, D.: 3D reinforced stitched carbon/epoxy laminates made by tailored fibre placement. Compos. A **31**(6), 571–581 (2000)
6. Guinovart-Diaz, R., Bravo-Castillero, J., Rodriguez-Ramos, R., Martinez-Rosado, R., Serrania, F., Navarrete, M.: Modelling of elastic transversely isotropic composite using the asymptotic homogenization method. Some comparisons with other models. Mater. Lett. **56**, 889–894 (2002)
7. Takano, N., Okuno, Y.: Three-scale finite element analysis of heterogeneous media by asymptotic homogenization and mesh superposition methods. Int. J. Solids Struct. **41**, 4121–4135 (2004)
8. Böhm, H.J.: A short introduction to basic aspects of continuum micromechanics. ILSB Report 206, Vienna (2007)

9. Chen, C.M., Kikuchi, N., Rostam-Abadi, F.: An enhanced asymptotic homogenization method of the static and dynamics of elastic composite laminates. Comput. Struct. **82**, 373–382 (2004)
10. Hassani, B., Hinton, E.: A review of homogenization and topology optimization I—homogenization theory for media with periodic structure. Comput. Struct. **69**, 707–717 (1998)
11. Hassani, B., Hinton, E.: A review of homogenization and topology optimization II—analytical and numerical solution of homogenization equations. Comput. Struct. **69**, 719–738 (1998)
12. Lukkassen, D., Persson, L., Wall, P.: Some engineering and mathematical aspects on the homogenization method. Compos. Eng. **5**, 519–531 (1995)
13. Zhu, H., Sankar, B.V., Marrey, R.V.: Evaluation of failure criteria for fiber composites using finite element micromechanics. J. Compos. Mater. **32**, 766–782 (1998)
14. Tsai, S.W., Wu, E.M.: A general theory of strength for anisotropic materials. J. Compos. Mater. **5**, 58–80 (1971)
15. White, S.R., Hahn, H.T.: Process modeling of composite materials: residual stress development during cure. Part I. Model formulation. J. Compos. Mater. **26**, 2402 (1992)
16. Karkanas, P.I., Partridge, I.K.: Modelling the cure of a commercial epoxy resin for applications in resin transfer moulding. Polym. Int. **41**, 183–191 (1996)
17. Johnston, A.A.: An Integrated Model of the Development of Process-Induced Deformation in Autoclave Processing of Composite Structures. Doctoral thesis, University of British Columbia (1997)
18. Svanberg, J.M.: Predictions of Manufacturing Induced Shape Distortions—High Performance Thermoset Composites. Doctoral thesis, Lulea University of Technology (2002)
19. Ruiz, E., Trochu, F.: Numerical analysis of cure temperature and internal stresses in thin and thick RTM parts. Compos. A **36**, 806–826 (2005)
20. Wille, T., Freund, S., Hein, R.: Comparison of Methods to Predict Process Induced Residual Stresses and Distortions of CFRP Components. In: NAFEMS Seminar: Progress in Simulating Composites, Wiesbaden (2011)

Chapter 16
Innovative Testing Methods on Specimen and Component Level

Falk Odermann and Tobias Wille

Abstract In general, tests can be divided into four categories: parameter estimation (e.g. material strength), phenomenological investigation, validation and qualification. According to this classification tests are carried out on a structural or component level and on a coupon level. For structural testing a Buckling Test Facility, a Variable Component Test Facility and a thermo-mechanical test field are described. Furthermore, information is given on specimen level tests with devices for standard test machines: Stringer Pull-off Device and 3D-Biax Device.

16.1 Buckling Test Facility

Buckling tests represent a specific topic in the whole area of testing. Especially investigations of thin walled structures, which are sensitive to imperfections, require sophisticated test facilities, [1, 2]. In order to serve the whole application range from basic research to industrial qualification, e.g. of stiffened or unstiffened cylindrical shells or panels, the Buckling Test Facility can be operated in four different configurations as described in the following sections.

F. Odermann (✉) · T. Wille
Institute of Composite Structures and Adaptive Systems,
Lilienthalplatz 7, 38108, Braunschweig, Germany
e-mail: falk.odermann@dlr.de

T. Wille
e-mail: tobias.wille@dlr.de

M. Wiedemann and M. Sinapius (eds.), *Adaptive, Tolerant and Efficient Composite Structures*, Research Topics in Aerospace, DOI: 10.1007/978-3-642-29190-6_16,

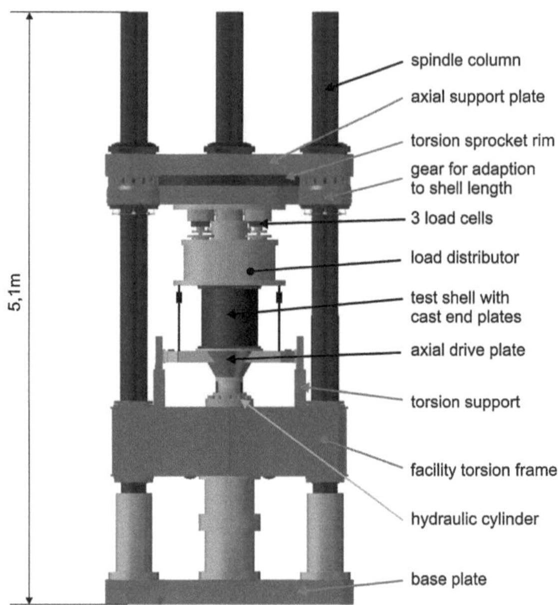

Fig. 16.1 Buckling test facility of DLR (configuration axial compression)

16.1.1 Static Axial Loading for Cylinders and Panels

The simplest kinds of all buckling tests are static tests under axial compression loading. Their results produce the base of experience for all the other buckling tests. In this configuration the Buckling Test Facility is used for phenomenological research and validation of numerical models since it was designed to meet these demands. By way of example, thin walled unstiffened and stiffened cylindrical shells, representative for space structures, and stiffened panels, representative for parts of aircraft fuselages, are specimens for investigation. Testing of thin walled structures requires a smooth and evenly distributed load introduction. In particular often very low prescribed displacements (e.g. fractions of a millimetre) are to be applied at high loads. Thus very small tolerances have to be ensured for the load introduction.

The general configuration of the facility is described in Fig. 16.1.

Loads are applied by the vertical hydraulic servo cylinder, equipped with a two stage servo valve with low volume flow. That valve allows precise control of axial movement. Load is directed through the axial drive plate to the specimen with cast end plates. At the top of the specimen the load distributor directs the load to three load cells and the enclosed axial support plate. During test a clamping mechanism fixes the axial support plate at the columns. Cylindrical shells and panels are coupled to the machine by means of equalizing layer and endplates to ensure a well defined contact between the specimen and the machine. First the specimen is cast into the end plates. Then between axial drive plate and load distributor this

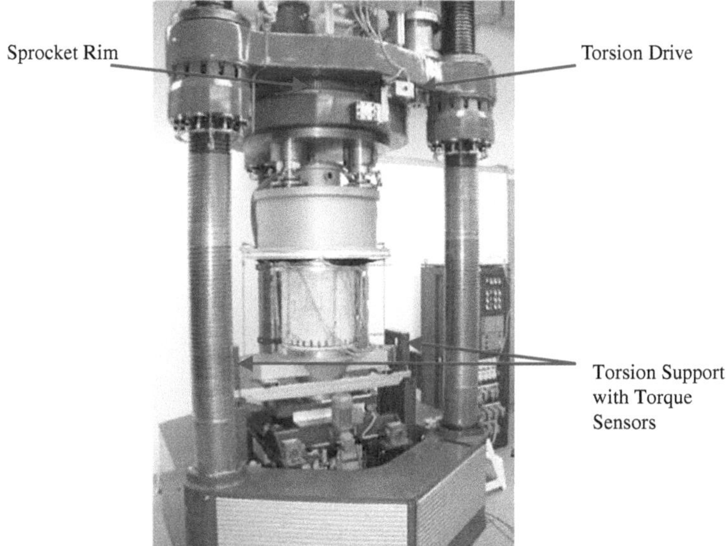

Fig. 16.2 Buckling test facility with torsion drive at the top and with unstiffened cylindrical composite shell

unit is set into the facility with an equalizing layer. Thus uniform connection between specimen and test facility is ensured. Panels are cast into boxes where this unit is set again between the axial drive plate and load the distributor by means of equalizing layers. Two displacement transducers besides the specimen are mounted between axial drive plate and load distributor. Additionally spatial displacements can be measured using digital image correlation system ARAMIS (GOM). Thermography is used for damage detection.

16.1.2 Static Axial Loads Combined with Torsion for Cylinders

Stiffened and unstiffened cylindrical shells can be statically tested with torsion loads in combination with axial compression. For this purpose a sprocket rim is used to rotate the top plate. The rim is driven by a sprocket wheel and a gearbox with an electric motor at the top of the axial support plate (Fig. 16.2).The moment of torsion can be measured with the help of sensors at the two lateral torsion supports. These supports contain deformable parts with strain gauges and have been calibrated before starting the test. The angle of torsion is measured via linear displacement transducers. Thus, axial load carrying capabilities under superposed torsion can be investigated. This test setup s used for phenomenological research as well as the validation of numerical models.

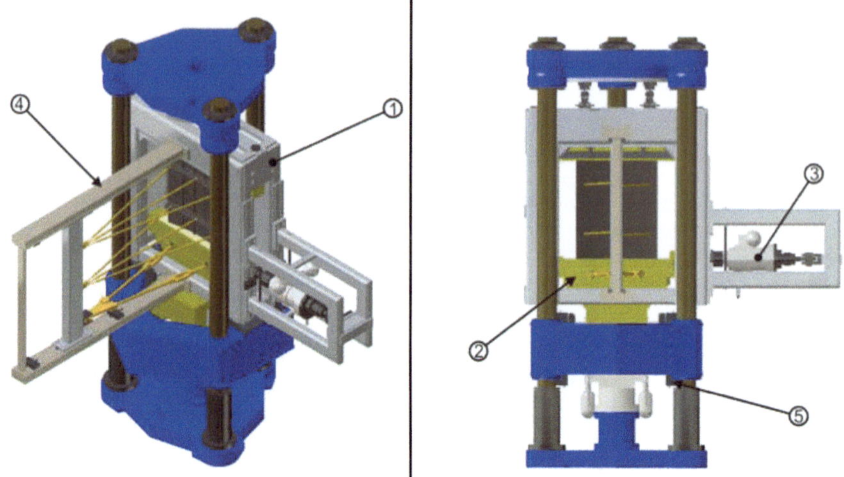

Fig. 16.3 Buckling test facility with compression shear test device

16.1.3 Combined Axial Compression Shear Test Device for Curved Stiffened Panels

In order to investigate the stability behavior of curved stiffened panels under combined axially compressive and in-plane shear loading, a special test device is needed. Different principles of shear-compression-test devices are currently applied [3, 4]. These devices apply loads with different kinematics according to the respective specimen geometry and specimen size. They are built as stand alone machines and need adequate space in a test hall. In contrast to that, a new test device was developed in order to extend the already described buckling test facility to the capability of in-plane shear loading in combination with compressive loads for curved stiffened panels.

The approach to investigate buckling under shear loading with free longitudinal edges of the panel was chosen to ensure constant boundary conditions during the test. Hence best comparability between boundary conditions of experiments and those in numerical models can be achieved within a validation process. The overall concept was based on the results of the numerical simulations and the given possibilities and limitations by the existing buckling test facility. In Fig 16.3 the new test device is shown as a CAD-model. The test device consists of a shear frame (1), a load transmission unit (2), a shear load cylinder bracket (3), the extension arm for the guiding rods (4) and an adapter for the existing buckling test facility (5), which decouples the axial movement from the horizontal one.

The test panel is cast into panel boxes. The upper box is clamped to the shear frame during testing. The lower box is affiliated with the load transmission. The compressive forces from the axial servo-cylinder are applied by way of a bearing

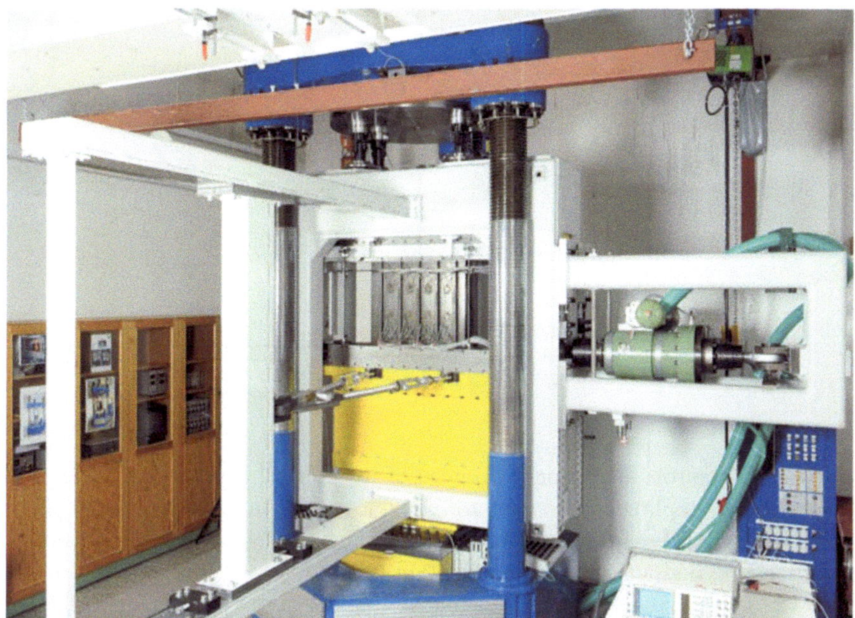

Fig. 16.4 Extended buckling test facility with compression shear test device

into the load transmission unit. The bearing guides the lower load transmission with its box parallel to the upper box. This way the upper and lower panel edges are kept parallel. The shear load is applied by the horizontal cylinder which is mounted in the cylinder bracket sideways to the lower clamping box. In order to allow the lower end of the panel to move according to its radius, the lower clamping box and the frames of the panel are guided through rods. These are mounted on the long extension arm and positioned at a common vertical axis. The axial load is measured with four load cells placed between shear frame and axial support plate. The shear force is measured with one load cell at the horizontal shear cylinder.

In Fig. 16.4 a photo of the new test device within the buckling test facility is shown. The extended facility allows the experimental investigation of the shear-compression-behavior of panels with curvatures representative for typical aircraft fuselage and wing structures.

16.1.4 Dynamic Loading of Cylinders

The Buckling Test Facility has got the capability to apply dynamic axial loads on panels and cylindrical shells. For this configuration a larger three-stage servo valve with shorter response time and higher volume flow, compared to those used with

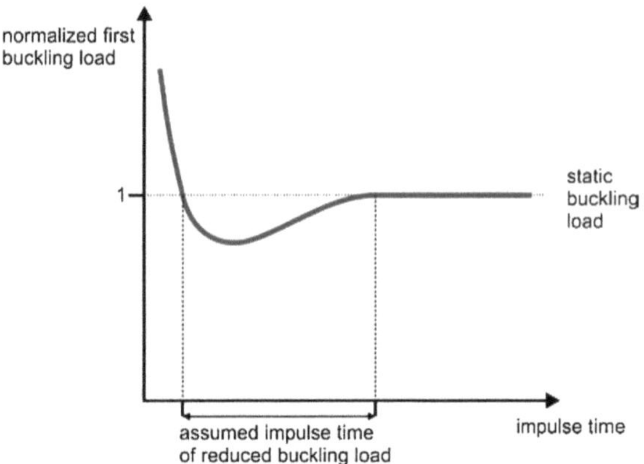

Fig. 16.5 Dynamic buckling load

static tests, controls the piston movement of the vertical hydraulic cylinder. Thus displacement amplitudes of 1 mm are possible with a frequency of up to 50 Hz. Tests with impulsive loading are conducted to investigate the buckling behavior of unstiffened cylindrical shells as a function of the load velocity or impulse time (Fig 16.5). It is assumed that an area of impulse times exists where the first buckling load of unstiffened cylinders is reduced. Some of these sophisticated experiments with phenomenological investigation were done and will be continued with extended devices.

16.2 Variable Component Test Facility

In fields of structural tests different structural elements and components of representative aircraft and vehicle parts have to be tested with regard to static strength and cyclic fatigue strength. These tests are used to validate and qualify manufactured parts. Those parts are tested in original size or scaled down size. To achieve this a variable test stand of adequate size is built up. This test facility consists of a universal mounting plate which is set on a fundament and suspended by springs of pressurized air. At top side of this plate equally spaced T-slots are milled to mount individually columns, angle plates or test frames for bearing specimens and load actuators. So it is possible to build up variable test stands for individual purposes. For example testing of frames from aircraft fuselages can be conducted by using angle plates to mount hydraulic servo cylinders and bearings (Fig. 16.6a). By using the same servo hydraulic cylinders, but different angle plates, beam or box like structures can also be tested with regard to torsion and/or bending (Fig. 16.6b).

In addition, a variable test stand gives the possibility to build up complete test facilities for time limited test procedures to overcome shortcomings of other test facilities which are already in use.

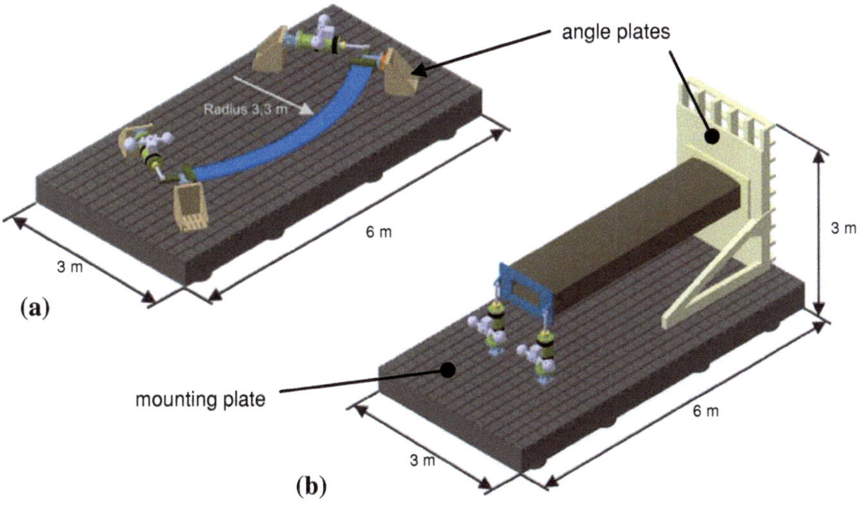

Fig. 16.6 Draft of variable test stand for frame test (**a**) and structural beams tests (**b**)

16.3 Test Devices for Standard Testing Machines

For standard testing machines different test devices have been developed to be used for new scientific investigations beside common known standard tests. Thus a standard test machine can supply tests for investigating special phenomena or to test specimens which are more complex than standard ones.

16.3.1 Stringer Pull-off Device

The omega stringer geometry became popular for modern CFRP-fuselage designs and provides an improved lightweight potential. In order to realize a reliable airworthiness of composite stringers, a novel test has been developed to experimentally analyze the skin-stringer bonding. A small bonding interface between stringer and skin significantly reduces the structural weight but affects the reliability of strength. Furthermore the bonding strength is influenced by the manufacturing process. The joint strength between the composite stringer feet and the skin can be analyzed by pulling off the stringer from the composite skin with the test device shown in Fig. 16.7. The orientation of the specimen can vary between 0° and 45° angles. Tests in the 0° orientation apply mainly peel stresses, whereas orientations between 0° and 45° apply a combination of peel and transversal shear stresses.

Fig. 16.7 Stringer Pull-off device with stringer skin specimen

16.3.2 3D-Biax Device

Even lightweight structures contain thick walled laminates to carry high and concentrated loads. Therefore material properties which are normal to the laminate layup are becoming interesting and have to be measured. According to the requirements to investigate the interlaminar properties of fiber reinforced composites the 3D-Biax Test Device [5] has been developed (Fig. 16.8a). It is based on commonly known methods, such as the ones for shear testing, [6, 7]. It is used for static testing of thick laminate specimens. Those specimens contain a large number of laminate layers and have a special shape (Fig. 16.8b). Load introduction is carried out through form fitting elements corresponding to this shape. A variable angle setup gives the possibility to test specimens with interlaminar tension, compression, shear or combined loads. According to the availability of standard test machines the device can be mounted to static as well as to dynamic test machines. Thus the 3D-Biax device offers great potential to investigate material properties and failure mechanisms with regard to the thickness of laminates.

16.4 Thermo-Mechanical Test Field

In many cases structures are exposed to harsh environmental conditions in addition to mechanical loads. Since the mechanical properties of composite materials deteriorate at elevated temperatures, structural failure can occur even at low level

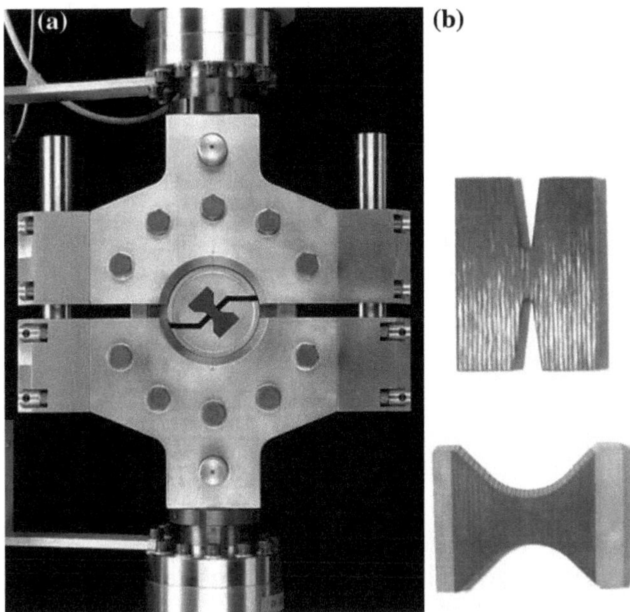

Fig. 16.8 3D-Biax Device for determining interlaminar properties of thick laminates (**a**), specimens (**b**)

mechanical loading. Therefore it is mandatory to accurately predict the temperatures of a structure throughout its complete mission. Computational and experimental examination methods are the basis for such predictions.

On the one hand the thermo-mechanical test field offers the wide potential from determining single parameters like heat transfer coefficients up to the thermal behavior of entire structures. On the other hand its well-known test conditions allow for the verification and validation of simulation tools and results.

The thermo-mechanical test field comprises the THERMEX (thermo-mechanical) test facility and the HRC (High Radiation Compartment) test facility. In order to satisfy particular project requirements, special test set-ups are additionally designed and *built*.

16.4.1 Thermo-Mechanical Test Facility THERMEX

The THERMEX test facility (Fig. 16.9) allows for thermo-mechanical testing of structures like flat plates, curved panels or complex structures with a maximum size of 1 × 0.8 m.

Power controlled infrared heating lamps with a maximum electrical power of 2 × 85 KW facilitate a tailored heating of the test structures up to 1,200°C.

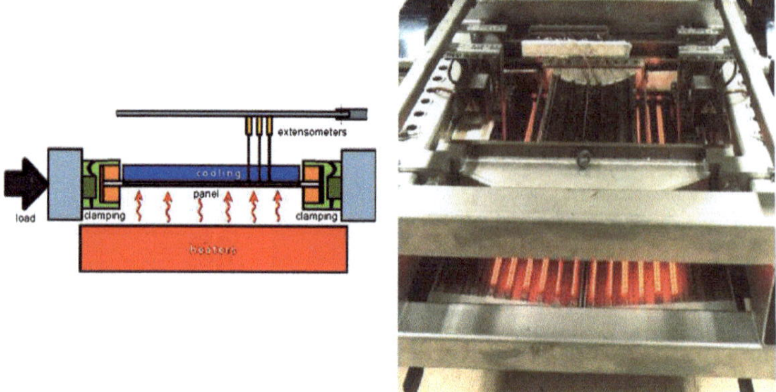

Fig. 16.9 Thermo-mechanical test facility THERMEX

Simultaneously time controlled tensile and compression loads up to 400 kN can be applied. In order to increase higher temperature gradients, parts of the structures can be cooled using liquid nitrogen.

The structural behavior at discrete points (e.g. temperature and strain) is recorded with a data acquisition system offering 64 channels. Additionally local temperature distributions can be documented by means of an infrared camera.

The fields of application comprise investigations of the thermo-mechanical behavior of e.g. hot primary structures, thermal protection systems or nozzle structures all made of conventional or fiber reinforced materials, [8, 9].

16.4.2 High Radiation Compartment

Usually the three heat transfer mechanisms conduction, convection and radiation appear in an overlaid manner. In most engineering applications a lack of clarity exists about their effective, relative contributions to the global heat balance. For a precise analysis within a structural design phase as well as for verification and validation purposes these contributions need to be quantified. Therefore the High Radiation Compartment (HRC) has been developed.

The HRC (Fig. 16.10) contains an insulated test box of $1 \times 0.3 \times 0.3$ m which is integrated in a vacuum chamber. A specific feature of the HRC is a temperature controlled copper plate heater enabling to reach temperatures of 500°C within a short period of time (<2 min). Due to a special surface finish with an emissivity of $\varepsilon = 0.96$ a high radiant flux can be generated. Radiation heat exchange measurements can be conducted within a pressure range of 0.1 mbar $\leq$ p ≤ 1 bar and in an atmosphere of selected gases. This allows for a specific investigation of convective and radiative heat transport fractions. The used insulation material,

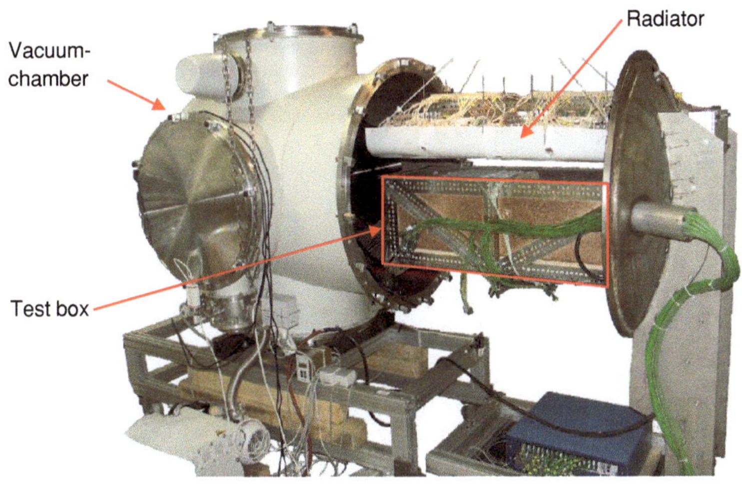

Fig. 16.10 High Radiation Compartment (HRC)

which prevents thermal losses and protects certain structural areas against radiation, also allows the investigation of specific conduction problems like bolted joints, spot welded or glued thermal contact areas, [10, 11].

References

1. Degenhardt, R., et al.: Experiments on buckling and postbuckling of thin-walled CFRP structures using advanced measurement systems. Int. J. Struct. Stab. Dyn. **7**(2), 337–358 (2007)
2. Zimmermann, R., Klein, H., Kling, A.: Buckling and postbuckling of stringer stiffened fibre composite curved panels–Tests and computations. Compos. Struct. **73**, 150–161(2006). Elsevier Ltd, Amsterdam doi:10.1016/j.compstruct.2005.11.050
3. Lovejoy, A.E.: Configuration and Sizing of a Test Fixture for Panels Under Combined Loads. NASA/CR-2006-214520 (2006)
4. Romeo, G., Frulla, G.: Nonlinear analysis of anisotropic plates with initial imperfections and various boundary conditions subjected to combined biaxial compression and shear loads. Int. J. Solids Struct. **31**(6), 763–783 (1994)
5. Hartung, D.: Materialverhalten von Faserverbundwerkstoffen unter dreidimensionalen Belastungen, Dissertation, Forschungsbericht DLR (2009)
6. Arcan, M., Hashin, Z., Voloshin, A.: A method to produce uniform plane–stress states with applications to fibre-reinforced materials. Exp. Mech. **18**, 141–146 (1977)
7. Iosipescu, N.: New accurate procedure for single shear testing of metals. J. Mater. Sci. **2**, 537–566 (1967)
8. Petersen, D., Klein, H., Hildebrand, B.: Recent Thermo-mechanical Test Series and Innovative Test Procedures. In: Proceedings, European Conference on Spacecraft Structures, Materials and Mechanical Testing, pp. 409–414. ESA (2001)

9. Petersen, D., Klein, H., Schmidt, K.: DLR THERMEX-B Test Facility for Cryogenic Tank Wall Structures. In: AIAA 8th International Space Planes and Hypersonic Systems and Technologies Conference, Norfolk, VA, USA, 27–30 April, 1998, pp. 21–30 (1998)
10. Teßmer, J., Spröwitz, T., Wille, T.: Thermal Analysis of Hybrid Composite Structures. In: Proceedings 25th Congress of International Council of the Aeronautical Science, Hamburg, Germany (2006)
11. Wille, T.: Simulation based Optimization of Production Processes using the example of Infrared Heating. Dissertation. DLR-Forschungsbericht. DLR-FB 2010-36 (2010)

Part III
Composite Design

Chapter 17
Compliant Aggregation of Functionalities

Christian Hühne, Erik Kappel and Daniel Stefaniak

Abstract The aggregation of functionalities offers additional benefits to the customers such as reduced weight, reduced life cycle costs and an increased range of applications. For a compliant aggregation of functionalities according to given requirements clear instructions on how to conduct lightweight design are essential, but often not available today. High performance lightweight structures are made from carbon fiber reinforced plastics increasingly. Due to the specific composite manufacturing process four different levels of function-integration are conceivable. The pre-fabrics or components of the composite can include smart materials with enhanced functionalities. The structure design can better exploit the composite potentials of anisotropic material properties. Passive components integrated into the structure provide additional functionalities as for example de-icing and lighting protection. In adaptive systems active elements significantly improves the ability of the structure to adapt changing environmental conditions. The development of the potentials resulting from the compliant aggregation of functionalities is presented in this chapter.

17.1 Motivation and Definition

The integration of function in the context of design issues describes the aim to aggregate multiply useful functionalities in as few parts as possible [1, 2]. Integrated structures lead to reduced assembly and joining efforts and in many cases to a much better and efficient use of the material of the structure. The customer's

C. Hühne (✉) · E. Kappel · D. Stefaniak
Institut für Faserverbundleichtbu und Adaptronik, Deutsches Zentrum für Luft- und Raumfahrt, Lilienthalplatz 7, 38108, Braunschweig, Germany
e-mail: christian.huehne@dlr.de

M. Wiedemann and M. Sinapius (eds.), *Adaptive, Tolerant and Efficient Composite Structures*, Research Topics in Aerospace, DOI: 10.1007/978-3-642-29190-6_17,
© Springer-Verlag Berlin Heidelberg 2013

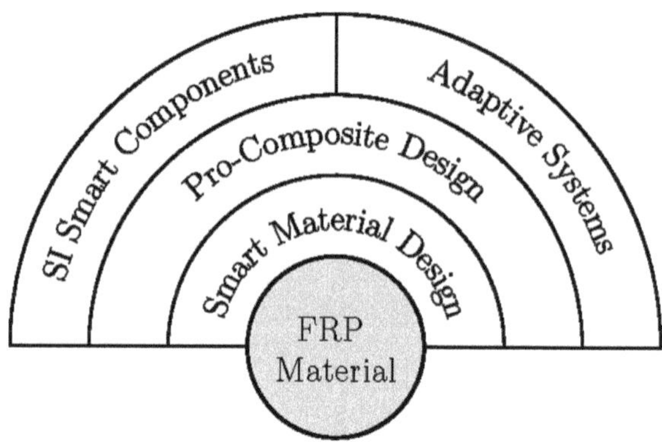

Fig. 17.1 Compliant aggregation of functionalities

value is not directly given by the number of integrated functionalities but the aggregation of functionalities offers additional benefits such as reduced weight, reduced life cycle costs and an increased range of applications.

Although the integration of multiple functionalities is part of almost each design recommendation, clear instructions on how to conduct lightweight design are hardly given. Focusing on fiber reinforced plastics (FRP) the aggregation of multiple functions is gaining more importance. Regarding high-performance structures, mainly carbon fiber reinforced plastics (CFRP) are used. After substituting metal components by composite structures in recent years, the integration of multiple functions is the next challenge for engineers striving for high-efficient structures.

Due to the characteristics of the composite manufacturing processes four different levels of function-integration are conceivable as depicted in Fig. 17.1. Those levels are smart material design, pro-composite design, the implementation of structure integrated (SI) smart components and adaptive systems. While the first two approaches can be applied in a regular manufacturing process without additional efforts, the third one demands additional equipment such as for example piezoelectric actuators, light-emitting diodes (LED) lights or embedded deicing elements. The full structural application approach demands appropriate kinematics and therefore an adapted design, as the whole structure is deformed.

A composite wing that fulfills aerodynamic requirements for natural laminar flow (NLF) is a convenient example to work out the different levels of the integration of functions. Figure 17.2 depicts such a wing structure with various integrated functionalities. Thus in the following, the single levels are briefly introduced and the general idea of each approach is presented.

17.2 Smart Material Design

The adaptation of material properties according to given requirements is summarized by the generic term smart material design. The requirements might be completely different with respect to the aspired aim. The improvement of the fire,

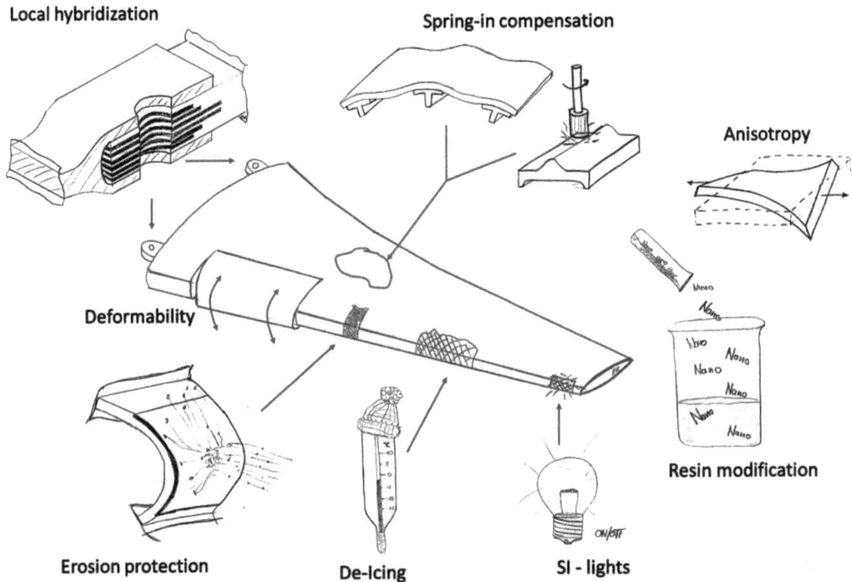

Fig. 17.2 Composite wing—compliant aggregation of multiple functions

smoke and toxicity (FST) properties of resin due to modification on the nano level as well as the enhancement of the bearing strength of holes due to local CFRP-metal hybridization are conceivable examples of smart material design. The design of those smart materials allows the structure to fulfil additional requirements.

17.2.1 Resin Modification

Although the specific composite properties are outstanding they can be further improved by the use of nanoparticles. A modification at nano level allows target-oriented amelioration of the macro-scale properties whereas the kind of improvement depends on the used particles and their properties [3].

In the context of the aggregation of function, the use of resins modified by nanoparticles allow a significant enhancement of FST properties. Thermal and electrical conductivity as well as chemical fire resistance can be affected by the choice of appropriate particles. With respect to the later described inherent deformations due to production which are mainly driven by the composite's anisotropy, modification with nanoparticles promises the reduction of the resin's thermal expansion and chemical shrinkage. That leads to a reduction of induced deformations during the manufacturing process. Adjusting the thermal expansion properties by using nanoparticles allows the design of thermally stable composites, which leads to a considerable improvement of the performance because fewer thermal stresses have to be considered within the design.

Fig. 17.3 Superimposing material specific advantages—CFRP-Metal hybridization (*left*) local hybridization, (*right*) surface protection

17.2.2 Fiber Metal Laminates

The development of fiber metal laminates is basically motivated by the idea to combine the advantages of both constituents. The combination of these materials can be performed for an entire part or only a part of it as well as for the entire laminate or only its surface. Therefore a distinction must be made with respect to the materials application, either as structural material or as local reinforcement or as surface protection. Figure 17.3 shows selected applications of CFRP-metal hybridization.

17.2.3 Structural Material

Glass laminate aluminium reinforced epoxy (GLARE) represent the best-known, successful application of FRP-metal hybrids, which is applied as structural material for the upper fuselage section of the Airbus A380. GLARE shows outstanding fatigue properties, an excellent impact resistance and a good residual strength. Further improved properties are the flame resistance and the corrosion behaviour [4, 5].

A new approach for high performance structural applications uses a new hybrid laminate lay-up with a metal layer thickness of less than 0.08 mm and a metal volume fraction of less than 8%. Therein the metal layers replace ±45° and 90° plies. Hence, stiffness and strength in 0° direction are not reduced in comparison to the use of varying fiber directions. In the context of function integration metal layers generate improved damage tolerance properties as they deflect interlaminar fracture into zones with increased delamination surfaces. They serve as crack arrest layers and their energy dissipation is elevated as a consequence of the increased number of interfaces within the laminate.

Beside enhanced transverse stiffness and strength properties compared to unidirectional laminates, CFRP-UD/steel-laminates show higher values for residual compression strength in compression-after-impact (CAI) tests and demonstrate up to 65% higher elastic modulus than common multi-axial CFRP reference specimens [6].

17.2.4 Local Reinforcement

In addition to their weight saving potential, composites enable the application of high integral design philosophy avoiding structural interconnections. However, the size of composite structures is often limited due to higher tooling complexity, rising manufacturing risks, restrictions of logistics and handling as well as inspection, maintenance and repair needs. Therefore bolted connections are unavoidable for aerospace applications. However, fiber laminates, especially highly orthotropic ones, are characterized by a low bearing strength and low shear load capabilities.

The usual practice to account for these low bearing and shear capabilities of the fiber laminates is to locally build up the joining area. However, the resulting thickened laminate leads to eccentricities in the load path and consequently to secondary local stresses and bending moments, especially for single-sided laminate thickening. This local thickening increases the complexity of all the adjacent parts resulting in larger metallic fittings and longer fasteners, further increasing the connection's weight and reducing the joint efficiency.

The local hybridization approach uses a fiber-metal laminate within the fastener areas as outlined in Chap. 19. By the gradual substitution of composite plies by high-strength steel or titanium foils as it can be seen in Fig. 17.3, the remaining continuous CFRP plies guarantee the load carrying capability and ensure that the fatigue behaviour is dominated by the outstanding properties of the carbon fibers.

At the same time the metallic fraction within the hybrid increases the bearing and shear strength capabilities as well as notched tension and compression strength, which are essential properties to obtain a high bolted joint efficiency. This approach eliminates the unwanted thickening, avoids the consequence of secondary bending moments and reduces the size and weight of fittings and fasteners significantly [7, 8].

17.2.5 Abrasion Protection

In many areas FRP do not meet the abrasion requirements even though their specific properties would justify their application for static or dynamic loading. For example, the wing's leading edge demands for abrasion resistant materials in combination with high surface quality requirements. As a pure composite surface would not be able to sustain the environmental loads of a leading edge, a hybrid design promises sufficient durability. Thus, very thin metal foils on the surface of the composite leading edge provide an abrasion resistant outer layer which can be applied only to parts of the wing without creating steps or gaps as it can be seen in Fig. 17.3. This example represents a compliant integration of an additional function without affecting the performance of the composite part.

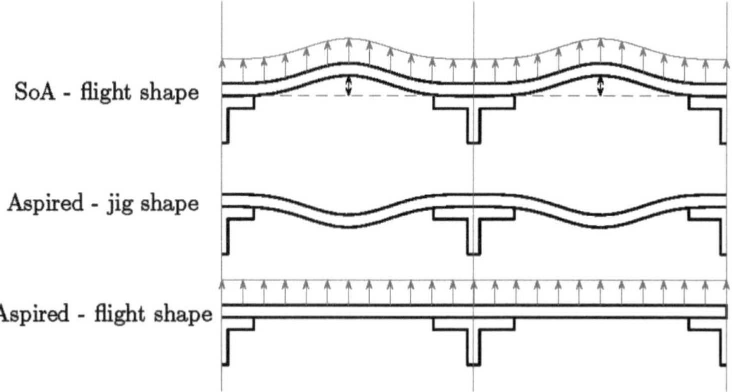

Fig. 17.4 Skin waviness due to aerodynamic loads and spring-in deformations

17.3 Pro-Composite Design

CFRP usually show anisotropic material behavior due to the combination of fibers in different orientations. That particular characteristic symbolizes strength and weakness of composite material simultaneously. Target-oriented interaction between design and analysis allow the adaptation of the structural behavior, as geometry and lay-up can be optimized according to given requirements [9]. Whereas structure tailoring is a key function for high-performance composite parts on the one hand the underlying material anisotropy is responsible for unwanted deformations due to the curing process with a resin system in production on the other hand. Those deviations are often referred to as spring-in deformations as outlined in Chap. 22. Commonly, when net-shape manufacturing is aspired, compensation strategies are necessary to counteract the spring-in difficulties. However, detailed and verified knowledge of deformations induced by the material anisotropy and the production process in combination with robust compensation strategies allow further integration of structural functions. Tailoring of the as-built geometry by use of shape compensation strategies gives further freedom within the design. In the context of enhanced wing concepts, NLF is regarded as one key issue in order to reduce aerodynamic drag significantly. Consequently, new and more stringent manufacturing tolerances for the geometry are required to achieve this goal. It is crucial to avoid waviness as well as steps and gaps of upper wing panels in order to achieve surface properties required for NLF. The target is to optimize the structural performance and shape of an upper wing cover for cruise loads, in order to receive NLF at this condition.

Within state of art (SoA) manufacturing it is still a challenge to predict spring-in deformations. According to Fig. 17.4, SoA flight shape—even if globally considered in the production jig-shape—shows certain local skin waviness induced by aerodynamic loads and as well deformations due to the production process.

Whereas the aspired flight shape shows no local waviness under aerodynamic loads the corresponding jig-shape is given by a superposition of compensated deformations due to production and inverted aerodynamic waviness. Regarding effective directions, aerodynamic waviness opposes the direction of the deformations due to production. Thus, utilizing those unavoidable but adjustable deformations to equalize the waviness due to aerodynamic loads is a promising approach to achieve the aspired flight shape. Fortunately, experimental investigations reveal that magnitudes of both effects are in the same range. With the use of commercial finite element tools relatively precise calculations of panel waviness can be achieved in coupled CSM-CFD simulations.

However, the simulation of the manufacturing process is rather well developed. In order to define tooling that compensates for such induced deformations, simple straightforward strategies mainly focus on part deformations and reveal promising results for extruded angled profiles. Therein, it is assumed that a defined composite material shows reproducible characteristics for one chosen production process cycle. Material parameters derived from experimental findings on a coupon test level are applicable to structures on a macro level [10, 11]. Stringer stiffened panels, however, show a rather complex deformation behavior because spring-in deformations are superposed with deformations induced by a thermal mismatch of stringer feet and skin during the curing process. Especially for T-stringers the effect of the filler geometry and the filler material needs further investigations. The aforementioned easy-to-use procedures are generally suitable for the application to more complex problems but they have to be validated experimentally. Fig. 17.5 shows the aspired procedure to define tooling geometries that compensate the fabrication induced deformations. Consequently a pro-composite design in this example enables the production of an upper wing cover suitable for NLF and thus for a further enhanced performance of the structure.

17.4 Structure Integrated Smart Components

The conversion of an electrical potential into heat is realized within resistance heating devices, as they are applied in de-icing applications for example. The piezoelectric and the inverse piezoelectric effect are exploited to transfer mechanical strain into an electric potential and vice versa. Generally piezoelectric ceramics can be used as actuation and as sensing elements for example in active vibration control, active structure acoustics control as shown in Chap. 35 or in the field of structure health monitoring as outlined in Chap. 37.

17.4.1 De-Icing

The icing of wings is still a certain risk in modern aviation. Based on an electro-thermal principle a de-icing system has been developed which is suitable for the integration in current composite wings [12]. An electro-conductive carbon-

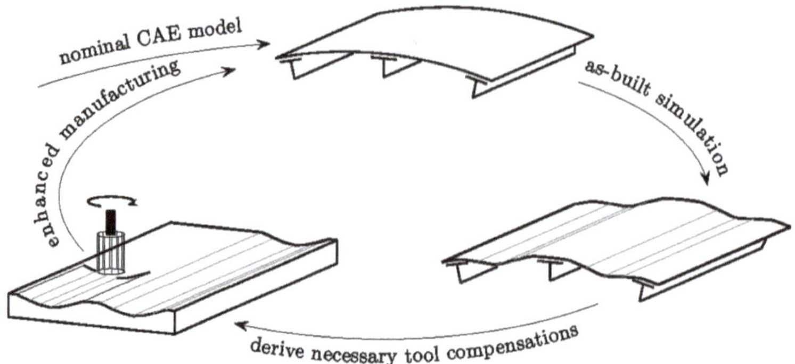

Fig. 17.5 Tool compensation process supported by numerical analysis and effective compensation strategies. Jig shape is adapted accounting for aerodynamic loads and fabrication induced deformations

Fig. 17.6 Infrared camera picture of a structure integrated De-Icing system

fiber-fabric has been integrated directly into the laminate and subsequently contacted electrically. The integration has been done in one-shot together with the production of CFRP parts and demonstrates the integration of effective function into the structure. Experimental validation obtains a homogeneous heating with a distinct borderline as illustrated in Fig. 17.6.

Due to the comparatively high energy consumption of such systems, heated areas are separated into smaller segments, which, on the other hand, demand more complex control. Experiments prove the functionality of de-icing systems integrated into the structure.

17.4.2 Structure Integrated Lighting System

Beside de-icing systems, structure integrated lights are another example of the integration of function [13] that have already been realized in demonstrator structures. With regard to the requirements of a composite wing such as a step- and gapless design of the outer skin, those direct integrated LEDs are suitable as positioning lights at the wing's leading edge. Figure 17.7 shows a generic, single curved CFRP structure with an integrated LED device.

Fig. 17.7 Structure integrated (*SI*) light systems

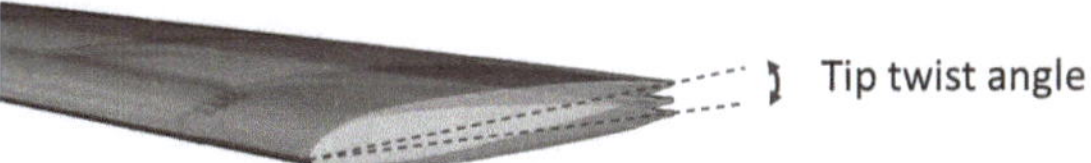

Fig. 17.8 Tip twist rotation induced by surface actuators

Within the automotive industry, light systems have gained an increasing importance in the last years. Beside the improved illumination, the design of illumination systems inside and outside the car has become a key-issue for car manufacturers or their suppliers. The automobile manufactures Audi for example utilizes the OLED (organic LED) technology in order to realize new illumination systems for interior or rear light applications [14]. That underlines the demand for new integrated light technologies for automotive applications.

17.4.3 Actuator Induced Morphing

Whereas incorporated thermal de-icing and lighting systems do not affect the geometrical component shape the use of actively controllable elements enables the shape control or morphing of composite structures as outlined in Chap. 32.

Nevertheless, the integration of piezoelectric actuators is accompanied by certain barriers, which directly affect the structural performance. Generally, opposing demands are competing: on the one hand, maximal actuation with maximal stroke is aspired, on the other hand, structural strength and durability should be identical to passive, non-actuated structures. Furthermore, operational parameters like voltage of piezoelectric actuators should be of a manageable magnitude. Thus, monolithic piezoelectric ceramics are hardly applicable as they demand high operational voltages. Therefore multi-layer actuators are the elements of choice as they generate comparable forces with significantly lower operational voltages.

Piezoelectric actuators are successfully applied to helicopter blades (model scale) in order to reduce rotor noise, fuselage vibration and fuel consumption [15]. With respect to the considered composite wing, integrated actuators are not expected to deform the entire wing, but can be utilized to realize actuated flaps [16] for example. Due to the direct integration of piezoelectric actuators into the laminate, the morphing of the blades can be realized without friction and wear (Fig. 17.8), as it would be the case for common mechanical components such as gears and levers.

Fig. 17.9 Morphing leading edge high lift device

The integration of those flat piezoelectric actuators into the laminate represents a disturbance of the structure itself. Due to the relative large actuator thickness, the mechanical performance is reduced because fiber waviness or neat resin areas support the initiation and growth of cracks. The integration of function requires a multi-objective optimization of the primary load carrying capability and additional functionalities.

17.5 Adaptive Systems

As mentioned above, a composite wing suitable for NLF requires a step- and gapless design of the outer skin surface as outlined in Chap. 31. A morphing high lift system at the leading edge of the wing is one alternative to conventional high lift devices such as slats which can satisfy the requirement of even and undisturbed surfaces.

Analogously to the SI actuator induced morphing the requirement for maximum morph-ability and maximum strength are competing. In contrast to the afore-mentioned embedded elements adaptive systems require a full adaptation of the carrying substructure. In order to get an efficient system structural stiffness of morphing areas must be small to give the required flexibility to the actuated degrees of freedom. As indicated in Fig. 17.9 morphing is support by the structural shape, whereas structural resistance to aero loads remains constant during the morphing process.

In order to control the shape of the morphing structures with respect to aero-dynamic loads appropriate kinematics are used which ensure that allowable strain limits are not exceeded for the whole laminate. Due to the large deformations of the leading edge skin carbon composites are hardly applicable due to their small yield strength. Thus, glass fiber material is utilized for a demonstrator structure as it is depicted in Fig. 17.9.

To sum up, the key in the context of a morphing high lift system at the leading edge of a wing is to design a skin which is stiff enough to carry aerodynamic forces on the one hand and on the other hand to ensure flexibility, which is necessary to sustain the change of structure and the related strain level without damages [16, 17].

References

1. Roth, K.: Konstruieren mit Konstruktionskatalogen: Band 1: Konstruktionslehre. 3. Auflage, erweitert und neu gestaltet. Springer, Berlin (2000) ISBN 3-540-67142
2. Fritsch, M.: Zur integralen Funktionsausnutzung von Bauelementen. Feingerätetechnik Technisch-wissenschaftliche Zeitschrift für Feinmechanik, Optik und Meßtechnik 16, Heft 9 (1967) ISSN 0014-9683, S.402-404
3. Arlt, C.: Wirkungsweisen nanoskaliger Böhmite in einem Polymer und seinem Kohlenstofffaserverbund unter Druckbelastung, Dissertation, Magdeburg (2011)
4. Vlot, A., Gunnink, J.G.: Fiber Metal Laminates an Introduction. Kluwer Academic Publishers, Dordrecht (2001)
5. Wu, G., Wang, J.M.: The mechanical behavior of glare laminates for aircraft structures. JOM J Miner. Met. Mater. Soc. 57(1), 72–79 (2005) DOI: 10.1007/s11837-005-0067-4
6. Stefaniak, D., Fink, A., Kolesnikov, B., Hühne C.: Improving the mechanical properties of CFRP by metal-hybridization. In: International Conference on Composite Structures ICCS16, Porto, June 2011
7. Kolesnikov, B., Herbeck, L., Fink, A.: CFRP/titanium hybrid material for improving composite bolted joints. Compos. Struct. 83, 368–380 (2008)
8. Fink, A., Camanho, P.P., Andrés, J.M., Pfeiffer, E., Obst, A.: Hybrid CFRP/titanium bolted joints: Performance assessment and application to a spacecraft payload adaptor. Compos. Sci. Technol. 70, 305–317 (2010)
9. Spröwitz, T., Hühne, C., Kappel, E.: Thermal Aspects for Composite Structures—From Manufacturing to In-Service Predictions. In: CEAS 2009, 26–29 Oct 2009, Manchester (2009)
10. Kappel, E., Stefaniak, D., Hühne, C.: A semi-analytical simulation strategy and its application to warpage of autoclave-processed CFRP parts. Compos. Part A Appl. Sci. Manuf. 42(12), 1985–1994 (2011)
11. Kappel, E., Stefaniak, D., Hühne, C.: Kompensation faserverbundspezifischer Fertigungsdeformationen im Werkzeug—Ein semi-analytischer Ansatz. In: NAFEMS Online Magazin 03/2011, Ausgabe 20
12. Jürgenhake, C.: Strukturintegrierte Leiterbahnen, Diplomarbeit, IB 131-2009/33
13. Franken, H.: Entwicklung eines Faserverbundkonzeptes einer strukturintegrierten Widerstandsflächenheizung für Enteisungssysteme und beheizbare Werkzeuge. IB 131-2004/08, DLR Braunschweig (2004)
14. Audi MediaInfo. Tomorrow's lighting technologies, Press Release http://www.audiusanews.com/print.do?id=2690
15. Monner, H.P., Opitz, S., Riemenschneider, J., Wierach, P.: Evolution of Active Twist Rotor Designs at DLR. In: 49th AIAA/ASME/ASCE/AHS/ASC Structures, Structural Dynamics, and Materials, Schaumburg, IL, 7–10 April 2008
16. Monner, H.P., Sinapius, M., Opitz, S.: DLR's morphing activities within the European network. In: AVT—Symposium on Morphing Vehicles, Evora, Portugal, pp. 1–31 (2009)
17. Monner, H.P., Kintscher, M., Lorkowski, T., Storm, S.: Design of a smart droop nose as leading edge high lift system for transportation aircrafts. In: AIAA Conference, Palm Springs, USA (2009)

Chapter 18
Boom Concept for Gossamer Deployable Space Structures

Marco Straubel and Michael Sinapius

Abstract Deployable structures are necessary to realize large but weight-efficient space systems. DLR provides a deployable mast that can be used either at once to realize e.g. long dipole antennas of some ten meters or to setup structures that use this mast as basic building block structure. This section shall, therefore, enable a basic insight on the concept and of the resulting challenges. Moreover, a deployment test series under weightlessness is presented and evaluated to show possible concepts of deployment control and demonstrate the potentials.

18.1 Large Gossamer Space Structures

Gossamer structures are optimized for their intended task. They provide only the required amount of stiffness. If the considered mission requires no significant acceleration for rapid repositioning or reorientation of the satellite, the required stiffness is very low and the resulting structures are very filigree, thin or *gossamer*. The following sections give a brief introduction in concepts for deployable structures.

18.1.1 Exemplary Deployable Space Structures

The currently available deployment systems can be basically clustered in three different groups: the pantographic systems, the elastically deformed systems and the inflatable systems.

M. Straubel (✉) · M. Sinapius
Institute of Composite Structures and Adaptive Systems, Composite Design,
Deutsches Zentrum für Luft- und Raumfahrt e.V. (German Aerospace Center),
Lilienthalplatz 7, 38108, Braunschweig, Germany
e-mail: marco.straubel@dlr.de

M. Wiedemann and M. Sinapius (eds.), *Adaptive, Tolerant and Efficient Composite Structures*, Research Topics in Aerospace, DOI: 10.1007/978-3-642-29190-6_18,
© Springer-Verlag Berlin Heidelberg 2013

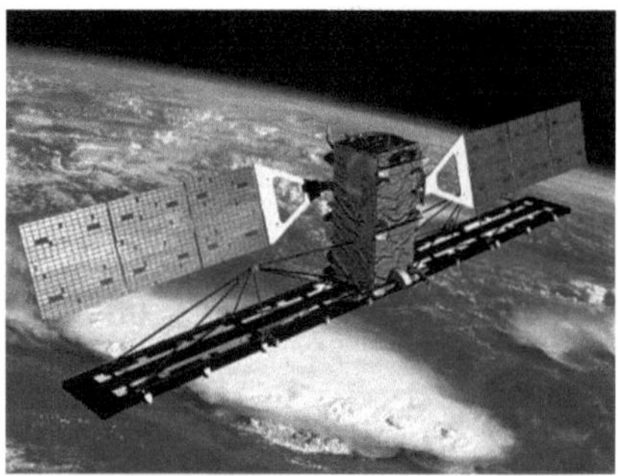

Fig. 18.1 RadarSat II with deployed SAR antenna and solar array (Courtesy of Canadian Space Agency)

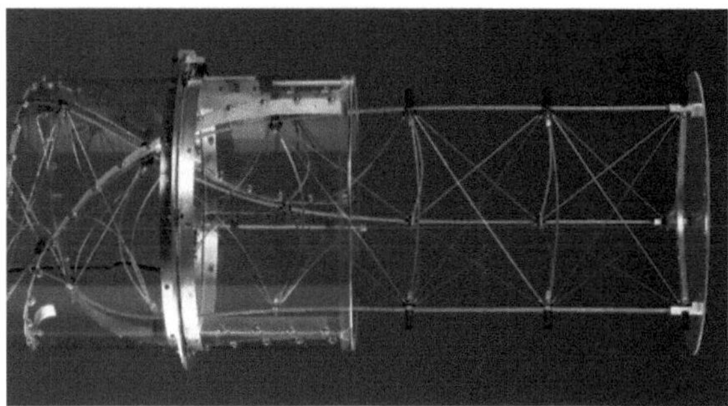

Fig. 18.2 CoilABLE boom (Courtesy of ATK)

Pantographic or manifold structures are made of stiff parts that are interconnected with hinges. The hinges are, in most cases, spring driven and use a locking mechanism that fixes the hinge after reaching its final position. The start of the deployment is commonly triggered by pyro-mechanical cutters. Motor actuated mechanisms are only used if the deployment needs to be reversible or the deployment velocity has to be controlled for technical reasons (e.g. requirement from Attitude and Orbit Control System (AOCS)). Figure 18.1 shows the classical application of such structures. Canada's RadarSat II uses one deployable radar antennas and two unfurlable solar arrays.

Another space proven system is the CoilABLE Mast from the US company ATK. The mast can be packed by twisting it around the length axis. As the boom is under pretension in its compressed state, the deployment needs to be controlled by a rotating nut (Fig. 18.2). Other truss concepts are Northrop–Grumman's *AstroMast,* ATK's *FAST* and *ADAM* masts, and the *TriLok* mast [1].

Figure 18.3 depicts a further concept for deployable structures. The shown flexible hinge element is able to perform large deflections without relying on classical hinge concepts, involving friction affected bearings, and features a high robustness against small distortions of the packed and deployed structure due to thermo-mechanical effects.

In combination with struts it can be used to design various one- and two-dimensional structures [2].

Fig. 18.3 Quadrilateral reticulations in (**a**) deployed and, (**b**) stowed configuration (Courtesy of CSA Engineering Inc.)

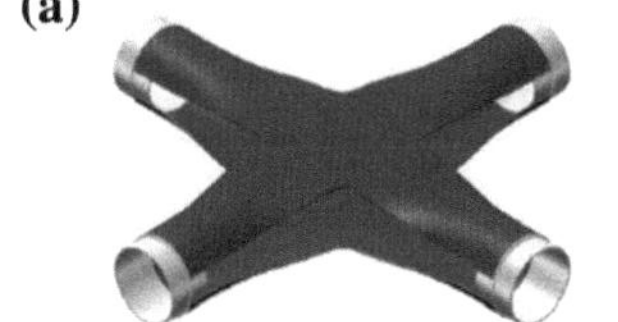

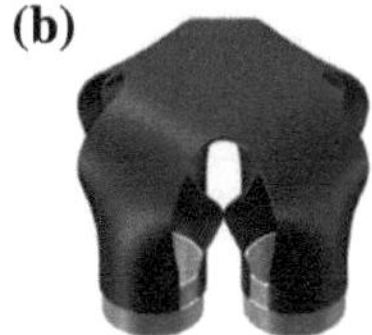

Fig. 18.4 Inflatable boom during folding (Courtesy of ASTRIUM)

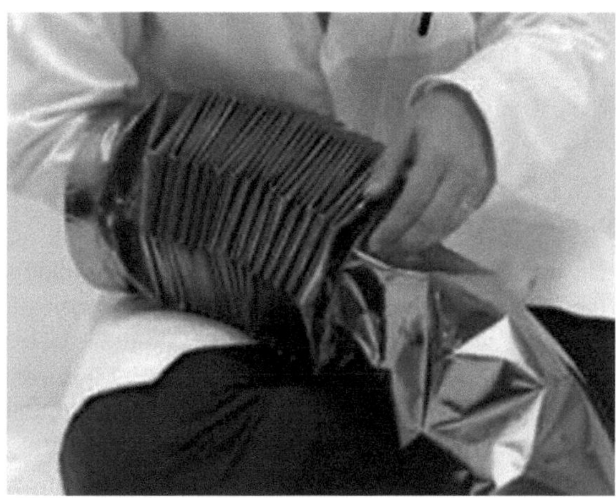

The third basic deployment concept is the inflatable one. It is characterized by a low material thicknesses and low weight (Fig. 18.4). However, an inflation system is required that contains either gas tanks with valves or pyro-active gas generators that have to be respected in the mass calculation. If inflatable structures need to maintain their stiffness over a longer time period, the gas inflation system needs to provide the pressure over the entire service time or the structure needs to be cured or *rigidised* in any kind after successful deployment. Different kinds of rigidization have been investigated: The curing of a resin impregnated composite by UV-light [3], by implemented resistive heaters [4] or the rigidising of an aluminium layer by stretching the material beyond its yielding point [5]. All rigidising concepts have been tested and successfully verified. However, the curing of composites under space conditions is a very inhomogeneous process that leads therefore to asymmetric curing which can lead to unintended distortions of the cured structure. Thus, they can be only recommended for structures with medium to low accuracy requirements.

Table 18.1 provides an assessment of the different deployment techniques depending on some characteristic parameters. As is shown in this table the appropriate deployment concept depends on the mission requirements. Stiffer structures are heavier and need more stowage volume but provide a higher geometrical accuracy.

Table 18.1 Trade off between deployment types

	Pantographic structures	Elastically deformed structures	Inflatable structures
Stiffness	++	+	+
Mass	o	++	+
Stowage volume	o	+	++
Shape accuracy	++	++	-
Controllability of deployment	++	++	+
Flight heritage	++	+	+

++ very good; + good; o moderate; - poor

18.1.2 Challenges and Needs

Despite of the high flight heritage of pantographic systems their numerous hinges increase the system complexity and thus the probability of deployment failures. In order to address this aspect ECSS requires each mechanism to be operated at minimum 50 times on-ground to verify its lifetime capability [6].

Thus, also a mechanism that needs to be actuated only once in space has to prove its capability to satisfy this test requirement.

Apart from the mechanisms, also the whole structure needs to be tested in order to validate the resonance frequencies in deployed state. Moreover, deployment tests under representative mechanical and thermal environment need to be performed. Depending on the system stiffness and size such tests require complex gravity compensation equipment and the use of cost-intensive thermal-vacuum chambers. Future large structures may become even so large that any ground testing will become impossible (e.g. 100×100 m solar sails). Thus, new concepts of verification need to be developed that use sub-scale tests to tune scalable finite element models and predict the behaviour of the final full scale structure only by simulation.

Concluding this, the need for a system complexity reduction as well as a reliable verification method is noticed.

18.1.3 Applications

Almost every huge space structure could be build using deployable concepts. However, this section will concentrate on 4 representative and most promising ones: Solar Sails, Antennas, Solar Arrays, and Deorbiting devices.

Solar sails are an alternative propellantless propulsion system utilizing the Sun's radiation pressure. At Sun-Earth distance a pressure of maximal 9.1 µPa can be used to accelerate a spacecraft. As this pressure is extremely small, the spacecraft needs to have a low mass-to-sail area ratio to generate a significant "thrust".

Fig. 18.5 On-ground deployment demonstration of a 20 × 20 m solar sail (ESA contract, DLR delivers booms and sails)

DLR has demonstrated in 1999 the deployment of such a solar sail by use of the introduced booms. During an on-ground demonstration a 400 m^2 solar sail was deployed autonomously from the centre module (Fig. 18.5). The deployment mechanism uses a boom root deployment.

Space antennas for radio communication and radar surveillance are a second application for deployment techniques. So far most deployable antennas consist of folded arrays like manifold structures as previously shown in Fig. 18.1. These structures were flown many times and have proven their reliability but have the disadvantage of a high mass.

To evaluate the lightweight potential of deployable gossamer SAR antennas, DLR and ESA established cooperation for research. The mechanical part of a deployable SAR antenna has been developed [7] in the study. The final design provides an antenna aperture of 40 m^2 with a complete mass of less than 60 kg. Figure 18.6 shows the result. The comparable antenna of RADARSAT-2 has a mass of 750 kg. Although RADARSAT-2 has a 4 times higher operational radar frequency and, therefore, increased shape accuracy requirements, this comparison shows the mass saving potentials of huge gossamer antennas.

Another application for gossamer structures is deorbiting. The increasing population of space debris is an upcoming problem. Out-of-service satellites as well as upper stages from launchers generate a risky environment for new satellites passing or sharing the same orbits. One possibility to decrease the amount of debris is a passive deorbiting of space systems by using the aerodynamic drag within lower earth orbits. Therefore, the very thin residual atmosphere is used to generate a constant decelerating force. To use this effect, large surfaces need to be deployed to increase the area-to-mass ratio (Fig. 18.7).

The last but not least application for deployable structures considers solar arrays for power generation in space. Conventional systems use foldable rigid planes or thin carrier foils that are fixed at deployable frames (Fig. 18.8). Deployable systems like the above introduced antenna could be a gossamer

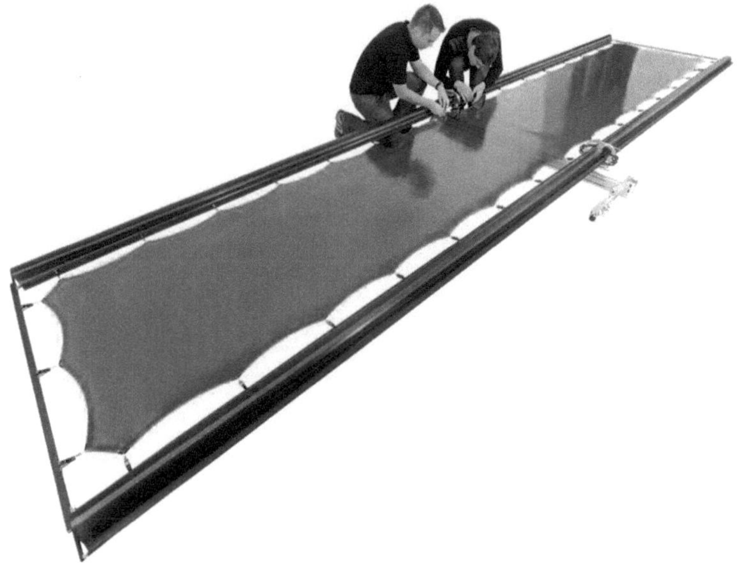

Fig. 18.6 1:3 sub scale model of a deployable SAR antenna (model size 6 × 1.4 m)

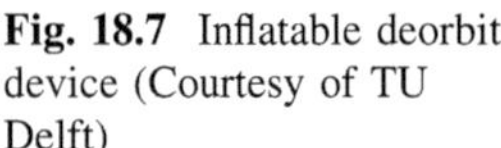

Fig. 18.7 Inflatable deorbit device (Courtesy of TU Delft)

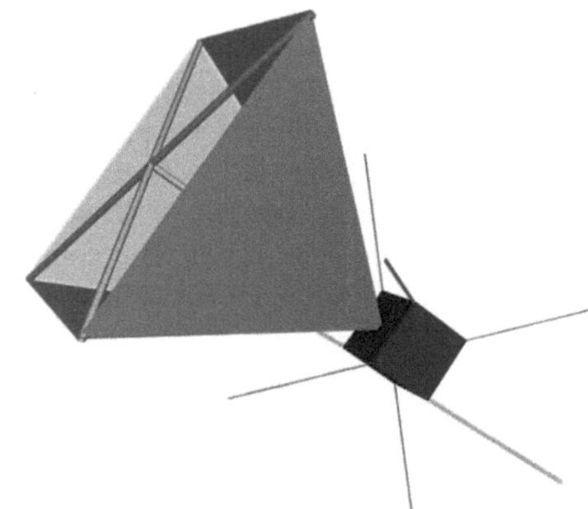

alternative to those conventional systems. The recent developments on flexible thin film solar cells support this trend.

18.2 DLR's Deployable Boom

The following sections introduce DLR's concept for deployable gossamer boom being as building block for versatile deployable space structures.

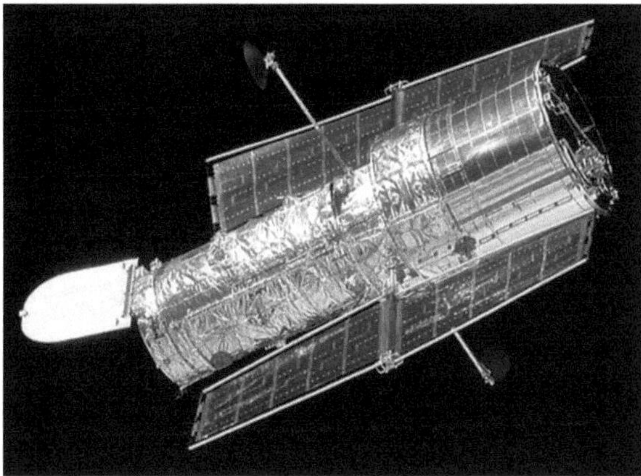

Fig. 18.8 Hubble space telescope with deployed solar arrays (Courtesy of NASA)

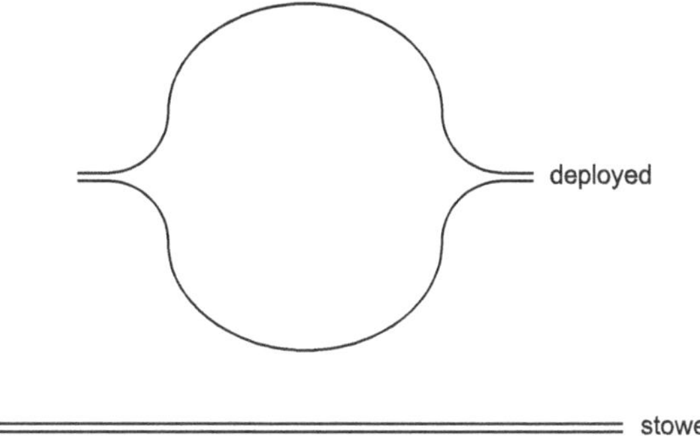

Fig. 18.9 Sketch of the booms cross section in deployed and stowed configuration

18.2.1 Concept

The basic concept of DLR's coilable composite lightweight booms for deployable space applications is quite simple: Two half shells of the boom are separately manufactured from a very thin CFRP laminate (0.1 mm thick) and are co-bonded afterwards. The cross section of the resulting boom is displayed in Fig. 18.9.

The upper sketch shows the cross sections shape in the unstressed, deployed configuration. Due to external forces, the boom can be flattened. Thereby, the bending stiffness in on direction decreases significantly and the boom can be coiled around a carrier hub.

18.2.2 Mechanical and Thermal Properties

During the last years of research a sophisticated boom version for the solar sail application as well as a low cost version has been developed. The related properties of both setups are listed in Table 18.2. However, these are only two possible

Table 18.2 Properties of two different boom configurations [8, 16]

	High performance boom	Low cost boom
Layer setup	Up to 12 layers of unidirectional prepreg per half shell	One layer of 0/90 prepreg per half shell
Layer thickness	0.1 mm	0.1 mm
Manufacturing	Autoclave based (expensive, restricts max. boom length)	Out of autoclave (use of modular curing tent)
Specific costs (incl. manpower, material; excl. mould)	1 k€/m	0.1 k€/m
Specific mass	100 g/m	62 g/m
Bending stiffness about X-Axis	5300 Nm^2	2,661 Nm^2
Bending stiffness about Y-Axis	3,850 Nm^2	3,325 Nm^2
Max. bending moment about X-Axis	±84 Nm	±13.24 Nm
Max. bending moment about Y-Axis	±52 Nm	±23.91 Nm
1st bending mode of 5 m long sample	3.01 Hz	3.26 Hz
CTE in longitudinal direction	$\leq0.5 \times 10^{-6}$ 1/K	$\leq2 \times 10^{-6}$ 1/K

configurations. The provision of tailored booms in relation to the given requirements is one advantage of this boom concept.

For further details on thermo-mechanical effects please refer to the doctoral thesis of Christoph Sickinger [8] or some previously published papers [9–12].

18.2.3 Deployment Control

Comparable to a coiled strip of sheet metal, the coiled boom intends to deploy itself by its own internal stored elastic strain energy from the moment the restraining force disappears. During on-ground tests, this self deployment is continuously and well directed as it is stabilized due to friction between the different layers enforced by gravity. But simulations and test had shown that the behaviour in weightlessness is more complex. Figure 18.10 shows the simulated deployment behaviour. The deployment process is very chaotic and shifts to a discrete amount of partial deployments.

As this deployment is not predictable and not controllable, it is not a suitable process for the majority of space applications. Thus, different concepts to control the deployment and results from test under artificial weightlessness are given hereafter.

The first concept is the *Electric Root Deployment*. The basic idea is to wrap the boom around a motor driven core that is located in the spacecraft. Like a cable winch the boom is deployed by the driven central hub that extracts the boom in a controlled and continuous manner from the module and guarantees a smooth deployment. This concept has been developed by the German companies Kayser-Threde and INVENT within the first Solar Sail study [17].

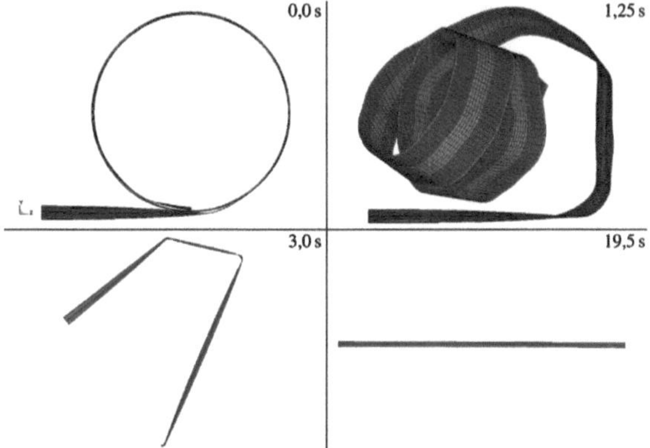

Fig. 18.10 Simulation of boom deployment within 0 g environment [8]

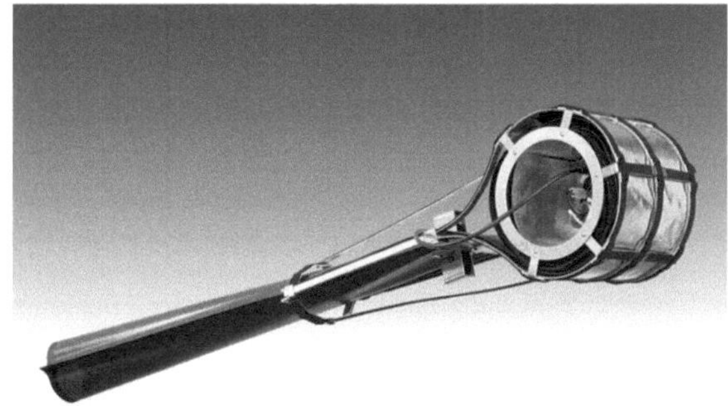

Fig. 18.11 Electrically driven tip reel deployment mechanism

One advantage of this concept is the possibility to coil more than one boom on the same cylinder. The past Solar Sail design includes a deployable cross like structure of four booms that are simultaneously deployed from one central hub. This concept is very weight efficient but has one disadvantage: The transition zone, required by the boom to evolve its full cross section and stiffness, is located next to the central core. Thus, the weakest part of the boom meets the most loaded region, the boom root.

The second concept, the *Electric Tip Deployment*, uses a comparable approach as the root deployment but the deployment location is changed. To counteract the collocation of highest load and lowest stiffness the weak transition zone is transferred from the root to the tip. This is realised by the design of the mechanism visible in Fig. 18.11. It contains a cylindrical part for storage of the coiled boom and is equipped with two wheels that are driven by a battery supplied electric motor.

In contrast to root deployment the mechanism detaches oneself from the main module after activation of the motor. The partly deployed end of the boom—as seen left hand in Fig. 18.11—is fixed at the centre module.

In relation to the previously introduced concept, one mechanism deploys only one boom. Hence, in case of deployment of multiple booms the specific weight per deployed boom is higher than in the other concept. But as the deployment

Fig. 18.12 Hybrid
deployable/inflatable boom

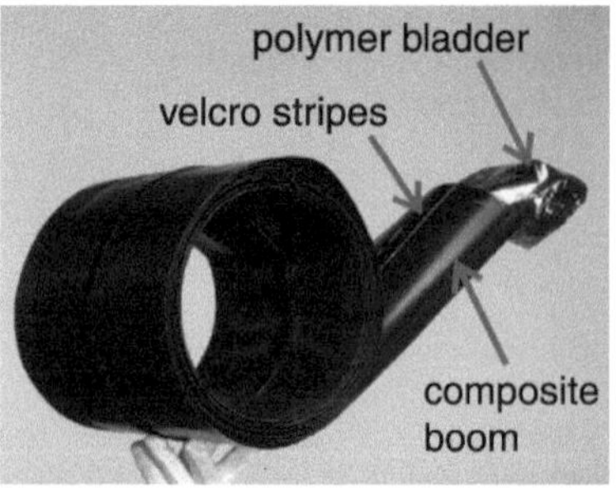

mechanisms are jettisoned after deployment, the total mass of the deployed structure is lower than of the one utilizing the root deployment principle. Thus, the electric tip deployment is the supreme solution for deployment of solar sails where the ratio between sail area and mass is proportional to the acceleration achieved by the solar radiation.

The third concept, the *Inflating Tip Deployment*, realises a completely other deployment principle and comes up without any electric motors or moveable parts. It consists mainly of two basic elements that control the deployment. As shown in Fig. 18.12 the boom is equipped with a Velcro layer on each side.

During coiling the hook part of one boom side locks into the loop part of the other boom side and prevents a self deployment of the packed mast. To deploy it anyhow, a gastight polymer hose of only 12 μm material thickness is inserted into the hollow boom already upon manufacturing. Once pressurised the bladder acts as pneumatic actuator that deploys the boom by locally breaking the Velcro connections.

Depending on the level of controllability the needed inflation gas can be provided by conventional gas tanks in combination with controllable valves or special gas generators that produce gas by decomposition of solid materials.

Compared to the other two concepts the inflating tip deployment requires the application and insertion of extra material to the booms which increases the specific mass per length by around 20–50%. However, the needed mass and the complexity of the deployment supporting device in the main module are reduced.

At least, all three introduced deployment control concepts have their pros and cons. The optimum concept depends on the desired application.

18.3 Tests

To verify and to attest the necessity of the presented deployment control concepts, a test campaign for two concepts under weightlessness has been planned and finally conducted in February 2009. Within DLR funding the entire experiment area of the test plane Airbus A300 ZERO-G of the French company NOVESPACE has been allocated to test deployable booms of realistic size under zero-g conditions.

18.3.1 Objectives

The objectives of the experiment were defined as followed:

- Demonstrate chaotic boom deployment without deployment control concept to verify necessity of deployment control concepts,
- Show directness of unfolding process using inflating tip deployment concept,
- Test repeatability of deployment using inflating tip deployment concept by multiple use of the same boom specimen,
- Test controllability of inflating tip deployment by decrease or total stop of gas supply during deployment process,
- Show directness of uncoiling process by electric tip deployment concept.

18.3.2 Test Procedure

To test the inflating tip deployment, a test rack has been designed to provide control on the necessary pressurized gas and to monitor the gas flow and pressure. Thereby, the gas pressure was controlled by the test PC that manages the gas delivering pump. Thus, different pressure characteristics were programmed to test the deployment at high and low speed as well as the possibility to stop and restart the deployment.

The chaotic deployment as well as the electric tip deployment were operated from a persons hands to avoid deployment of the boom inside of the test rack and reduce the complexity of the test rack to a minimum level.

Furthermore the total number of 8 high resolution cameras was used to observe the deployment. The time offset of all cameras, the test rack and the aircrafts own data recording equipment has been identified to synchronize the data sets afterwards.

18.3.3 Test Results

As presented in Fig. 18.13 the deployment of the uncontrolled boom was analogue to the simulation shown above in Fig. 18.10. The process is hardly directed and very discontinuously.

Figure 18.14 displays one frame of the experiment records that has been created for each of the 26 experiments. It contains the synchronized video frames of all eight cameras, the pressure and mass flow rate date as recorded from the test rack and the information on 0 g quality and flight altitude as provide by the aircrafts board system. In addition the white ball at the coiling hub has been tracked by

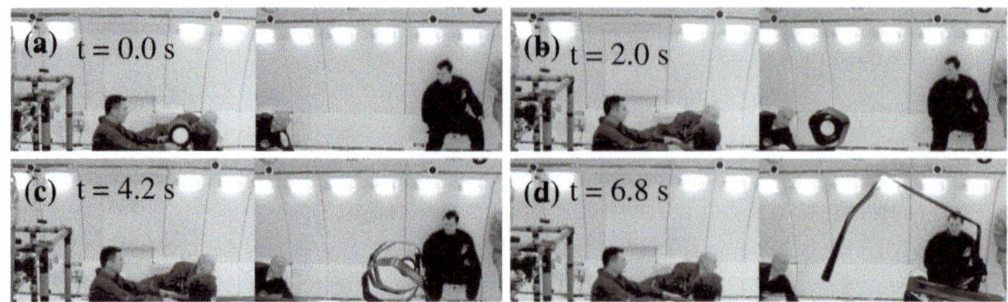

Fig. 18.13 Chaotic deployment of a boom without deployment controlling concept under 0 g

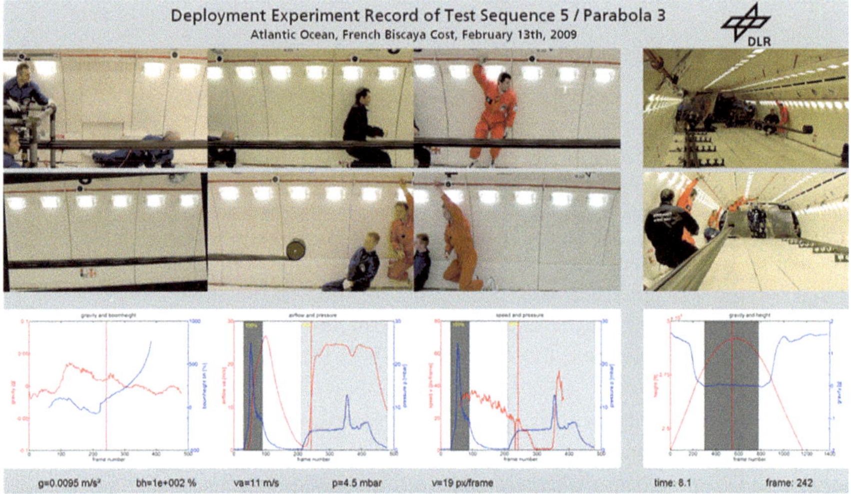

Fig. 18.14 Exemplary experiment record of inflation driven deployment [15]

image processing software to generate values on the hub's horizontal deployment velocity and height over ground. Concentrating on the content of the frame it shows the successful deployment of a 14 m long CFRP boom by inflation. The electric tip deployment mechanism performs best. During all five experiments it deploys an 8 m boom with high repeatability and without system related difficulties.

Concluding the presented test the following statements can be formulated: Deployment without any control mechanism is possible but not recommended. Inflation controlled deployment works very well. However, degradation effects of the Velcro tape have been observed after the same boom has been deployed up to five times (decreased deployment inhibiting forces). Furthermore, the inflation driven deployment process seems to be very demanding for the boom material. The choice of the Velcro thickness and characteristic breaking force is important to ensure no damage of the boom during deployment.

The electrical driven mechanism is the best directed deployment principle. It handles the boom very carefully and causes therefore no material damage during any test. The drawbacks of the principle are the high weight as well as the complexity of the mechanism. Furthermore, it considers the jettisoning of the mechanism at the end of the deployment process which will cause space debris. Therefore, is will not be suitable for earth orbit missions but only for interplanetary and deep space missions. For further details please refer to [13].

18.4 Conclusion and Perspectives

In this section some general applications of deployable space structures have been discussed and challenging aspects have been named. Furthermore, the basic concept of DLR's deployable CFRP masts as well as promising test results was presented.

Therefore, the demand for gossamer structures in general as well as approaches for their realisation have been introduced and discussed.

As a future perspective, the recently started GOSSAMER project [14] needs to be mentioned. It pursues a new on-orbit verification approach that will verify the deployment technologies for solar sail propulsion, thin film solar array technology and deorbiting structures. Therefore, three missions of increasing complexity and size are planned.

The addressed orbit will vary from mission to mission from very low Earth orbits to orbits beyond GEO. Hence, the scientific and monetary value will rise from one mission to the following which enables the sponsoring agencies to *buy* the next mission only if the previous less-expensive mission had been a success.

The funding for the first sub mission GOSSAMER-1 by DLR, ESA and EU is almost committed. It will demonstrate an on-orbit deployment of a 5×5 m solar sail. The launch is scheduled for mid 2014.

References

1. Jenkins, C.H.: Recent advances in gossamer spacecraft. In: AIAA, 2006, vol. 212, Progress in Astronautics and Aeronautics Series, 212, iSBN-10: 1-56347-777-7 ISBN-13: 978-1-56347-777-5
2. Footdale, J.N., Murphey, T.W.: Deployable structures with quadrilateral reticulations. In: 50th AIAA/ASME/ASCE/AHS/ASC Structures, Structural Dynamics, and Materials Conference, Palm Springs, CA, USA, 4–7 May 2009
3. Allred, R., Hoyt, A., McElroy P.: Uv rigidizable carbon-reinforced isogrid inflatable booms. In: 43rd AIAA/ASME/ASCE/AHS/ASC Structures, Structural Dynamics, and Materials Conference, Denver, Colorado, USA, 22–25 April 2002
4. Sandy, C.: Next generation space telescope inflatable sunshield development. In: 2000 IEEE Aerospace Conference Proceedings, vol. 6, Proceedings Paper, pp. 505–519, 18–25 Mar 2000

5. Fang, H., Lou, M., Huang, J.: Design and development of an inflatable relectarry antenna. JPL, Technical Report, 15 May 2002, IPN Progress Report 42–149
6. ECSS, ECSS-E-ST-33-01C—Mechnisms, European Cooperation for Space Standardization Std., Rev. ECSS-E-ST-33-01C, 06 Mar 2009
7. Straubel, M., Sickinger, C., Langlois, S.: Trade-off on large deployable membrane antennas. In: 30th ESA Antenna Workshop, Noordwijk, The Netherlands, 27–30 May 2008
8. Sickinger, C.: Verifikation entfaltbarer composite-booms für gossamer-raumfahrtsysteme. Dissertation, Technische Universität Carolo-Wilhemina zu Braunschweig, publisher: Shaker, ISBN: 978-3-8322-8049-9, (Mar 2009)
9. Sickinger, C.: Analysis of thermally induced disturbances of a gossamer composite boom. In: Boom 4th European Workshop on Inflatable Space Structures, 16–18 June 2008
10. Sickinger, C., Herbeck, L., Ströhlein, T., Torrez-Torres, J.: Lightweight deployable booms: Design, manufacture, verification, and smart materials application. In: 55th International Astronautical Congress, IAF/IAA/IISL, Ed., 04–08 Oct 2004
11. Sickinger, C., Herbeck, L., Ströhlein, T., Torrez-Torres, J.: Deployable structurs in space. In: 12th National SAMPE-Symposium Faserverbundstrukturen, Braunschweig, 2–3 Mar 2006
12. Sickinger, C., Assing, H., Köke, H., Straubel, M.: Verification methology for self-deploying support frames, In: 1st CEAS European Air and Space Conference, Berlin, Germany, 11–13 Sep 2007
13. Straubel, M., Sinapius, M., Langlois, S.: On-ground rigidised, deployable masts for large gossamer space structures. In: European Conference on Spacecraft Structures, Materials & Mechanical Testing, Toulouse, France, 15–17 Sep 2009
14. Geppert, U., Biering, B., Lura, F., Block J., Straubel, M., Reinhard, R.: The 3-step dlr-esa gossamer road to solar sailing. Advances in Space Research, vol. In Press, Corrected Proof, 2010
15. Bock, M.: Synchronisation und Aufbereitung von Kamerabildern und Messdaten des CFK-Mast-Entfaltungsversuchs unter Schwerelosigkeit diploma thesis—Otto-von-Guericke Universität Magdeburg, 02 Aug 2009
16. Hillebrandt, M.: Experimentelle Verifikation von Low-Cost-CfK-Booms (engl. experimental verification of low-cost-CFRP-booms), Aug 2009, student research project, supervised by DLR and University of Braunschweig
17. Leipold, M., Runge, H., Sickinger, C.: Large SAR membrane antennas with lightweight deployable booms. In: 28th ESA Antenna Workshop on Space Antenna Systems and Technologies. DCR Institute of Composite Structures and Adaptive Systems and Kayser-Threde (2005)

Chapter 19
Local Metal Hybridization of Composite Bolted Joints

Axel Fink

Abstract Composite technologies have been proven to be advantageous in allowing for the development of aircraft and spacecraft structures which feature highly integral design concepts. However, structural joining using conventional mechanical fastening techniques still remains an indispensable issue within the design of advanced composite structures. Crucial challenges facing structural joining are the inherent complexity of the stress state at the bolt locations on the one hand and the multifaceted fracture mechanics of composite material on the other. The aerospace industry's increasing requirement for weight reductions and a more efficient use of composites demands not only an accurate understanding of this material's mechanics and its damage behavior at composite joints but the development of advanced joining techniques as well in order to be able to fully exploit the outstanding capabilities of composite material. The use of a local hybridization with metal represents a suitable and technologically feasible means to increase the mechanical efficiency of highly loaded composite bolted joints which allows for a significant improvement of the overall structural efficiency of real composite structures. This chapter presents the hybrid reinforcement concept and addresses some fundamental topics of this technology's mechanical behavior.

A. Fink (✉)
Institute of Composite Structures and Adaptive Systems,
German Aerospace Center (DLR e.V.), Lilienthalplatz 7, 38108, Braunschweig, Germany
e-mail: christian.huehne@dlr.de

M. Wiedemann and M. Sinapius (eds.), *Adaptive, Tolerant and Efficient Composite Structures*, Research Topics in Aerospace, DOI: 10.1007/978-3-642-29190-6_19,
© Springer-Verlag Berlin Heidelberg 2013

19.1 Local Hybridization

From the very beginning of the history of technology the invention of efficient, strong and durable joining techniques has been representing a mandatory and challenging task always going along with the development of new tools, devices and structures. Though often being attributed to a secondary role, the joining techniques became decisive in terms of the structural functionality of most of the technical inventions, ranging from the prehistoric stone axe to the latest spacecraft launchers.

Modern aircraft and spacecraft structures are still characterized by the use of structural joining. Even for advanced composite structures, mechanical fastening with rivets and bolts remain the most important structural joining technique with regard to its reliability, detachability, inspectability, repairability and robustness. Joints represent a structural discontinuity which results in a distortion of the load paths hence exciting local stress intensities which develop to an inherent source of damage or even structural failure. Numerous historical and recent incidents attest the important role of joints in terms of a safe operation of transport vehicles, irrespective of the structural material used. The loss of the composite vertical tail plane of the American Airline A300–600 [1], the cracked Southwest Airline 737 fuselage [2], but even the sinking of the RMS Titanic [3] are only some examples related to the natural weakness of joints.

Metals are prone to develop crack nucleations at the stress intensity locations of bolt loaded holes under cyclic loading. This leads to crack formation, its propagation with considerable crack growth rates and finally to a residual strength dominated by the material's fracture toughness. While metals standout through static notched strength, which is insensitive to stress intensities, they feature important drawbacks in terms of fatigue resistance. The fatigue behavior of metal can be considerably improved by a laminated hybridization with fiber reinforced plastic layers—taking advantage of the so called *crack bridging effect* provided by the fibers—which has led to the development of FMLs (fiber metal laminates) for fatigue critical applications. Glare (aluminum hybrid laminate) [4] and HTCL (titanium hybrid laminate) [5] are some of the most relevant examples. For joints and especially for the highly fatigue critical lug joints of wing-fuselage attachments, Glare has been demonstrated to show considerable fatigue resistance and damage tolerance allowing for a higher stress level or longer inspection periods of the joint [6].

Contrary to metals, composite material features a pronounced notch sensitivity which especially develops under quasistatic loading. However, composite bolted joints stand out due to their excellent fatigue resistance and a residual strength that often equals or exceeds the quasi-static strength as a consequence of inherent material softening mechanisms. This behavior of composite material shall be considered as the first paradox with regard to the metal's behavior. Figure 19.1 shows the joint efficiency—ratio of joint strength to unnotched strength—of single-row (1RBJ) and three-tandem-row (3RBJ) bolted joints in terms of their quasistatic strength as a function of the diameter-to-width ratio. Due to their notched sensitivity composites reach maximum efficiencies of 30% to 40%,

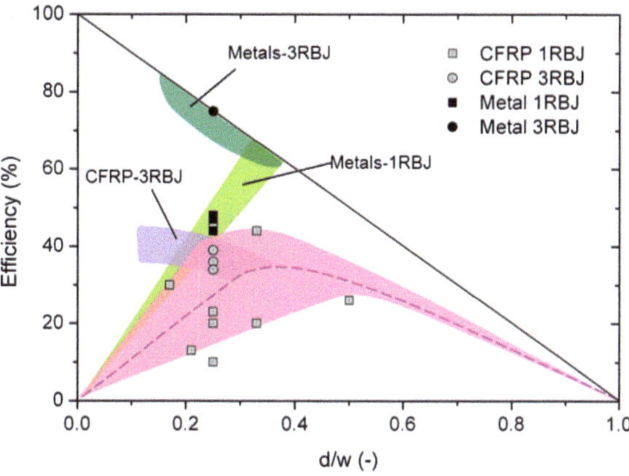

Fig. 19.1 Efficiency of composite and metal single-row and multi-row bolted joints

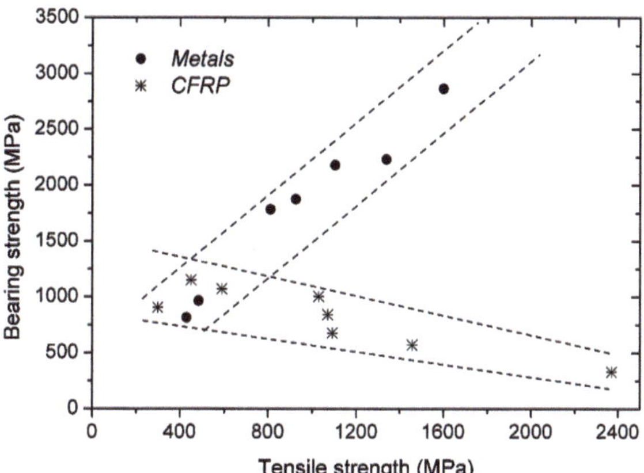

Fig. 19.2 Bearing strength against the unnotched tensile strength of composites and metals

multi-row joints being only slightly more efficient than optimal single-row joints. Metals, however, offer efficiencies in the range of 80%.

What might be understood as the second paradox of composites in relation to their bolted joint behaviour is described in Fig. 19.2 showing the bearing strength—one important factor to attain high joint strengths, especially for single-row configurations—in relation to the material's unnotched quasistatic strength. Metals exhibit an increase of bearing strength with larger tensile strengths. The bearing strength of composites, on the contrary, decreases with rising material strength, which means that the material's directional performance on the one hand and the joint performance on the other are two opposing phenomena. This discrepancy affects the lightweight potentials of composite materials within a real structural context.

In analogy to the improvement of monolithic metal, the hybridization of composite laminates is deemed an effective means to overcome the inherent deficiencies of composite laminates. With regard to the complex stress state at a bolt loaded hole and the corresponding simultaneous demand for sufficient bearing, shear and notched strength capabilities, the laminate's hybridization with

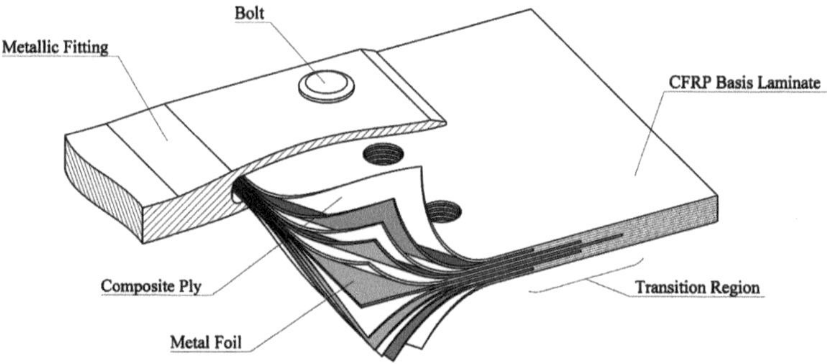

Fig. 19.3 The local reinforcement approach by means of laminate hybridization

high-strength metal takes advantage of its favourable isotropic properties in order
to improve the bolted joint performance.

The local hybridization technique has hence been proposed and developed as a
bolted joint reinforcement approach. This technique entails the local substitution
of composite plies by metal sheets or foils of corresponding thickness within the
bolted joint region, as depicted in Fig. 19.3, without the implementation of any
interfacial adhesive. Continuous plies, which should contribute most to the total
load carrying of the laminate, are allocated adjacent to each embedded foil, hence
acting as an adhesive interlayer. The ply substitution technique allows for reducing
or, in the optimal case, eliminating the need for local thickening. This effect avoids
local eccentricities and their associated secondary stresses and reduces fastener
grip lengths and the peripheric elements' geometric complexity.

The reinforcement with metal has often been addressed as the embodiment of
massive metal doublers on top of or within the composite laminate. The use of
thick plates or sheets is deemed, however, inefficient due to high interlaminar
stresses at free edges and crack surfaces, the limited effect of fiber bridging, the
low amount of interfaces, the pronounced delamination risk and the occurrence of
irregularities at the transition region which all translate to moderate strength,
damage tolerance and fatigue resistance capabilities.

The key for a high performance hybrid laminate is found in a high discretization
of the lamination—a principle that generally applies for all multiphase laminated
materials—, which can be especially accomplished for prepreg laminates by
means of ply-wise hybridization. The maximum metal content amounts in that
case to about 50%.

19.2 Improvement of Bearing Strength

With regard to their galvanic compatibility, their low CTE (coefficient of thermal
expansion) mismatch and high specific strengths, titanium alloys are deemed the
best material choice for the hybridization of carbon fiber reinforced plastics. Apart

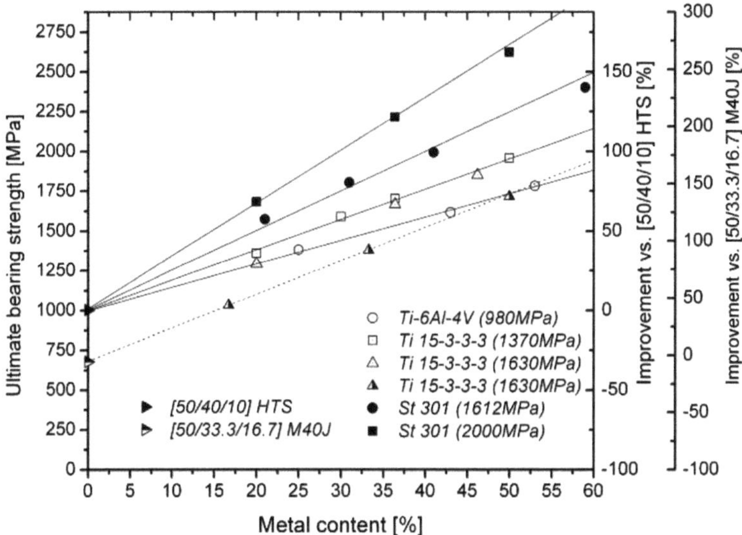

Fig. 19.4 Bearing strength improvement of a [50/40/10] and a [50/33.3/16.7] composite laminate with HTS and M40J fibers resp

from the obvious but not trivial requirement for a strong and durable bond between the metal and the composite constituent, the metal's stiffness and the strength are two essential factors defining its reinforcement capabilities. The magnitude of reinforcement is moreover conditioned by the hybrid laminate's metal content and its laminate ply configuration.

The improvement of the ultimate bearing strength of two exemplary composite laminates is presented in Fig. 19.4 as a function of the metal content and the metal strength. The reinforcement effect increases with rising metal content and strength when substituting all but the main load carrying 0°-plies. However, the impact of rising titanium strength is limited, providing no further increase of the laminate bearing strength beyond tensile strengths of 1370 MPa. This effect, which is also present in terms of the offset bearing yield strength, is a consequence of the damage initiation and accumulation within the composite plies which remains unavoidable irrespective of the titanium strength [7, 8]. Only a simultaneous increase of the metal modulus allows for the development of the advantageous effect of rising metal strength [7], which is best achievable by the use of steel as the reinforcement material. Despite the considerable thermal residual stresses excited by a larger CTE mismatch, steel hybrid laminates reach exceptionally high bearing strengths in comparison to titanium hybrid laminates, even in the case of equal metal strengths and metal contents. This means in turn, that less metal is required to achieve the same reinforcement effect, which translates to less metal integration efforts and costs and a larger hybridization flexibility. The reinforcement effect results to be more pronounced for less strong carbon fibers—M40 J against HTS fibers in the present case—as well as for laminates with a higher level

Fig. 19.5 Ultimate bearing strength (UB) and 0.5%-offset bearing strength (OB) of a titanium hybrid laminate with a varying content (X%) of 0, 90 or ±45° plies (*R* remaining content)

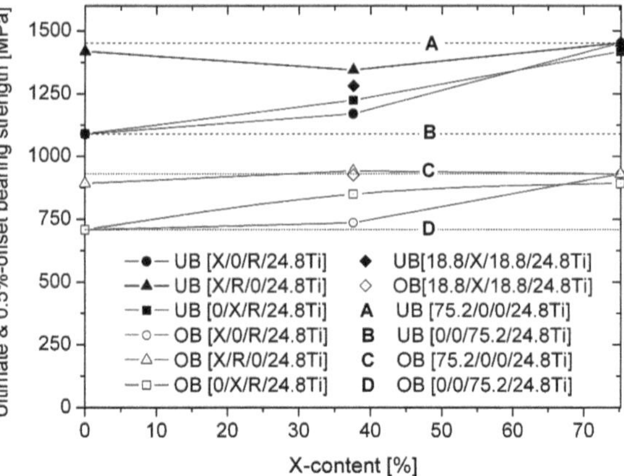

of orthotropy, as presented in Fig 19.5 on the bases of a titanium hybrid laminate with the moderate metal content of about 25% and varying composite ply orientations. In terms of both, ultimate bearing strength and 0.5%-offset bearing strength, 0° and ±45° plies show nearly the same impact on the hybrid laminate's bearing capabilities, whereas a 90° ply content tends to reduce them. With rising metal content, the metal constituent becomes more predominant and the influence of the orientations of the composite plies diminishes. However, the fracture behavior still remains dominated by the material's laminated configuration and its specific damage mechanisms [7], which include the deterioration of the composite constituent, ply delamination and the related out-of-plane deformation, and finally lead to a sudden loss of load bearing capabilities. Only for high metal contents beyond 50% a positive effect of 90° plies is present [9], which is considered a consequence of their stabilizing effect against out-of-plane deformation.

19.3 Reinforcement of Bolted Joints

For multi-row bolted joints, the transfer load is shared—in an unbalanced fashion for more than two rows—among the bolt-rows. Hence, the laminate's bearing strength and shear strength, although still important, play a secondary role in terms of the ultimate joint strength. Regarding the by-pass and transfer-load interactions at the critical outer bolt rows, the laminate's notched strength represents the predominant material property attaining high ultimate joint strengths. The joint strength of a three-row bolted joint (Fig. 19.6), with a width-to-diameter ratio and pitch-to diameter-ratio of 4 and 3 respectively, is presented in Fig. 19.7 as a function of the metal content for a hybrid laminate composed of 0° plies and titanium (Ti-6Al-4 V) or steel (301).

The strength of a single-row joint with equal width-to-diameter ratio and an edge-to-diameter ratio of 3 is included for comparison. For a three-row bolted

Fig. 19.6 Three-row bolted joint on a CFRP/steel laminate

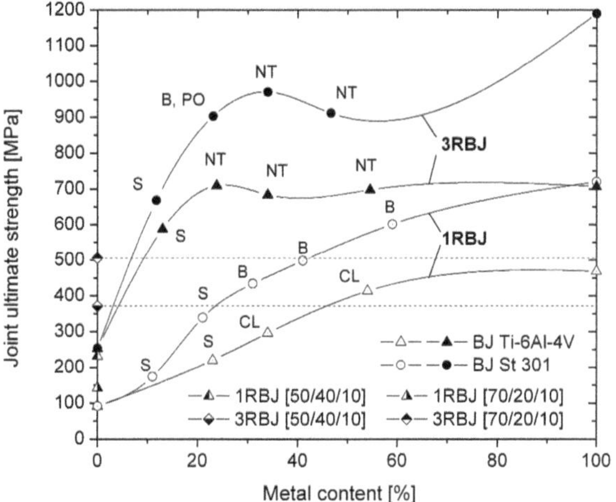

Fig. 19.7 Strength of a three-row (3RBJ) and single-row (1RBJ) bolted joint of a 0°/metal hybrid laminate vs. the metal content. *B* bearing failure; *S* shear-out failure; *NT* net-tension failure; *CL* cleavage failure; *PO* pull-out failure

joint, a relatively small metal content is necessary to provide a unidirectional laminate with a sufficient shear and bearing strength in order to initiate a net-tension failure and develop maximum load capabilities. Due to the lower strengths of metal, a higher hybridization degree reduces the notched strength, hence offering no further reinforcement capabilities. Due to a 100% load transfer, a single-row bolted joint demands a higher hybridization degree for achieving adequate shear and bearing strengths. Nevertheless, single-row bolted joints with metal contents between 20 and 40% are capable to reach the joint strengths of three-row-bolted joints of typical composite laminates. The resulting weight saving potential is significant and obvious.

With regard to the predominant notched strength at multi-row bolted joints, the hybridization results in greater efficiency for less orthotropic, and hence lower-strength, composite laminates. On the contrary, the highest reinforcement effects at

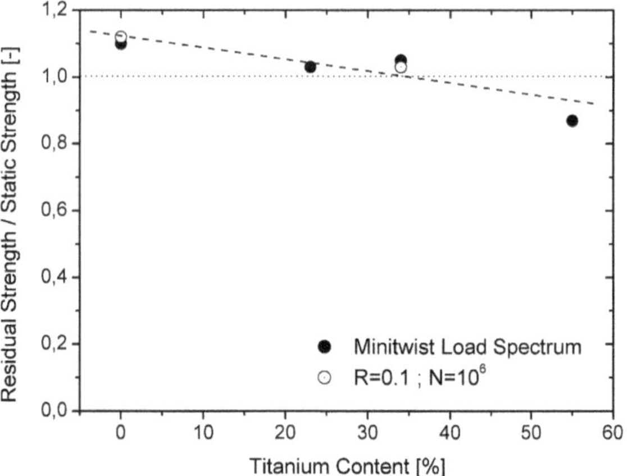

Fig. 19.8 Residual strength after MINITWIST spectrum loading and R = 0.1 cyclic loading

single-row bolted joints, demanding high bearing strengths, are reached for highly orthotropic laminates.

The metal hybridization evidently improves the quasistatic strength of bolted joints. For joints on monolithic composite laminates, the fatigue behaviour is usually not an issue due to the low strains demanded from static requirements on the one hand and the favourable fatigue resistance on the other. For hybrid materials, on the contrary, it gains in importance with regard to the presence of a metal constituent, especially for aircraft applications. Provided the case of adequate interface bond strength and material's configuration, low thermal residual stresses and proper laminate discretization, the fatigue resistance is however deemed to be positively influenced by the crack bridging effect, well known from typical FML applications. After being subjected to a MINITWIST spectrum loading during five aircraft lives, no cracks or delaminations are present for bolted joints on hybrid titanium material irrespective of the metal content. The residual strength is depicted in Fig. 19.8 which demonstrates no apparent loss of static strength capabilities for lower titanium contents. The residual strength slightly decreases with rising metal content [10]. Crack formation within the titanium plies and a corresponding delamination is nevertheless observed for hybrid titanium laminates under a quite severe cyclic loading with a loading ratio of R = 0.1 and an amplitude loading of 66% of the quasistatic strength [11]. Specimens with lower titanium contents survive one million cycles without rupture and reach residual strengths similar to the quasistatic strengths.

19.4 The Transition Region

In order to exploit the full reinforcement capabilities of the hybridization, the transition region from pure composite to hybrid laminate has to reach equal or better joint efficiencies in comparison to the bolted joint. Two mechanisms

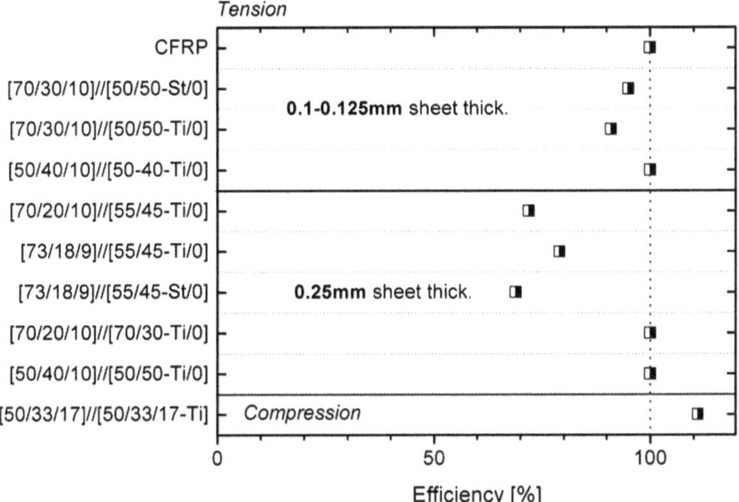

Fig. 19.9 Strength efficiency of exemplary transition regions with different metal contents and sheet thicknesses under tension and compression loading

determine the strength limitations of the transition region: the delamination of both the interrupted composite plies and the metal sheets on the one hand and the overstressing of adjacent continuous plies at each substitution point on the other [8, 11]. The total effective stiffness—thickness times modulus—of the metal sheets and the interlaminar fracture toughness represent two key issues affecting the delamination resistance. The interlaminar toughness is dependent on the matrix properties, the fiber-resin interaction and the strength of the adhesive interface between the metal and the resin. High bond strength—and durability—is achieved by specific metal surface treatments [12]. Special testing methods have been developed to substantiate the critical energy release rate of hybrid laminates as a measure of the interlaminar fracture toughness [13]. It is essential to account for the interlaminar fracture toughness as a function of the laminate thickness [7] when designing the transition region. The use of thin and compliant metal foils, as well as a staggered configuration of the substitution points enables a smooth longitudinal and flexural stiffness change. With regard to their large modulus and CTE, steel foils are much more prone to delamination than titanium foils. An overview of the quasistatic strength of different transition region configurations is presented in Fig. 19.9 in terms of their efficiency (ratio of the transition region's ultimate strength to the composite laminate ultimate strength). For transitions with no interruption of 0° layers, exceptionally high joint efficiencies of almost 100% are achieved under tension loading irrespective of the foil thickness. A substitution of load carrying 0° plies leads to a reduction of the load capability, which is more pronounced for thicker metal sheets. Nevertheless the efficiencies are within a range of 70% to 90% which is considerably larger than the bolted joint's efficiency. Whilst the titanium and the steel hybrid laminates with thin foils show a fiber failure due to overstressing, the final failure of steel laminates with thicker foils is triggered by delamination. The compression strength denotes no strength loss in comparison to the reference composite laminate, the slight exceedance of a 100% efficiency being deemed associated to the known peculiarities of compression testing.

The material and transition region's configuration factors influencing the static behavior are also determining in terms of fatigue behavior. Thicker titanium sheets are prone to develop cracks—depending on the load level and the fatigue loading type—which are however effectively bridged, under the formation of localized, delaminations, by the adjacent continuous $0°$ plies. The substitution of $0°$ plies turn out to be critical, arising as potential source of delamination at the ply substitution, points especially under repeated compression loading [11]. Despite the crack and delamination formation, well-designed transition regions offer residual tensile strengths comparable to the quasistatic strengths.

19.5 Exemplary Applications

The load capabilities of hybrid single bolted joints are superior to multi-row composite bolted joints. Despite similar or even lower specific strengths (ratio of strength to laminate density), the local hybridization offers drastic weight saving potentials due to the reduction of the amount of bolt rows and fasteners, the elimination of pad-ups, the diminution of overlap lengths and the simplification of connecting fittings and peripheric elements. Hence, the weight efficiency of a reinforcement technique mainly results from secondary mass savings as a consequence of the large absolute—and not specific—joint strength capabilities.

Hence, the local titanium hybridization enables to introduce a single row bolted joint instead of a baseline double row bolted joint at the upper interface root joint of a payload adaptor. This in turn reduces the amount of fasteners and the weight of the aluminium connecting ring. This enables a possible mass reduction of 25% of the additional joint mass which translates, in this special case, to a corresponding payload mass increase [13].

Another example represents the intersegmental joints' reinforcement of the future Ariane 5 composite booster case. The extremely high axial force flow resulting from the operational inner pressure demands a massive reinforcement of the composite laminate at the bolted intersegmental joints. The baseline design hence features a double-row bolted joint and a considerable local laminate thickening. By introducing a local titanium or steel hybridization of the composite laminate the double-row design can be substituted by a single-row design while eliminating any local thickening. Despite of an increase of the laminate's density, the reduction of the heavy steel connecting rings and the amount of steel bolts result in an overall reduction of the additional joint mass between 20 and 30%. Moreover, the reduction of the hoop plies and the new ring design enables the alleviation of the

Fig. 19.10 Transition region (CFRP/steel, resin infusion)

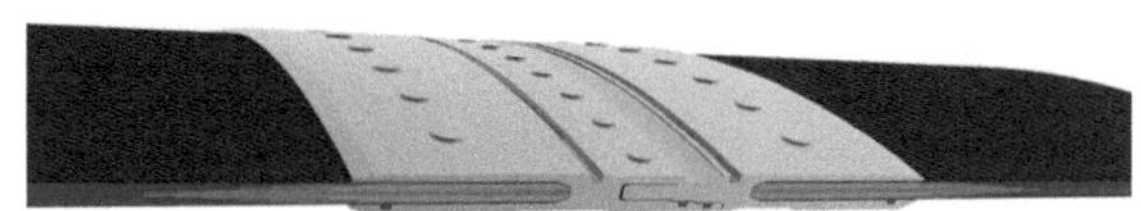

Fig. 19.11 Conceptual joint design

radial constraint at each intersegment joint and the related secondary stresses [14]. The compatibility of the hybrid technology with resin infusion techniques was successfully demonstrated for this application [15]. The transition region and the conceptual joint design are shown in Figs. 19.10 and 19.11 respectively.

References

1. Raju, I., Glaessgen, E., Mason, B., Krishnamurthy, T., Davila, C.: NASA Structural analysis report on the American airlines flight 587 Accident—Local Analysis of the Right Rear Lug. AIAA Paper 2005–2255, NASA Langley Research Center (2005)
2. Gründer, M.: Risse im Blech. Flugrevue **6**(2011), 88–89 (2011)
3. Woytowich, R.: Riveted hull joint design in RMS titanic and other pre-world war I ships. Mar. Technol. **40**, 82–92 (2003)
4. Vlot, A., Gunnik, W.: Fibre metal laminates, an introduction. Kluwer Academic Publishers, Dordrecht (2001)
5. Johnson, W.S., Cobb, T.: Hybrid titanium composite laminates: a new aerospace material. In: 21st Annual Adhesion Society Meeting, NASA Langley Research Center (1998)
6. Schijve, J.: Fatigue of structures and materials. Springer, Berlin (2009)
7. Fink, A., Camanho, P.: Reinforcement of composite bolted joints by means of local metal hybridization. In: Composite Joints and Connections, Woodhead Publishing Ltd. (2011)
8. Camanho, P., Fink, A., Obst, A., Pimenta, S.: Hybrid titanium-CFRP laminates for high performance bolted joints. Compos. A **40**, 1826–1837 (2009)
9. Fink, A., Kolesnikov, B., Herbeck, L.: Effizienzsteigerung von Hochleistungsfaserverbundstrukturen durch lokale CFK/Metall Verstärkung, Deutscher Luft- und Raumfahrtkongress DGLR, Braunschweig, Nov. 2006 (2006)
10. Kolesnikov, B., Herbeck, L., Fink, A.: CFRP/Titanium hybrid material for improving composite bolted joints. Compos. Struct. **83**, 368–380 (2008)
11. Fink, A.: Lokale Metall-Hybridisierung zur Effizienzsteigerung von Hochlastfügestellen in FAserverbundstrukturen, DLR Forschungsbericht 2010–2014 (2010)

12. Fink, A., DuMars, B. Development of an improved surface preparation for titanium bonding and titanium graphite laminates for aircraft and space vehicle applications. AIRTEC 5th International Conference, Frankfurt (2010)
13. Fink, A., Camanho, P., Andrés, J.M., Pfeiffer, E., Obst, A.: Hybrid CFRP/titanium bolted joints: performance assessment and application to a payload adaptor. Compos. Sci. Technol. **70**, 305–317 (2010)
14. Fink, A., Camanho, P., Canay, M., Obst, A.: Increase of bolted joint performance by means of local laminate hybridization. In: 1st CEAS European Air and Space Conference, Berlin (2007)
15. Fink, A., Camanho, P., Andrés, J.M., Pfeiffer, E., Obst, A.: CFRP/metal hybrid material for improving composite bolted joints. In: 11th European Conference on Spacecraft Structures, Materials and Mechanical Testing, Toulouse (2009)

Chapter 20
Payload Adapter Made from Fiber-Metal-Laminate Struts

Boris Kolesnikov, Daniel Stefaniak, Johannes Wölper
and Christian Hühne

Abstract In comparison to other transport systems, launch vehicles are characterized by relatively light but extremely valuable payloads. The launcher's upper stage structures, e.g. payload adapter and fairing, offer the highest weight saving potential. An effective weight reduction can only be achieved by the combined utilization of high performance materials and adapted construction methods. To improve the structures damage tolerance a new hybrid lay-up has been developed, which combines the properties of both, steel and carbon fiber reinforced plastics (CFRP). This chapter presents a preliminary design of a payload adapter as a framework, which is based on the high performance material properties of uni-directional CFRP-steel-laminates, offering a considerable weight saving potential.

20.1 State-of-the-Art Construction Technologies for Payload Adapters

Previous adapters used in launcher structures are made of metal, fiber reinforced plastic (FRP) or a combination of both materials. The adapter 1666A of the 'ARIANE 4', for example is an aluminium semi-monocoque construction, Fig. 20.1a, whereas CFRP is used for the adapter 1,194 V, Fig. 20.1b. The latter is designed as a sandwich-construction with aluminium frames. Similar material combinations can be found in the 'ARIANE 5', in which the VEB (Vehicle Equipment Bay) type A is made of CFRP as sandwich, Fig. 20.2a. Monolithic CFRP adapters are also used for this launcher, Fig. 20.2b.

B. Kolesnikov (✉) · D. Stefaniak · J. Wölper · C. Hühne
Institute of Composite Structures and Adaptive Systems, Composite Design,
Deutsches Zentrum für Luft- und Raumfahrt e.V. (German Aerospace Center),
Lilienthalplatz 7, 38108, Braunschweig, Germany
e-mail: boris.kolesnikov@dlr.de

M. Wiedemann and M. Sinapius (eds.), *Adaptive, Tolerant and Efficient Composite Structures*, Research Topics in Aerospace, DOI: 10.1007/978-3-642-29190-6_20,
© Springer-Verlag Berlin Heidelberg 2013

Fig. 20.1 a Adapter 1666A
and **b** 1,194 V of 'ARIANE
4' [1, 21]

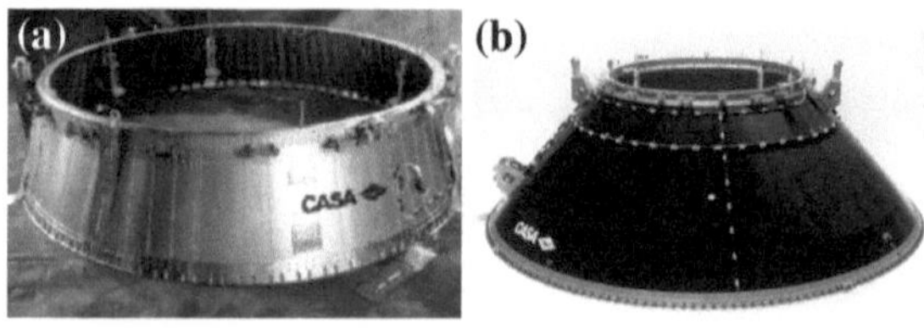

Fig. 20.2 a VEB structure
type A with adapter and
b monolithic CFRP adapter
of 'ARIANE 5' [1, 21]

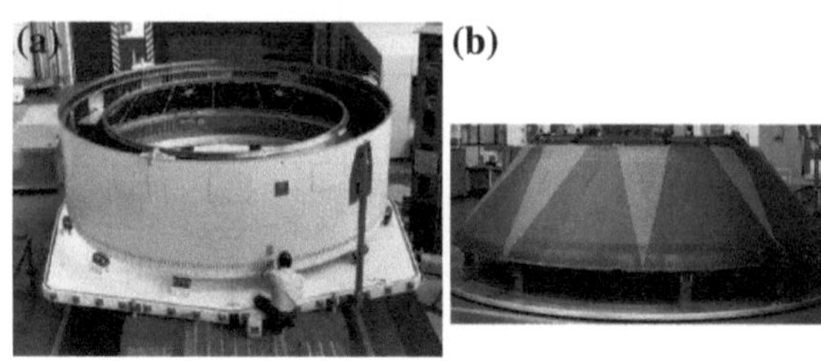

Fig. 20.3 a CFRP-lattice
structure adapter and
b aluminium adapter of
'Proton-M' [2]

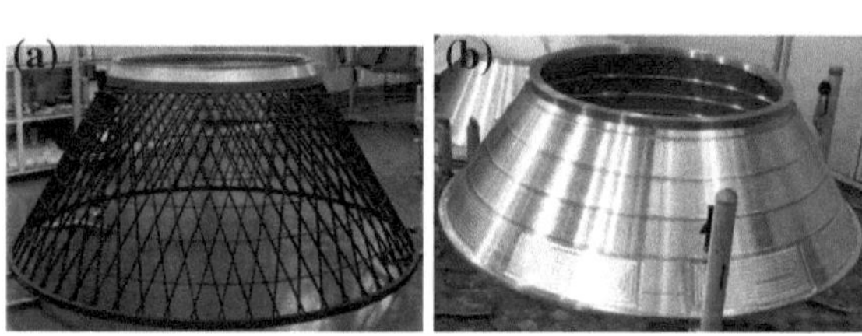

CFRP-sandwich structures are also used in the Russian launcher 'Soyuz' [1].
For many years, CFRP-lattice-structures for adapters have been successfully used
in Russian aerospace [2, 3–5]. Figure 20.3 shows a (a) CFRP lattice-structure
adapter and (b) aluminium adapter with a variable wall thickness of the Russian
launcher 'Proton-M'. Weight- and cost savings up to 60% can be achieved by
utilizing a CFRP-lattice structure compared to a metallic [2].

However, in all these examples metal and CFRP are combined on a structural
level and not on a material scale. Fiber metal laminates have not been found in
adapters for launcher structures so far.

20.2 Current Fiber Metal Laminates

Research on fiber metal laminates has been done for more than 40 years to
improve the material performance of the individual constituents [6, 7]. The well-
known material GLARE (**Gla**ss Fiber **Re**inforced Aluminium) is a combination of
layers of glass fiber reinforced plastic (GFRP) which is used in the upper fuselage
of the 'AIRBUS A380'. The use of glass fibers considerably reduces the risk of
fatigue compared to monolithic aluminium [7].

Local reinforcement of joining areas is another important discipline for fiber
metal laminates [6, 8–11]. Fiber laminates, especially highly orthotropic laminates,

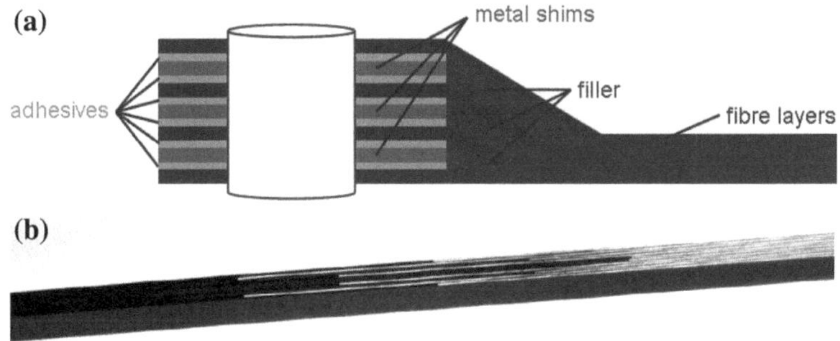

Fig. 20.4 **a** Transition zones with and **b** without eccentricity [11]

are characterized by low bearing and shear capabilities. Therefore, a local laminate built-up at the joining area is necessary to increase the load capacity of the laminate, Fig. 20.4a. Secondary stresses and eccentricities due to thickening on one side are the unfavourable consequences.

Another solution to improve the low bearing strength of fiber laminates is the local reinforcement technique characterized by the gradual substitution of specific composite plies by high-strength metal foils, cf. Fig. 20.4b. This approach eliminates any laminate thickening and secondary stress and consequently reduces the size and weight of fastening elements [11]. The transition region between the pure composite and the metal reinforced coupling region reaches a coupling efficiency (ratio of transition strength to strength of basic composite) of up to 100% [8, 10–12]. This leads to an increase of the specific strength of the bolted joining of up to 41% compared to pure CFRP [11].

20.3 Framework Design for an Upper Stage Adapter

In the following a VEB-structure as a V-ring construction with an adapter in framework design is investigated. As a consequence of the framework design the struts, serving as main structural elements, are only loaded in longitudinal direction. This allows a unidirectional alignment of all carbon fibers as long as damage tolerance needs are sufficiently satisfied. Thus, struts with quadratic and circular cross sections are investigated (Fig. 20.5).

20.4 Fiber Metal Laminates Increase Degree Capacity Utilization of CFRP-Strut

In a variety of aerospace applications, the industry's increasing requirements for higher structural efficiency compete with the fundamental requirements for damage tolerance. Especially, when high specific uniaxial mechanical properties are desired,

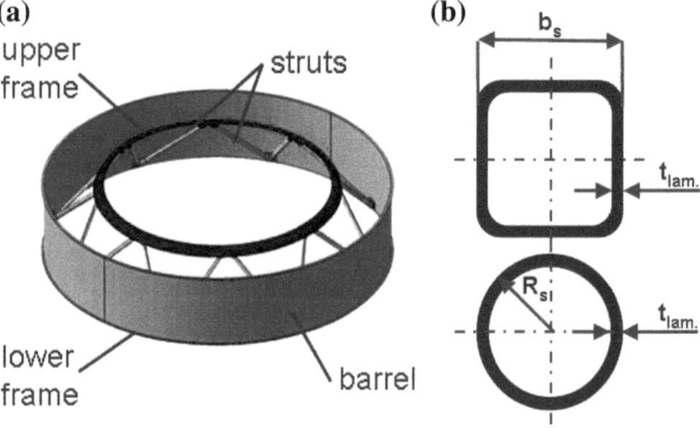

Fig. 20.5 **a** V-ring with adapter as framework design using struts with **b** quadratic and circular cross section

notch and impact sensitivity properties drastically limit the fiber fraction in load direction since laminates are created by stacking sequences with various orientations. Additionally, the thickness of equally orientated layers is limited to reduce crack distribution.

As a result, stiffness and strength per unit weight of the laminate for a given direction are lower than the corresponding values of a unidirectional composite. Particularly buckling endangered longerons and struts with high stiffness requirements suffer a loss of their lightweight potential due to the reduced residual strength capability.

Conventional fiber metal laminates, for example those described in the patents of Kolesnikov [12] and Westre [13] contain metal-layers with a thickness of 0.08–1.0 mm and therefore relatively high metal fractions impair the weight efficiency of the laminate.

Recognizing these limitations, a new laminate lay-up has been developed to improve the CFRP performance [14]. This new laminate consists of metal layers with a thickness of less than 0.08 mm and unidirectionally aligned fiber layers which are stacked as alternating layers of metal and fiber. Hence, stiffness and strength in the 0°-direction are not reduced in comparison to the use of variant fiber directions. The metal layers deflect inter-fiber-fracture in zones of delamination; they serve as crack arrest layers and the energy dissipation is elevated due to the increase of the delamination area as a consequence of an increased number of interfaces within the laminate. Transverse stiffness and strength are increased compared to unidirectional laminates while the specific stiffness in the 0°-direction is higher compared to common multi-axial CFRP laminates (Fig. 20.6).

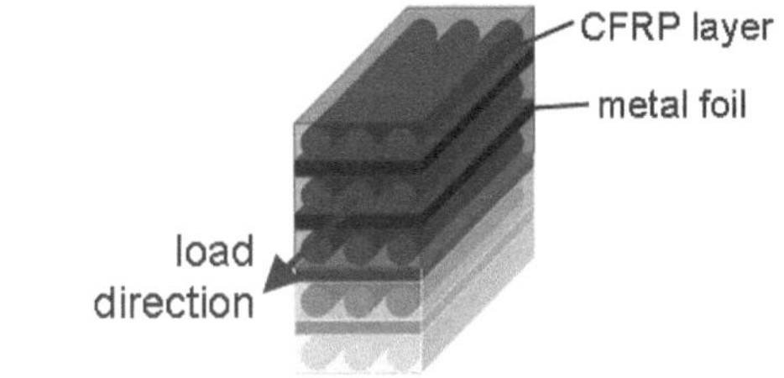

Fig. 20.6 Unidirectional fiber metal laminate [22]

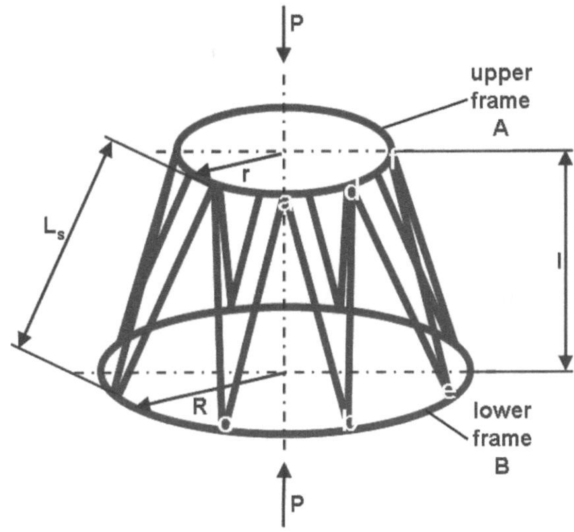

Fig. 20.7 Geometrical relations for framework adapter

20.5 Analytical Preliminary Design of Framework-Design

The local and global buckling of the structure is estimated for the preliminary design and the bending moment in the frame is investigated. Based on these results, different adapter configurations are compared.

20.5.1 Geometrical Relationships of Struts in a Conical Framework

Figure 20.7 shows the proposed conical framework structure [15] for the adapter. The length L_s shown in Fig. 20.7 is determined by

$$L_S = \sqrt{R^2 + r^2 + l^2 - 2Rr\cos(\pi/n)} \qquad (20.1)$$

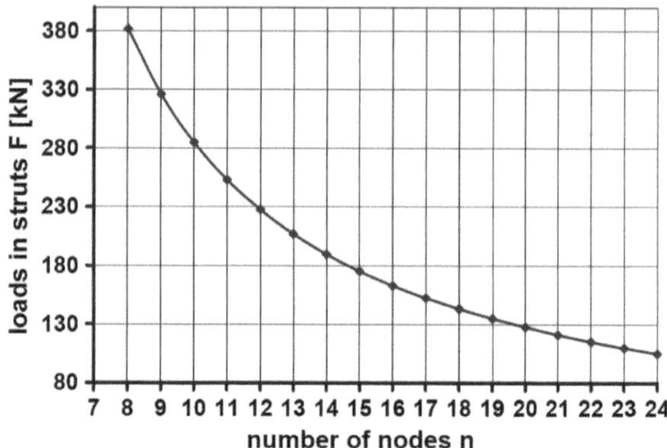

Fig. 20.8 Strut loads as a function of the number of nodes

The normal forces in the struts F of a conical framework due to the load P are

$$F = \frac{PL_S}{n2l} \tag{20.2}$$

The graph in Fig. 20.8 shows the relation between strut forces and the number of nodes for a given load case. Increasing the number of nodes from 8 (equals 16 struts) to 24 (equals 48 struts) reduces the normal strut forces by a factor of 3.64.

20.5.2 Estimation of Local and Global Buckling Stress of Struts

The global and local buckling stresses of struts with a square cross section, as shown in Fig. 20.5(b), are estimated as follows [16, 17]:

$$\sigma_{buckl.global} = \frac{\pi^2 E_1 I_{\min}}{A_S(vL_S)^2} \tag{20.3}$$

$$\sigma_{buckl.local} = \frac{\pi^2 t_{lam.}^2}{6b^2}\left(\sqrt{\tilde{E}_1 \tilde{E}_2} + \tilde{E}_1\mu_2 + 2G_{12}\right) \tag{20.4}$$

$$\tilde{E}_1 = \frac{E_1}{(1 - \mu_1\mu_2)} \tag{20.5}$$

$$\tilde{E}_2 = \frac{E_2}{(1 - \mu_1\mu_2)} \tag{20.6}$$

For struts with a circular cross section, the global and the local buckling stresses according to [15, 18, 19] are

$$\sigma_{buckl.global} = v\frac{\pi^2}{2}E_1\left(\frac{R_S}{L_S}\right)^2 \tag{20.7}$$

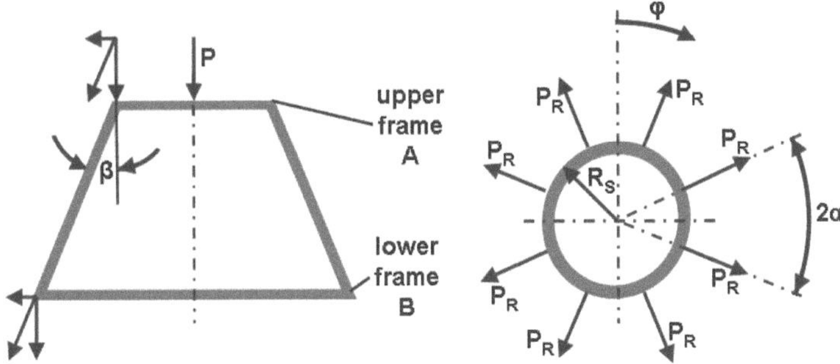

Fig. 20.9 Radial loads in adapter frames

$$\sigma_{buckl.local} = \frac{1}{\sqrt{3(1-\mu_1\mu_2)}}\sqrt{E_1 E_2}\frac{t_{lam.}}{R_S}\sqrt{\frac{\sqrt{E_2/E_1}+2G_{12}(1-\mu_1\mu_2)/E_1+\mu_2}{2\sqrt{E_2/E_1}+E_2/G_{12}-2\mu_2}}$$

$$(20.8)$$

The effective buckling length factor $v = 1$ that, represents a simple support of a bar is used for the calculation of the global buckling stress with Eqs. (20.3) and (20.7). The non-axially symmetric buckling-stress is estimated with Eq. (20.8).

20.5.3 Radial Loads in Frames

As a consequence of the conical adapter shape, radial loads P_r (Fig. 20.9) act in the nodes a, b, c, d etc. (Fig. 20.7).

The radial loads P_r are given by

$$P_r = \frac{P}{n}\tan\beta \tag{20.9}$$

The relation of radial loads P_r in frame B as a function of the number of nodes is given in Fig. 20.10. Increasing the number of nodes from 8 (equals 16 struts) to 24 (equals 48 struts) reduces the radial loads P_r by factor 3.

20.5.4 Maximum Bending Moment in Frames

The radial loads P_r generate a bending load in frame B. According to [16] and [20] for $0 \le \varphi \le \alpha$ (see Fig. 20.9) the bending moments in frame B are estimated by:

$$M = \frac{P_r R}{2}\left(\frac{\cos(\varphi)}{\sin(\alpha)}-\frac{1}{\alpha}\right) \tag{20.10}$$

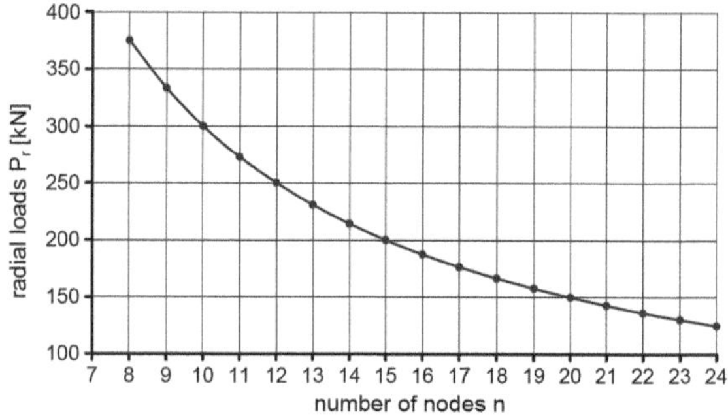

Fig. 20.10 Radial loads in frame B as a function of the number of nodes

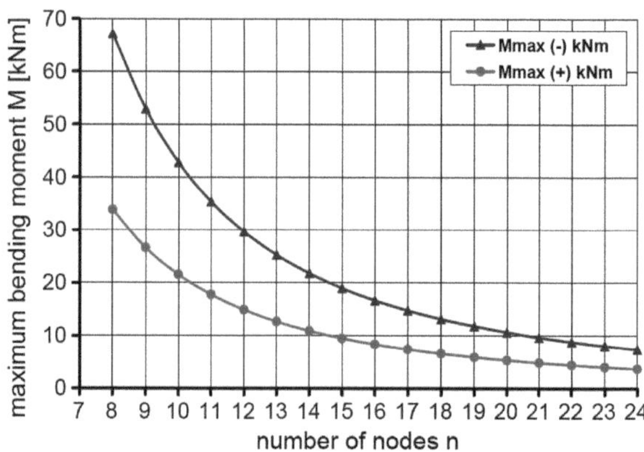

Fig. 20.11 Maximum bending moments in frame B as a function of the number of nodes

in which the angle α is

$$\alpha = \frac{\pi}{n} \tag{20.11}$$

The positive maximum bending moments in frame B occur at angles $\varphi = 0,\ 2\alpha,\ 4\alpha\ \dots$:

$$M_{\text{max}}^{(+)} = \frac{P_r R}{2}\left(\frac{1}{\sin(\alpha)} - \frac{1}{\alpha}\right) \tag{20.12}$$

The negative maximum bending moments in frame B prevail at angles $\varphi = 1\alpha,\ 3\alpha,\ 5\alpha\ \dots$:

$$M_{\text{max}}^{(-)} = \frac{P_r R}{2}\left(\cot(\alpha) - \frac{1}{\alpha}\right) \tag{20.13}$$

The maximum bending moments in frame B as a function of the number of nodes are given in Fig. 20.11. Increasing the number of nodes from 8 (equals 16 struts) to 24 (equals 48 struts) reduces the maximum bending moment almost by a factor of 7.

Table 20.1 Weight reduction compared to CFRP-sandwich reference; V_M: metal volume fraction

Configuration	1	2	3	4
Description	CFRP 76.5/ 23.5/	CFRP-titanium	CFRP 75/25/0	CFRP-steel
Metal volume fraction (%)	0	1.74	0	3.1
Cross sections struts (mm)	Quadratic 49 × 49	Quadratic 45 × 45	Circular radius: 28.2	Circular radius: 30.1
Thickness laminate (mm)	2.125	2.29	2	1.29
Weight reduction (%)	25.2	24.5	29.4	36.9

20.5.5 Weight Saving Potential of Framework Configurations

Detailed investigations prove that the total weight is nearly independent of the number of nodes for the considered node numbers in the previous calculations. However, based on the considerable reduction of the bending moments in frame B for increased node numbers, $n = 22$ nodes are selected for further investigations. A further increase of the node numbers seems unnecessary since the maximum bending moment asymptotically approaches a lower limit.

The weight reduction of the struts is regarded as a primary optimization parameter. Struts of pure CFRP and unidirectional CFRP-steel-laminates are investigated based on a classical optimization method. Laminate thickness, strut diameter and lay-up are systematically varied to reach a similar safety margin against compression failure as well as local and global buckling in the struts.

Therefore, two different unidirectional fiber metal laminates (CFRP-titanium and CFRP-steel) with following parameters were investigated:

- thickness of CFRP-layer (prepreg): $t_{CFRP} = 0.125$ mm
- thickness of titanium- and steel-layer: $t_{titan} = t_{steel} = 0.01$ mm
- fiber orientation in CFRP layer: exclusively in the direction of the applied load (unidirectional)
- metal volume fraction V_M: 1.43% up to 7%

Following the described optimization method, the total weight of the adapter was calculated for different configurations of the derived bending loads in frames A and B using the given material parameters. The four most promising configurations are listed in Table 20.1, whereas the weight reduction is given in comparison to a reference CFRP-sandwich construction. Therein no coupling elements are considered, neither for the framework nor for the sandwich design.

Fig. 20.12 External load
application

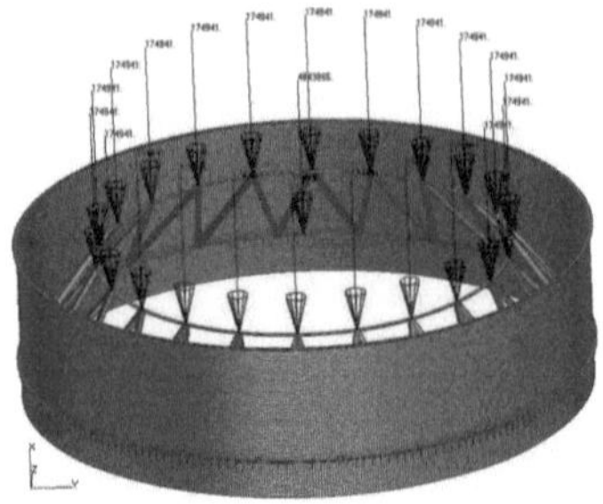

Fig. 20.13 (**a**) Damage after
30 J impact; UD (**b**), [62.5/
25/12.5] (**c**); [92.5/7.5St/0]

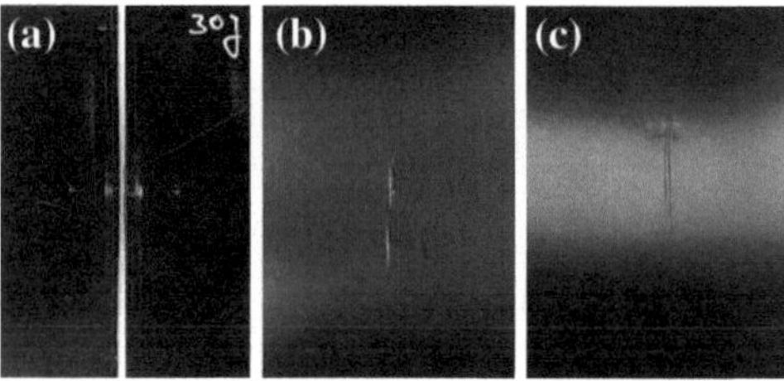

20.6 FEM Analysis for Preferred Framework Configuration

The intention of the FEM-calculation is the validation of the analytical solutions
for the framework and the sandwich design of the adapter. MSC software PA-
TRAN (pre- and post-processing) and NASTRAN (Solver) was used for the
simulations.

The 44 struts of one of the framework configurations are modelled with beam-
elements. The connection between beam and barrel is modelled with RBE3 elements. The loads are applied to the nodes according to Fig. 20.12.

The FEM-results show a good agreement with the analytical solutions.

20.7 Experimental Investigation of Unidirectional CFRP-Steel-Laminates

Compression-After-Impact (CAI) examinations with 30 J impact were performed
at CFRP-UD-laminates, CFRP-62.5/25/12.5-laminates and CFRP-UD/steel-lami-
nates with different metal volume fractions using metal foils of 0.05 mm thickness.
The UD-laminates showed catastrophic failure after impact, whereas the residual
strength after impact could be tested for the other lay-ups, see Fig. 20.13.

The CFRP-UD/steel-laminates showed higher values for residual compression
strength and demonstrated an elastic modulus increased by 65% compared to the
CFRP-62.5/25/12.5 reference laminate (Fig. 20.14).

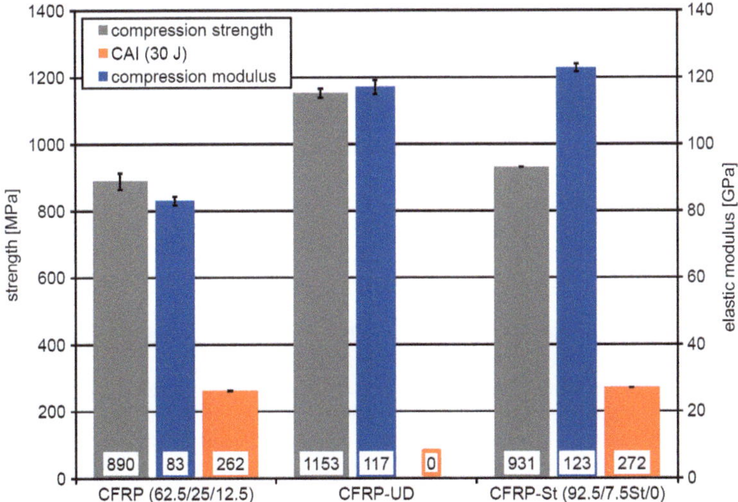

Fig. 20.14 Compression properties

20.8 Conclusion

Advantages in weight efficiency of an adapter in framework design compared to a sandwich design could be shown analytically. The results were validated by a numerical simulation showing very similar values. The weight efficiency can be improved by using unidirectional CFRP-steel-laminates. However, damage tolerance properties have to be proven for these novel materials. Experiments for a steel thickness of 0.05 mm showed promising results, but thinner foils offer a higher potential for the regarded adapter.

References

1. Gómez-Molinero, V.: History and lessons learnt from the development of mechanical systems for different launch vehicles. In: 1st CEAS European Air and Space Conference, CEAS-2007-303
2. Vasiliev, V., Barynin, V., Rasin, A., Petrokovskii, S., Khalimanovich, V.: Anisogrid composite lattice structures—development and space applicatioins. In: 11th European Conference on "Spacecraft Structures, Materials and Mechanical Testing", Toulouse, France (2009)
3. Vasiliev, V., Barynin, V., Rasin, A.: Anisogrid lattice structures—survey of development and application. Compos. Struct. **54**, 361–370 (2001)
4. Rasin, A., Vasiliev, V.: Development of composite anisogrid spacecraft attach fitting. In: 11th European Conference on Composite Materials, Rhodos, Greece (2004)
5. Vasiliev, V., Rasin, A.: Anisogrid composite lattice structures for spacecraft and aircraft application. Compos. Struct. **76**, 182–189 (2006)

6. Nadler, M.A., Yoshino, S.Y., Darms, F.J.: Boron/epoxy support strut for non-integral cryogenic tankage. In: Materials and Processes, 15th SAMPE-Symposium, Los Angeles (1969)
7. Vlot, A., Gunnink, J.G.: Fiber metal laminates an introduction. Kluwer Academic Publishers, Dordrecht (2001)
8. Fink, A.: Local metal hybridization increasing the efficiency of highly loaded composite bolted joints. Dissertation, University Braunschweig, Germany (2010)
9. Worobeij, W.W., Sirotkin, O.S.: Joints for FRP constructions' (russian). Maschinostroenie, Leningrad (1985)
10. Kolesnikov, B., Herbeck, L., Fink, A.: Fortschrittliche Verbindungstechnikenvon Faserverbunden. In: DGLR-congress, Dresden, Germany, vol II, September, pp. 1419–1428 (2004)
11. Kolesnikov, B., Herbeck, L., Fink, A.: CFRP/titanium hybrid material for improving composite bolted joints. Compos. Struct. **83**, 368–380 (2008)
12. Kolesnikov, B., Wilmes, H., Herrmann, A.S., Pabsch, A.: Verbundmaterial mit einem verstärkten Verbindungsbereich. European patent EP 1 082 217 B1, 2002
13. Westre, W.N.: u. a. "Titan-Polymer hybrid Laminate". Patent DE 697 34 616 T2, 2005
14. Kolesnikov, B., Fink, A., Hühne, C., Stefaniak, D., Borgwardt, H.: Strukturelement aus einem Hybridlaminat. patent application DE 10 2010 035 324.8-16, 2010
15. Balabuch, L.I., Alfutov, N.A., Usükin, W.I.: 'Structural mechanics for rockets' (russian). Wysschaja schkola, Moskau (1984)
16. Pisarenko, G.S., Yakowlew, A.P., Matweew, W.W.: 'Reference book for strength of materials' (russian). Naukova Dumka, Kiev (1988)
17. Vasiliev, V.V.: Mechanics of Composite Structures. Taylor & Francis, London (1993)
18. Structural Materials Handbook, Vol. 1—Polymer Composites: Section VI—Design of Structures, Chapter 25, Design of Struts, ESA PSS-03-203 Issue 1. Noordwijk, The Netherlands (1994)
19. Beloserov, L.G., Kireev, B.A.: FRP under mechanical and thermal load (russian). Phismatgis, Moskau (2003)
20. Roark, R.J.: Formulas for Stress and Strain, 3rd edn. Mcgraw-Hill book company, Inc, New York (1954)
21. Gómez-Molinero, V.: General view of the spacesystem structures evolution and furure chellenges, European Conference on Spacecraft Structures. In: Materials & Mechanical Testing 2005 Noordwijk, The Netherlands, 10–12 May 2005
22. Stefaniak, D., Fink, A., Kolesnikov, B., Hühne, C.: Improving the mechanical properties of CFRP by metal-hybridization. In: International Conference on Composite Structures ICCS16, Porto, June 2011

Chapter 21
About the Spring-In Phenomenon: Quantifying the Effects of Thermal Expansion and Chemical Shrinkage

Erik Kappel, Daniel Stefaniak and Christian Hühne

Abstract A straightforward approach to predict spring-in deformations of angled composite parts is presented. Therefore, a proposal by Radford is extended in order to calculate the spring-in contribution due to chemical shrinkage. For this, the volumetric shrinkage of neat thermoset resin, which is in the range of 2–7%, is transferred to equivalent strains on ply level assuming no shrinkage in fiber direction. As the fiber volume fraction (FVF) affects mechanical and chemical properties significantly, the spring-in angle is affected as well. Therefore, the numerical investigation accounts for the spring-in angle and its thermal and chemical contributions depending on the FVF. Classical laminate theory (CLT) is utilized to homogenize layup expansion and shrinkage properties. For validation purposes, model predictions are compared with measurement results gained from one manufactured test specimen. Good agreement between analytical and experimental results is found. Furthermore, the chemical contribution of the total spring-in angle $\Delta\varphi$ turned out to be significantly larger than the thermal contribution.

21.1 Problem's Topicality and Influence Nowadays

The propagation of carbon fiber reinforced plastics in civil aircraft since 1975 has been accompanied by remarkable effort in modeling the specific material behavior of composites. Considering rapidly expanding quantities for modern aircraft, the efficient manufacturing processes gain more importance within the process-chain. Fabrication with high dimensional fidelity is one essential key issue in order to

E. Kappel (✉) · D. Stefaniak · C. Hühne
Institut für Faserverbundleichtbu und Adaptronik, Deutsches Zentrum für Luft- und Raumfahrt, Lilienthalplatz 7, 38108, Braunschweig, Germany
e-mail: erik.kappel@dlr.de

M. Wiedemann and M. Sinapius (eds.), *Adaptive, Tolerant and Efficient Composite Structures*, Research Topics in Aerospace, DOI: 10.1007/978-3-642-29190-6_21,
© Springer-Verlag Berlin Heidelberg 2013

achieve a cost-efficient overall concept. First-Time-Right fabrication must be aspired as manual rework is commonly cost intensive and time consuming. Due to the complex non-isotropic material properties of composites and to common high-temperature and high-pressure manufacturing processes, carbon fiber reinforced plastics tend to show undesirable deviations from the nominal geometry after manufacturing. Those deviations, often referred to as spring-in, are detrimental to achieving a cost-efficient process-chain, as they increase assembly costs due to additional efforts as for example shimming.

Apart from detailed numerical models, which account for the entire complexity of the manufacturing process there is a strong desire to have easy to use pre-dimensioning tools, which are able to predict expected manufacturing deformations with sufficient accuracy as it is desirable at the tool design for example. RADFORD [6] designated thermal expansion and chemical shrinkage as the main driver for manufacturing-induced deformations. Therefore, the present study quantifies thermal and chemical fraction of the spring-in deformation by extending RADFORD's approach [6]. As expansion and mechanical properties depend significantly on the FVF, the spring-in angle and its contributions are calculated as a function of the FVF within this study. Comparison of geometrical measurements of a fabricated test specimen with the predicted spring-in angle $\Delta\varphi$ and its fractions shows promising accuracy.

21.2 Sources of Spring-In Deformations

Carbon fiber reinforced plastics (CFRP) are commonly composed of isotropic resin and orthotropic fibers. Within current manufacturing processes elevated temperatures and high pressure are applied in order to accelerate the manufacturing process and to reduce the void content. The orthotropic mechanical and thermal properties of carbon fibers in combination with the curing process of the resin result in complex physical interactions during the entire manufacturing process. Whereas the fiber properties tend to show no strong temperature dependency, the resin properties change significantly during processing. Within the composite heat transfer mechanisms are acting caused by thermal boundary conditions due to autoclave heating for example, and due to internal heat generation, induced by the ongoing chemical reaction of the resin. Furthermore, the rheological properties of the resin change starting with a viscous behavior in the liquid state, running through a visco-elastic behavior in the rubbery state and ending up with an elastic behavior in the cured rigid state. A unidirectional ply/laminate shows transversal isotropic material properties, resin properties being dominant in the plane perpendicular to the fiber orientation. Within the aircraft industry, multi-angle laminates are common in order to account for the acting stress-states. In general, those layups show orthotropic material properties, however, the in-plane and out-of-plane properties differ significantly as the former are dominated by the fiber properties and the latter are resin dominated. This is shown exemplarily in

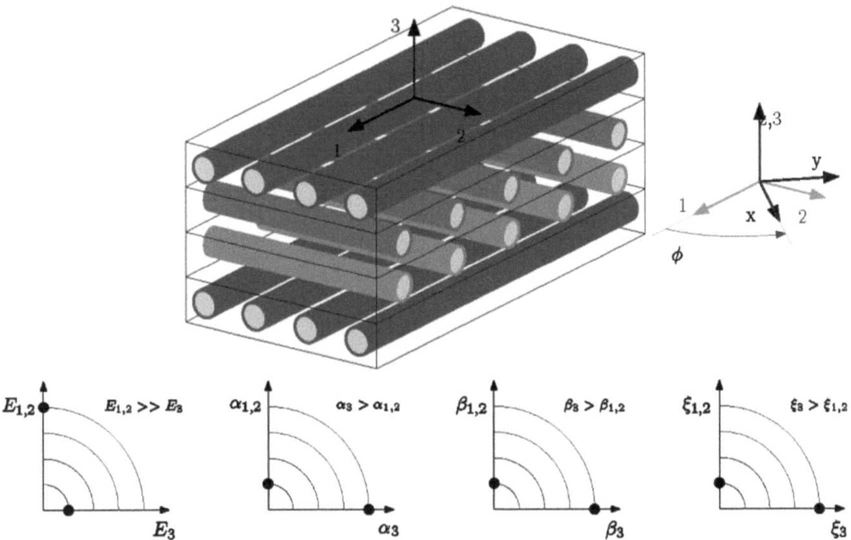

Fig. 21.1 Orthotropic properties of a $[0,90]_s$ laminate—E_i denotes the orthotropic Young's moduli, α_i denotes coefficient of thermal expansion, β_i coefficient of moisture expansion, ξ_j denotes the effective chemical shrinkage, respectively

Fig. 21.1 for a $[0,90]_s$ layup. Therein, the orthotropic Young's moduli E_i, the coefficients of thermal expansion (CTE) α_i, the moisture swelling coefficients β_i and the effective chemical shrinkage coefficients ξ_j are depicted in polar-plots schematically.

Those orthotropic properties do not induce shape changes in flat areas during the manufacturing process but they affect curved sections. RADFORD [6] developed an analytical relation between the aspired reference angle $\tilde{\varphi}$ and the as-built angle φ as a function of the tangential and radial expansion, as shown in Fig. 21.2. Consequently, the spring-in angle $\Delta\varphi$ is defined as the difference between the as-built angle and the reference angle. Regarding thermal effects, application of an isothermal load to a curved laminate with consideration of the CTEs depicted in Fig. 21.1 a large change in thickness direction is obtained but no corresponding change in arc length.

Without allowing shear deformations an angle-change is the consequence. It should be noted that the index k within Eq. 21.1 represents thermal expansion, moisture swelling and chemical shrinkage, which are handled in exactly the same manner. Investigating the consequences of those part deviations, it must be distinguished between direct and indirect deformations. Figure 21.3 depicts direct deformations as they occur on profile structures and indirect ones, which are common for monolithic fabricated stringer-skin sections for example.

Within the present article direct shape changes are investigated utilizing classical laminate theory (CLT). It is focused on symmetric multi-angle aircraft

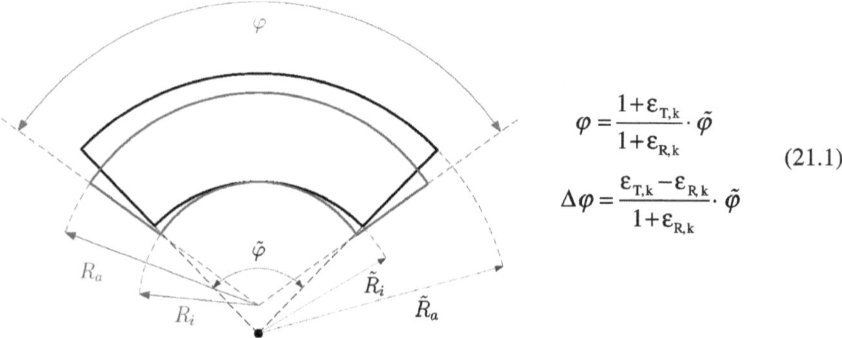

$$\varphi = \frac{1+\varepsilon_{T,k}}{1+\varepsilon_{R,k}} \cdot \tilde{\varphi}$$

$$\Delta\varphi = \frac{\varepsilon_{T,k}-\varepsilon_{R,k}}{1+\varepsilon_{R,k}} \cdot \tilde{\varphi} \qquad (21.1)$$

Fig. 21.2 Change in angle due to orthotropic expansion properties according to RADFORD [6]

laminates consisting of ±45, 90 and 0° plies. For validation purposes, numerical and experimental results obtained with selected CFRP test specimens are compared.

21.3 Analytical Investigation of the Spring-In Effect and its Contributions

As aforementioned, manufacturing of CFRP is affected by a multitude of different parameters. The fiber volume fraction (FVF) is one essential measurand in order to assess the performance of a fabricated composite part. As the mechanical properties depend on the FVF, the spring-in behavior is affected as well. According to RADFORD [6], thermal expansion and chemical shrinkage are the main contributors to the spring-in deformations. SVANBERG [9] for example, stated a volumetric shrinkage of neat resin of about 5.4%. Commonly 2–7% cure shrinkage are typical for thermoset resins. Contributions due to moisture effects are neglected within the present study, hence, they are hardly present right at the end of the manufacturing process. According to SPRÖWITZ et al., [8] the volume change of a reference volume with the edge lengths a, b and c with orthotropic material properties is given by $\Delta V = \varepsilon_a\varepsilon_b\varepsilon_c + \varepsilon_a\varepsilon_b + \varepsilon_a\varepsilon_c + \varepsilon_b\varepsilon_c + \varepsilon_a + \varepsilon_b + \varepsilon_c$.

As chemical shrinkage occurs in the resin solely and with respect to the high fiber stiffness, it is assumed that strain due to chemical shrinkage occurs perpendicular to the fibers only. Thus, strains in fiber direction ε_a are assumed to be zero. Introducing $\varepsilon^* = \varepsilon_b = \varepsilon_c$ for strains in the resin-dominated direction, the volume change due to chemical shrinkage can be written as follows.

$$\Delta V = (\varepsilon^*)^2 + 2\varepsilon^* \qquad (21.2)$$

That assumption corresponds to results presented by ERSOY [2], who reported that in the manufacturing process unidirectional (UD) laminates reveal significantly

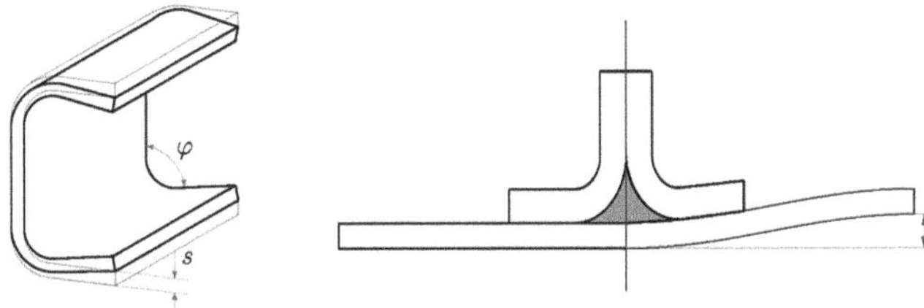

Fig. 21.3 Direct shape changes of a C-profile (*left*), Indirect changes of a monolithically fabricated stringer-skin section (*right*)

smaller through-thickness strains than cross-ply laminates. ERSOY [2] traced that back to the fact that due to the bi-axial fiber orientation of the cross ply laminate resin shrinkage must occur mainly in thickness direction. Solving the Eq. 21.2 for ε^* gives strains perpendicular to the fibers as a function of the volume change ΔV. Regarding a UD ply, the resin content is given by 1-FVF multiplied with the reference volume $\tilde{V}$. Considering that, effective strains are obtained to $\varepsilon^*_{\text{eff}}$.

$$\varepsilon^*_{\text{eff}} = -1 + \sqrt{1 + (1 - FVF)\Delta V} \longrightarrow \qquad \alpha_{\text{eq}} = \frac{\varepsilon^*_{\text{eff}}}{\Delta T} \tag{21.3}$$

Dividing by the cool-down temperature ΔT allows direct comparison of thermal and chemical parameters. Within a small Matlab routine Radford's approach has been implemented in order to calculate the spring-in angle and its fractions. The input parameters are ply thickness t_k, resin shrinkage ΔV, nominal section angle $\tilde{\varphi}$, cool down temperature ΔT and an arbitrary layup such as $[45, -45, \ldots]_s$. Based on orthotropic fiber and isotropic resin properties, engineering constants and thermal expansion coefficients are calculated as a function of FVF. Therein, rules of mixture proposed by SCHÜRMANN [8] are utilized.

$$\underline{\alpha} = \{\alpha_1 \quad \alpha_2 \quad 0\}^{\text{T}}, \quad \underline{\alpha}_{\text{eq}} = \frac{1}{\Delta T} \cdot \{0 \quad \varepsilon^*_{\text{eff}} \quad 0\}^{\text{T}} \tag{21.4}$$

As Radford's approach is applicable only for homogeneous orthotropic materials, expansion properties are homogenized, as shown in Eq. 21.5.

$$\left\{ \begin{array}{c} \overline{\underline{\alpha}}_{\text{th}} \\ \overline{\underline{\alpha}}_{\text{ch}} \end{array} \right\}_{\text{lam}} = \underline{\underline{A}}^{-1} \cdot \sum_{k=1}^{N} \underline{\underline{\overline{Q}}}_k \cdot t_k \cdot \left\{ \begin{array}{c} \overline{\underline{\alpha}} \\ \overline{\underline{\alpha}}_{\text{eq}} \end{array} \right\}_k \tag{21.5}$$

Therein $\overline{\underline{\alpha}}_{\text{th}}$ and $\overline{\underline{\alpha}}_{\text{ch}}$ correspond to the vectors of homogenized thermal and chemical expansion in global coordinates, $\underline{A}$ represents the extensional stiffness matrix, $\underline{\underline{\overline{Q}}}_k$ represents the lamina stiffness matrix of the k-th ply and represents the thickness of the k-th ply. It should be noted that both consist of in-plane properties only. However, Radford's formula is based on the plane-stress (PS) assumption. Thus, its application demands knowledge of expansion properties in thickness direction. Analogous to the transverse direction of a UD ply the out-of-plane

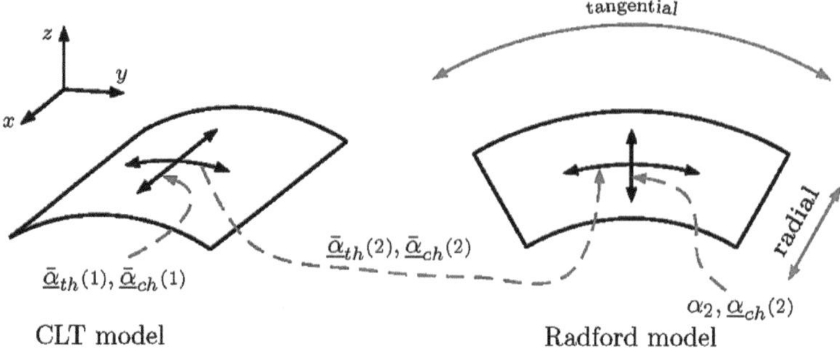

Fig. 21.4 Sources of the parameters applied in Radford's approach

Fig. 21.5 Spring-in angle as a function of FVF for a quasi-isotropic laminate [45,−45,90,0]$_s$, *gray* area represesnts estimated FVF of fabricated specimens, *blue* area represents unrealistic and undesirable FVFs

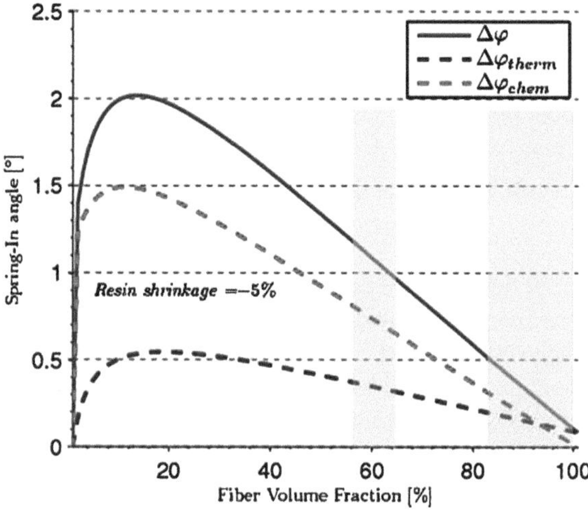

properties of the laminate are resin dominated. Thus, it is assumed that the laminate expansion in thickness direction equals the expansion of the UD plies perpendicular to the fibers. Figure 21.4 depicts that schematically.

Applying the spring-in angle calculation according to Eq. 21.1, while tangential expansion properties are taken from the CLT model and radial properties in the Radford model are taken from the local ply properties shown in Eq. 21.4, allows the calculation of the corresponding spring-in angle. According to the variety of conceivable layups, a global statement about thermal and chemical contributions of the spring-in behavior is not possible. However, Fig. 21.5 depicts it for a quasi-isotropic [45,−45,90,0]$_s$ layup exemplarily. Therein the spring-in angle $\Delta\varphi$ corresponds to the sum of thermal contribution $\Delta\varphi_{therm}$ and chemical contribution $\Delta\varphi_{chem}$.

Coefficients of thermal expansion (CTE) of the fibers and the resin utilized in the present study are taken from Hexcel® data sheet, ANGHELESCU [1] and PELLEGRINO [5] ($\alpha_1 = -0.63 \times 10^{-6} 1/K$, $\alpha_2 = 7.2 \times 10^{-6} 1/K$ and $\alpha_{res} = 65 \times 10^{-6} 1/K$) Inverse rule of mixture, as shown by SCHÜRMANN [7], is applied to determine lamina CTEs as a function of FVF. Regarding realistic FVF

 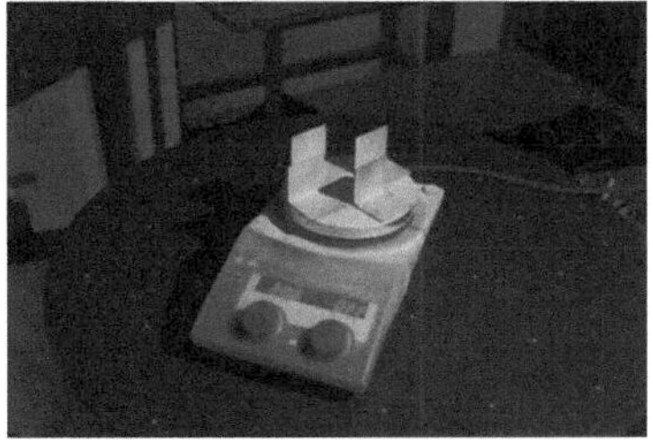

Fig. 21.6 L-profile test specimens (*left*), Hot-plate experiment (*right*)

Table 21.1 Difference of the spring-in angle between room temperature (25°C) hot conditions (~ 180°C)

Layup	$\Delta\varphi$ (T = 25°C)	$\Delta\varphi$ (T $\approx$ 180°C)	$\Delta\varphi_{\text{therm}}$ (°)	Difference (%)
$[45,-45,90,0]_s$	1.36	0.82	0.54	39.7

of 40% up to 80% the chemical fraction decreases from 71 to 61%, whereby the total spring-in angle decreases from 1.6 to 0.6°. A FVF of 60%, as it is aspired in common aircraft composite parts, corresponds to a spring-in angle of 1.1° with a chemical contribution of 67%, which is as twice as high as the thermal contribution.

21.4 Experimental Investigations

To verify the analytical approach test specimens have been manufactured using a 8552/AS4 prepreg system. Dimensions and shape of the L-profile test specimens are similar to those used by KLEINEBERG [4]. Detailed information about the process boundary conditions such as bagging, cure cycle and tool properties can be found at KAPPEL et al. [3]. The manufactured specimens have been measured using a full-field GOM Atos system in order to determine fabrication induced deformations. While chemical shrinkage is irreversible, thermal expansion is an elastic phenomenon. Assuming that vitrification of the resin took place at T = 180° at the manufacturing process, this temperature is necessary to remove thermo-elastic strains induced by the cool-down from curing to room temperature.

For the application of corresponding thermal loads, specimens have been placed onto a controllable hot-plate, as shown in Fig. 21.6. The temperature of the hot plate has been set to 200°C in order to compensate loss of temperature due to convection and radiation. After 10 min on the hot plate, specimens have been measured and actual angles have been evaluated using the GOM Inspect software.

The results of the experiment are summarized in Table 21.1. The manufactured test specimen showed a spring-in angle at room temperature (T = 25°C) of 1.36°. For a temperature of approximately 180°C a spring-in angle of 0.82° was obtained.

That complies to a thermal contribution of the spring-in angle of 0.54°, which is 39.7% of the total angle $\Delta\varphi$. FVF of the fabricated specimen is approximately 60% as the manufacturer's recommended cure cycle (MRCC) has been used. Comparing measurement results with the numerical results, depicted in Fig. 21.5, shows reasonably good agreement.

Considering that neat resin shrinkage has only been estimated with 5%, that the specimen temperature during the measurement was approximately 180°C and that process and tool conditions as well as tool-part interactions as described by TWIGG et al. [10] and KAPPEL et al. [3] remain unaccounted in Radford's approach, the results agree surprisingly well.

21.5 Conclusions

Radford's approach has been extended in order to calculate spring-in deformations due to chemical shrinkage. Therefore, the volumetric shrinkage of neat resin has been transferred to an equivalent shrinkage in the resin dominated direction, assuming no chemical shrinkage in fiber direction. CLT is used to homogenize layup expansion properties, handling chemical expansion analogously to thermal expansion. Subsequently, Radford's approach is utilized accounting for thermal and chemical fractions of the spring-in deformation. The FVF is one essential parameter for composite parts, as it affects stiffness and expansion properties significantly. Consequently, the spring-in angle $\Delta\varphi$ and its fractions has been calculated as a function of FVF within this study. For verification purposes, the extended approach has been experimentally validated with one L-profile test specimen fabricated out of 8552/AS4 prepreg material. Reheating the parts using a hot-plate allows to blank out the thermo-elastic contribution of the spring-in angle, which is induced by the cool-down during manufacturing. Using a full-field measurement system, the thermal fraction of the spring-in angle of a $[45,-45,90,0]_s$ specimen could be obtained to nearly 40% of the total spring-in angle $\Delta\varphi$, which is reasonably close to the numerical prediction. Nevertheless, further specimens with different lay-ups are necessary to validate the applicability of the presented procedure more thoroughly. Although slight uncertainties remain due to its simplicity, the proposed strategy delivers promising results for estimating the total spring-in angle and its thermal and chemical contributions.

References

1. Anghelescu, M.S., Alam, M.K.: Carbon foam tooling for aerospace composites. SAMPE Conference Cincinnati, Ohio (2007)
2. Ersoy, N., Tugutlu, M.: Cure kinetics modeling and cure shrinkage behavior of a thermosetting composite. Polym. Eng. Sci. **50**(1), 84–92 (2010)
3. Kappel, E., Stefaniak, D., Hühne, C.: A semi-analytical simulation strategy and its application to warpage of autoclave-processed CFRP parts. Compos. Part A, **42**(12), 1985–1994 (2011)

4. Kleineberg, M.: Präzisionsfertigung komplexer CFK-Profile am Beispiel Rumpfspant, TU Carolo-Wilhelmina zu Braunschweig (2008)
5. Pellegrino, S.: Ultra-thin Carbon Fiber Composites: Constitutive Modeling and Applications to Deployable Structures. Lectures 1–2, California Institute of Technology (2009)
6. Radford, D.W.: Shape Stability in Composites. Rensselaer Polytechnic Institute, Troy (1987)
7. Schürmann, H.: Konstruieren mit Faser-Kunststoff-Verbunden. Springer, ISBN 3-540-40283-7 (2005)
8. Spröwitz, T., Hühne, C., Kappel, E.: Thermal aspects for composite structures—from manufacturing to in-service predictions. In: CEAS 2009, 26–29 Oct. 2009, Manchester, UK, (2009)
9. Svanberg, J.M., Altkvist, C., Nyman, T.: Prediction of shape distortions for a curved composite C-spar. J. Reinf. Plast. Compos. **24**, 323 (2005)
10. Twigg, G., Poursatip, A., Fernlund, G.: Tool–part interaction in composites processing. Part I: experimental investigation and analytical model. Compos. Part A Appl. Sci. Manuf. **35**, 121–133 (2004)

Chapter 22
Carbon Fiber Composite B-Rib for a Next Generation Car

Jörg Nickel and Christian Hühne

Abstract Increasing environmental, economical, and social issues force future car concepts to strive for maximum efficiency. As metal designs have reached a very high level of maturity, further potentials are seen especially with extremely lightweight (carbon) fiber reinforced composites (CFRP) exhibiting high strength and stiffness, which are advantageously integrated into multi-material designs. This section focuses on improvements in weight reduction, safety and modularisation strategies for future cars. A novel Rib and Space-Frame concept incorporating carbon fiber composites is presented. Herein, an essential component replacing the former B-pillar is the B-rib, which uses a novel mechanical principle to meet the side impact crash requirements. Furthermore, using the Rib and Space-Frame concept as an example, the entire process chain of fiber composites—from materials to design, sizing, prototype manufacture, and testing—is being presented also opening up perspectives for future mass production strategies.

22.1 Challenges of Future Individual Mobility

Three central challenges arise from the tense relation between the demands for—and the negative effects of—mobility: securing mobility for people and goods, protecting the environment and preserving resources, and improving safety and security without neglecting cost efficiency. Subsequently the demands for future vehicles by far exceed traditional requirements in automotive design [1–3]. A major

J. Nickel (✉) · C. Hühne
Institute of Composite Structures and Adaptive Systems, Composite Design, Deutsches
Zentrum für Luft- und Raumfahrt e.V. (German Aerospace Center), Lilienthalplatz 7, 38108,
Braunschweig, Germany
e-mail: Joerg.nickel@dlr.de

M. Wiedemann and M. Sinapius (eds.), *Adaptive, Tolerant and Efficient Composite Structures*, Research Topics in Aerospace, DOI: 10.1007/978-3-642-29190-6_22,

objective is to minimise fuel consumption and CO_2 emissions e.g. by reduced driving resistance [2]. The driving resistance comprises four relevant aspects:

- **Rolling resistance** F_{RR} resulting from the friction between the wheels and the ground.
- **Grade resistance** F_{GR} dependant on the inclination of the ground.
- **Acceleration resistance** F_{AR} induced by the mass moments of inertia of the rotationally and transversally accelerated masses of the vehicle.
- **Aerodynamic drag** F_{AD} considering the air drag coefficient, the frontal area, and the current velocity.

In any driving situation the summed resistances are in equilibrium with the output of the engine at the driving wheels. This relationship is given as the sum of driving resistances (22.1).

$$M_m \cdot \frac{i_{tot}}{r} \cdot \eta_{tot} = \underbrace{m \cdot g \cdot f \cdot \cos\alpha}_{F_{RR}} + \underbrace{m \cdot g \cdot \sin\alpha}_{F_{GR}} + \underbrace{e \cdot m \cdot \ddot{x}}_{F_{AR}} + \underbrace{c_W \cdot A \cdot \frac{\rho_L}{2} \cdot v^2}_{F_{AD}}$$

$$(22.1)$$

Variable	Definition	Unit
M_m	Wheel hub torque	(Nm)
i_{tot}	Overall gear transmission ratio of powertrain	[1]
r	Dynamic wheel diameter	(m)
η_{tot}	Gross degree of efficiency of powertrain	[1]
m	GVWR = gross vehicle weight rating	(kg)
g	Gravity acceleration	9.81 (m/s^2)
f	Rolling drag coefficient	(–)
α	Angle of elevation	(°)
e	Mass factor (for calculation see Ref. [4])	(–)
$\ddot{x}$	Vehicle acceleration	(m/s^2)
c_W	Air drag coefficient	(–)
A	Vehicle frontal area	(m^2)
ρ_L	Density of ambient atmosphere	(kg/m^3)
v	Driving velocity	(m/s)

This correlation reveals the strong influence of the vehicle mass which occurs as a factor in three of the four terms of the above-mentioned equation. In a steady driving condition on even ground, the influence of the mass decreases to a minimum as only the rolling resistance and the aerodynamic drag are relevant. Under discontinuous driving conditions or when the velocity is frequently changing, as for e.g. urban traffic or short trips, the mass impact is dominant. Here, a mass reduction can result in a significant decrease of fuel consumption, i.e. 100 kg less weight usually results in an approximate fuel reduction of 0.5 l/100 km.

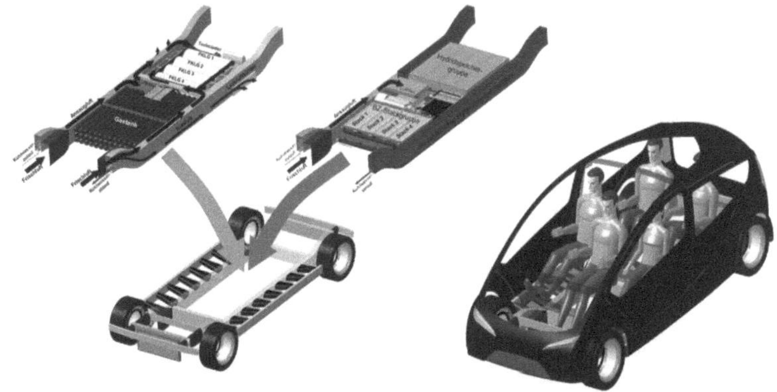

Fig. 22.1 Vehicle concept with intrusion-resistant containment (DLR courtesy)

As the car body has an average share of 25% of the overall weight, potentials for lightweight design are mainly seen here.

Further issues arise from the major trend of individualisation. The effects are reduced specific quantities per car variation [5, 6] and demands for additional options of modularisation and derivatisation, also of drive-train, car body and structural components. These aspects have to be considered in future vehicle concepts.

22.2 Novel Vehicle Concept

A novel upper middle class vehicle structure, that is fiber composite intensive in the Rib and Space Frame Design, is seen as one solution to the challenges of future personal mobility, [1], Fig. 22.1. Beside its modular construction incorporating a CFRP intensive Multi Material Design [7], it is characterised by the simple integration of alternative drive-train concepts into the intrusion-resistant floor area, which is protected by side crash compartments [8]. These would generally absorb the forces of side and pole impacts. The basic Space Frame structure, which consists of ring frames instead of A, B, and C pillars, is fitted onto the floor structure and combined with geometrically simple profiles and highly integral cast metal joints.

22.2.1 Composite B-Rib: An Essential Component of the Vehicle Concept

In the overall development, the focus of this Rib and Space Frame Design is initially on the ribs with their higher design and dimensioning complexity. The B-rib, which replaces a standard automobile B-pillar, was selected as an

Fig. 22.2 B-rib position in vehicle concept

example [8]. The ribs are made of carbon fiber composites. Carbon fiber reinforced polymers (CFRP) are relatively expensive (around 25–50 €/kg), but offer the required stiffness and energy absorption levels while remaining lightweight [7, 8].

Basically the B-rib has to fulfil side impact conditions according to Euro-NCAP (50 km/h, 950 kg, 300 mm height of load introduction) [8]. For certification in the US, the more ambitious American IIHS requirements (50 km/h, 1,500 kg, 379 mm height of load introduction) are valid. Following concept development, topology optimisation, construction and structural and dynamic analyses, prototypes successfully meet these requirements.

22.2.2 *Functional Principle of the Composite B-Rib under Side Impact*

The most significant load cases for the design of the floor assembly arise from both the side and pole impact test procedures. The components of the drive unit must not be forced against an incompressible object, as this would cause the acceleration values for the occupants to increase too sharply.

Therefore a chassis design of the car body is developed that forms an outer zone acting as a "crash compartment". This zone extends from the sidewall to the continuous side members, which are used for their suitability in lightweight construction.

The B-rib, as one essential component of the Rib and Space-Frame concept, Fig. 22.2, uses a novel mechanical principle to meet the side impact crash requirements as shown in the equivalent mechanical diagram, Fig. 22.3.

When a side impact occurs, the rib in the vicinity of the roof beam acts as a deformable joint, Fig. 22.4. By contrast, the continuous side member, Fig. 22.5 between the roof beam and the rocker panel, is designed to be extremely stiff, so as to provide the best possible protection to the occupants in the head and shoulder region matching the side impact criteria of IIHS. The resulting rotation about the

Fig. 22.3 Equivalent mechanical diagram of B-rib

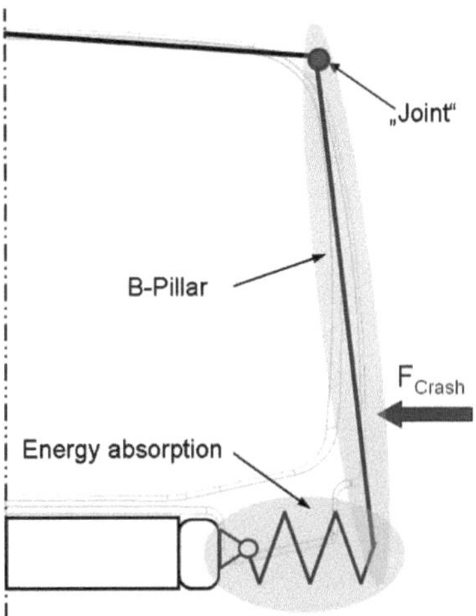

Fig. 22.4 Functional area of B-rib: deformable joint

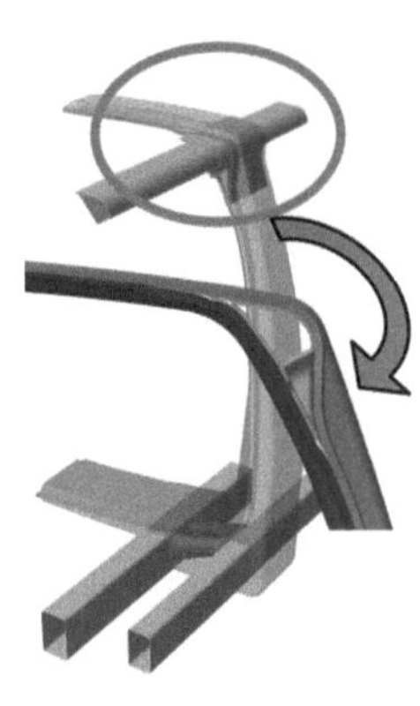

Fig. 22.5 Functional area of B-rib: continuous side member

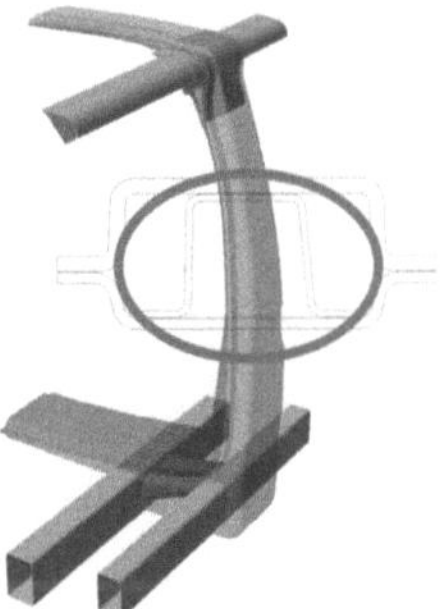

deformable joint transfers the crash energy to the crash cone elements, Fig. 22.6 in the floor region. These are located in the area between the rocker panel and the continuous side member and absorb energy through a crushing action.

Fig. 22.6 Functional area of B-rib: crash cone element

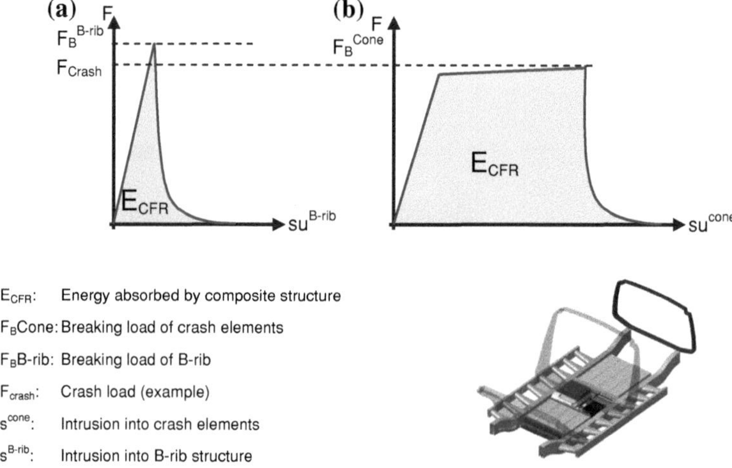

Fig. 22.7 Energy dissipation in B-rib (**a**) and crash cone (**b**) under side impact

The B-rib is designed to show a breaking load $F_{B\ B\text{-}rib}$, which is significantly higher than any occurring side impact crash load F_{Crash}, whereas the breaking load of the crash cone elements $F_{B\ Cone}$ is dimensioned to the load level of the side impact F_{Crash}. So the crash cones will always work for energy dissipation and provide secure functionality, Fig. 22.7.

The lightweight potential of the B-rib is shown through a weight reduction of up to 35% with significantly increased safety and overall performance compared to the steel reference structure of a compact class vehicle.

22.3 Challenges in Design and Manufacture

The development of the composite B-rib covers the entire process chain. Herein are included conceptual and detailed design, static and dynamic calculation, sizing, and simulation, and lay-up design. Manufacturing aspects such as design of prototypical autoclave manufacturing concepts and manufacture of prototypes are also addressed. Validation tests are included as well.

The frame is constructed from two shells with internal Ω-stiffening, Fig. 22.5. The internal Ω-profile has to be produced within strict wall thickness tolerances in order to ensure the best fit.

Analytical examination, topology optimisation, and static and dynamic numerical calculations yield results for required wall thickness, fiber orientation and the amount of fibers in various directions. The values in brackets describe the percentage of fiber volume content in the $(0°/\pm45°/90°)$-direction.

Outer shell: A high wall thickness is required for load introduction and stability. For that reason and also for introduction of direct side impact, a high volume content of $0°$-oriented fibers is needed in the compression-loaded area of the side beam represented by a fiber orientation share of (65/25/10).

Inner shell: As the inner shell only takes tension loads during side impact, a lower wall thickness is sufficient. The share of fiber orientation, with a high volume content of $0°$-oriented fibers in the tension-loaded area, is similar to that of the outer shell: (65/25/10).

Internal Ω-stiffening: The wall thickness is determined by the requirements from local buckling of the web and a high share of $0°$ and $\pm45°$-oriented fibers due to combined bending and shear loads: (50/40/10).

The specifications arising from the design and calculation have to be realised in a manufacturing-oriented and quality-controlled manner. The prototypical tool and manufacturing concept has to guarantee the required component quality at maintainable cost and effort. As a consequence single-sided aluminium tools and a liquid composite moulding process (LCM) using an autoclave are preferred. Right from the beginning, the design of the B-rib considered aspects of series production. In order to achieve attractive cycle times and to meet industrial requirements, resin transfer moulding (RTM) procedures with highly reactive resins are options for future manufacturing strategies.

The detailed design, manufacturing simulation and prototype manufacture include the following steps. First, based on the calculated results from the topology optimisation, the final geometry of the B-Rib is defined. Subsequently the lay-up and the ply-book can be derived as a basic document for manufacture.

Using this lay-up sequences and ply-book as an input for the draping simulations, the optimal and wrinkle-free fiber lay-up based on the applied preforms (satin weave offering best drapability) can be determined. The results enable generating the flat patterns needed for numerical controlled cutting of the preforms, Fig. 22.8.

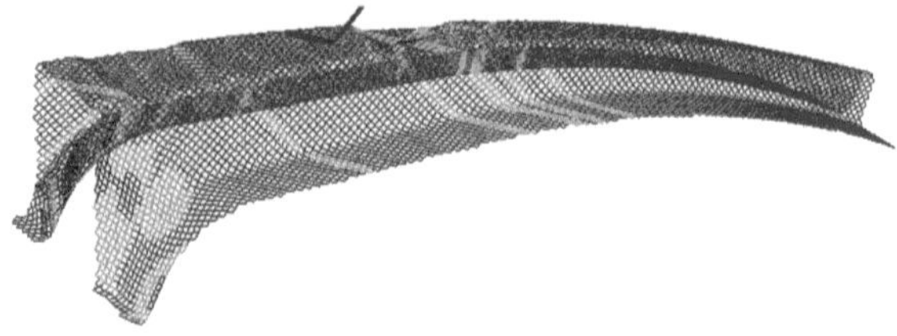

Fig. 22.8 Draping simulation of the internal Ω-profile located in the side area of the B-rib

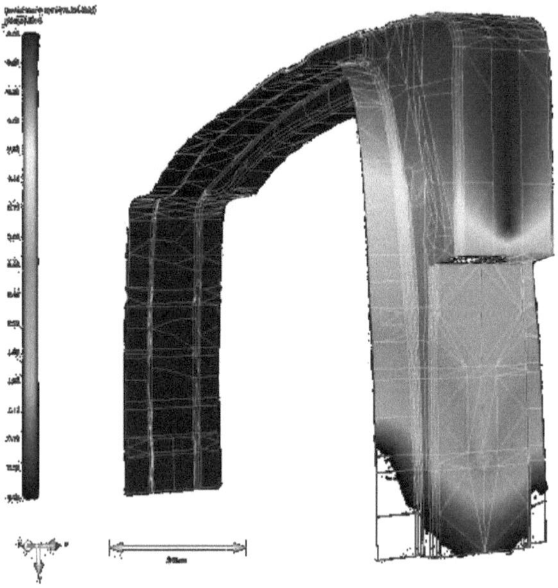

Fig. 22.9 Flow simulation of the outer shell of the B-rib

A flow simulation has to be performed in order to determine the best-suited injection management of the tool and to evaluate and to verify the manufacturing concept, Fig. 22.9.

Another essential step lies in the development of suitable tool concepts for prototype manufacture, Fig. 22.10. For the outer shell with its complex geometry and undercuts, a positive mould is best suited. For producing the inner shell, a negative mould is used. For prototype manufacture a manageable effort is essential. As a consequence single-sided aluminium tools and a liquid composite moulding process (LCM) using an autoclave are preferred.

Many processes are available to manufacture CFRP components. Infusion technology involving modern preform processes offer an attractive, automatable and low-cost manufacturing method for complex components. In contrast to pre-preg processes with resin-impregnated fiber layers, infusion processes involve injecting dry semi-finished fiber products with a low-viscosity resin. These allow complex geometries to be achieved through cost-reducing textile technologies such as sewing and draping. SLI (Single-Line-Injection) is suitable for the frames

Fig. 22.10 Prototype B-rib (half, rendering) and autoclave tool concept

Fig. 22.11 Functional principle of ultrasonic thickness measurement

considered here [7]. This only requires a solid tool half ("open mould") which gives the shape. The other mould half is replaced by an air-tight membrane. The required consolidation pressure is applied by an autoclave. The pressurised injection allows even low-permeability preforms to be soaked with resin and large flow paths to be achieved without flow promoters. Not using flow-promoting manufacturing aids allow a very high surface quality even on the membrane side by caul plates.

By means of tool-integrated ultra-sound sensors, the wall thickness during the manufacturing process can be monitored. Controlling the injection pressure, the quantity of resin in the component can be adjusted with corresponding effects on the fiber volume content and the thickness of the component, Figs. 22.11 and 22.12. The required sharp tolerances during production are guaranteed by applying ultrasonic thickness measurement and control [9, 10].

Using a pasty two-component adhesive the single components are joined together to the prototype B-rib, Figs. 22.13, 22.14 and 22.15. For increased buckling stability, side zones are filled with foam. After attaching the fittings, e.g. door hinges etc., the prototype B-rib is ready for further validation testing.

Fig. 22.12 Internal Ω-profile tool with vacuum assembly and integrated sensors

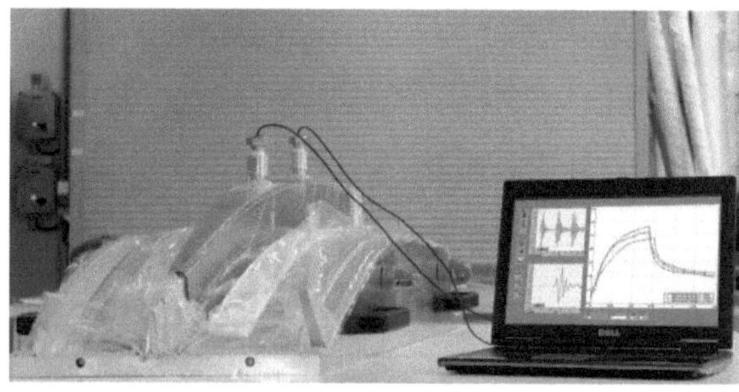

Fig. 22.13 Side view of carbon fiber composite B-rib

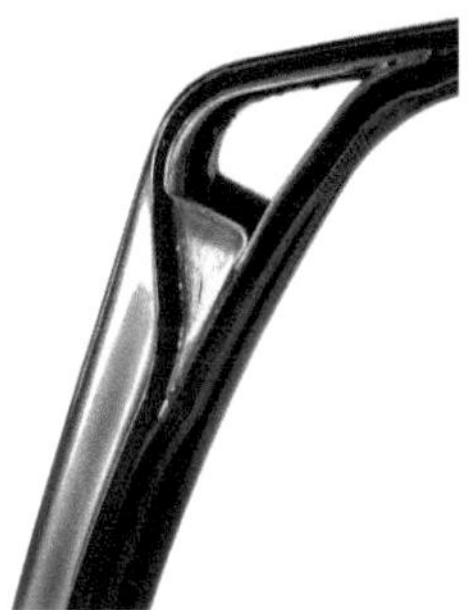

Fig. 22.14 Carbon fiber composite B-rib

Fig. 22.15 Cross section of carbon fiber composite B-rib

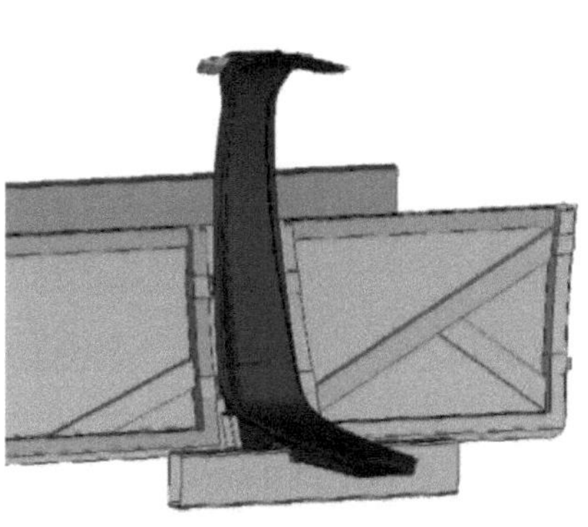

Fig. 22.16 Crash simulation and validation test of carbon fiber composite B-rib

22.4 Validation by Means of Static and Dynamic Tests

The structural concept of the B-Rib design and the manufactured prototypes have to meet static and dynamic requirements according to legal regulations. For dynamic side impact testing, there are mainly two important regulations that have to be fulfilled. In Europe, this is currently the Euro-NCAP regulation demanding a side impact velocity of 50 km/h and a mass of the impactor of 950 kg, introduced at a height of 300 mm [8]. In the US, the more ambitious American IIHS requirements at 50 km/h impact velocity, impactor mass of 1,500 kg at 379 mm height of load introduction are valid.

The designed and manufactured B-Rib meets both, the Euro-NCAP requirements and the American IIHS side impact demands, Fig. 22.16 proving the theoretically calculated and expected behaviour, Fig. 22.7.

22.5 Conclusion & Perspectives

The development of the composite B-rib covers the entire process chain. Herein are included conceptual and detailed design, static and dynamic calculation, sizing, and simulation, and lay-up design. Manufacturing aspects such as design of prototypical autoclave manufacturing concepts and manufacture of prototypes are also addressed. Validation tests are included as well.

The lightweight potential of the B-rib is shown through a weight reduction of up to 35% with significantly increased safety and overall performance compared to the steel reference structure of a compact class vehicle.

Experiences reveal the inherent potentials and interferences between a pro-composite, i.e. fiber composite optimised component design, and cost-efficient and cost-effective manufacturing technologies, like Resin Transfer Moulding (RTM) and their variants, when seeking an optimum solution in terms of weight and cost ready for mass production [11].

References

1. Friedrich, H.E., Kopp, G.: Leichtbau und Modulbauweisen für zukünftige Fahr-zeugkonzepte, Stuttgarter Symposium, (2007)
2. Haldenwanger, H.G.: Zum Einsatz alternativer Werkstoffe und Verfahren im kon-zeptionellen Leichtbau von PKW-Karosserien, TU Dresden, Ph.D-thesis, (1997)
3. Krusche, T., Leyers, J., Oehmke, T., Parr, T.: Bewertung von Modularisierungs-strategien für unterschiedliche Fahrzeugkonzepte am Beispiel des Vorderwagens. ATZ **10**, 928–933 (2004)
4. Robert Bosch GmbH (Hrsg.): Kraftfahrtechnisches Taschenbuch, 25. Auflage, Friedr. Vieweg & Sohn Verlag, GWV Fachverlage GmbH. Wiesbaden Oktober 2003, ISBN 3-528-23876-3 (2003)
5. Kroker, J.: Zukünftige Ansätze für modulare Karosseriestrukturen. Faszination Karosseriebau, Braunschweig (2005)
6. Schöll, R., Friedrich, H.E., Kopp, Gu., Kopp, Ge.: Innovative Fahrzeugstruktur in Spant- und Space-Frame-Bauweise. ATZ (111. Jahrgang), pp.52-58. Vieweg + Teubner Verlag. ISSN 0001-2785 10810, (2009)
7. Herrmann, A.S., Sigle, C.: Das Single-Line-Injection-Verfahren zur Herstellung von Hochleistungsverbunden. DGLR-Jahrestagung, S. 1–7 (1999)
8. Schöll, R., Erdl, G., Friedrich, H.E.: Innovative Fahrzeugleichtbaustruktur in Spant- und Space-Frame-Bauweise. Konstruktion Zeitschrift für Produktentwicklung und Ingenieur-Werkstoffe, pp 37–38. Springer VDI Verlag, ISSN: 0720-5953, (2009)
9. Mc Hugh, J.: Ultrasound Technique for the Dynamic Mechanical Analysis of Polymers. Berlin, Technische Hochschule, Dissertation, (2007)
10. Stahl, A., Nickel, J., Kühn, M., Liebers, N.: Qualitätsgesicherte Fertigung von Faserkunststoffverbund-Bauteilen. ATZproduktion 05–06, 2010, Nov. 2010, 3. Jahrgang, pp. 48–52. Springer Automotive Media, Springer Fachmedien Wiesbaden (2010)
11. Nickel, J., Schöll, R., Friedrich, H.E., Wiedemann, M., Hühne, Ch.: B-Säulen-Ringspant in CFK-intensiver Bauweise für ein Multimaterialauto. ATZ-Fachtagung Werkstoffe im Automobilbau, 18–19 May 2011, Stuttgart, (2011)

Chapter 23
Automated Scarfing Process for Bonded Composite Repairs

Dirk Holzhüter, Alexander Pototzky, Christian Hühne
and Michael Sinapius

Abstract Today's bonded composite repairs rely heavily on manual grinding. The human influence on scarf tolerances and consequently on assembly and structural performance can be reduced by automating the process of scarf manufacturing. A process based on contact free surface scanning, surface reconstruction, automated repair design and automated milling of the repair scarf is presented. A machine and software design for validation purposes is described. Several repair specific design considerations relevant for the construction of a mobile scarfing machine are discussed. The redesign of a standard 3-axis milling machine to a mobile automated scarfing unit is presented and the architecture of the associated software framework is described. An outlook to future validation steps is given.

23.1 Motivation

The steady increase of CFRP-Structures in modern aircraft will reach a new dimension with the entry into service of the Boeing 787. The extensive use of Carbon Fiber Reinforced Plastics (CFRP), e.g. in fuselage and wing shells, marks a major step forward in the application of modern composite materials. Those structures tend to be damaged more often compared to composite structures on classic aircraft because of their location. Typical threats for those aircraft structures are environmental hazards like lightning and bird strike, airport operations (e.g. service vehicles), maintenance and pilot handling (e.g. tail strike) [1].

D. Holzhüter (✉) · A. Pototzky · C. Hühne · M. Sinapius
Institute of Composite Structures and Adaptive Systems, Composite Design,
Deutsches Zentrum für Luft- und Raumfahrt e.V. (German Aerospace Center),
Lilienthalplatz 7, 38108, Braunschweig, Germany
e-mail: dirk.holzhueter@dlr.de

M. Wiedemann and M. Sinapius (eds.), *Adaptive, Tolerant and Efficient Composite Structures*, Research Topics in Aerospace, DOI: 10.1007/978-3-642-29190-6_23,
© Springer-Verlag Berlin Heidelberg 2013

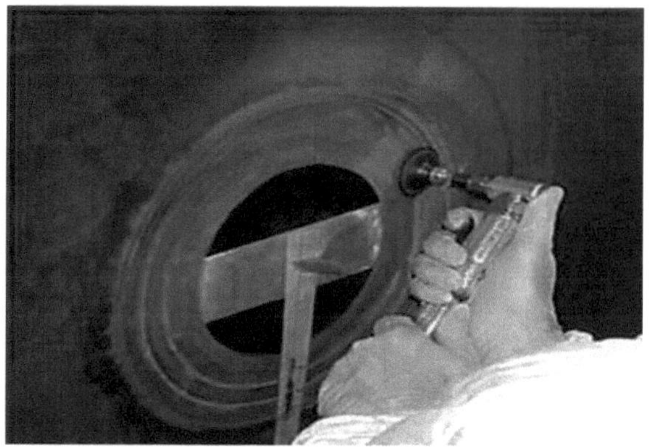

Fig. 23.1 Manual scarfing by angular grinding [7]

Repairs of composite aircraft structures are already of great economic importance. Lufthansa Airlines carried out 1,600 repairs on composite structures in 2006 (Fig. 23.1) [2]. A standard repair procedure on composite components is the bonded scarf patch repair [3]. The damage size will be determined by a NDT procedure, e.g. ultra sonic or thermography. The damage is commonly removed by hand drilling or grinding. The paint is locally removed down to the first ply. The typical procedure of preparing the bonding surface is to continuously scarf the structure around the damage in a circle (Fig. 23.1) or to manufacture a stepped contour [4]. Typical scarf ratios are in the range of 1:20–1:60 of the local structural thickness. The scarf is manufactured by manually grinding ply by ply by using an angular grinding tool. The craftsman orients himself at the ply pattern to identify the local depth. A circular template is used to get the correct diameter. It is essential for the mechanical performance of the repair to fabricate the scarf in a small tolerance. The grinding activates the surface by removing the resin from the fibers and breaking open superficial chemical bonds for the following bonding process. The manufacture of the scarf can be quite complicated if the repair surface is curved or if local reinforcements, edges, thickness changes or bolted joints are located within the scarf area. Manually grinding a scarf is a slow process which can take several hours. The quality of the procedure also depends strongly on the experience of the technical personnel. Many different procedures are known for the patch manufacturing. The Co-Bonding of wet prepreg by using a film adhesive is a wide spread repair process among the industry. The wet lay-up is still flexible and compensates even large tolerances. The curing temperature for the repair lay-up as well as the adhesive is restricted to 130°C. Local heat sinks on the parent structure restrict the use of higher ranges. The out-of-autoclave curing condition requires special repair pre-pregs. The alternative process of secondary bonding a cured patch requires a separate tooling. The tooling is directly moulded in the scarf by using a low temperature prepreg. This female tooling is moulded a second time to produce a male tool which is used in an autoclave to manufacture the proper patch at original processing conditions. Due to manufacturing distortions it is hardly possible to use a film adhesive for the bonding. Normally a paste adhesive is better suited to compensate the tolerances although both types of adhesives are used today. The process is much more time consuming compared to a wet lay-up process, but produces a better mechanical performance. The cured patch repair is mainly restricted to military aircraft while wet lay-up is a preferred process for helicopters.

Up to now the bonded composite repair is a highly manual process. Especially the manufacturing of the scarf is a process with a huge human factor which may be minimized by the automation of this process step.

23.2 The Automatic Scarfing Process

The state of the art scarfing procedure has a great human impact on manufactured tolerances which influences the mechanical performance as well as assembly tolerances of the repair. The automation of this process is a major step forward in supporting the certification of bonded repairs for primary aircraft structures.

The manual grinding is therefore replaced by a numerical controlled (NC) milling of the scarf. A major requirement of the automatic milling process is the need for the local geometry information, mainly the surface contour and thickness distribution. In 2006 Baker [5] described a CNC milling of a scarf contour. The surface geometry has been acquired by contacting the surface with a probe. The data was used afterwards for the CNC programming. In 2009 Wittingham [6] describes the usage of a surface profiling equipment and 5 axis milling to produce scarfs of repair specimen. Up to now no mobile process for the machining of 3D scarf repairs is known.

The proposed process chain automatic scarfing process starts with acquiring the geometry surrounding the repair (Fig. 23.1). The specified region surrounding the damage is scanned and local surface heights are saved to a point cloud data format. The data is automatically used to virtually reconstruct the surface by a NURBS (Non-Uniform B-Spline) approximation of the point cloud generating a repair surface for a CAD repair design (Fig. 23.2). A software framework marshalling a CAD-kernel automatically generates a scarf design depending on user specific information like the scarf diameter, the scarf ratio, and the scarf depth (Fig. 23.3). The design is directly based on the reconstructed surface. The CAD geometry may also be used in the future to generate cutting patterns for prepreg or wet lay-up patch repairs as well as for a finite element modelling of the repair. The CAD model is mainly used for outlining a milling strategy and to simulate the milling process offline (Fig. 23.4). A milling strategy is than simulated and compiled into a standard command code for CNC devices (G-Code). The generated command file is loaded into the control software of the mill and used for the machining process (Fig. 23.5). The milling of the scarf depends on the control commands which have been derived from the surface reconstruction. It is therefore essential that the machine does not move between the scanning process and the machining. The machine needs to be fixated directly onto the repair surface in order to assure

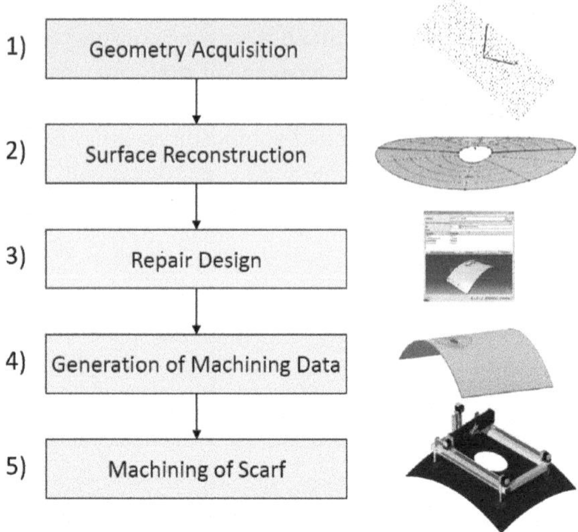

Fig. 23.2 Automatic scarfing process chain

this. This also solves problems with induced machining vibrations. Many rotor-crafts are equipped with rubber gears which tend to vibrate easily. A surface mounted solution is preferable in these cases. A mounting via suction cups is a common solution for surface mounted machines.

The number of needed control axis depends on the required repair geometry. A 3-axis portal is only capable of machining a continuous scarf because of the curvature of the repair scarf. The direction of the machining tool cannot be normal to the surface which is necessary to machine the rims of a stepped repair. A 5-axis portal machine or a 7-axis serial robot would be necessary to manufacture both stepped and continuous scarfs which allow of aligning the tool normal to the surface. Experimental comparison between stepped and scarf repairs showed a superiority of the scarf repair in static tensile testing which supports the sufficiency of a 3-axis solution for future repair applications.

23.3 Machine Design for the Automatic Scarfing Process

The Automatic Scarf Concept has been realized in a specific machine design in order to validate concept assumptions and prove concept feasibility. Figure 23.3 visualizes the realized machine which has been derived from the requirements. The basis of the machine is a customized 3-axis High-Z 720/T from CNC-STEP with a machining area of $720 \times 400 \times 110$ mm. The machining area has been chosen to manufacture typical scarf ratios of up to 1:40 in a 3 mm laminate for an assumed damage of 50 mm. The configuration is geometrically able to machine a continuous scarf but not a stepped one as outlined for 3-axis devices in the

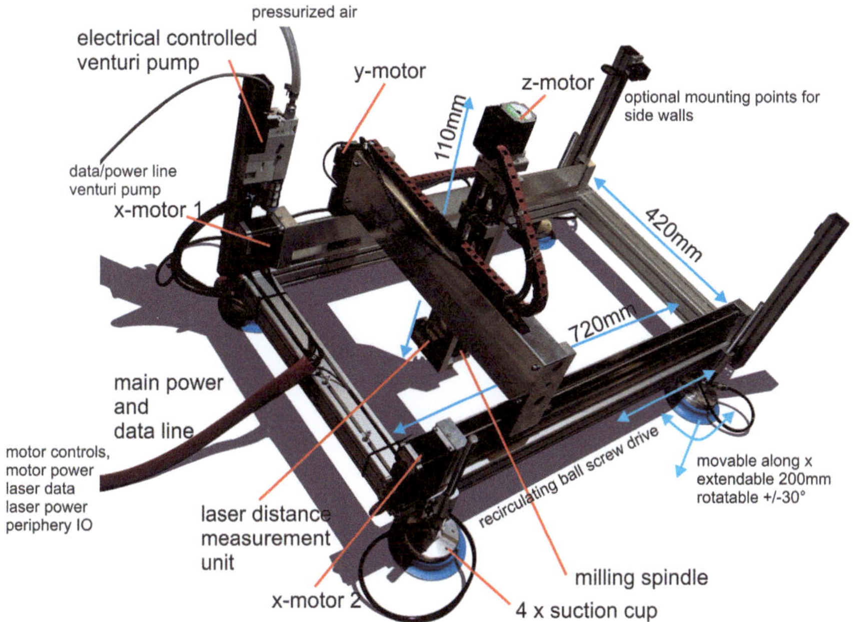

Fig. 23.3 Machine realized for validation of automated scarfing process

Fig. 23.4 Dependency of machining time of scarf ratio and structural thickness

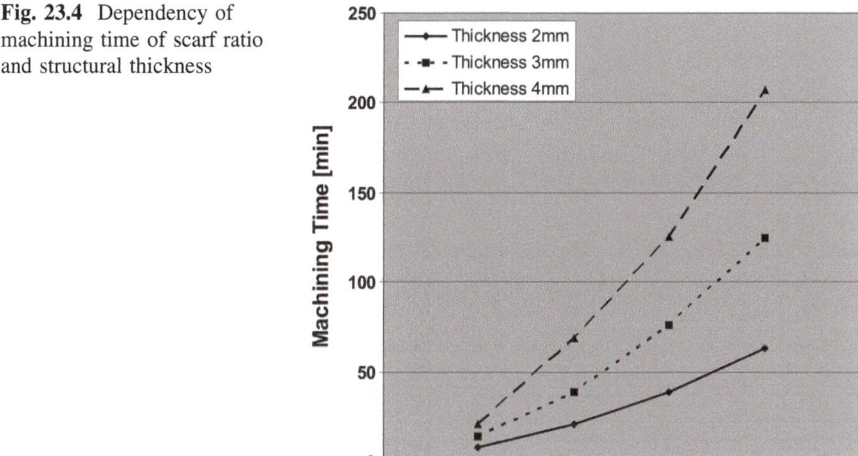

previous paragraph. Two stepping motors for the x-axis, one for the y-axis, and one for the z-axis each drive a recirculating ball screw with a feed rate of 0,00625 mm per 1/8 step. The maximum motor rates have been experimentally

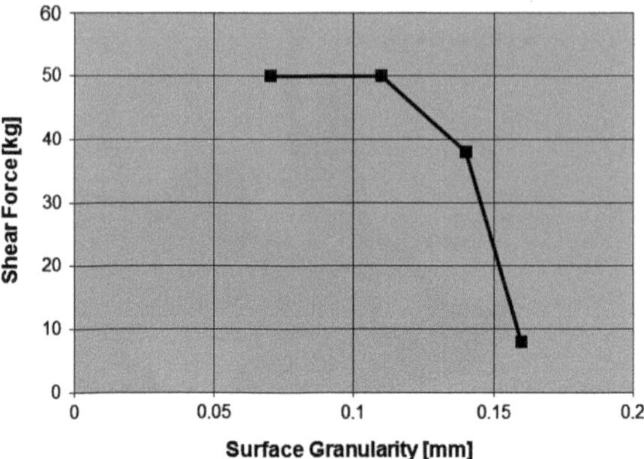

Fig. 23.5 Shear force of one suction cup for different surface granularities

tested to identify suitable machining speeds. The machining time is scaled by the feed rate linearly and by thickness and scarf ratio non-linearly. Figure 23.4 visualizes the dependency of machining time on scarf ratio and thickness. The data has been generated by simulating the machining of different circular repair scarfs. Assuming a machining speed of 1200 mm/min a standard repair with a scarf ratio of 1:20 and 3 mm laminate thickness would be machined in 40 min.

The needed machining precision is derived from the tolerance of the scarf contour. A deviation of ± 0.1 mm of scarf contour results in a maximum failure of 4 mm in repair radius (assumed scarf ratio of 1:20). The maximum failure is scaled by scarf ratio linearly but not by thickness. This means in the given example case that the scarf ratio is reduced by 4% at maximum which seems still acceptable.

The machine's reproducible precision is specified by the manufacturer at 0.01 mm. The milling spindle has standard rating of 1000 W and is able to cover turn rates of 5,000 up to 25,000 U/min. It holds tooling diameters of 1–8 mm. The machine is controlled manually by software inputs or automatically by scripted commands (G-code). A separate hardware motor controller converts digital inputs into control signals for the stepping motors and is able to control all three axis simultaneously. The virgin machine weighs approximately 30 kg.

The machines first modification is a 3 mW 660 nm wavelength laser distance measurement unit. The laser safety class is 3R which allows the usage of the laser without safety restrictions. It is directly mounted besides the milling spindle and aligned with the machines z-axis. It measures the relative height between the laser diode and a surface beneath. The maximum precision of the laser is specified at ± 0.03 mm. Experimental validation showed a maximum error of ± 0.02 mm by analyzing 400 measurements. The laser system is synchronized with the step signals of the machines x-motors by the synchronization unit and is DC-isolated by the isolation unit. The synchronization unit consists mainly of an integrated controller (IC) equipped with an internal counter. The IC can be programmed to trigger a laser measurement at a specific step count of the x-axis. Data communication is handled by RS232 serial protocol and transmitted via a hardware

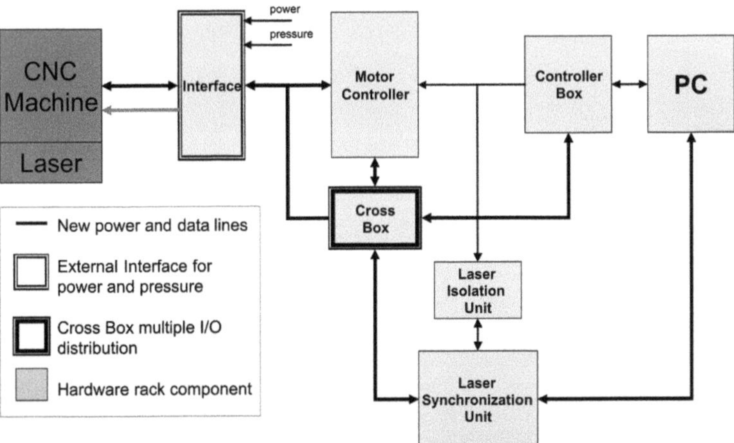

Fig. 23.6 Schematic of redesigned electrics

FTDI$^©$ RS232/USB converter to a PC. The synchronization allows the calculation of the local x and y position of the machine if proper zeroing of the machine axis has been done. The point laser combined with the hardware provides a very simple hardware implementation to measure a local surface contour and synchronize this with the actual machining position. The scanning is restricted to line patterns due to the fact that the stepping and directional signals of the x-axis motor controller is solely used to trigger a measurement. The offset between milling tool and laser is measured and compensated before machining. The position of a small step in a template tool is measured and touched by the tool allowing a maximum precision of approximately 0.08 mm.

The second main modification is the replacement of the standard mounting points by four suction cups. An electrical controlled venturi pump is operated by pressurized air and generates an absolute pressure of 0.1 bar. Pressurized air is chosen as the main vacuum source because it is commonly available in repair shops or easily produced by a portable compressor. Each suction cup withstands a shear force of ~ 90 kg, which is more than the machine's total weight. The pump was chosen because of its fail safe design. An internal valve closes in case of pressure or electrical loss, which locks the suction cups for a sufficient amount of time to secure machine and operating personnel. Each suction cup adjusts to a local structural curvature of about $\pm 15°$. The mounting points are rotatable for about $\pm 30°$ and movable in z and x direction. The system has been tested on different composite surfaces (typical peel ply and tooling surfaces) and different types of sandpaper granularity. Results are given in Fig. 23.5.

The third modification is a complete rewiring of the electrical interfaces. Figure 23.6 shows the schematic of the electrical design. The machine has been divided into two units, the machining unit and the control unit. Both units are connected with only one electrical and one pressure line allowing a fast disconnect

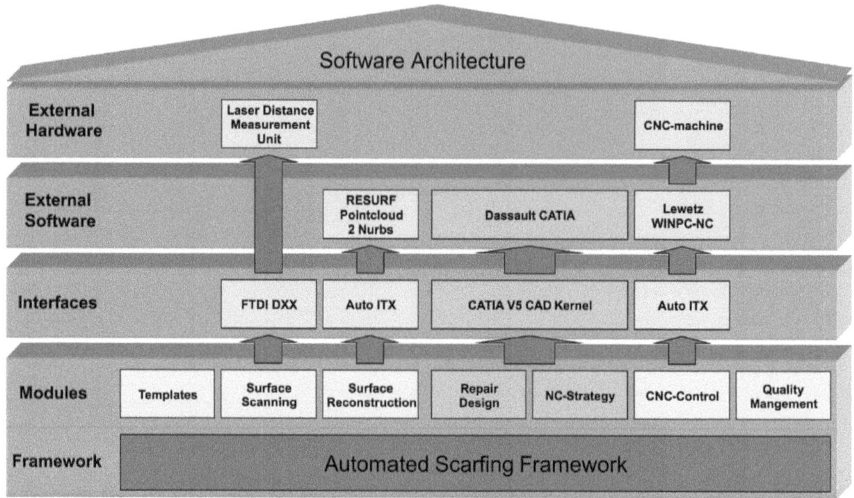

Fig. 23.7 Software schematic

of the machine. A relocation test has been performed in approximately 15 min. The electrical line contains the motor controller wiring, laser power and data transfer, and periphery I/O lines. All commands and data input/outputs are controlled from a portable PC. Everything has been mounted into a movable board for laboratory use only.

23.4 Software Framework for the Automated Scarfing Concept

A major aim of the automated scarfing concept is keeping the process simple in order to prevent damages from structure and personnel. This is the topmost priority of the developed software framework for command and control of the machining process. Figure 23.7 shows the frameworks schematic. The software is developed in Microsoft Visual C# object oriented programming, which allows easy driver integration for periphery devices like the laser unit and the venturi pump. The whole software process has been split into multiple software modules to keep the software flexible for future developments.

The template module allows the choice and definition of standard repair scenarios. This includes repair e.g. geometry and machining strategy. This allows standardizing the process and reducing human influence due to erroneous inputs. The second module controls the scanning process of the repair surface. The user is able to vary the scanning resolution in machines x and y direction. The software communicates with the laser via a FTDI[©] USB/RS232 interface driver. It also

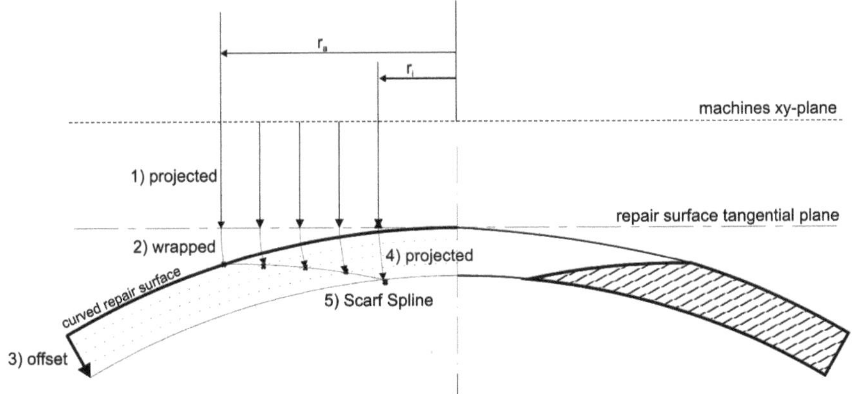

Fig. 23.8 Design of curved scarf contour

generates a suitable G-code command script for the scanning process, which is automatically loaded by the machines control program. The machine scans the repair surface in lines and returns the local z values of the laser, which is saved in xyz-ascii data file. The data is processed by the surface reconstruction module. A proper 3D modelling of the repair design needs a G0, G1 and G2 continuous surface. Due to the noise in the laser measurement it is necessary to do a surface approximation. A 3D NURBS fitting of the point data was chosen. The algorithm for this fitting is implemented by using a 3rd party software (Pointcloud2NURBS$^{©}$ by Resurf$^{©}$) which is a low cost tool. The output is an .igs data file. The software is marshalled via the AutoITX$^{©}$ interface driver allowing an easy integration of the reconstruction software. The repair design module controls the construction parameters of the repair design. It is possible to generate circular as well as elliptical scarf designs. Scarf ratios and repair diameters are used to virtually construct the repair design based on the reconstructed surface. The framework uses Dassault CATIA$^{©}$ V5R18 via its API (Application Programming Interface). CATIA$^{©}$ is a standard tool within aerospace industry and provides the geometry operations to construct the repair design. A cost efficient alternative is the OPENCASCADE$^{©}$ geometry library, which is a freeware tool. Figure 23.8 shows the design of the scarf contour. The inner and outer repair contours are defined in the machines xy-plane and projected (1) along the machines z-axis onto a tangential repair plane. These projections are wrapped over the repair surface (2) and projected in local normal direction (3) onto offset surfaces (4) to construct local support points. These support points are connected by a spline function to define the 2D scarf contour (5). This scheme is used to design a spline every 30°. These splines are used as guides to loft the scarf surface. The repair design is a parametric script which allows changing of scarf ratio, repair diameters and layout (circular or elliptic). For load optimized repair designs it is possible to define varying scarf ratios for the elliptic layout.

Module number five is used to define machining specific parameters and to generate the machining code. Especially tool diameters, machining feed and turn rates can be chosen. Additionally the machining strategy can be changed. NC-strategy influences scarf tolerances due to the fact that local laminate stresses relax during machining. In most cases a machining on nominal value without any roughening is chosen to prevent the part from deforming. This strategy wears the tool and introduces great machining forces. The module provides options to choose the standard or alternative strategies e.g. separate roughening and finishing. The sixth module controls the command software of the milling machine. This is done again via AutoITX. The milling process can be started and stopped from this module. The last module supports quality management processes. The geometry data is saved and a report is generated. The report contains a summary of manufacturing parameters and chosen options.

The developed modular software framework provides a command and control environment for the complete machining and design process of a 3D curved repair scarf.

23.5 Summary and Outlook

The bonded composite repair process depends on the quality of the repair scarf. The automation of the repair scarf manufacturing is an important step forward in improving repair quality and reproducibility. A process based on contact free laser scanning, surface reconstruction, CAD-design and mobile CNC milling has been developed and realized in an integrated machine design. The machine as well as the software design has been outlined. The validation of the concept still remains open for the future. Especially comparing manually manufactured scarfs to the automatically manufactured ones the next step is validating the concept. The machine has been designed to achieve the necessary repair tolerances. This has to be demonstrated in different typical repair situations, e.g. circular repairs of curved panels. The process gives a new design freedom to the repair engineer. Especially load optimized repairs with varying scarf ratios or freeform scarf contours will be possible. Both options may improve robustness of the repair design and further improve mechanical performance of the bonded repairs.

A major future aspect to consider is the influence of grinding on the bonding characteristic of the surface. For certifying an automatic milling process as bonded repair preparation it is necessary to prove that the milled surface has the same bonding performance as a grinded one. A manual grinding after the milling would negate every advantage of the automatic scarfing because it would be done manually. Therefore it is essential to provide sufficient data to show compliance of bonding characteristics between grinded and milled repair surfaces.

References

1. Armstrong, K.B., Bevan, L.G., Cole, W.F.: Care and Repair of Advanced Composites: SAE International (2005)
2. Sauer, C.: Lufthansa perspective on applications & field experiences for composite airframe structures. In: 3rd FAA Workshop for Composite Damage Tolerance & Maintenance, Lufthansa Technik: Tokyo, Japan, (2009)
3. Baker, A.A., et al.: Composite materials for aircraft structures, American Institute of Aeronautics and Astronautics, (2004)
4. Niu, M.C.-Y.: Composite airframe structures : Practical design information ands data 1st ed, Hong Kong: Hong Kong Conmilit Press, p. 664, (1992)
5. Baker, A.S.: Defence, and D.I.V. technology organisation victoria air vehicles. Development of a hard-patch approach for scarf repair of composite structure. Available from: http://handle.dtic.mil/100.2/ADA458447 (2006)
6. Whittingham, B., et al.: Micrographic studies on adhesively bonded scarf repairs to thick composite aircraft structure. Compos. A Appl. Sci. Manuf. **40**(9), 1419–1432 (2009)
7. Boeing, Boeing Patch Repair, SB09_compositesthematerials_i.jpg, Editor, http://www.compositesworld.com/articles/composites-repair (2009). p. A repair tecshnician scarfs the area around a damagse cutout prior to application of a composite patch

Part IV
Composite Technology

Chapter 24
Self-Controlled Composite Processing

Markus Kleineberg

Abstract Technical properties of composite materials highly depend on their distribution of fibre and matrix components. To combine these two components in the best possible configuration a variety of technologies has been developed. However a remaining challenge is that all semi finished products have decisive tolerances that may interact during production and lower the level of structural performance. To avoid extended safety margins, a variety of strategies has been developed to maximize reproducibility in the crucial production steps by detecting and compensating dominating tolerance issues in the process chain.

24.1 Introduction

The manufacturing procedure for composite components is different from a typical production scenario for metallic structures. Metallic structures are usually created from given materials with defined mechanical properties. Typical technologies to create the final structure are machining, welding or casting. To manufacture continuous fiber reinforced composite components the final material composition and the geometrical shape are both defined in the manufacturing procedure by embedding the reinforcing fiber architecture in a polymer matrix.

To ensure a reproducible component quality and performance all relevant component parameters concerning fiber architecture, matrix chemistry, fiber to matrix interface and also geometrical shape and surface properties need to be controlled in the manufacturing procedure. To maximise process efficiency it is

M. Kleineberg (✉)
Institute of Composite Structures and Adaptive Systems, Deutsches Zentrum für Luft- und Raumfahrt, Lilienthalplatz 7, 38108, Braunschweig, Germany
e-mail: markus.kleineberg@dlr.de

M. Wiedemann and M. Sinapius (eds.), *Adaptive, Tolerant and Efficient Composite Structures*, Research Topics in Aerospace, DOI: 10.1007/978-3-642-29190-6_24,

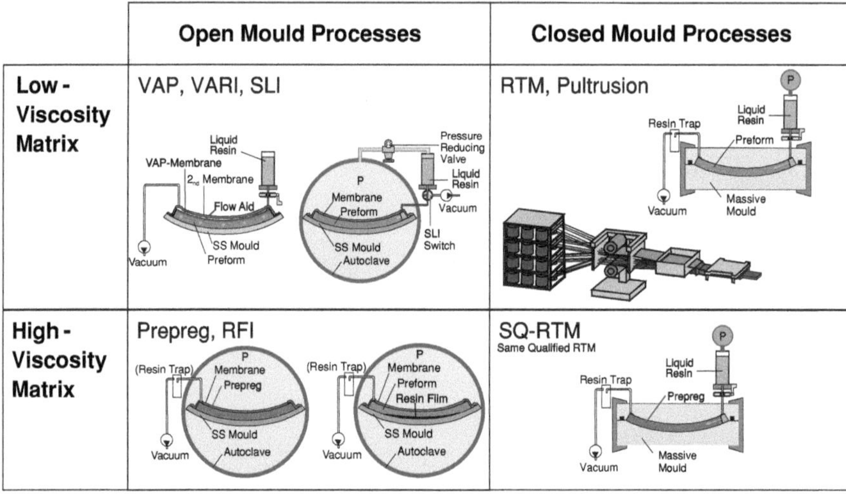

Fig. 24.1 Processing options

crucial to establish an adequate process maturity that takes into account a certain bandwidth of material and environmental tolerances.

Today's composite component production scenarios are often dominated by the manual interaction of skilled and experienced technicians that follow a predefined production procedure where the required lay-up and processing conditions are provided. Since all used semi-finished products and technical equipments have certain tolerances the [1] production procedure is typically set-up in a highly conservative way in order to ensure the required quality of the product even under worst case conditions.

To compete with highly productive manufacturing concepts for metallic structures that have proved their efficiency in numerous applications it is essential to find technical solutions that enable future composite production concepts to work close to their theoretical limits [2].

In contrast to today's composite production scenarios where manly incoming inspections are carried out to check basic properties of the semi-finished products [3] the aim of the "Self-Controlled Composite Processing" strategy is to measure, check and adapt all relevant production parameters throughout the complete manufacturing chain from lay-up concepts up to infusion and curing.

24.2 Process Chain

To characterise composite manufacturing processes they can be subdivided from a matrix point of view into low and high viscosity approaches and from a moulding point of view into open and closed mould concepts (Fig. 24.1).

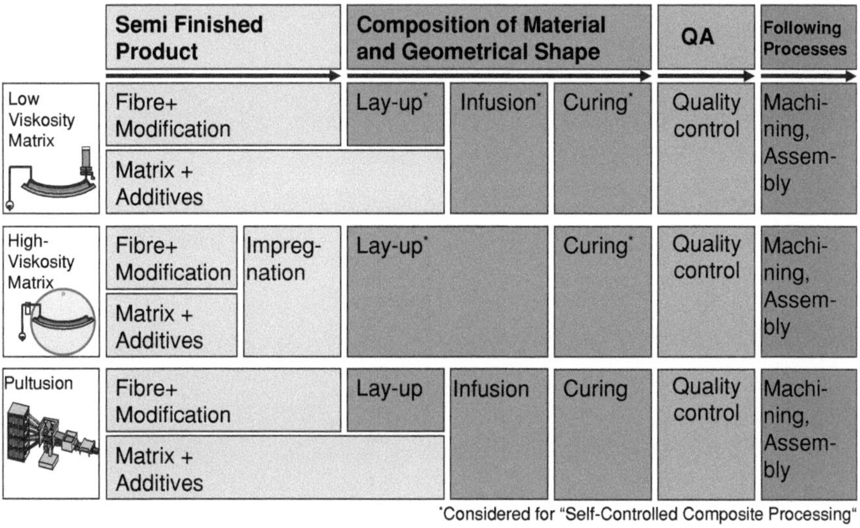

	Semi Finished Product		Composition of Material and Geometrical Shape			QA	Following Processes
Low Viskosity Matrix	Fibre+ Modification		Lay-up*	Infusion*	Curing*	Quality control	Machining, Assembly
	Matrix + Additives						
High-Viskosity Matrix	Fibre+ Modification	Impregnation	Lay-up*		Curing*	Quality control	Machining, Assembly
	Matrix + Additives						
Pultusion	Fibre+ Modification		Lay-up	Infusion	Curing	Quality control	Machining, Assembly
	Matrix + Additives						

*Considered for "Self-Controlled Composite Processing"

Fig. 24.2 Process chains

Each combination has its unique advantages and disadvantages which means that the driving production boundary conditions like the size of parts and complexity and also the required production rate need to be taken into account to identify the most efficient production concept [4].

Since ensuring the correct material composition and the correct geometrical shape are the most critical elements of the composite process chain, especially for complex composite components, these aspects are in the focus of Self-Controlled Composite Processing strategies (Fig. 24.2). Pultrusion is a highly effective approach for straight, continuous composite products but the required control concepts are highly specialised and the resulting product is usually handled like a semi finished product in metallic designs.

24.3 Options for Self-Controlled Lay-up/Preforming Processes

With respect to the lay-up procedure Self-Controlled Composite Processing can be a viable option in automated preform procedures in order to generate net shaped fiber architectures with variable geometry (see Chap. 25: Continuous Preform with variable height adjustment). Defects that need to be monitored in this production step are gaps within the plies, frayed cutting edges and also deviations in the resulting fiber angles due to draping problems (Fig. 24.3) [5]. Fiber angle deviations will primarily have an impact on the resulting component performance but they also affect the permeability of the preform which in turn could cause severe problems

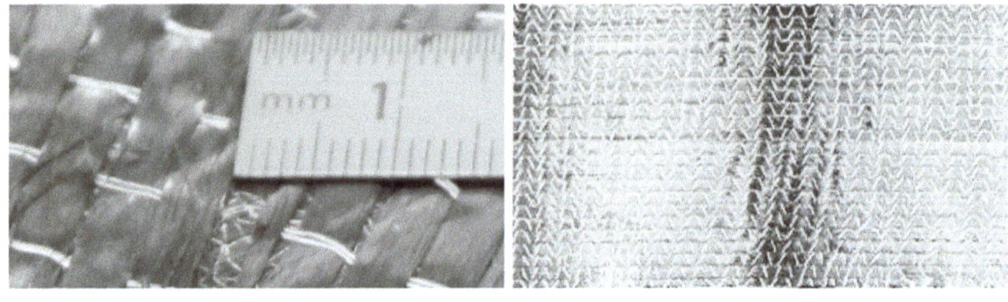

Gaps in Preform   Fiber Angle Deviation

Fig. 24.3 Typical preform problems

Topology Analyses
with Laser Stripe Sensor

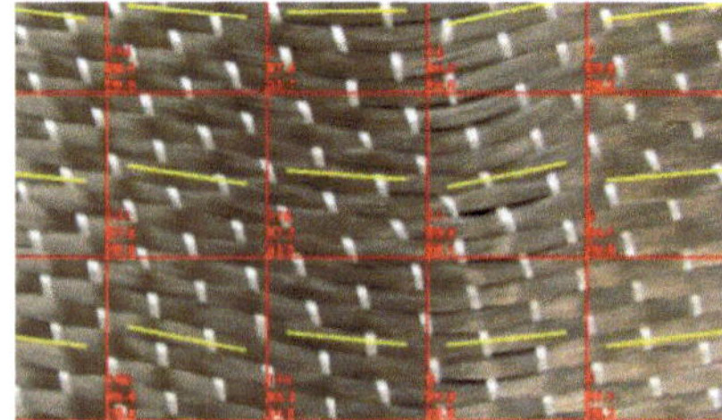

Grey Scale Value Analyses
with CCD Sensors

Fig. 24.4 Preforming process control sensors

during the following infusion step. Gaps in the preform and an insufficient geometrical preform accuracy are even more critical for a reproducible infusion process because they have the ability to disturb the flow front formation which in turn may lead to a significant quality reduction due to entrapped voids [6]. Especially closed mould processes are critical because the related massive moulds do not have the ability to compensate geometrical preform deviations as it is the case in more flexible open mould processes (see Chap. 26: Sensitivity Analysis Of Influencing Factors On Impregnation Process Of Closed Mould RTM).

To control the lay-up/preforming process special optical sensors can be applied that either identify the topology of the preform or derive fiber angle deviations through a Grey Scale Value analysis (Fig. 24.4) [5].

The required step towards Self-Controlled Composite Processing would be to feed the results back into the preform process e.g. by adapting the preform manipulator settings or by adjusting the cutter devices. Identified faulty materials may be sorted out automatically to maximise the component performance level.

When all preform parameters have been adjusted the complete stacking needs to be consolidated e.g. by activation of a binder. To do this it is possible to apply an electromagnetic field which induces eddy currents within a multiaxial carbon fiber preform (see Chap. 27: Induction technique in manufacturing preforms) [7, 8].

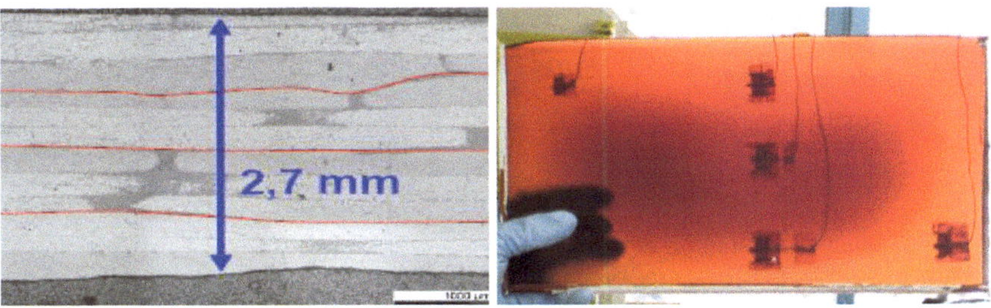

<table>
<tr><td>Fiber Content/Thickness Tolerances</td><td>Exothermal Overheating</td></tr>
</table>

Fig. 24.5 Typical infusion/curing problems

Since the warming of the preform is dependent on the fiber architecture (formation of electrical loops) both the electrical energy required by the inductor and also the heat distribution in the preform can be analysed (e.g. with an IR sensor) and used for process control purposes.

24.4 Options for Self-Controlled Infusion/Curing Processes

The final embedding of the fiber architecture in a void free and sufficiently cured matrix as well as assuring the required geometrical shape tolerances are the dominating quality aspects of the infusion and curing step. The fiber content of a laminate or the dedicated laminate thickness (Fig. 24.5) are also important quality aspects that directly determine the performance of the manufactured composite component. The combination of dry fiber preforms and low viscosity resin offers the opportunity to adjust the fiber content during the process by controlling the amount of resin in the laminate before the curing is initiated. The fiber content of high viscosity prepreg laminates can only be adjusted by controlled bleeding and is therefore much more difficult. On the other hand prepreg laminates can easily be toughened because the viscosity is less important for the process. To ensure an accurate degree of cure both overheating e.g. due to uncontrolled exothermal reactions (Fig. 24.5) and insufficient crosslinking e.g. due to inadequate activation energy has to be avoided [9].

To control the infusion/curing process ultrasonic sensors (Fig. 24.6) can identify both the crosslinking stage of the matrix and in open mould processes also the gap distance that can be used to determine the fiber content of the laminate [10]. To ensure a homogeneous temperature distribution infrared sensors (Fig. 24.6) have the ability to show areas with deviating temperatures. Once the heating and cooling behaviour is fully understood conventional thermocouples may be a cost effective alternative. Besides sluggish, convection based heating concepts it is also possible to provide the required activation energy in form of electromagnetic energy e.g. in the range of microwaves. The advantage would be that the energy can be provided directly and selectively to the matrix and other materials with an explicit dipole character.

Self-Controlled Processing during the infusion/curing step can be achieved for the fiber content by controlling the infusion pressure using the data of the gap

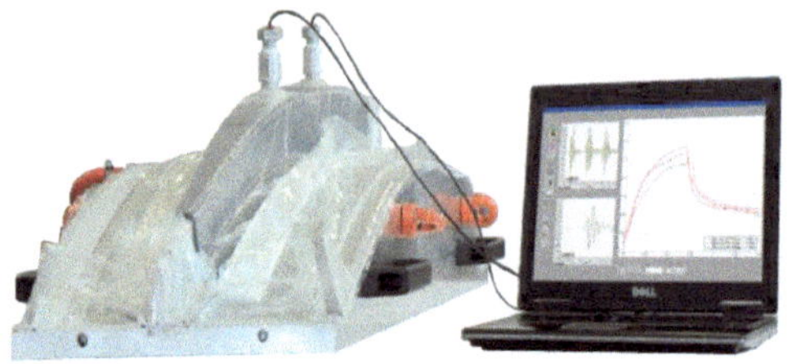

Laminate Thickness and Cure Monitoring
with Ultra Sonic Sensors

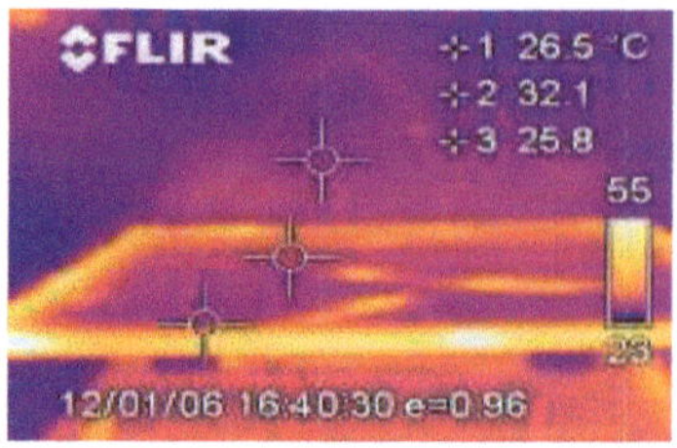

Temperature Distribution Monitoring
with Infra Red Sensors

Fig. 24.6 Infusion/curing process control sensors

measuring ultrasonic sensors (see Chap. 29: Interactive Manufacturing Process Parameter Control). The locally measured degree of cure in combination with information about the global temperature distribution can be used to adapt the processing parameters in a way that the process results are reproducible at the highest possible level. Combining prepreg and infusion processes (see Chap. 28: Combined Prepreg- and Injection Technology) as well as the selective application of energy can further extend processing opportunities by eliminating specific technology disadvantages.

References

1. Hong LI et al.: Probability-Based Modelling of Composites Manufacturing and its Application to Optimal Process Design; University of British Columbia; Revised February 25, 2002; Vancouver (2002)
2. Bauer, J: Increasing Productivity in Composite Manufacturing; NATO RTO AVT Panel spring symposium and specialists' meeting Loen, NORWAY (2001)
3. MIL-HDBK-17/3F (VOL. 3 OF 5), Department Of Defense Handbook: Composite Materials Handbook—Polymer Matrix Composites Materials Usage, Design And Analysis 17 Jun 2002 (2002)
4. Prof. Dr. Paolo Ermanni; Skript zur ETH-Vorlesung (151-0307-00L) Composites Technologien; Version 4.0; August 2007; Zürich (2007)
5. Dr. Andrea Miene; Bildanalytische Qualitätssicherung in der Preformfertigung Thementag: ZFP in der Produktion, Augsburg March (2010)
6. Chee, C.W.: Modelling the effects of textile preform architecture on permeability. Doctoral Thesis, October 2006; University of Nottingham (2006)
7. Ströhlein, T., et al.: Continuous inductive preforming for RTM-production; CFK Convention Stade; June 2008; Stade (2008)
8. Frauenhofer, M., et al.: Inductive preformings. In: 15th International Conference Mechanics of Composite Materials, Riga pp. 87–97 (2008)
9. Pullen et al., D.A.: Remote Cure Sensing of Polymer Composites; NATO RTO MEETING PROCEEDINGS 9, Intelligent Processing of High Performance Materials; May 1998, Brussels (1998)
10. David, D.S., Kim, R.S.: Ultrasonic cure monitoring of advanced composites. Sensor Rev. **19**(3), 187–191 (1999)

Chapter 25
Continuous Preforming with Variable Web Height Adjustment

Henrik Borgwardt

Abstract Preforming is required for complex shaped profiles manufactured in liquid composite moulding (LCM) processes. The stacking is made of dry fiber fabrics, which are infiltrated in a later process step. The complete stacking is called preform. The fixing of the textile layers can be realised through stitching or binder technology. The main disadvantage is the immense rate of manual work within the preform process. In consequence, the manufacturing is costly in terms of time and high effort for quality control. Automated preforming can reduce the costs by increasing the output and production rate while minimising waste. Preform profiles with variable outlines and non-extrudable sections are of particular interest for the aviation and automotive industry. The DLR Institute of Composite Structures and Adaptive Systems has developed an innovative device to overcome the previous limitations and to fulfil the industrial demands.

25.1 Introduction

Reducing fuel consumption and CO_2 emissions by 50% per passenger kilometer until 2020 (Vision 2020) is one of the targets established by the Advisory Council for Aeronautical Research in Europe (ACARE) [1]. About half of the demanded reduction shall be delivered by more efficient aircraft and therefore new and innovative concepts for aircraft are needed. A change of material from aluminium to lighter fiber reinforced polymers (FRP) in selected structures is one of the contributors to achieve the targets. Due to this, next generation aircraft are going to

H. Borgwardt (✉)
Institute of Composite Structures and Adaptive Systems, Deutsches Zentrum für Luft- und Raumfahrt, Lilienthalplatz 7, 38108, Braunschweig, Germany
e-mail: henrik.borgwardt@dlr.de

M. Wiedemann and M. Sinapius (eds.), *Adaptive, Tolerant and Efficient Composite Structures*, Research Topics in Aerospace, DOI: 10.1007/978-3-642-29190-6_25,
© Springer-Verlag Berlin Heidelberg 2013

feature a rising percentage of lighter and more efficient new structures made of carbon fiber reinforced polymers (CFRP). Additional the sharp rise in prices for fuel and the dramatically stiffening competition of airliners require drastic reductions in weight and costs for the next generation of aircraft.

One obstacle for an intensive use are the substantial material and manufacturing costs of light weight structures made of carbon fibers. For their composite airframe structure activities, aircraft manufacturers are currently using carbon prepreg material. Components made by this processes are characterised through excellent inter-laminar strength and constantly high fiber volume content. But on the other hand the use of prepregs is limited by gently curved surfaces. Additionally prepreg material is expensive due to the specific storage conditions at very low temperatures [2].

In contrast to the automotive industry aerospace components generally require better predictability of the mechanical properties. In low-cost markets, e.g. the automotive industry high production and output rates combined with low-cost manufacturing processes are required [3].

Dry fiber textiles with its ease handling, forming abilities and storage conditions are one instrument to growth of composites in engineering and automotive markets. Due to that there is an interest in dry fiber preforming technologies linked with low-cost manufacturing techniques such as liquid composite moulding (LCM), especially focused on the out-of-autoclave principles, resin transfer moulding (RTM) or vacuum assisted resin infusion (VARI) processes.

The textile preforming technology is subdivided into the preforming process and the consolidation with a matrix material. Through stacking single layers of dry fiber textiles within the preforming process the so-called preform is build up. To fix the single layers in the form two main technologies are currently applied: Stitching and binder technology [4]. Stitching is an established technology also well-known from the clothing industry that shows some disadvantages in the structural behaviour. With the binder technology, an auxiliary material with thermoplastic characteristics is used to bond the layers together by melting and consolidating again when cooling down. In opposite to resin impregnation the preform assembly is still performed manually. Preparing the whole lay-up and consolidating the complex preform is subject of manual labour still today.

In the recent past, handling systems for plane or gently curved surfaces have been developed. Cutting of layers by automated conveyer cutters and stacking with robots have been realised. For constant cross-sections like stringers or frames a few manufacturing processes are under development [4]. The novel approach is to manufacture the whole preform by continuously stacking several textile tapes together and subsequently and/or parallel undergoing a forming process. Textile tapes are continuously formed into preforms with this process similar to a pultrusion process. The functionality of this process has already been proven by several prototype devices. However, this technique is not applicable for profiles with variable outlines and non-extrudable cross-sections like aircraft frames or automotive space-frame profiles.

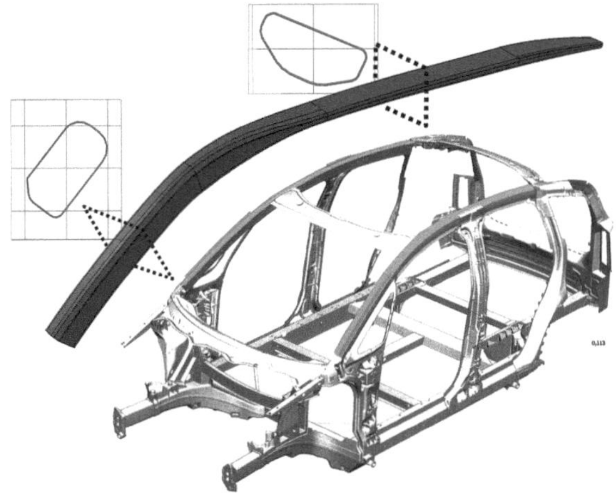

Fig. 25.1 Automotive space frame architecture (Source: Audi with additional elements by DLR)

25.2 Continuous Preforming with Variable Web Height Adjustment

Replacing manual work by automated preforming processes increases the output rate and the preform quality by improving the repeatability. Robustness comparable to well-established industrial metal working procedures is an essential requirement of composite manufacturing processes. Focusing on aerospace and automotive industries' future needs, profiles are essential construction elements. The established semi-monocoque design in aircraft technology and space-frame architectures (Fig. 25.1) in the automotive industry are based upon profiles. There is a wide field of possible construction elements, e.g. stringers (T, Ω) with constant and extrudable cross-sections, aircraft frames with various radii and variable web height and closed, curved, and space-frame profiles with variable cross-sections. The manufacturing of straight preform profiles is demonstrated by the German Aerospace Center (DLR) in the project MOJO [4]. Other initiated projects deals with continuous manufacturing of constant cross-section aircraft frames [5] and T-Stringers [6]. Straight outlines are the characteristic of both profile types addressed in the projects mentioned before. A matter of particular interest of the aviation and automotive industry are preform profiles with variable outlines and non-extrudable sections. Manual work is dominating the manufacture of profiles with variable contours (Fig. 25.2). With state-of-the-art preform production lines these geometries are difficult to realise.

The novel continuous preform production line (Fig. 25.3) is able to manufacture I-Beam profiles with variable web height (Fig. 25.4) by using high performance non-crimp fabric (NCF). The material storage is integrated in the machine. The NCF is delivered on rollers, which can easily be installed into the production line using a plug-and-play system. Inside and on one outside of the NCF-tapes a binder with thermoplastic behaviour is integrated. Rollers and adjustment units guide two

Fig. 25.2 Curved carbon fiber preform profiles

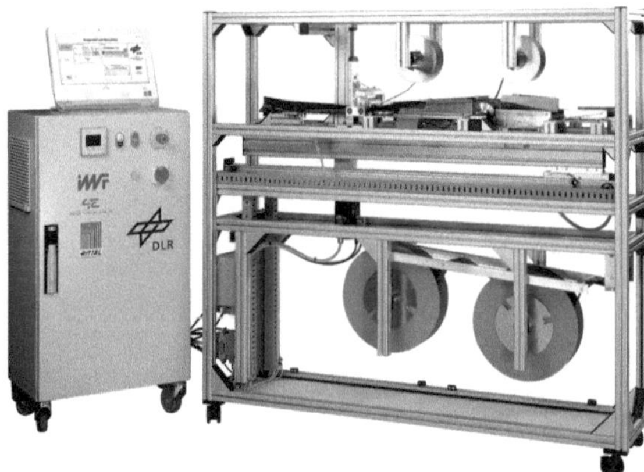

Fig. 25.3 Fully automated continuous preforming machine for profiles with variable web height

Fig. 25.4 Preform profile with variable web height adjustment

parallel tapes to the forming units. In the first forming step, the material is formed into an L-profile. This forming step is realised by sheet metal guides. These guides are combined with heating units, which warm up the binder inside the tapes. Two contoured rollers on each side perform the consolidation. Afterwards the tapes are formed into C -profiles and guided together. The binder activation is performed by bonded membrane heat elements. In the next step the two C-profiled tapes get connected on their webs with compaction rollers. In this way the I-Beam profiles are manufactured. Afterwards the formed fabric runs into a split roller stand. The upper roller stand is mounted on a vertical linear motion unit. The height of the I-Beam profile can be continuously adjusted by moving the upper roller stand. The height is continuously adjustable from nine to 100 mm. A distraction lance realises the actuation with a clamping device mounted on a horizontal linear axis. An integrated cutting unit assures the preform profile's net shape quality. The complete manufacturing procedure is computer controlled and monitored.

25.3 Evaluation of Performance

25.3.1 Performance

For the tests, NCF material with four layers (quadri-axial) with tricot stitching and an area weight of 512 g/m^2 is used. A Spunfab epoxy based binder is applied between all layers and on one top side of the material. The maximum machine speed is 400 mm/s. Limited by the local heat entry from heating units to the NCF material for binder activation, the preforming speed is 20 mm/s. The resulting preforming output rate for the above-mentioned I-beam profiles with curved outline, width 25.7 mm, mean thickness 0.5 mm of the flanges, and 1.0 mm in the web, having a mass of 176 g/m yields 12.2 kg/h. This means, the production time of one meter of net-shaped I-beam profile with gusset filler is 50 s. The production rate of the continuous preforming device compared to manual work with special preforming tools is approximately the same. But the variability and the option to manufacture endless profiles with variable web height are the essential advantages of the automated continuous preforming unit. The maximum profile length is limited by the size of the stored material rollers. Variations of the profile shape for this projected machine is limited by the range of vertical linear axis of 100 mm. With the parameters the preforming unit can produce variable profiles with web heights from 10 to 100 mm. Various profile shapes are demonstrated. Minimum length to web height variations are restricted by the relatively high bending stiffness of the used NCF material [7]. While performing the continuous preforming process with 20 mm/s, changes in the web height geometry from 1/3 (height/length variation) were demonstrated without wrinkling.

25.3.2 Preform Shape and Tolerance Capability

The length of the test profiles is in a range from 520 to 1211 mm. The height of the I-beam profiles is variable from 10 to 35 mm (Fig. 25.5).

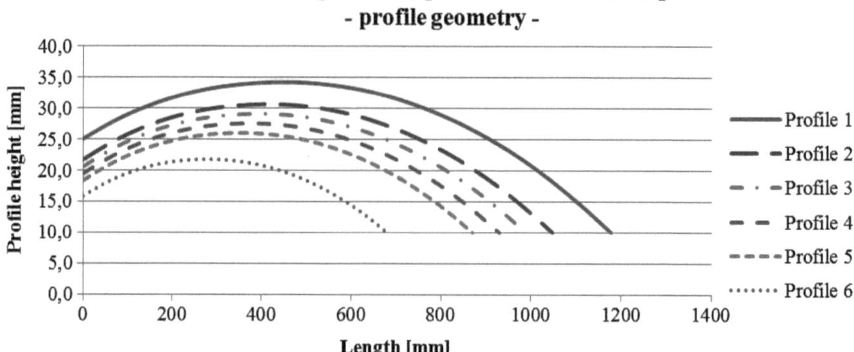

Fig. 25.5 Continuous preforming with variable web height—profile geometry

Fig. 25.6 Simulated fiber angle deformation ±45°

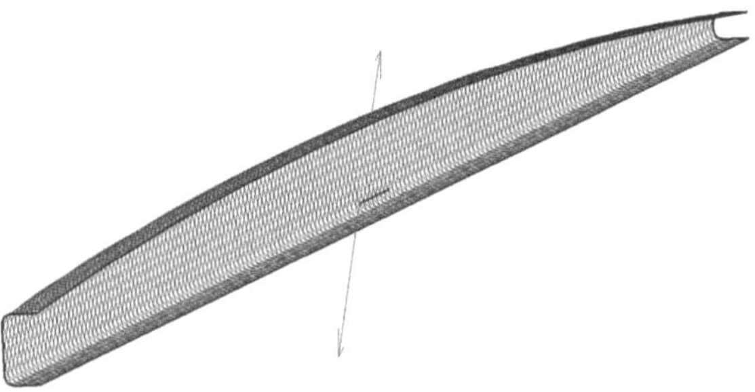

Fig. 25.7 Simulated fiber angle deformation 0/90°

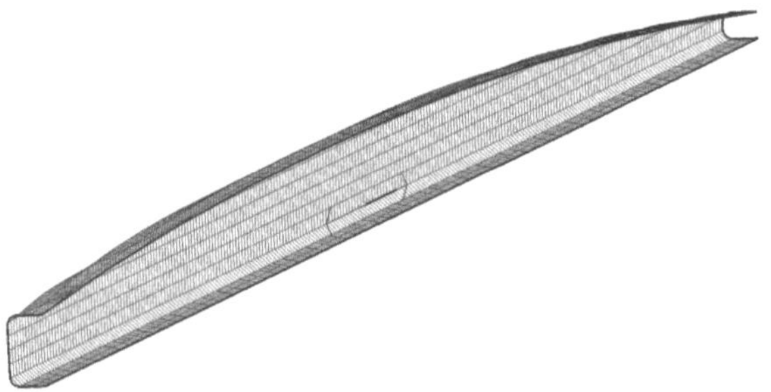

The vertical linear axis provides a tolerance of 0.05 mm. The accuracy of the preform profile web height is around −0.5 to +1 mm. While the web height is changing and a tape with constant width is used for the preforming process, the upper flanges width is changing. To get net shape preforms the edges of the upper flanges are continuously cut to a parallelised I-beam contour. The accuracy of the parallel upper flanges is ±0.4 mm. The radius between web and flange is 3 mm.

Any profile geometry has been examined in terms of producibility by means of draping simulation before. Draping simulation software tools can be used to simulate the draping process and such fabric shear behaviour of NCF generate 2D flat patterns of the draped fabrics. Draping simulation software predicts the flat pattern geometry and checks their producibility. Fiber angle deformation due to the preforming process was simulated with the 3D simulation software CATIA and FiberSim by Vistagy.

The Simulation did not reveal any deformation (Figs. 25.6 and 25.7) and deviation in fiber angles of more than five degrees, i.e. the profiles shall be preformed without any restrictions. However, current draping simulation tools are based on non-dynamic draping algorithms. Continuous processes cannot be simulated yet. This is objective of on-going research.

25.4 Conclusion

A further growth of composite applications in low-cost markets depends on the reduction of manufacturing costs. This can be achieved by automating the preforming process chain. Complete automation of the preform manufacture will result in improved repeatability and reduced cycle times. Especially for low-cost markets such as automotive industry fully automated production lines with high repeatability and robust processes are necessary [8]. One approach to efficient processes with reduced cycle times is the continuously automated preforming of dry fiber textile tapes into preform profiles.

A novel, low-cost preforming machine was developed for the manufacture of I-Beam profiles with variable web height adjustment. This new and highly flexible manufacturing device has demonstrated the continuous production of dry carbon fiber preform for aircraft and automotive parts. The realised preform production line offers a technology readiness level 4/5, i.e. the prototype is validated in a relevant environment. Various profiles with different length and curved outlines were produced and applied to a test configuration.

Further perspectives may be opened up by increased production rates through more efficient local heat entry by means of the induction technology [9, 10]. In this context options of various profile cross sections have to be examined.

References

1. ACARE: Addendum to the strategic research agenda, Advisory council for aeronautics research in Europe (2008)
2. Potluri, P., Atkinson, J.: Automated manufacture of composites: handling, measurement of properties and lay-up simulations. Compos. Part A **34** 493–501 (2003)
3. Friedrich, H.E., Kopp, G.: Leichtbau und Modulbauweisen für zukünftige Fahr zeugkonzepte. In: Stuttgarter Symposium (2007)
4. Cinquin, J., Voillaume, H., Stroehlein, T., Ruzek, R.: Modular joining of B-stage cured composite element with forming process and film adhesive for structural application, In: Proceedings of the ICCM-17 17th International Conference on Composite Materials, 27–31 July, Edinburgh, UK, (2009)

5. Brötje-Automation: Lösungen für anspruchsvolle Faserverbundteile., Kunststoffe; 5/2010; München, 2010, Carl Hanser Verlag (2010)
6. Kirst, P.: Untersuchungen zur Oberflächenwelligkeit von stringerversteiften CFK-Strukturen für Flugzeugflügel Oberschalen in Liquid Composite Moulding Verfahren (LCM). DLR, Braunschweig (2011)
7. Borgwardt, H., Ströhlein, T., Krzywinski, S.: Entwicklung einer Anlage zur kontinuierlichen Fertigung von Preformprofilen mit variabler Steghöhe aus Faserverbundwerkstoffen, Interner Bericht, IB131-2009/16, DLR, Braunschweig (2009)
8. Mills, A.: Automation of Carbon fibre preform manufacture for affordable aerospace appltications. Compos. Part A **32**, 955–962, (2001)
9. Benkowsky, G.: Induktionserwärmung: Härten, Glühen, Schmelzen, Löten, Schweißen. (5. Auflage) Berlin, (1990)
10. Bengtsson, P.: Rapid automated induction lamination (RAIL) of carbon fiber weave and thermoplastic film. Lulea University of Technology, Lulea (2006)

Chapter 26
Sensitivity Analysis of Influencing Factors on Impregnation Process of Closed Mould RTM

Martin Friedrich, Wibke Exner and Mathias Wietgrefe

Abstract This paper reports on different parameters that influence the closed mould resin transfer moulding (CM-RTM) process for fiber reinforced plastics. A sensitivity study of selective parameters is performed. This includes material parameters (i.e. viscosity, permeability), process parameters (i.e. temperature) and geometrical parameters (i.e. position of preform in the tool). Furthermore, fiber type and targeted fiber volume content (FVC) are considered to validate the full range of fiber reinforced plastics. As an example for the sensitivity study, the aeronautical carbon fabric G0926 and epoxy resin system RTM6 (both manufactured by Hexcel) are analyzed for targeted fiber volume contents in a range of $\sim 60\%$. The infiltration of a rectangular panel was simulated with the flow simulation software RTM-Worx by Polyworx. It was found that the infiltration of a simple geometry can differ by app. factor 3 in terms of duration, when only considering the tolerances of material and process parameters ("upper tolerance limit" vs. "lower tolerance limit" scenario). For different types of composite materials observed in this study, it can even go up to Factor 1,000. To achieve a reliable RTM process, these aspects—material types and range of tolerance—have to be considered.

26.1 Introduction

For many component shapes, the material group of fiber reinforced plastics exceeds the potential of lightweight metallic materials. Weight reductions of 25–30% compared to metals, are realistic [1]. To take advantage of this class of

M. Friedrich · W. Exner (✉) · M. Wietgrefe
Institute of Composite Structures and Adaptive Systems,
Deutsches Zentrum für Luft- und Raumfahrt,
Lilienthalplatz 7, 38108, Braunschweig, Germany
e-mail: wibke.exner@dlr.de

M. Wiedemann and M. Sinapius (eds.), *Adaptive, Tolerant and Efficient Composite Structures*, Research Topics in Aerospace, DOI: 10.1007/978-3-642-29190-6_26,
© Springer-Verlag Berlin Heidelberg 2013

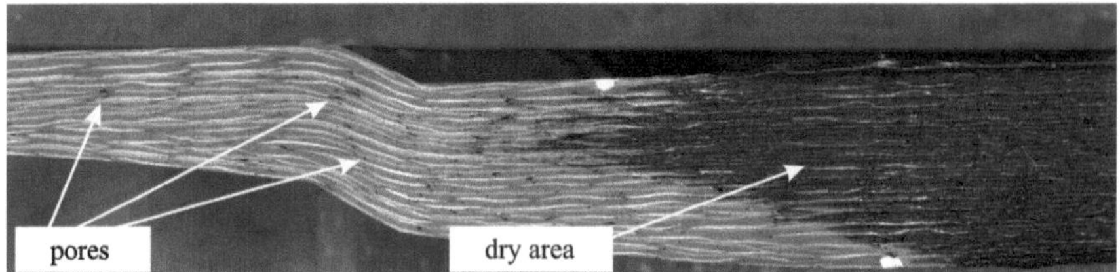

Fig. 26.1 Failure patterns of an incompletely impregnated component

materials in other areas, for example in vehicle construction, efficient manufacturing processes play a major role in reducing the costs of high-fiber composites to an acceptable level.

One of these efficient manufacturing processes is resin transfer moulding (RTM), in which, in contrast to the prepreg method, the impregnation of the fiber is carried out within the manufacturing process itself. This method is particularly suitable for large-scale production, since a fully enclosed, rigid mould with a geometrically defined component cavity and a fixed volume ("closed mould") is used. This type of processing thus allows the manufacturing of components with low tolerances and good surface finish on all sides [2].

The range of closed mould RTM parts is quite versatile. While some industries manufacture large RTM parts (e.g. automotive or marine industry), aeronautical RTM parts are comparably small. It has to be considered that the manufactured composites differ in their properties as well as the targeted fiber volume contents.

Therefore it is not feasible to use all kinds of composite materials to produce all the different types of RTM parts, as this could result in a large number of scrape components.

In contrast to the enormous advantages of the closed mould RTM processing at component level, many material and process variations as well as process tolerances have a greater influence on the quality of fiber impregnation during the production process, compared to the single-side membrane method.

Of all quality criteria of fiber composite components, the global and microscopic distribution of fibers and resin within a component are the most significant. Both have an influence on the overall structural integrity of a component as well as its fatigue performance. The main goal in the production of composite components therefore has to be the complete saturation of the fiber material with resin within the entire component volume. Figure 26.1 shows the types of failure patterns in case of an incompletely impregnated component: dry areas and pores.

The filling ratio of a component describes the degree of filling of a component with resin during the injection process. For a "good" component, a degree of filling near 100% is required. While "dry areas", depending on their extent, can be critical for the structural integrity of a component, pores (small and very small inclusions) cause a significant reduction in fatigue performance [3].

The aim of this paper is to give a sensitivity analysis of the closed mould RTM process to show the bandwidth of this process within the full range of composites

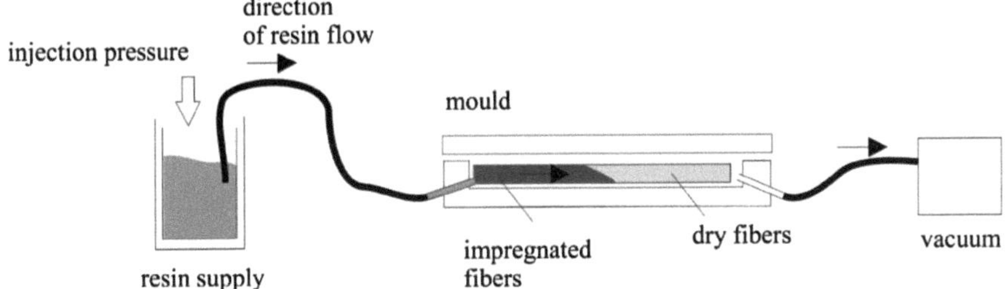

Fig. 26.2 Schematic description of resin flow in closed mould RTM

and exemplarily for one material configuration used for aeronautical RTM parts. Different influencing factors are being considered. This includes material parameters (i.e.: viscosity, permeability), process parameters (i.e.: temperature), and geometrical parameters (i.e.: position of preform in the tool). As an example the "best case" and "worst case" scenarios are validated for aeronautical composite materials via resin flow simulation. The knowledge and consideration of all influencing factors can help to avoid common mistakes and get a reliable RTM process with low scrap rates.

26.2 Resin Flow in Closed Mould RTM

While manufacturing the composite components using the closed mould RTM process, resin flows through tubes, channels in the mould (free cross-sectional areas), and through a porous medium, characterized here as the fiber reinforcement (Fig. 26.2).

In the following section the mathematical assumptions of the individual models are shown.

26.2.1 Resin Flow in Free Cross-Sectional Areas

26.2.1.1 Lines of Circular Cross Section

The flow rate of the resin in tubes can be calculated in an idealized way according to the law of Hagen-Poiseuille for circular cross sections.

$$\dot{V} = \frac{\pi \cdot r^4}{8 \cdot \eta} \cdot \frac{\Delta p}{l} \tag{26.1}$$

where, $\dot{V}$ is the volume flow rate through the tube (m³/s), r is the inside radius of the tube (m), l is the length of the tube (m), η is the dynamic viscosity of the flowing fluid (Pa s), Δp is the pressure difference between the beginning and end of the tube (Pa).

26.2.1.2 Lines of Rectangular Cross Section

The flow of the resin system within the rectangular tubes of the mould (including the flow in the border area between fibers and tool) follows, in an idealized way, the law of Hagen-Poiseuille for rectangular cross sections:

$$\dot{V} = \Delta p \cdot \frac{K \cdot \min(b,h)^3 \max(b,h)}{2\eta l} \tag{26.2}$$

$$K = 1 - \sum_{\infty}^{1} \frac{1}{(2n-1)^5} \cdot \frac{192}{\pi^5} \cdot \frac{\min(b,h)}{\max(b,h)} \tanh\left((2n-1)\frac{\pi}{2}\frac{\max(b,h)}{\min(b,h)}\right) \tag{26.3}$$

where, b is the width of the channel (m), h is the height of the channel (m).

26.2.2 Resin Flow in Porous Media

The flow of a fluid through a porous medium is described by Darcy's Law (26.4). In combination with the continuity Eq. 26.5 and neglecting gravitation, the basis of isothermal flow simulation of the RTM process (26.6) can be derived.

$$v = \frac{Q}{A} = -\frac{1}{\eta} \cdot k \cdot \nabla p \tag{26.4}$$

$$\nabla \underline{v} = 0 \tag{26.5}$$

$$\nabla \cdot \left(\frac{[k]}{\eta} \cdot \nabla p\right) = 0 \tag{26.6}$$

where, v is the fluid velocity, η is the resin viscosity, Q is the amount of fluid, k is the perform permeability, A is the surface and ∇p is the fluid pressure change. The flow process of the resin system through a fiberreinforcement can be simulated by computer modeling to predict dry-spot formation, incomplete mould filling and curing. Flow simulation software, therefore, become very important tools in mould designing and manufacturing, process configuration, reducing manufacturing costs, and improving quality. For the present study the flow simulation software RTM-Worx will be used.

Table 26.1 Tolerances of preform in the mould (as measured at DLR)

	Manual process (mm)	Automated process (mm)
$t_{cutting}$	± 1	± 0.2
$t_{handling}$ (single layer, position)	± 1	± 0.3
$t_{preforming}$	± 2	± 1
$t_{placing}$ (position)	± 0.5	± 0.3

26.3 Analysis of Influencing Factors

There are three main factors influencing the impregnation process in closed mould resin transfer moulding: a geometric driven factor, the permeability of the fibers to be impregnated and the viscosity of the resin system.

26.3.1 Geometric Influence

The addition and superimposition of cutting-, handling-, preforming and placing-tolerances during manufacture unavoidably leads to the formation of gaps between the fiber reinforcements (preform) placed into the mould and the side walls of the component cavity of the mould.

The possible width of this gap ($t_{channel}$) can, therefore, be determined according to:

$$t_{channel} = l_{component-cavity} - l_{fibre-material} + t_{tolerance} \qquad (26.7)$$

with

$$t_{tolerance} = t_{cutting} + t_{handling} + t_{preforming} + t_{placing} \qquad (26.8)$$

Typical values for these tolerances are provided in Table 26.1 based on measurements at DLR.

During injection, resin flow in these channels (gaps), as shown on Fig. 26.3, can not be prevented.

Depending on its magnitude, this resin flow can lead to premature confluence of flow fronts, and thus prevent a complete impregnation of the fibers in the component.

The magnitude of the effect of these resin runners depends mainly on geometric constraints on the interaction of the preform and the mould (channel cross section, see Eq. 26.7 in resin flow in open rectangular cross-sections).

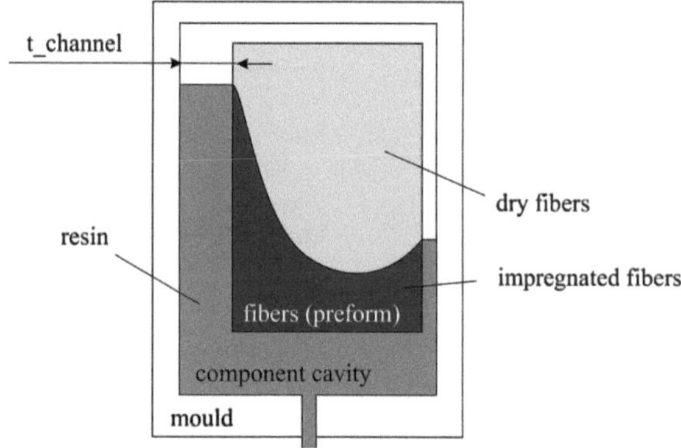

Fig. 26.3 Resin flow during injection

26.3.2 Permeability of Fiber Reinforcement

Permeability is a measure of the ability of a porous material to transmit fluids, and therefore the permeability tensor of fibrous reinforcements is one of the main material input parameters for macroscopic flow simulation in liquid composite moulding.

Even though the permeability tensor has importance for the simulation of flow processes in composite materials, there is no standardized measuring method. Hence, different setups/measurement devices have been developed especially for fibrous applications. The first step towards standardization of permeability measurement has been implemented by ONERA with participation of the DLR measurement device [4].

DLR uses a 2-D setup with a central injection system for the in-plane permeability measurement. The height of the cavity is defined by the thickness of a metal frame; therefore, the compaction of the reinforcement in the mould (i.e. the targeted fiber volume content) can precisely be adjusted and the measurements can, therefore, be representative for closed mould RTM. Pressure sensors are placed on the bottom part of the mould with the position defined only by the radius/distance from the centre. The direction of the sensors–they are placed on the x-axis, y-axis, and in a 45° angle–is not directly included in the calculations. Figure 26.4 shows the 2-D radial resin flow in carbon fiber reinforcement [5].

DLR processes its data with the 2-D solution of Darcy's law

$$K_e = \frac{Q\eta}{2\pi z} \cdot \frac{\ln(r_2/r_1)}{(P_1 - P_2)} \qquad (26.9)$$

where, K_e is the effective permeability between two sensors on one straight line (m^2), Q is the constant volume flow (m^3/s), n is the dynamic viscosity of the flowing fluid (Pa s), z is the laminate thickness (m), r_1, r_2 are radii of measurement points (m), P_1, P_2 are pressures at measurement points (Pa).

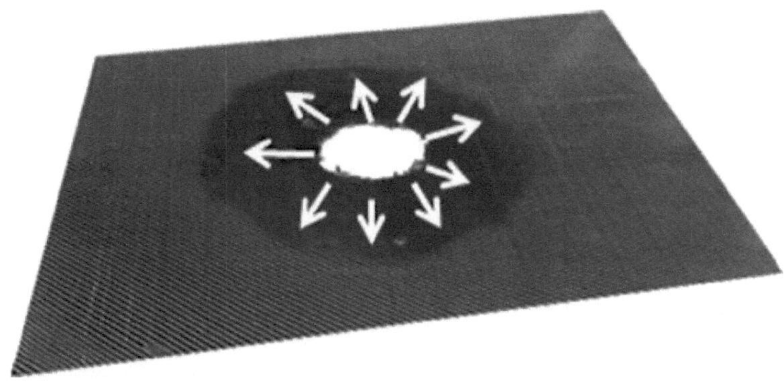

Fig. 26.4 2-D radial flow (DLR Bremen)

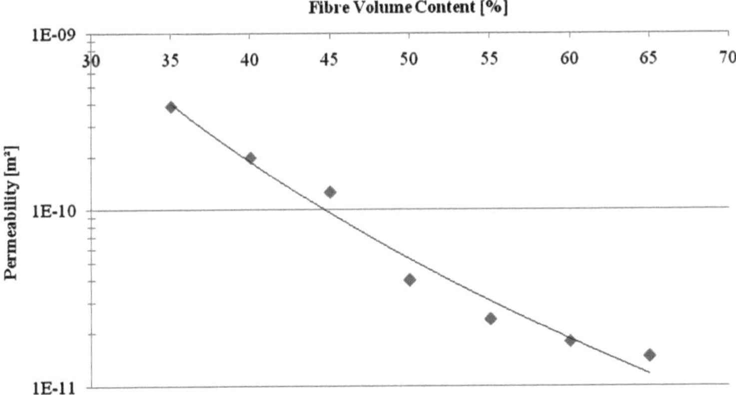

Fig. 26.5 Effective permeability of carbon fabric G0926 (Hexcel) as measured by DLR

Ten pressure sensors placed on the bottom half of the closed mould allow the calculation of 13 different K_e values due to the arrangement of the sensors. The average of all K_e values gives the total effective permeability.

The ratio K_x/K_y is calculated with the time at which the sensors in the respective directions have reached a certain pressure level. These calculations are performed by the software "PPerm" by Pole de Plasturgie de l'Est (manufacturer of the permeability measurement device). Ratio and total effective permeability lead to K_x and K_y values.

For the present study the focus is on the order of magnitude of the permeability value. Therefore, the difference of K_x and K_y (permeabilities in x- and y-direction) is excluded from the study and only the effective permeability K_e is considered.

Figure 26.5 shows the permeability of the carbon fabric G0926 by Hexcel which is used exemplarily in this study to show the permeability change depending on the fiber volume content.

Table 26.2 Viscosity of different resin systems at 25°C

Resin	Manufacturer	Mixing Ratio (mass fraction)	Viscosity (mPa·s) at 25°C
RTM6	Hexcel	–	70,000
Araldite LY556: Aradur 917: Accelerator DY070	Huntsman	100:90:1	900
SR8100:SD8822	Sicomin	100:31:00	300

26.3.3 Resin Properties

A major influence on the duration of the infiltration process is the viscosity of the resin system. The viscosity depends mainly on the type of resin system, the temperature, and the degree of cure. Table 26.2 shows the viscosity of three different fully uncured resin systems at 25°C.

Usually, a higher temperature leads to a lower viscosity of one resin, but also accelerates the chemical reaction of curing. Therefore, the infiltration window–the period of low viscosity–decreases. Usually, a viscosity lesser than 500 mPa·s is recommended. Operating experience at DLR shows that for aeronautical RTM parts, the initial viscosity should be lower than 100 mPa·s. Figure 26.6 shows exemplarily the isothermal viscosity of the resin system RTM6 for different temperatures.

26.4 Sensitivity Analysis

In the industrial production of composite parts using the closed mould RTM process, a statistics-driven summation and superimposition of the described effects can be observed. In order to determine the magnitude of the injection characteristics resulting from these effects, a two-part sensitivity analysis is performed:

- Part 1: impact of influencing factors within the range of composites
- Part 2: impact of tolerances within the same component

The geometry of the component chosen for this study is a rectangular plate. This is a simplification of the complex component shapes occurring in reality, but nevertheless it already reflects the impact of the influences in a good manner.

26.4.1 Impact of Influencing Factors Within the Range of Composites

The goal of the first part of this sensitivity study is to show the total bandwidth of the injection process cycle. For this purpose an analysis of the closed mold RTM manufacturing of two components with identical geometry and size dimensions,

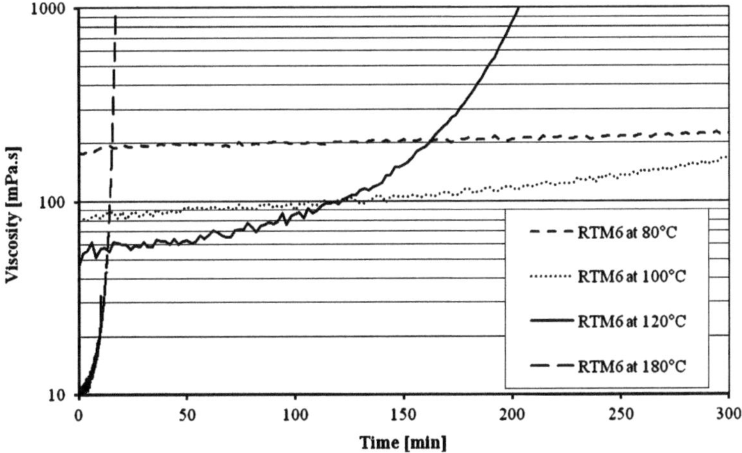

Fig. 26.6 Isothermal viscosity of RTM6 at different temperatures

but with large differences in the material and process parameters which influence the effects described above (lower and upper limit), is carried out.

In detail there is a variation of

- the fiber volume content (FVC) which influences the permeability (Sect. 26.3.2)
- the resin system and the associated injection temperature influencing the viscosity (Sect. 26.3.3) and
- the production type which is relevant for the gap size in the mould edges (Sect. 26.3.1).

Table 26.3 shows the material and process parameters for the lower limit (component 1) and upper limit (component 2) and the resulting behavior of the injection process.

The injection times are 7.4 s for component 1 and 8,770 s for component 2, each based on the boundary conditions specified in the Table. The included figure of the injection simulation shows the differences in the flow path of the resin system within the mould.

26.4.2 Impact of Tolerances Within the Same Component

While in Sect. 26.4.1, the material and process parameters were varied within a whole range of possible material and process parameters (for the closed mould RTM production of long fiber reinforced components), in the second part of this study, the process sensitivity has been analyzed by varying the material and process parameters on a scale as it occurs for one single component (component 3) within a mass production process.

Table 26.3 Comparison of lower and upper limit within the range of composites

	Component 1 (lower limit)	Component 2 (upper limit)
Component geometry	Flat panel	Flat panel
Length, width, thickness	$600 \times 400 \times 2$ (mm^3)	$600 \times 400 \times 2$ (mm^3)
Fiber reinforcement	G0926 (Hexcel)	G0926 (Hexcel)
Fiber volume content	30%	60%
=>permeability	9.67 e$-$10 m^2	0.184e$-$10 m^2
Resin system	RTM6 (Hexcel)	Sicomin SR8100, SD8822
Injection temperature	120°C	25°C
=>resin viscosity	27 m Pas	300 m Pas
Production type	Manual	Automated
=>gap width	Top: 0.0 mm, Bottom: 2.0 mm Left: 2.0 mm, right: 0.0 mm	Top: 0.8 mm, Bottom: 0.0 mm Left: 0.0 mm, right: 0.8 mm
Inlet/vent	Bottom center/top center	Bottom center, top center
Injection pressure	6 bar	6 bar
Impregnation progress (blue => red)		
Infiltration time	7.4 s	8770.0 Sec (146 min)

The following parameters were varied in a manner arising solely by the tolerances of material and process parameters:

- the FVC which influences the permeability (Sect. 26.3.2) and its variations results due to the given constant volume of the closed mould, from tolerance of the fiber areal weight of the fiber reinforcement.
- the injection temperature influencing directly the viscosity of the resin system (Sect. 26.3.3)
- the position within the tolerance filed of the production type which is relevant for the gap size in the mould edges (Sect. 26.3.1).

Table 26.4 shows the injection behavior for the lower and upper tolerance limit of the component 3.

Table 26.4 Comparison of difference tolerance limits within the same component

	Component 3 (tolerance limit combination A)	Component 3 (tolerance limit combination B)
Component geometry	Flat panel	Flat panel
Length, width, thickness	$600 \times 400 \times 2$ mm^3	$600 \times 400 \times 2$ mm^3
Fiber reinforcement	G0926 (Hexcel)	G0926 (Hexcel)
Fiber volume content	57.5% (370 g/m^2 − 15 g/m^2)	62.3% (370 g/m^2 + 15 g/m^2)
⇒permeability	0.225e−10 m^2	0.149e−10 m^2
Resin system	RTM6 (Hexcel)	RTM6 (Hexcel)
Injection temperature	125°C (120°C + 5°C)	115°C (120°C − 5°C)
⇒resin viscosity	23 m Pas	31 m Pas
Production type	Automated (upper tolerance limits)	Automated (lower tolerance limits)
⇒gap width	Top: 0 mm, Bottom: 0.6 mm Left: 0.6 mm, right: 0.0 mm	Top: 0.3 mm, Bottom: 0.0 mm Left: 0.0 mm, right: 0.3 mm
Inlet/vent	Bottom center/top center	Bottom center, top center
Injection pressure	6 bar	6 bar
Impregnation progress (blue => red)		
Infiltration time	203.0 s (3.38 min)	599.0 Sekunden (9.98 min)

The injection time for component 3 process variant 1 (tolerance limit combination A) is 203 s with the boundary conditions specified in the table, while the injection time for the component process variant 2 (tolerance limit combination B) and the designated boundary conditions is 599 s. As shown in the included simulation figure, there is also a difference in the flow path of the resin system within the mould.

26.5 Conclusion

This study shows the causes for the large impact of material and the process parameters on the impregnation process, as it can be observed during manufacturing practice. Particularly significant is the magnitude of these influences, when looking at the entire spectrum of the fiber composite component production by closed moulding RTM (Sect. 26.4.1).

The magnitude of the differences in the injection time for components with identical geometry can be a factor of larger than 1,000 by using the same injection pressure, only depending on what kind of fiber and resin material is used and how the fibers are processed (cutting, handling).

Besides that, it was shown that there are, even within the recurring production process within a mass production for a particular fiber composite component, process and material variations (resulting solely from tolerances) present, which strongly influence the impregnation of the component (Sect. 26.4.2).

The injection time varies only due to the process tolerances for this sample component by factor 3 and the flow path of the resin in the mould is also different.

Especially, the latter decides whether, through the appropriate location of the vent, the component becomes completely saturated or there are dry areas remaining within the component.

For a cost efficient production of fiber composite materials, the highest process reliability by minimizing the impacts of these influencing factors of the manufacturing process, is one of the key factors for preventing expensive scrap parts. To achieve this high level of process reliability, there are (in addition to the obligation to adjust the process parameters to the materials used) two possible approaches according to the results obtained:

1. Minimization of the influencing factors by significantly cutting tolerances of today's conventional materials and processes (weights, temperatures, cutting variations etc.).
2. Using a very robustly designed injection process which minimizes the effects of the described influence factors.

While the first approach almost automatically leads to a drastic increase in material and process costs, and thus undermines the approach to achieve cost reduction through high-process reliability, the second approach of using a robust injection process can be considered as a very effective way to achieve both process reliability and process cost reduction.

Consequently, this will result in a cost reduction for the entire component.

Further studies concerning the set-up of robust injection moulding processes are currently being performed by the authors of this article.

References

1. Flemming, M., Ziegmann, G., Roth, S.: Faserverbundbauweisen: Fertigungsverfahren mit duroplastischer matrix. Springer, Berlin (1999)
2. Kruckenberg, T., Paton, R.: Resin transfer moulding for aerospace structures. Springer, The Netherlands (1998)

3. Chambers, A.R., Earl, J.S., Squires, C.A., Suhot, M.A.: The effect of voids on the flexural fatigue performance of unidirectional carbon fibre composites. Int. J. Fatigue **28**, 1389–1398 (2006)
4. Arbter, R., et al.: Experimental determination of the permeability of textiles: A benchmark exercise. Compos. Part A **42**(9), 1157–1168 (2011)
5. Weitzenböck, J.R., Shenoi, R.A., Wilson, P.A.: Radial flow permeability measurement. Compos. Part A **30**(6), 781–796 (1999)

Chapter 27
Inductive Preforming

Dipl-Ing Tobias Ströhlein

Abstract The most time consuming step in manufacturing of high performance composites usually is the manually driven preforming step. Individual layers are draped in their formed position and fixed to each other by activating a binder. The innovative technology based on inductive heating presented in this chapter speeds up this process. As many parameters affect the inductive heating rate, a simple analytical model based on empirical data will be introduced. With this model, the resulting temperature can be predicted for given process parameters. Conversely, the best fitting process parameters can be found given a required heating temperature (depends on the binder).

27.1 Introduction

Over the past decade, the manufacturing of fiber-reinforced parts has shifted from mainly manual production to automated production [1, 2]. Reliable automated equipment is now available [3], but the processing time for high performance components is still poor. Consequently, the production rate is low.

The liquid-composite-moulding (LCM) technology nowadays offers comparable mechanical properties to Prepreg at lower material and processing costs [2, 4]. Since the used dry fabrics are more flexible and easier to drape, more complex highly integral designs are feasible [1]. This will result in a reduction of assembly costs. However, the preforming step is still the main factor driving the cost up in

D.-I. Tobias Ströhlein (✉)
Institute of Composite Structures and Adaptive Systems, Deutsches Zentrum für Luft- und Raumfahrt, Lilienthalplatz 7, 38108, Braunschweig, Germany
e-mail: christian.huehne@dlr.de

M. Wiedemann and M. Sinapius (eds.), *Adaptive, Tolerant and Efficient Composite Structures*, Research Topics in Aerospace, DOI: 10.1007/978-3-642-29190-6_27,
© Springer-Verlag Berlin Heidelberg 2013

the production of composite [1, 2]. During this step, plane dry fabrics are cut, draped into a 3D shape, compacted and fixed in their preliminary position. For the fixation of the carbon layers, low content of adhesive (binder) with thermoplastic characteristic is used. The binder can be activated by increasing the temperature above the melting point. The state of the art method is using convective or conductive heating techniques in combination with an oven, an autoclave or a draping-iron [1, 5, 6]. The main disadvantage is the low heat transfer rate, which results in a long process time. The only heat transfer path is through the preform surface into the layup. Within the preform itself, the energy must be transported by heat conduction, which is time consuming, especially for thick preforms.

An alternative preform technique based on electromagnetic energy transfer is introduced and the research results are presented. The results allow to develop an analytical model that predicts the resulting temperature and the process parameters.

27.2 Inductive Heating Mechanism

Inductive heating for metal components is well known and has been used in heat treatments for over 100 years [7]. Over the past decade, research investigated the application of the well known inductive heating to thermoplastic composites. The goal is to melt, weld or bend the parts by heating the thermoplastic matrix. Although the studies showed quite promising results with respect to an increased heating rate, the costs in terms of needed equipment as well as the effort to predict the results are quite high. For prepreg composites and simple inductor coils, some models predict the resulting temperature (e.g. [8–11]). However, these models are not valid for dry fabrics or for composite optimized inductors powered by low frequencies.

In comparison to the induction technique for metals, where most of the energy is transferred to the surface, very different volumetric heat mechanisms occur in composites [12–14].

In general, inductive heating involves a frequency generator that transforms the supplied frequency and current (50 Hz/230 V in Europe) to the target frequency and current in order to charge an attached inductor-coil. Due to the alternating current, an electromagnetic field occurs around the windings of the coil. Eddy currents are induced in any conductive material within the alternating magnetic field. For carbon fibers within this magnetic field, a current is induced along the fibers. Due to the electrical resistance, the induced energy is transformed into heat energy, which melts the binder, ref. Fig. 27.1. The induction of electrical energy along the fibers works fine at high frequencies [15, 16]. Unfortunately, high frequency equipment is expensive and inflexible with regard to the distance between the inductor and the frequency generator. Alternatively, energy can also be transferred at low frequency, with the transformer principle, into "global loops". These arise automatically in preforms as long as layers with different fiber orientation are used [15, 17].

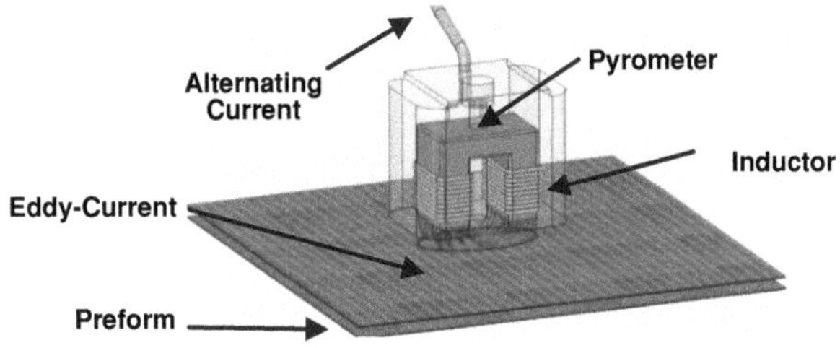

Fig. 27.1 Principle of inductive heating

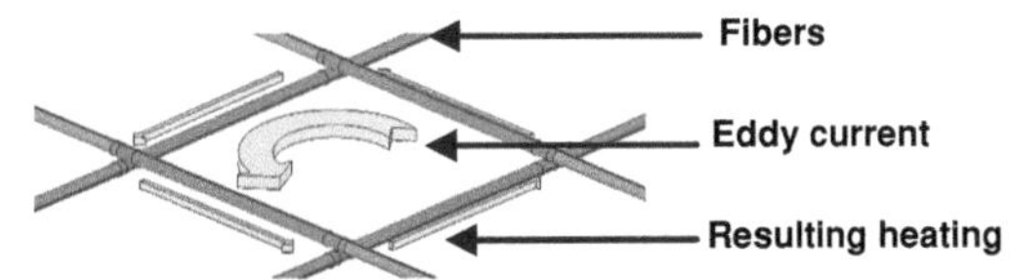

Fig. 27.2 Principle of eddy current in a global loop

Three different heat generation mechanisms have been identified [15]:

- Fiber Heating by Joule Loss
- Junction Heating by dielectric hysteresis
- Junction Heating due to contact resistance between fibers

Fink [8] found that the kind of heat mechanism mainly depends on the distance between the carbon fiber layers with different orientation. In case of a preform, the distance is zero; therefore, the fiber heating by Joule Loss is dominant. Fiber heating by Joule Loss means, that the fibers of different orientation form global loops with a homogeneous electrical resistance (Fig. 27.2). As opposed to the other two effects, no local (over-) heating at junctions is expected. This is of great importance, because the technology should never affect the quality of the preform.

Nowadays, the potential of inductive preforming has already been validated for several industrial applications such as Wing-spars, continuous production of complex H-shaped bars, and doorframes for aircraft. The equipment has been further adapted to composite applications. The following Fig. 27.3 illustrates the potential decrease in heating time for a thick preform.

In the past, the ideal parameters for an efficient and stable process had to be found by trial-and-error. There is no known mathematical model that predicts the needed power and application time in order to generate the heat for a specific binder activation temperature within a complex preform. In the following sections a method is presented to develop such a model for a wide range of applications.

27.3 Method to Identify the Parameter Influence

To identify the influencing parameters, the basic physics of the induction technique—independent of the application—has to be analysed.

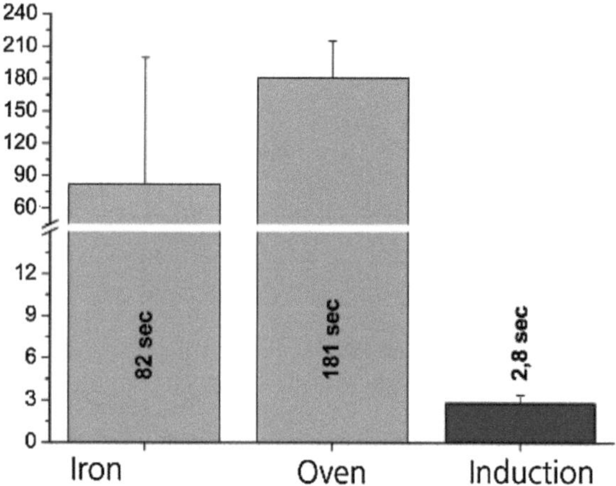

Fig. 27.3 Heating rate of different heating techniques for standard frame lay-up with 20 layers of NCF material

The induced voltage u_{ind} can be calculated with the following formula:

$$u_{ind} = 2 * \Pi * f * \mu * H * A \tag{27.1}$$

The induced voltage depends on the frequency f, the electrical permeability μ, the field intensity H and the heated area A. The field intensity H can further be expressed through the Biot-Savart law:

$$dH(r) = \frac{I}{4\pi} \frac{d\mathbf{l} \times \mathbf{e}_r}{r^2}; \quad \mathbf{r} = r\mathbf{e}_r \tag{27.2}$$

The field intensity differential, and therefore the induced voltage, depends on the current I, the cross product of the lead length differential and the unit distance vector to the lead $d\mathbf{l} \times \mathbf{e}_r$ and is inversely proportional to the square distance to the lead r^2. The electrical permeability μ depends on the electrical material-properties of the preform. It varies with the resistance between the fibers as well as the kind and amount of used (isolating) binder.

After the main parameters have been identified, their effect on typical preforms must be analysed to determine their mathematical relationship.

Therefore, a parameter study with more than 300 variations has been performed. For each single influencing parameter, first a series of experiments have been performed, where only one specific parameter is varied. In the second step, two parameters are varied in combination to find possible cross influences. The studies reveal that the distance from the inductor to the preform has a large effect on the electromagnetic field, and therefore on the transferred energy and heat. By varying the thickness of the preform, the distance from any additional layer to the inductor also changes.

A stepped test preform has been created where nine different thicknesses can be tested at nine different positions, while all other parameters, like compaction pressure as well as all other material parameters, stayed constant.

It was decided to measure the thermal response of the induced power in a homogeneous preform indirectly. Two optical heat measurement devices were used, a pyrometer and an infrared thermo camera. The pyrometer is directly integrated into the inductor to measure the surface temperature of one side. The temperature of the opposite (free side) of the preform is measured with the infrared thermo camera. The highest generated temperatures as well as the starting temperatures are recorded to obtain the heating temperature for the analysis. The heating temperature is defined as the difference between the starting temperature and the final temperature at the end of the induction process.

To identify the influence of material properties (binder, kind of fabric, layup) a second test has been set up with UD-fabric, woven fabric and non-crimp fabrics (NCF). For each sub-test, a comparable 2 mm thick stacking made of each material was created to identify the raw material effect. Similarly, the layup has been varied from a biaxial to a quasi isotrop layup while varying the binder content from zero to the standard value.

27.4 Parameter Influence

As starting point for the model, a reference parameter set need to be defined:

Equipment		
Frequency generator		EW5 (iff GmbH)
Inductor		U7050 (iff GmbH)
Reference values		
Power (set output)	$P_{ref} =$	2.25 kW
Frequency	$f_{ref} =$	15 kHz
Activation time	$t_{ref} =$	1 Sec.
Compaction pressure	$p_{ref} =$	1,000 mbar
Reference preform		2 mm NCF [8(+45°,−45°)]

The above-defined reference results in a heating temperature ΔT_{ref} of 120°K. This value is needed in the subsequent formulas. In the following sections, the influence of each parameter on this reference condition is analysed.

27.4.1 Preform Thickness

For any kind of heat transfer, the resulting temperature difference ΔT depends on the heat energy Q, specific heat capacity cp as well as the heated mass m.

$$Q = (A * d * \rho) * cp * \Delta T \tag{27.3}$$

In case of a preform with varying thickness d, the mass will vary linearly in correlation with d, as the area A and density ρ are constant. By increasing the mass within the electromagnetic field, the heat energy Q stays constant, while the heating temperature ΔT drops. As approximation, the relation of the change in thickness d to the resulting temperature drop ΔT_d can be represented mathematically by:

$$\Delta T_d = -0.02 * \Delta T_{ref} * d \tag{27.4}$$

The parameter ΔT_d is dependent on the value of the reference temperature difference ΔT_{ref} and the thickness of the investigated preform d.

27.4.2 Material Parameters

The tests validate the assumption that the heating temperature depends directly on the contact between fibers in different directions. In case of UD-raw material, the coupling rate is 52% of that for the NCF reference material, and for woven fabrics it is 119%. In the model, these values are represented by the parameter M_{Typ}, giving the temperature difference in accordance to the reference temperature ΔT_{ref} by:

$$\Delta T_{Typ} = \Delta T_{ref} * (M_{Typ} - 1) \tag{27.5}$$

It was found, that the tested binder (EPR05311) with 5 g/m^2 has a constant effect on the heating temperature. Independent of the layup, raw material and process parameters, the heating rate is lower by $\sim 3\%$ when the binder material is used ($M_{Binder} = 0.97$) than when the binder is not used. This is reflected in the equation below:

$$\Delta T_{Binder} = \Delta T_{ref} \left(M_{Binder} - 1 \right) \tag{27.6}$$

The layup of the preform influences the number of junctions available to form a global loop, which results in varying electrical resistance in the loop. With a biaxial layup of (0/90°) or of (+45°/−45°), a global loop compatible with the inductor flux needs less junctions giving a higher heating temperature ($M_{Stacking} = 1$) than a (0/45°) layup ($M_{Stacking} = 0.88$) with non-orthogonal fiber to fiber crossing. This effect can be taken into account by:

$$\Delta T_{Layup} = \Delta T_{ref} \left(M_{Stacking} - 1 \right) \tag{27.7}$$

With these three equations (Eqs. 27.5–27.7), which are based on empirical data, the material dependent influence can be accounted for with a standard deviation of 3 K.

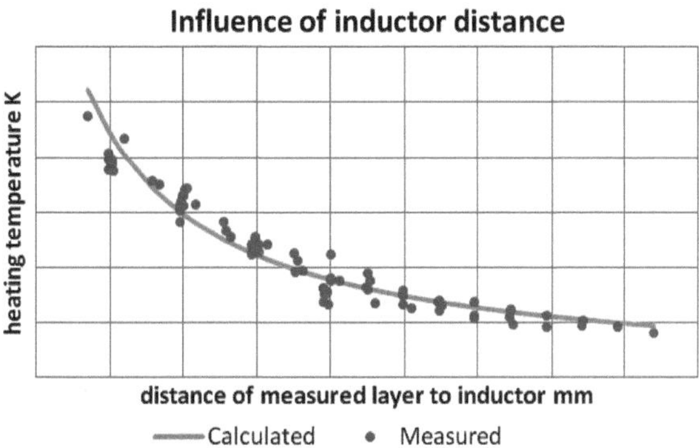

Fig. 27.4 Influence of inductor distance to the heating temperature

27.4.3 Compaction Pressure

The analysis shows, that the heat rate increases with the compaction pressure. By increasing the compaction pressure, the resistance between the single layers will be reduced and the heating temperature increased. The highest heating rate is reached when all fibers are in direct contact with each other. For the model, this effect can be taken into account by:

$$\Delta T_{Comp.} = \Delta T_{Ref}\left(\left(0.046 * \ln\left(\frac{p}{mbar}\right) + 0.67\text{K}\right) - 1\right) \qquad (27.8)$$

The reference Temperature (ΔT_{Ref}) is measured at 1,000 mbar compaction pressure resulting in $\Delta T_{Comp.}$ of ~ 0 K. Any lower compaction pressure results in a negative $\Delta T_{Comp.}$, indicating that the heating temperature decreases following Eq. 27.8.

27.4.4 Distance to the Inductor

As shown in Eq. 27.2, the distance d from specific layers to the inductor is inversely proportional to on the heating temperature and has a large influence on it. This correlation can be validated quantitatively by analysing the experimental results, which leads to the following approximation:

$$\Delta T_Z = \text{T}_{Ref} * \left(\frac{3}{d} - 1\right) \qquad (27.9)$$

This formula predicts the resulting heating temperature influence with a medial standard deviation of 2.17 K.

The graph in Fig. 27.4 shows the measured values as well as the predicted curve featured by Eq. 27.9.

27.4.5 Frequency

An often citied quadratic influence of the frequency [12, 16, 18] on the induced power was not found in the experiments. For all experiments, a frequency generator EW5 from IFF GmbH has been used with an advanced control unit. Herewith, it is possible to set the output frequency as well as the output power independently of each other. At constant output power, no appreciable change in the resulting temperatures was measured, while the whole frequency range of 10–30 kHz was tested in 2.5 kHz steps.

Only under certain conditions, such as at very low power set to very high frequency, a drop instead of an expected rise in temperature was measured. This effect depends on the control characteristic of the frequency generator and not on electromagnetic coupling effects. A frequency effect will, therefore, not be taken into account for the model.

27.4.6 Activation Time and Power Level

By increasing the activation time t or the output power p, Q will increase linearly giving the following equations:

$$\Delta T_{Power} = \Delta T_{Ref} * \left(\frac{P}{P_{Ref}} - 1 \right) \tag{27.10}$$

$$\Delta T_{Time} = \Delta T_{Ref} * \left(\frac{t}{t_{Ref}} - 1 \right) \tag{27.11}$$

27.5 Mathematical Model to Predict Resulting Heat at Each Layer

The resulting heating temperature ΔT can now be calculated for each layer intersection by adding the results of Eqs. 27.3–27.11 to the reference temperature ΔT_{Ref}:

$$\Delta T = \Delta T_{ref} + \Delta T_d + \Delta T_{Typ} + \Delta T_{Binder} + \Delta T_{Layup} + \Delta T_{Comp.} + \Delta T_z + \Delta T_{Power}$$
$$+ \Delta T_{Time}$$

$$\tag{27.12}$$

To validate the model, the measured temperature differences resulting from more than 300 experiments are compared to the calculated temperatures.

The standard deviation measured and the predicted values are below 5 K. By taking also thermal effects like the conduction through the thickness and the radiation heat loss into account, the standard deviation can be lowered to 3 K. As long as the activation time is low (0.5–1.5 s), the mentioned thermodynamic effects are found to be small in comparison to the total heating temperature and are therefore negligible.

With Eq. 27.12 it is now possible to predict the heating temperature between all layers. By calculating the resulting temperatures for each layer-to-layer interface line by line in a spreadsheet, the machine parameters can be varied until all heating temperatures at any position are within the defined process window.

27.6 Conclusion

A preform technology based on inductive heating has been presented and its main heating principle has been identified. The potential to boost the time consuming preforming step has already been shown. However, a method to predict the resulting heating temperature was still missing. The method enables find optimal parameters for efficient serial production.

This drawback has been solved by identifying the major parameters and analysing their quantitative influence on the heating. Taking this information into account, a mathematical model is now available for prediction of each individual temperature in any layer crossing in order to find and validate the optimal machine parameters.

References

1. Herbeck, L., Sickinger, C., Kleineberg, M., Riedel, U.: Faserverbundstrukturen—Von der Idee bis zum Prototypen. CCG-Seminar. Braunschweig, 31 Jan 2006 (2006)
2. Kleineberg, M.: Präzisionsfertigung komplexer CFK-Profile am Beispiel Rumpfspant. (Dissertation/Technische Universität Carolo-Wilhelmina zu Braunschweig) Braunschweig, (2009)
3. Borgwardt, H.: Kontinuierliches Preforming mit automatisierter variabler Höhenanpassung. AVK (vor Bewertungskommite) Frankfurt, 8 Jul 2009 (2009)
4. Lowe, J.: Kohlenstofffasern der nächsten Generation. Sampe-Deutschland. Berlin, 18 Feb 2009 (2009)
5. Ermanni, P.: Composites Technologien. Zürich, (2007)
6. Flemming, M., Ziegmann, G., Roth, S.: Faserverbundbauweisen—Fertigungsverfahren mit duroplastischer Matrix. Berlin, (1999)
7. Rudnev, V., Loveless, D., Cook, R., Black, M.: Handbook of induction heating. New York, (2003)
8. Fink, B.K., McCullough, R.L., Gillespie, J.W.: A local theory of heating in cross-ply carbon-fiber thermoplastic composites by magnetic induction. Polym. Eng. Sci. 32(5), 357–369 (1992)

9. Fink, B.K., McCullough, R.L., Gillespie, J.W.: Experimental verification of models for induction heating of continuous-carbon-fiber composites. Polym. Compos. **17**(2), 198–209 (1996)
10. Kim, H.J., Yarlagadda, S., Shevchenko, N.B., Fink, B.K., Gillespie, J.W.: Development of a numerical model to predict in-plane heat generation patterns during induction processing of carbon fiber-reinforced prepreg stacks. J. Compos. Mater. **37**(16), 1461–1483 (2003)
11. Lin, W., Miller, A., Buneman, O.: Predictive capabilities of an induction heating model for complex-shape graphite fiber/polymermatrix composites. In: 24th International SAMPE conference. Toronto, 20 Oct 1992 (1992)
12. Yarlagadda, S., Kim, H.J., Gillespie, J.W., Shevchenko, N.B., Fink, B.K.: A study on the induction heating of conductive fiber reinforced composites. J. Compos. Mater. **36**(4), 401–421 (2002)
13. Johnson, R.J., Pitchumani, R.: Enhancement of flow in VARTM using localized induction heating. Compos. Sci. Technol. **63**(15), 2201–2215 (2003)
14. Suwanwatana, W., Yarlagadda, S., Gillespie, J.W.: Hysteresis heating based induction bonding of thermoplastic composites. Compos. Sci. Technol. **66**(11–12), 1713–1723 (2006)
15. Ahmed, T.J., Stavrov, D., Bersee, H.E.N., Beukers, A.: Induction welding of thermoplastic composites–an overview. Compos. A Appl. Sci. Manuf. **37**(10), 1638–1651 (2006)
16. Frauenhofer, M.: Schnellhärtung struktureller Verbundklebungen mittels elektromagnetischer Wechselfelder. TU-Braunschweig) Braunschweig, 5 Mar 2010 (2010)
17. Rudolf, R., Mitschang, P., Neitzel, M.: Welding of high-performance thermoplastic composites. Polym. Polym. Comp. **7**(5), S.309–S.315 (1999)
18. Fink, B.K.: Heating of continuous-carbon-fiber thermoplastic-matrix composites by magnetic induction. Delaware, 692 p (1991)

Chapter 28
Combined Prepreg and Resin Infusion Technologies

Novel Integral Manufacturing Processes for Cost-Efficient CFRP Components

Robert Kaps and Martin Wiedemann

Abstract This chapter presents a novel manufacturing technology for the production of integrated structures made of carbon fiber composite materials. This manufacturing process is able to considerably reduce the production time of large assemblies for primary structural applications. It is based on a combination of the prepreg and the resin infusion technology, both of which are already established in industrial aerospace production. Experimental studies have been carried out to demonstrate the utility of this process. These investigations focused on the characterization of the generated contact zone through the use of two matrix systems as well as its mechanical properties.

28.1 Introduction

The material costs of aerospace composites are considerably higher than those of the currently used lightweight metal alloys. These higher costs can be compensated by the application of integrated manufacturing technologies that reduce the number of production steps associated with the assembly of complex structural components. The reproducibility and robustness of these integral technologies are in substantial demand since the rejection costs for large structural component are considerable.

Just the handling and positioning of semi-finished fiber components actually accounts for approximately 30% of the cost of structural components made of resin

R. Kaps (✉) · M. Wiedemann
Institute of Composite Structures and Adaptive Systems, Deutsches Zentrum für Luft- und Raumfahrt, Lilienthalplatz 7, 38108, Braunschweig, Germany
e-mail: Robert.Kaps@dlr.de

M. Wiedemann
e-mail: Martin.Wiedemann@dlr.de

M. Wiedemann and M. Sinapius (eds.), *Adaptive, Tolerant and Efficient Composite Structures*, Research Topics in Aerospace, DOI: 10.1007/978-3-642-29190-6_28,
© Springer-Verlag Berlin Heidelberg 2013

infusion and up to 45% of the cost of prepreg structures, depending on their geometric complexity [1].

The use of CNC tape-laying machines can clearly simplify the lay-up of prepreg materials in the production of simple geometric structures. However, this is usually not feasible in the case of complex and integral structural components such as stiffeners, frames and load transmission structures. If mechanical stability and weight requirements demand the production of such complex components by means of the prepreg technology, the costs increase significantly. A promising approach to counteract this trend is to combine the prepreg and infusion technologies into one single process. This methodology is presented on the following pages.

28.1.1 Prepreg Technology

The manufacture of fiber composite materials using woven or unidirectional (UD) fiber layers pre-impregnated with matrix resin (prepreg) has become the most widespread production process in the aerospace industry. The characteristics of the prepreg-based manufacturing process are a high degree of process reliability and the high quality of the components [2]. The excellent mechanical properties of prepreg composites result from the straight lay up of the carbon filaments using UD tapes and toughener-modified matrix resins. These features lead to the best ratio of weight to mechanical properties feasible today for primary aerospace structures.

28.1.2 Infusion Technology

Liquid resin infusion (LRI) technology is a term applied for a group of processes used to manufacture fiber composites from dry fiber preforms that involve infusion of liquid matrix materials followed by subsequent curing.

The particular advantage of the LRI processes is the ability to comparatively freely combine fiber and matrix materials, which allows for high flexibility in the manufacturing process. Additionally, the use of semi-finished products, like non-crimp fabrics (NCF), substantially reduces the preforming efforts especially for complex preform geometries.

Unlike prepreg resins, infusion resins require lower viscosity in order to completely impregnate the dry fiber material within the available process window. However, the lower viscosity concurrently leads to an increased brittleness of the resin matrix. Consequently, the mechanical strength values attainable in the components with the LRI processes are, in some cases, lower than the values attainable by the prepreg technology [3]. Another problem associated with the infusion resins is the high volume shrinkage and the related spring-in effects [4]. Technologically critical can be the prevention of dry spots as a result of inhomogeneous soaking of the fibers in large thin shells.

28.1.3 Integrated Technologies

Aside from joining together already cured composite components, aircraft manufacturers have been applying the so-called co-bonding processes in the production of composite rudder units for years [5]. In this process, individual subassemblies manufactured or preformed by different technologies are joined by means of sequential curing. However, the matrix systems of the subassemblies do not cross-link to each other in this process since the matrix of one joining partner is already cured.

A specific process using combined semi-finished parts is presented by Ermanni [6]. In this process, uncured prepreg structures are placed on wet-wound carbon fiber components. The assembly is then jointly vacuum-bagged and cured. This process can be used, for example, for a wet-wound fuselage barrel with prepreg stringer reinforcements. These type of processes that assemble non-cured parts are called co-curing technologies.

Another process using the co-curing technology is described in [7]. Dry fiber preforms are placed on large prepreg shell structures. A process called resin film infusion (RFI) is used for the impregnation of the dry preforms. The resin film fractions are placed inside the vacuum bag together with the dry fiber preforms. The heating phase liquefies the resin film, wetting the dry fiber preforms. To obtain optimal compatibility, the resin film and the prepreg matrix should preferably consist of the same resin system. However, the flow range of the resin is limited (a couple of centimeters) in the dry fabrics. Especially in the case of complex structures and high resin viscosities, multiple resin reservoirs are necessary, which make the lay-up more difficult. It also increases the risk of creating defects since the wetting of the fibers by the resin is difficult to control.

28.2 Combined Prepreg and Resin Infusion Technology

A component manufactured by the combined prepreg and resin infusion (CPI) technology consists of a prepreg and an infused component, whereby the infusion resin is supplied from outside through the vacuum bagging of the component. Thus, the application of this manufacturing process allows on one hand the creation of simple geometries, exposed to high stress, solely made from prepreg materials. On the other hand, the reinforcement or force-guiding structural elements with challenging geometries can be implemented by dry fiber preforms with much less manufacturing effort. The subsequent autoclave-based process includes the resin infusion into dry fibers and joint curing of the entire structure. The result is a complex composite component possessing the favorable mechanical properties of prepreg as well as the cost efficiency of liquid resin infused (LRI) structures.

Since the two combined processes have opposing requirements with regard to the viscosity of the matrix material, the use of two different resin systems in the CPI manufacturing is unavoidable. This leads to the formation of a potentially critical contact area between the two resin systems. Consequently, comprehensive investigations of the transition zone have been carried out by means of a selected pair of materials. The optimization of the process control and the acceptance of the new process in industrial-scale production are further demands of the investigations.

28.2.1 Effects in the Transition of Prepreg to Infusion Resin

The combination of two systems of semi finished products in one single process and the combination of two different uncured resin systems need to be analyzed in order to avoid any degradation of the material properties in the contact zone.

(a) Chemical compatibility

To ensure the compatibility of the prepreg and the infusion resin, the chemical composition of both should preferably be similar. The chosen infusion resin is a tetraglycidyl-methylene-dianiline (TGMDA) with two aromatic amin hardener components 4,4'-Methylen-bis(2,6-diethylanilin) M-DEA and 4,4'-Methylen-bis(2-isopropyl-6-methylanilin) M-MIPA [8]. This resin, called RTM6, is supplied by Hexcel Composites and is developed in accordance with requirements of the aerospace composites industry. The viscosity at the infusion temperature of 120°C is about 33 mPas. The prepreg resin system basically consists of the same epoxy polymer TGMDA but with a DDS (4, 4'-diaminodiphenyl sulfone) hardener agent.

In fact, based on the similarity of the basic epoxies RTM6 and Hexply6376, it is assumed that both matrix resins will be highly compatible in the contact zone and form a highly stressable compound.

A good test to validate this assumption is the measurement of the surface Young's modulus of the resin in the contact zone. The Young's modulus of a specimen surface with the two neat resin systems in contact was tested in Fig. 28.1 by micro indentation. The modulus ranges from the nominal value for the prepreg resin to the value of the infusion resin with no significant sign of a discontinuity. A significant discontinuity would be an indication of a disturbed ratio of the reactive components and thus an incompatibility of the resin systems [9].

(b) Fluorescence microscopy of the transition zone

Prepreg materials are characterized by the non-hardened matrix-components possessing a certain fraction of excess resin matrix. In regular production, the excess resin is taken by the bleeder ply or peel ply [10]. As a result, the dry fiber layers of the CPI assembly may take up a small amount of prepreg resin. This take up is driven by the pressure difference between the vacuum applied to the dry fibers and the resin pressure of the prepreg generated through the compaction pressure of the autoclave. The capillary forces between the dry fibers generate an additional force to disperse the prepreg resin. It can be expected that both effects will shift the resin transition zone towards the area of dry fibers.

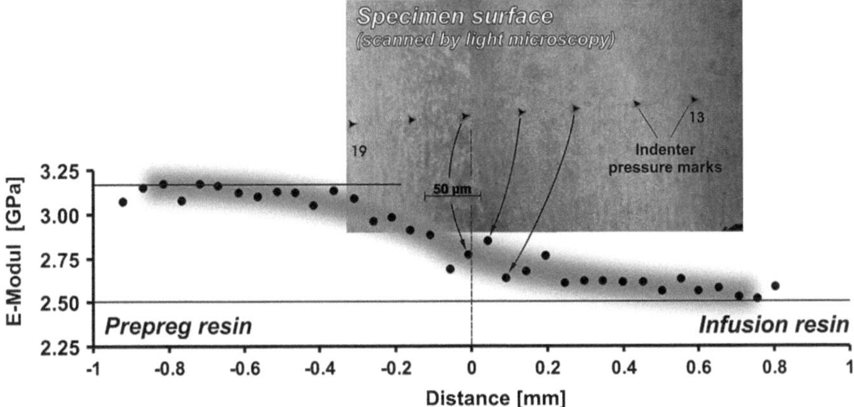

Fig. 28.1 Indentation of a neat resin specimen of the two adjacent resin systems by micro indentation

For an assessment of the matrix distribution in sample materials, micrographs were analyzed. The infusion resin was colored by a fluorescent dye to allow differentiation between the matrix resins. This method can be used to analyze the impact of various selected process parameters on the spatial distribution of the matrix resins.

The fluorescent dye's accurate representation of the location of the infusion resin was proved by analyzing a suitable sample both with the fluorescent microscope and the energy dispersive X-ray (EDX) analysis that detected the sulfur allocation of the DDS hardener agent [11].

The temperature level and duration of the dwell time just before the infusion of RTM6 emerge as the major process parameters in the CPI production process. The temperature of the dwell time has an impact on the viscosity of the prepreg resin and, therefore, on the migration distance of the resin into the dry fibers. Prolonging the dwell time increases the extent of the distribution of the prepreg resin into the dry fiber material.

Figure 28.2 (left) shows a sample manufactured with a dwell time of 90 min at 120°C that was implemented prior to the infusion of the dye-containing resin. Compared to Fig. 28.2 (right), in which the duration was 30 min at 90°C, a clearly more extensive distribution of the prepreg resin into the former dry fiber material is evident. Inspection of the matrix transition in Fig. 28.2 (left) shows that colored resin areas completely surrounded by the prepreg resin system are notable. These "infusion resin islands" are basically generated by capillary forces within the rovings of the unidirectional (UD) woven fabric, whereby the areas surrounding the rovings remain rather dry and are subsequently filled with infusion resin.

(c) Simulation of the CPI process for a generic laminate

Mild bleeding into the dry fibers can have a positive effect on the mechanical properties of the contact zone by creating a smooth transition between the two matrix systems. Moreover, this effect can prevent a mechanically critical border of the component (attachment of stringer to the skin, e.g.) from coinciding with the

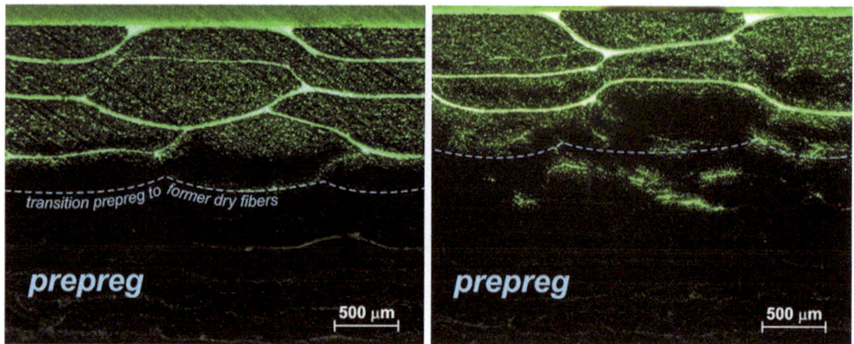

Fig. 28.2 CPI samples manufactured at dwell time parameters of 120°C/90 min (*left*) and 90°C/ 30 min (*right*)

transition zone of the two matrix systems. The positioning of the toughened prepreg resin in that region will also enhance the strength of the joining.

On the other hand bleeding can lead to exceeding the limit of the industry standard for the fiber volume fraction V_f. This limit is actually $V_f = 60 \pm 4\%$. To analyze the controllability of the maximum fiber volume fractions by the CPI process parameters, a through thickness simulation of a generic CPI-laminate was developed.

The physically basis of this simulation is the description of a stationary flow of a fluid through a porous media presented in Eq. 28.1 (Darcy's Law):

$$u_D = -\frac{K}{\eta}\frac{\Delta p}{L} \tag{28.1}$$

where u_D represents the Darcian velocity of the fluid, K the permeability, η the viscosity of the fluid, L the distance and Δp the pressure difference between two points. Adapted to a constant volume completely filled with variable fractions of fibers and resin, the following leading differential equation describes the time dependent change of the fiber volume fraction V_f:

$$\frac{\partial V_f}{\partial t} = \frac{1}{2}\nabla\left(\frac{\overline{K(V_f)}}{\eta}\nabla\sigma_f(V_f)\right) \tag{28.2}$$

where t represents the time, and σ_f the relaxation pressure of the fiber bed [12]. In the simulation the differential equation is solved by the finite difference method for the thickness direction of a generic CPI-laminate.

Figure 28.3 shows the result of simulated CPI processes with two sets of parameters with opposite effects to the prepreg bleeding. The first diagram shows the local fiber volume fractions at 90°C process temperature and 15 min dwell time prior to the infusion process. The second one displays a process time of 120°C and 30 min dwell time. The lower part of the laminate represents the Prepreg layers, the upper one the infusion laminate.

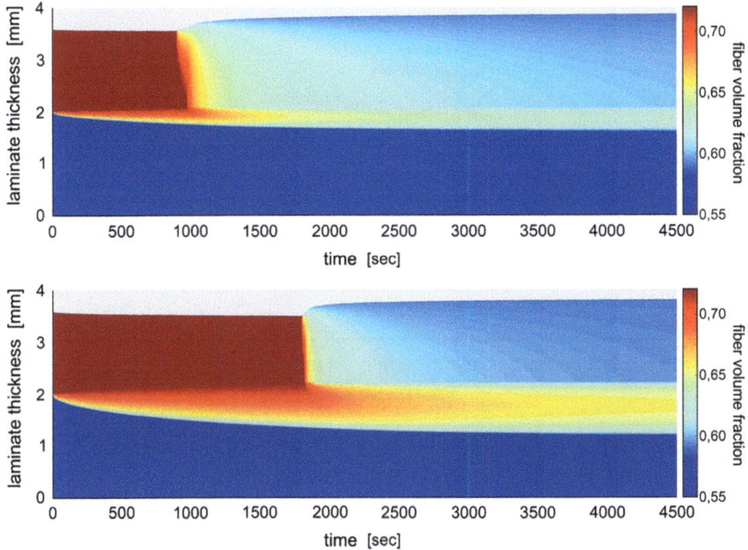

Fig. 28.3 Through thickness simulation of a 4 mm CPI-laminat with two sets of process parameters: 90°C process temperature and 15 min dwell time (*upper diagram, #1*), 120°C process temperature and 30 min dwell time (*lower diagram, #2*)

At the end of the upper process in Fig. 28.3, the infusion process was able to restore the former high fiber volume fractions resulting from the bleeding effect back to $V_{f,\,max\,end\,\#1} = 63{,}05\%$. The final maximum fiber volume fraction in the lower process exceeds the standards limit with $V_{f,\,max\,end\,\#2} = 65{,}7\%$. This high fiber volume fraction is a result of the higher dwell time and thus a wider distribution of the prepreg resin into the dry fiber layers. A secondary effect is a higher grade of reaction that leads to a higher viscosity of the prepreg resin at the end of the dwell time. The dimension of the interaction zone is qualitatively comparable to the microscopic results in Fig. 28.2.

For a large number of parameter sets, the maximum fiber volume fraction was calculated to identify the process window for certain laminate thicknesses. The results are shown in Fig. 28.4 with a diagram including all sets of process parameter with a maximum allowed fiber volume fraction of $V_{f,\,max} = 64\%$. Each line is a function of the process temperature and the dwell time. As shown in the diagram, practicable process windows exist for laminates equal or larger 1 mm thickness (e.g. temperature $\leq 108°C$ and dwell time ≤ 30 min).

28.2.2 Mechanical Tests

The successful introduction of the CPI technology will ultimately depend on whether the mechanical properties of the joined components will not weaken the overall composite structure compared to standard bonding technologies.

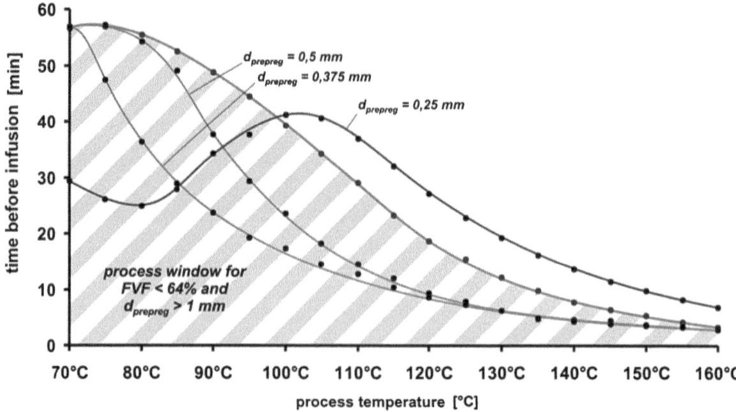

Fig. 28.4 Process window for maximum fiber volume fractions equal to the limit of $V_{f, max} = 64\%$ as a function of process temperature and dwell time

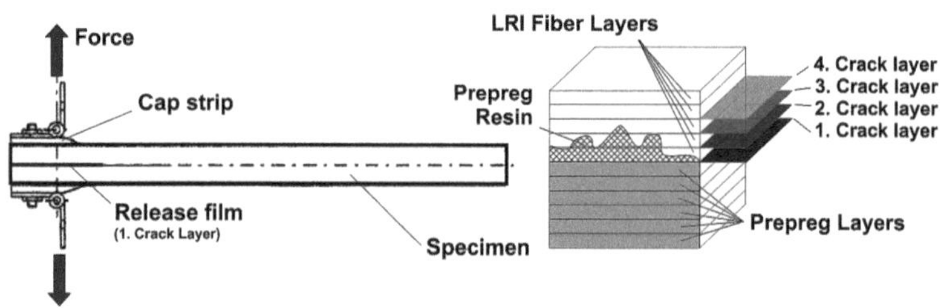

Fig. 28.5 Description of the peel samples and crack positions

The following test programs were carried out in order to demonstrate the mechanical performance of the contact zone.

The fracture toughness G_C is capable of reflecting changes in the mechanical strength properties in the transition zone when the process parameters, dwell time and dwell time temperature, are varied [13].

Peel test samples were prepared and tested to determine the fracture toughness applying an aerospace standard [14]. An initial crack, created by an inserted halogen film, was required for this type of sample. It had to be ensured that this crack coincided with a relevant interlaminar separation plane containing the corresponding matrix transition. Since it is impossible to exactly determine the actual position of the prepreg resin in the transition zone, the number of samples for each set of parameters is increased to reflect four different crack levels (Fig. 28.5).

The results of a sample series of pure infusion composites and pure prepreg composites, respectively, demonstrates the practically independence of G_C from

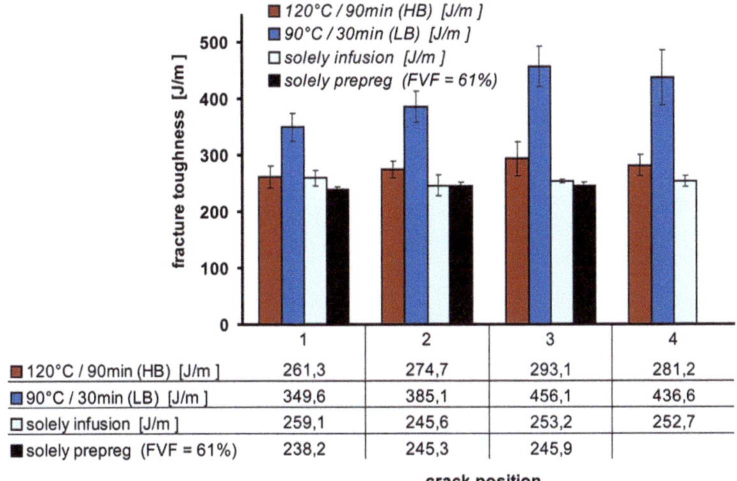

	1	2	3	4
120°C / 90min (HB) [J/m]	261,3	274,7	293,1	281,2
90°C / 30min (LB) [J/m]	349,6	385,1	456,1	436,6
solely infusion [J/m]	259,1	245,6	253,2	252,7
solely prepreg (FVF = 61%)	238,2	245,3	245,9	

Fig. 28.6 Fracture toughness G_{IC} in CPI-fabricated samples depending on the position of the initial crack layer

the actual level of crack layer. All samples yielded the sample fracture toughness values with a standard deviation of less than 2.5% (Fig. 28.6).

Preliminary rheological tests on the prepreg resin system Hexply6376 were carried out in order to determine the sensible ranges of temperature and dwell time to be varied in the experiments. This resulted in the use of two dwell time temperatures and durations in the investigation of the influence of process parameters on the fracture toughness: 120°C at a dwell time of 1.5 h and 90°C at a dwell time of 0.5 h. The comparison of the respective samples reveals that these parameters have an impact on the sample properties. The fracture toughness of the 90°C/30 min dwell time samples is approx. 30% higher than that of the 120°C/90 min samples.

The distribution of fracture toughness within the different layers of both CPI samples series reveals a peak of fracture toughness exactly at the crack level at which the matrix transition zone is expected to be. The relatively low values in the direction of the prepreg at crack layer no. 1 were confirmed in tests on pure prepreg material. The average fiber volume fraction determined in these tests by the analysis of photomicrographs was about 60–63%.

The influence of the process parameters on the distribution of fracture toughness is shown in Fig. 28.6. The overall magnitude of the toughness values in the 90°C/30 min series of samples is higher than that of the 120°C/90 min dwell time samples. A possible explanation is the ratio of the relative gelling times of the two matrix systems. At a dwell time temperature of 120°C at 90 min, the cross-linking of the prepreg resin progresses significantly beyond the level found at lower temperatures and shorter dwell times. One possible consequence might be less bonding to the subsequently infused resin and, therefore, slightly lower fracture toughness. In both sample series the maximum fracture toughness appears in crack layer number 3. This region represents the mean level of matrix transition compared with the micrograph analyses [11]. It can be concluded that the peel samples, as well, show that the resin transition zone has no weakening effect in a bonding area of a CPI manufactured component.

28.2.3 Sample Structures

Technology sample structures are used to demonstrate the feasibility of manufacturing and the testing of the production methods. A further task is to prove the successful implementation of mechanical concepts. Examples for each of these purposes are presented.

(a) Stringer-reinforced panel

A promising application of the CPI technology is the usage of complex LRI components as reinforcing elements on prepreg skins.

Such structures are always needed where thin-walled shells have to bear huge bending and/or torsion loads, like fuselage shells or wing covers. Their thin and lightweight structure makes such components sensitive to buckling and require suitable reinforcements like stringers and frames. Planar or slightly curved prepreg shells with a sufficiently large radius can be laid-up automatically by CNC-controlled machines. This leads to a significant reduction in the cost of the shell material lay-up. The stringers of the skin panel sample presented in this chapter are provided in the so-called omega shape.

Layers of woven fabric including binder were laid up on a ROHACELL foam core. A vacuum-assisted precompaction process and subsequent trimming produced easy-to-handle dry fiber preforms. Stringers with this design are inherently stable and do not require tools for their assembly or fixation.

After the lay-up of the 2 mm prepreg skin, the stringers were fixed in their proper position, whereby repositioning was possible without any difficulties. Provision with covering plates, including the infusion line, was followed by the vacuum bagging. The finished shell sample is shown in Fig. 28.7. Ultrasound scans revealed that the attachment zones of the stringer foot to the prepreg shell were free of defects and free of pores.

(b) Fracture experiments on spars

Preservation of the favorable properties of the semi finished materials combined to a structural element is the basic requirement for the use of the CPI technology. For this purpose, a three-cell experimental spar was designed to allow a direct comparison in fracture tests between a spar made solely by resin infusion and a spar made with prepreg fractions [15].

The concept of the three-cell spar allows the forces from the thrust of the web to be coupled into the spar caps in multiple places. Tensile and compressive forces are picked up mainly by the caps of the spar. High-stiffness and high-strength

Fig. 28.7 Finished stringer-reinforced prepreg shell made by the CPI fabrication technology

Fig. 28.8 Manufacture of the spar with prepreg belts (*left*); spar with final wrapping in the tool (*right*)

materials are preferably used in the flanges. These areas are particularly well suited for accentuating the favorable properties of UD carbon fiber prepregs. For this reason, an experimental spar each with infused and with prepreg caps was designed and then exposed to load until fracture occurred.

Both tested spars were equipped with the identically sized webs made of dry braided fiber preforms. The structure of the spar with prepreg flanges and web cells is shown in Fig. 28.8 (left). The right part of Fig. 28.8 shows the finished preform with its final layer of braided fabric tube in the tooling (no lid yet) ready for vacuum bagging and processing.

The fracture experiments are successful with regard to the nominal loads of 15 kN (Fig. 28.9). A breaking load of 15.27 kN was achieved with the pure infused spar and 16.24 kN with the CPI spar. In both cases the starting point for the collapse of the spar was the failure of the compression loaded flange. However, unlike the infused spar cap "wet web" löschen, the prepreg flange remained attached to the web cells over the entire length of the spar except for the fracture zone. The better attachment of the prepreg spar flange to the web cells can be explained by the improved mechanical properties of the toughness-modified

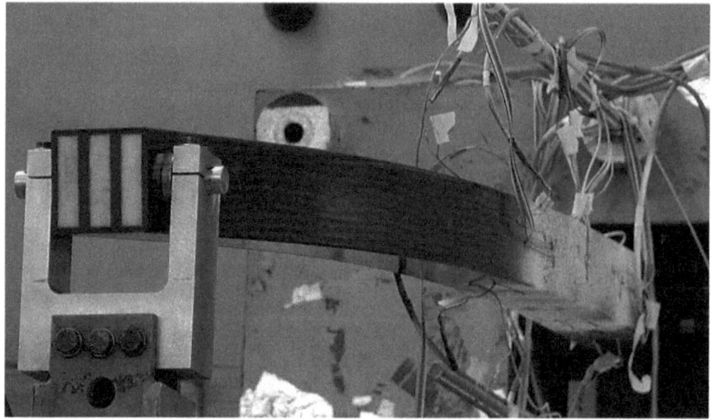

Fig. 28.9 Fracture test of the CPI-spar

prepreg resin that penetrated from the flange into the contact zone during the dwell time of the CPI process. These observations and the fact that the calculated fracture loads were actually attained, demonstrate the successful applicability of the CPI fabrication technique. The desired combination of the favorable properties of the semi-finished elements is indeed preserved in the finished component.

28.3 Conclusions

The suitability of the CPI technology and its variants for the production of primary aerospace components has been demonstrated by a large variety of applications. For this purpose, components and samples manufactured with this technology were systematically tested for their performance using current technological and material parameters. The investigations focused on the compatibility of the matrix materials, characterization of the transition zone, determination of the mechanical properties as well as the manufacture and testing of entire assemblies. The feasibility was tested in several samples as a basis to demonstrate applicability of the CPI technology.

The visual characterization of the transition zone, using dyed infusion resin, allows to study the impregnation behavior of the prepreg resin in the dry fibers and the identification of temperature, time and capillarity as the driving force. The determination of the fracture toughness was used in order to assess the influence of the main process parameters. The duration and the temperature of the dwell time just before the infusion of the resin into the dry fibers have an impact on the spread of the prepreg resins in the contact zone. A shorter dwell time with a lower temperature generate a higher main level of fracture toughness.

Comparative fracture tests with large spars made using both the CPI technology with prepreg flanges and in pure LRI technology allowed to demonstrate that the combination of favorable mechanical and handling properties of the semi-finished parts are retained in a complex CPI structural element.

Sample structures were used to study the manufacturing of complex components applying the prepreg component as a shell element with infusion filled dry fiber reinforcements elements.

It can be concluded that the presented CPI process is suitable for the manufacture of highly loaded CFRP components. Compared to bonding and welding technologies, the co-curing class CPI technology does not need any additional manufacturing steps and is free of problems concerning the fitting accuracy of the parts to be assembled. Furthermore, the use of infused fiber preforms in combination with prepreg preforms has the ability to significantly reduce efforts to produce large integral assemblies for primary structural applications.

References

1. Kleineberg, M.: Cost Effective Series Production of Composite Structures. Paris Air Show Le Bourget (2001)
2. Bergmann, H.W.: Konstruktionsgrundlagen für Faserverbundbauteile. Springer, New York (1992)
3. Bibo, G.A., et al.: Mechanical characterisation of glass- and carbon-fibre-reinforced composites made with non-crimp fabrics. Compos. Sci. Technol. **57**, 1221–1241 (1997)
4. Nuri, E., et al.: Development of spring-in angle during cure of a thermosetting composite. Composites: Part A **36**, 1700–1706 (2005)
5. Patent EP 1231046 A3, Verfahren zur Herstellung von Elementen aus Verbundmaterial durch Co-Bonding Technik (2002)
6. Ermanni, P.A.: Die kombinierte Nasswickel- und Prepregbauweise—Ein Verfahren zur wirtschaftlichen Herstellung von CFK-Flugzeugrumpf-strukturen. Dissertation ETH Nr. 9339, Zürich (1990)
7. Patent DE 19915083 C1, Verfahren zur Erstellung faserverstärkter Kunststoffbauteile mit nicht-vollständig abwickelbarer Geometrie (2000)
8. Patent DE_10156123 B4 (2003)
9. Munz, M.: Evidence for a three-zone interphase with complex elastic–plastic behaviour: nanoindentation study of an epoxy/thermoplastic composite. Appl. Phys. **39**, 4044–4058 (2006)
10. Hexcel Composites: Prepreg Technolgy. Publication No. FGU 017 Duxford (1997)
11. Janusz, A.: Erschließung von Methoden zur Analyse der Matrixverteilung in Faserverbundwerkstoffen mit unterschiedlichen Matrixsystemen. Diploma Thesis, Technische Universität Braunschweig (2006)
12. Kaps, R.: Kombinierte Prepreg- und Infusionstechnologie für integrale Faserverbundstrukturen. Dissertation, Technische Universität Braunschweig (2011)
13. Schulz, M.: Festigkeitsuntersuchungen von CFK Proben in hybrider Prepreg-/Nasstechnologie. Thesis, Technische Universität Dresden (2005)
14. AITM 1-0005 Iss2. Interlaminar fracture toughness energy Mode I—G_{IC}; Airbus Standard (1994)
15. Steeger, S.: Bauweisenvergleich von in Nass- und Hybridtechnologie gefertigten CFK-Biegeträgern mit Hilfe der Durchführung und Auswertung von statischen Bruchversuchen. Diploma Thesis, Technische Universität Braunschweig (2005)

Chapter 29
Interactive Manufacturing Process Parameter Control

Markus Kleineberg, Nico Liebers and Michael Kühn

Abstract The most critical step within the process chain is the curing of the matrix because here the different semi-finished products are transformed into a completely new compound material. In case of thermoset matrix systems all geometrical characteristics of the produced composite component are frozen as well. Tolerances in the chemical composition of the matrix or simple aging aspects can influence the crosslinking reaction while typical weight tolerances of the fibre product may affect the fibre content, the laminate thickness or even the global geometrical shape of the composite component through the so called "Spring-In" phenomenon. Interactively controlling the curing step is a promising approach to solve most of the above mentioned problems.

29.1 Introduction

Along with the growing demand for continuous fiber reinforced composite structures for future energy efficient aircrafts and other vehicles there is an emerging need for highly productive manufacturing strategies. The related high rate production processes will benefit from a significantly increased level of automation to increase the production speed and reproducibility. Looking at the process itself it will be necessary to reduce infiltration and curing times as much as possible to optimise the efficiency of the process. Furthermore high rate production processes have to be absolutely reliable to minimize quality control efforts and the scrap rate.

M. Kleineberg (✉) · N. Liebers · M. Kühn
Institute of Composite Structures and Adaptive Systems, Deutsches Zentrum für Luft- und Raumfahrtz, Lilienthalplatz 7, 38108, Braunschweig, Germany
e-mail: markus.kleineberg@dlr.de

M. Wiedemann and M. Sinapius (eds.), *Adaptive, Tolerant and Efficient Composite Structures*, Research Topics in Aerospace, DOI: 10.1007/978-3-642-29190-6_29,

Today the production processes for high performance CFRP components are restricted because all crucial parameters like for example pressure, cure temperature, dwell time, heat-up and cool-down rates are specified and fixed. In order to meet the demands mentioned above a dynamic control of all those process parameter based on an online condition monitoring of the important laminate properties should be a viable option. A highly promising research approach in this respect is the observation of the laminate throughout the complete process via ultrasonic sensors. Such sensors have to be compatible to the conditions of the process and the derived signal has to be reliable enough to correct the process boundary conditions as required. Compared to the well-known dielectric cure sensors not only the surface but also the local volume of the laminate is measured by ultrasonic sensors. This in turn means that not only resin flow fronts and the state of cure at the sensor surface can be measured but also volumetric conditions like the local laminate thickness and the degree of cure over laminate thickness. Possibly also void content can be analysed and utilised to optimise the process parameters.

29.2 Typical Production Processes for Composite Structures

A variety of different processes is available to manufacture CFRP components. Besides thermoplastic and thermoset matrix systems it is also important to distinguish between high viscosity resin systems typically used for prepreg processes and low viscosity resin systems typically used for LCM (Liquid Composite Moulding) processes. Depending on the application open moulds with a variable cavity or closed moulds with a geometrically constant cavity are viable. Open mould concepts are based on a rigid mould side which is sealed with an elastic membrane, Fig. 29.1b. By varying the pressure on the membrane it is possible to adjust the fiber content or respectively the thickness of the laminate especially for LCM Processes. To demould a thermoset component a minimum degree of cure has to be ensured. For thermoplastic components it is necessary to cool down below a temperature threshold specific to the matrix before demoulding. Elevated pressure can be used in combination with closed moulds (Fig. 29.1a) and low viscosity resin systems to accelerate the infusion process and to reduce the volume of possible voids. For open mould concepts an autoclave can be used to increase the pressure on the membrane, Fig. 29.1b.

29.3 Crucial Manufacturing Process Parameters

To evaluate the quality of a high performance laminate matrix the aspects related to the architecture and the orientation of the fibers have to be analysed. Destructive test methods in combination with a concept to assure quality are typically used to

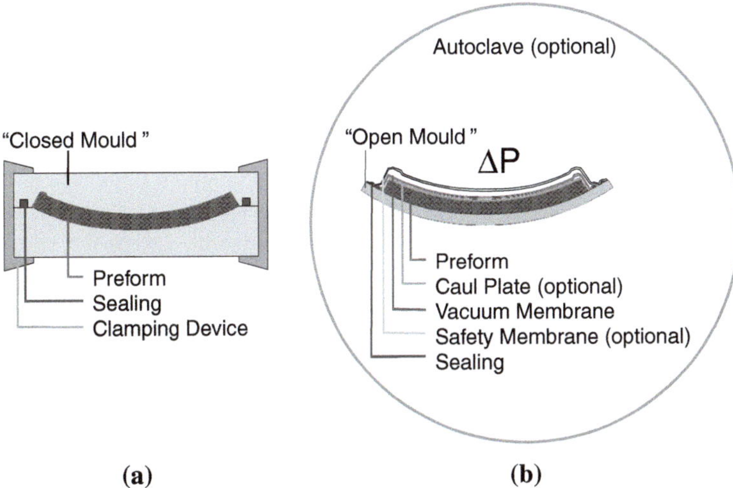

Fig. 29.1 Closed (**a**) and open (**b**) mould concepts

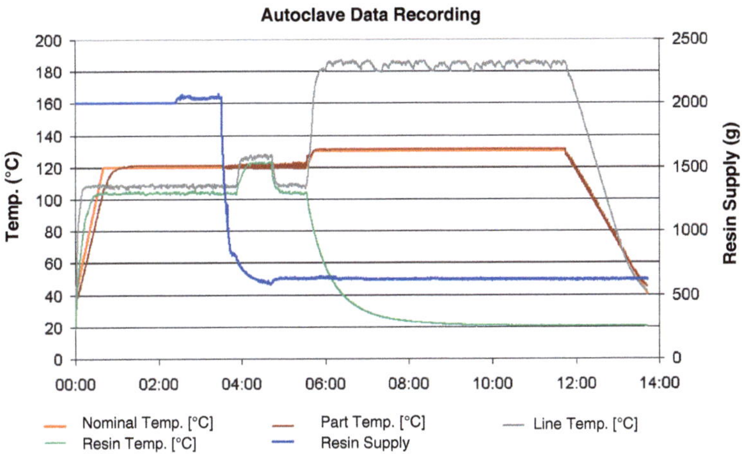

Fig. 29.2 Typical process data recording

guarantee the correct architecture and orientation of the fibers. Matrix related quality assurance is usually based on process parameter data recordings of time, temperature and pressure (Fig. 29.2) and a final ultrasonic and optical inspection. If the process parameter data recordings are within an acceptable tolerance bandwidth one assumes that the real properties of laminate like fiber volume content, degree of cure and void content are almost alright. Finally one relies on the final ultrasonic inspections.

The drawback of this approach is that faulty composite components will not be identified until at the end of the production chain. In addition to the enormous effort for the time consuming and expensive final inspection all components are led through the production chain no matter whether they are correctly processed or not. During the infusion of dry semi-finished fiber products in an open mould process, one particular problem is the scatter of the laminate thickness as a result of areal weight variations in the semi-finished product. Furthermore the history of the resin system has an impact on the gelation behaviour and on the viscosity of the resin which in turn significantly affects the processing window. To make sure that most of these uncertainties are taken into consideration the pre-defined processing conditions are highly conservative. The related additional costs for the prolonged use of manufacturing equipment and relatively high rejection rates are being accepted because other cost drivers dominate the labour intensive classical manufacturing process for high performance composite components.

29.4 Interactive Manufacturing Process Control Using Ultrasound

Sensor-guided controls which optimise both productivity and quality can be a short term solution for the dilemma mentioned above. One of the most important measurement methods in this context is the ultrasound technology which provides indications of the injection and resin curing progress, about void formations and the laminate thickness [1]. This sensors, which can be integrated into the manufacturing equipment, do not require any direct contact with the component as their ultrasound penetrates the tool wall. This means that the component surface is not impaired by sensor marks. The thickness of the laminate respectively the fiber volume content is an especially critical parameter in open mould injection. It shows high deviations from tolerance which can be brought down to a very low level of less than 100 µm with the aid of ultrasonic measuring systems. With this quality parameter in the new process control approach the injection process can be sensor-controlled via the differential pressure between the ambiance (autoclave) and the injection system. The ultrasonic signal is used in the next step in the process, the curing, to ascertain the completion of cross-linking and thus the time point of demoulding. Ultrasound transmission is used to monitor production, i.e. with a separate ultrasound sender and receiver, Fig. 29.3.

This has the advantage that it is a strong, easy to evaluate measurement signal which is easy to evaluate and insensitive to interference. The sensors are integrated into the mould and the vacuum assembly with specially developed adaptors.

In order to minimise the influence of the tool, a reference signal is recorded at each measuring point before measurement so that only the changes in the sound duration and signal amplitude are measured.

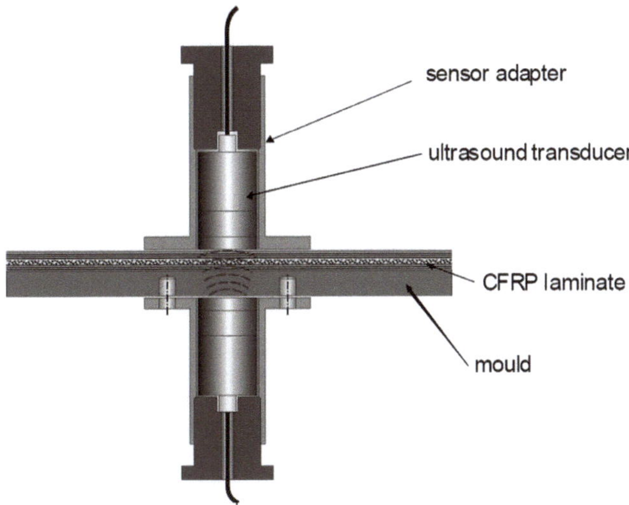

Fig. 29.3 Arrangement of US sensors

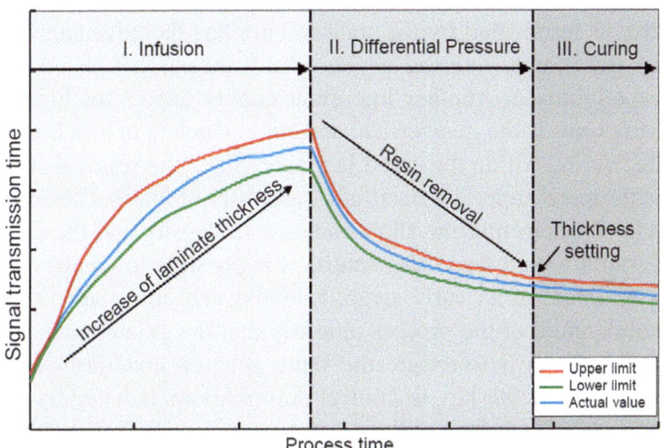

Fig. 29.4 Procedure for thickness adjustment

29.4.1 Interactive Thickness/Fiber Volume Content Control

As the signal transmission time is influenced by factors including the travelling path and the material and temperature-dependent sound velocity, the ascertainment of thickness always has to take place under the same process conditions.

In order to achieve a precise thickness setting, reference curves have to be recorded and the achieved thicknesses on the finished part have to be defined. The reference curves serve as models to control the process parameters and to precisely adjust the laminate thickness, Fig. 29.4.

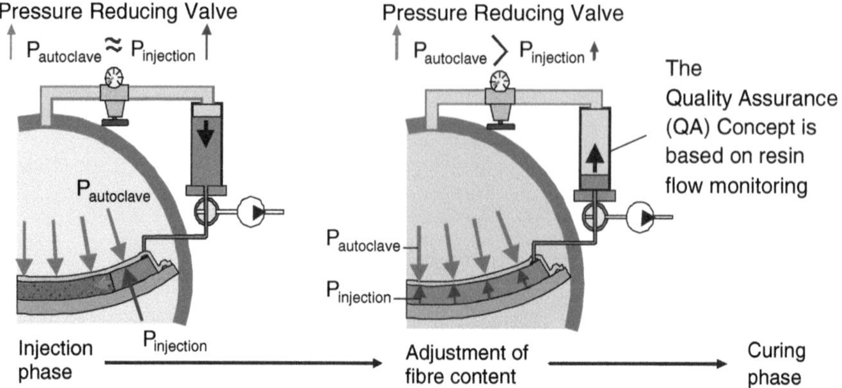

Fig. 29.5 Adjustment of fiber volume content with the SLI process [2]

29.4.2 Interactive Cure Control

A process that is terminated by the state of cure has the advantage that the cycle time can be significantly reduced which in turn means that also related machine costs can be minimised. Another important quality aspect for high performance composite structures is the geometrical accuracy which is in turn highly dependent on the residual stress within the cured laminate. Since the residual stress cannot be avoided it is far more important to reduce the scatter of internal stresses induced by the cure cycle. By detecting the slight increase in density and the related increase in speed of sound within the curing matrix it is possible to identify the beginning of the gelation phase at an early stage. Knowing the cure status it is possible to adjust the temperature of the process in a way that the gelation can be accelerated or delayed. The target is to ensure the same gelation conditions for every component because that is the key to control deformations induced by stress like for example "Spring-In" effects.

29.4.3 Interactive Void Content Control

Comparable to conventional ultrasonic inspections it is possible to identify possible void formations in the area of the sensor by analysing the modulated sensor signal during the infusion phase. As long as the resin is still in a liquid state it is possible to increase the process pressure in the case of detected void problems. By increasing the process pressure the void volume can be compressed (Boyle's/Mariotts's law) and the ability of the resin to absorb the gaseous void content is increased (Henry's law). The described procedure could be a valid counter measure against the contamination of resin batches with increased amounts of water or

Fig. 29.6 Omega shaped frame structure

solvents. The enhanced process maturity could be used for example to lower the resin specification requirements which in turn should lead to reduced resin costs.

29.5 Application Examples

Open mould LCM processes like e.g. the SLI (Single Line Injection) method offer the biggest potential for interactively controlled processes, especially if they are carried out in an autoclave at an elevated level of pressure (Fig. 29.5).

Prepreg processes have also been monitored successfully with ultrasonic sensors but the adjustment of the fiber volume content by controlled bleeding is highly dependent on the viscosity of the prepreg system. A serious equipment related problem is caused by the very demanding pressure and temperature conditions required for the manufacturing of high performance 180°C composite components. Furthermore the question had to be clarified how many sensors are necessary to get a reliable prediction of the final laminate parameters.

29.5.1 Manufacturing of Omega Shaped Frame Structures

To increase the potential of lightweight design structures the automotive industry has recently enhanced their efforts in the field of high performance composite structures. To obtain a better understanding of typical automotive boundary conditions an initiative has been started with focus on future car concepts [3], Fig. 29.6.

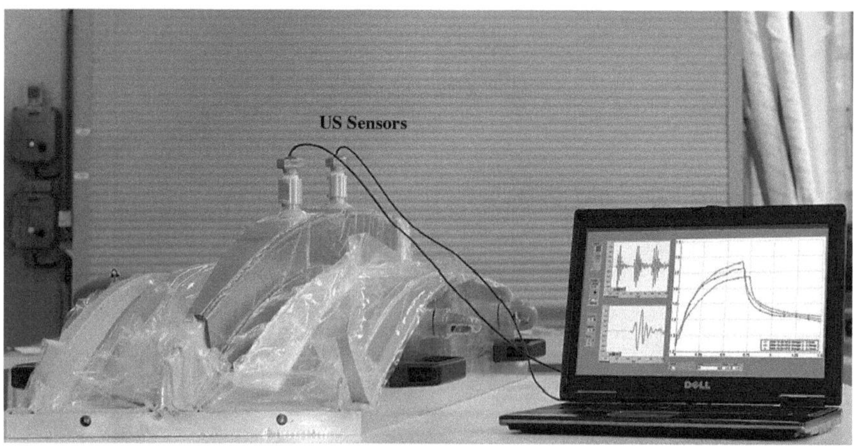

Fig. 29.7 Manufacturing of complex frame structures

Within this initiative one topic was the design and manufacturing of an innovative, crash optimised B-rib for future car concepts. The concept of this CFRP B-rib is being described more in detail in Chap. 23 in this book. For the development of a concept to assure mould quality a high degree of integration is taken into account right from the start. As the sensors do not deliver a signal until the cavity is filled with resin and both sensors are supposed to react at the same time, a flow simulation is carried out to find out the best positions for the sensors which are installed to check the advancing flow front during the infusion.

In the next step the position and size of the inflow channels integrated in the mould are varied in order to identify the best infusion procedure. Two ultrasonic sensors are screwed into the aluminium tool in such a way that the sensor faces are parallel to the mould's surface. After a reference signal has been recorded and the preform has been laid down, the second pair of sensors is placed directly opposite the sensors in the mould. In order to ensure an even surface and homogenous pressure distribution on the laminate, a caul plate is laid directly onto the perform (Fig. 29.7). During the first infusion the ultrasonic signals are recorded and the transmission times correlated with the component thickness. The recorded reference curve is consulted to set the component thickness in the following infusion (phase I, Fig. 29.4). After the infusion has been initiated, the actual ultrasonic signal curve is assimilated to the reference curve by regulating the differential pressure between the infusion system and the autoclave (phase II, Fig. 29.4). The fiber volume content is slightly reduced for a thicker component by infiltrating more resin; resin is pressed out to produce a thinner component. When the target thickness is reached, the infusion is stopped and the curing cycle initiated (phase III, Fig. 29.4).

Fig. 29.8 Manufacturing of coupon panels

29.5.2 *Manufacturing of Coupon Panels*

For the identification of viable process parameters for innovative fiber preforms and resin systems the proper monitoring of coupon panels manufacturing has a high priority. To reach the required close tolerances typical RTM (Resin Transfer Moulding) processes and rather small, closed moulds are used for the coupon panels manufacturing. To be more efficient especially in extensive qualification and screening programs the controlled LCM process have proved to be a highly competitive alternative. Tolerances below $\pm 0,1$ mm can be produced repeatedly for typical 4 mm coupon panels which means that the thickness accuracy is absolutely comparable to RTM manufacturing processes. As shown in Fig. 29.8 a single ultrasonic sensor is sufficient to get a good reference for the average thickness value of the coupon panel. By adjusting the difference in pressure between the autoclave and the infusion system the required nominal panel thickness is set.

29.6 Conclusion

Even though it has turned out to be quite tricky to extract the various information from the signal of the ultrasonic sensors especially the interactive thickness control option works fine. The robustness of the measurement equipment and the quality of the signal are at an acceptable level and the experiences especially with the manufacturing of coupon panels are positive.

References

1. Mc Hugh, J.: Ultrasound technique for the dynamic mechanical analysis of polymers. Dissertation, Technische Hochschule, Berlin (2007)
2. Herrmann, A. S., Sigle, C.: Das Single-Line-Injection-Verfahren zur Herstellung von Hochleistungsverbunden. In: DGLR-Jahrestagung, pp. S.1–S.7 (1999)
3. Schöll, R., Friedrich, H.: Innovative Fahrzeugstruktur in Spant- und Space-Frame- Bauweise. In: ATZ 01, pp. S.52–S.58 (2009)

Part V
Adaptronics

Chapter 30
Autonomous Composite Structures

Hans Peter Monner and Michael Rose

Abstract The vision behind so-called autonomous composite structures is the creation of a new class of composite lightweight structures with significantly enhanced capabilities with respect to traditional design. This includes health monitoring features to increase the maintainability and to enlarge the achievable design layouts as well as the incorporation of noise reduction and vibration control capabilities directly into the structure. The overall quantity to be minimized is the weight per surface ratio, which can only be put below a certain application dependent threshold value by active methods. Therefore three different key technologies have to be merged into one autonomous system: energy harvesting, smart structures and fiber composites. This section gives an overview about the requirements for current and future research to make this vision real and presents examples which demonstrate that some key aspects of autonomous composite structures are already realizable with "state of the art" techniques.

30.1 Limitations of Purely Passive Structural Design

In many applications and especially in the aerospace industry, composite structures are increasingly used to substitute other materials like aluminium in certain parts or whole assembly groups due to their excellent stiffness to weight ratio.

H. P. Monner (✉) · M. Rose
Institute of Composite Structures and Adaptive Systems,
German Aerospace Center DLR, Lilienthalplatz 7, 38108, Braunschweig, Germany
e-mail: hans.monner@dlr.de

M. Rose
e-mail: michael.rose@dlr.de

M. Wiedemann and M. Sinapius (eds.), *Adaptive, Tolerant and Efficient Composite Structures*, Research Topics in Aerospace, DOI: 10.1007/978-3-642-29190-6_30,
© Springer-Verlag Berlin Heidelberg 2013

Fig. 30.1 Technologies to be merged into an autonomous composite structure

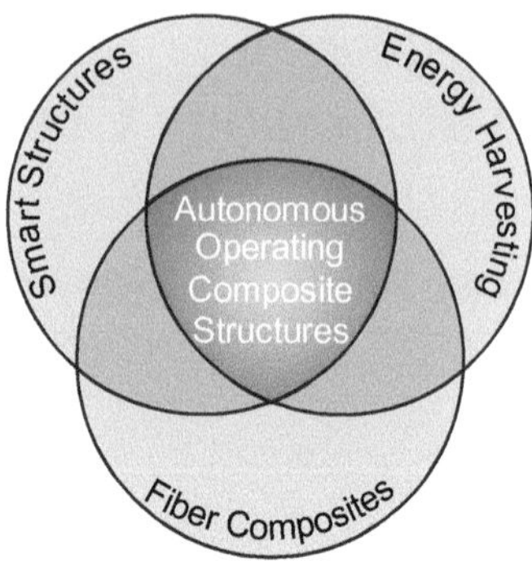

Prominent examples are given by the planes *A380* from Airbus and the *Dreamliner* from Boeing. A small reduction in aircraft weight has big impact on market value in the aerospace industry due to the leverage effect on fuel saving. But there are some challenging difficulties in every design of composite fiber based structures of reasonable complexity [1]:

1. Joining composite fiber components in an assembly task needs sophisticated techniques and these connections tend to weaken the structures or have a big weight penalty.
2. Damages due to surface impacts or internal delamination may severely degrade the structural load carrying capacity and are hard to detect by visual inspection.
3. Due to the high stiffness and low weight, large surfaces tend to vibrate and have low resistance against noise transmission.

The linkage problems can be reduced by manufacturing large composite fiber components like wings or complete sections of the fuselage in one piece using big autoclaves and sophisticated design tools for the fiber layout. But the other problems are inherent properties of these modern materials and there are no known solutions to solve them by purely passive design. The damage detection problem leads to so-called health monitoring systems, which are described in Sect. 30.3.

A crucial issue is the addiction of lightweight structures to radiate or transmit noise due to structure-borne sound or excitating noise e.g. at the outside of aircrafts. Broadband high frequency noise can be damped by suitable passive concepts with relatively low weight penalty, but the low frequency noise up to 500 Hz can not be treated in the same way, because absorbers with big masses would be needed to have suitable resonance frequencies. In this field, active or semi-active control concepts as described in Sect. 30.4 are quite promising to fill the frequency gap.

All these active strategies need suitable energy supply units to work properly. In the past, the needed energy was provided by external units with heavy weight and large energy consumption. A central point in autonomous system design is the integration of small energy supply units into the structure itself to provide the needed energy in a decentralized manner converting energy from the environment like heat, light or even vibrations.

In conclusion, to circumvent the limitations of purely passive designed structures, the vision is to establish a new class of composite lightweight structures, which have the capability to operate autonomously in a wanted manner, e.g. to reduce noise or to monitor the structure's health, and this only by using ambient energy. For this, three different key technologies have to be merged into one system: Energy harvesting, smart structures and fiber composites (Fig. 30.1). In the following an overview of these key technologies and how they have to interact with each other is given.

30.2 General Aspects of Smart Structures

The combination of fibers with a matrix is a successfully proven technology in various branches, giving the engineer a large amount of freedom in designing a lightweight structure. This reaches from fibers and matrices selection over a loadpath optimized design to the possibility to integrate additional functionality into the structure. Up today the latter is only merely addressed in existing industrial design processes but it is now recognized that a significant amount of lightweight potential lays in this approach: sensors, actuators, electronics, etc. become an integral part of the composite structure.

A smart structure involves five key elements: structural material, distributed actuators and sensors, control or analysis algorithms, and power conditioning electronics. With these components a smart structure has the capability to respond to changing environmental and operational conditions (such as vibrations, noise, and shape change) or to be used to monitor the structural health. In order to respond to changing environmental and operational conditions, microprocessors evaluate the responses of the sensors and use integrated control algorithms to command the actuators to apply localized strains/displacements/damping to alter the elasto–mechanical system response [2].

The actuators and sensors are highly integrated into the structure by surface bonding or embedding without causing any significant changes in the mass or structural stiffness of the system. In order to do so a fundamental part of smart structures are solid state actuators and sensors based on smart materials like piezoelectrics, magnetostrictives and electroactive polymers. Piezoelectrics are the most popular smart materials [3]. They react with deformation/strain when an electric field is applied and conversely produce voltage at the electrodes when strain is present. Therefore they are successfully used both as actuators and sensors. They are only capable to produce small strains directly, but the broad frequency range of these devices in combination of acceptable linear behavior in suitable operating conditions makes them the natural selection for actuators and sensors in most vibration/noise control smart structures. However, on the way to be autonomous the energy supply of a smart structure represents one of the most demanding tasks. The promising technologies developed within the emerging field of energy harvesting have the potential to close this gap.

30.3 Health Monitoring for Damage Detection

For structural health monitoring the actuators are used as transmitters for lamb waves, which propagate in the structure and are detected by the sensors. Microprocessors with specific embedded algorithms analyze the sensor signals to identify the structural integrity [4]. An integrated system for structural health monitoring allows an improved lightweight design, because composite structures can be realized that lacked the possibility of traditional visual inspection. Moreover, this system can be used for monitoring and damage detection of the structure.

Lamb waves can travel through large areas of components in contrast to longitudinal waves, which are only suitable for local testing with high resolution. For damage detection it is crucial to know how defects within the materials affect the propagation of wave modes and to select suitable ones based on this knowledge. The frequency range for lamb wave testing is between 10 kHz and 2 MHz, excitations are narrow banded modulated sinus signals or bursts.

30.4 Noise Reduction with Active Control

Active measures for noise cancellation allow avoiding the implementation of additional mass as damping material since especially CFRP structures have an acoustically critical weight to stiffness ratio. Low frequency noise can now be sufficiently addressed. There are mainly three approaches for noise and vibration cancellation based on structural concepts:

1. Active methods based on feed-forward control with reference signals.
2. Active methods based on feedback control algorithms.
3. Semi-active methods with hybrid electronic networks.

The first two strategies are discussed more thoroughly in Chaps. 34 and 36 whereas the third strategy is the focus of Chap. 33. The feed-forward control usually uses inherently stable FIR filter and can give very good performance if good reference signals are available. Good reference signals are characterized by the fact, that they have a good correlation to future excitations of the structure under investigation. This limits the applicability of feed-forward control to certain application scenarios. Excellent performance is given in situations where noise travels long distances before leaving the system like in air conditioning ducts or fabrication halls, polluting the environment with noise through chimneys. If such reference signals are not available, feedback algorithms based on H_2- or $H_\infty-$control design are a common choice to increase structural damping by active means [5, 6]. A fair balance regarding achievable performance and robustness has to be found always a posteriori. If system parameters drift (e.g. temperature changes, change of friction or change of system configuration like open/closed doors), the expectations on frequency dependent damping have to be lowered. Otherwise the closed control is prone to become unstable with very high oscillations or even system failure.

Both control types can be used in combination to benefit from each strength. Certain system excitations can be reduced by feed-forward control, which may also lead to improved stability of structures. The filter length of feed-forward FIR filters may also be substantially lowered if certain plant system poles can be stabilized by an inner feedback control.

Generally speaking the semi-active methods are inferior to full active ones with respect to achievable performance, because they can't realize frequency transfer functions with magnitudes bigger than one (at least if purely passive networks are simulated by the used hybrid networks). But there are several aspects which make them a good candidate for the control part of autonomous composite structures:

1. Low energy consumption.
2. High miniaturization possible.
3. Decentralized operation of several units possible.

The consumed energy is only needed to supply active components like operational amplifier with the necessary energy. Therefore lower requirements are needed for energy harvesting methods and this promises high miniaturization levels. Several hybrid networks have been tested in the past like tuned absorbers acting as conventional mass spring absorbers with less weight impact. This kind of absorber acts only locally within the frequency range. Other more sophisticated networks like the ones based on the simulation of negative capacities have a broad band damping effect. Due to the fact, that they are designed to increase the damping of the structure by electrical resistors in combination with a cancellation of the piezoelectric capacity, they are very robust and stable and many autonomous units with such kind of networks can easily be used in a decentralized fashion [7, 8].

30.5 Energy Harvesting

In many technical applications there is a considerable amount of ambient energy available. This ambient energy is typically present in the form of thermal energy, light (solar) energy, or mechanical energy. The process of capturing this energy from the environment or a surrounding system and converting it to perform any useful work is known as energy harvesting [9]. Possible power suppliers can be thermoelectric generators [10], mechanical vibration devices using piezoelectric transducers, wind turbines, and solar cells. A combination of several energy harvesting strategies in one device can be advantageous to increase the harvesting capabilities. Besides the energy converter a harvesting module that effectively captures, accumulates, stores, conditions and manages power for the device is required. Other forms of energy harvesting include the wireless energy transmission for SHM sensor nodes [11].

The miniaturization of the energy harvesting system is a crucial task to achieve a sufficient mass to energy conversion ratio. Significant progress in microelectronics allows a reduction in mass and volume as well as an increase in energy conversion.

This is of special importance when it comes to the system integration into fiber composites.

There are many other technologies for smart structures that have proven their potential also in combination with fiber composites. By replacing power conditioning hardware and cables with energy harvesting technology the overall system complexity and weight can further be reduced and the vision of an autonomous operating composite structure comes closer to reality.

References

1. Flemming, R., Roth, S.: Faserverbundbauweisen—Eigenschaften—mechanische, konstruktive, thermische, elektrische, ökologische, wirtschaftliche Aspekte, ISBN 3-540-00636-2. Springer, NewYork p. 601 (2003)
2. Monner, H.P.: Classic and emerging smart materials and their applications. In: RTO-AVT-141—Specialists' Meetings on Multifunctional Structures/Integration of Sensors and Antennas, Vilnius, Lithuania pp. 1–17 (2006)
3. Chopra, I.: Review of state of art of smart structures and integrated systems. AIAA J. **40**(11), (2002)
4. Hillger, W., Pfeiffer, U.: Structural health monitoring using lamb waves. In: 9th European Conference on Non-Destructive Testing (ECNDT), Berlin, 25–29 Sept p. 7 (2006)
5. Misol, M., Algermissen, S., Unruh, O., Haase, T., Pohl, M., Rose, M.: Active control of sound transmission through a curved carbon fiber reinforced plastic (CFRP) panel. In: Proceedings of ICAST2011, pp. 10–12. Corfu, Greece (2011)
6. Rose, M., Oliver, U., Haase, T.: Vibration control of stiffened plates with embedded cavities using flat piezoceramic devices. In: International Conference on Adaptive Structures ICAST, State College, Pennsylvania, 4–6 Oct 2010
7. Hagood, N.W., von Flotow, A.H.: Damping of structural vibrations with piezoelectric materials and passive electrical networks. J. Sound Vib. **146**(2), 243–268 (1991)
8. Pohl, M.: Noise and vibration attenuation of a circular saw blade with applied piezoceramic patches and negative capacitance shunt networks. ISMA Leuven, Belgium, pp. 411–424 (2010)
9. Galea, S.C., Van der Velden, S., Moss, S., Powlesland, I.: On the way to autonomy: the wireless-interrogated and self-powered smart patch system. Encycl. Struct. Health Monit. p. 22 (2009)
10. Inman, D.J., Sodano, H.A.: Energy harvesting using thermoelectric materials. Encycl. Struct. Health Monit. http://dx.doi.org/10.1002/9780470061626.shm102, doi:10.1002/9780470061626shm102 (2009)
11. Farinholt, K.M., Park, G., Farrar, C.R.: Energy harvesting and wireless energy transmission for SHM sensor nodes. Encycl. Struct. Health Monit. **9**, 269–280 (2009)

Chapter 31
Design of a Smart Leading Edge Device

Markus Kintscher and Martin Wiedemann

Abstract To make use of low-drag future generation wings with high aspect ratio and low sweep for natural laminar flow, new high lift devices have to be developed (ACARE (Addendum to the Strategic Research Agenda, 2008), Horstmann (TELFONA, Contribution to Laminar Wing Development for Future Transport Aircraft, 2006)). At the wing leading edge a smart e.g. morphing high lift device is being developed which provides a high-quality surface without gaps and steps. Due to the low maturity of morphing skins (Thill et al. (The Aeronautical Journal, 112:117–138)) the challenge of high strains has to be solved by an adequate design and sizing process. The presented design process comprises the requirements of a smart leading edge device, the structural pre-design and sizing of a full-scale leading edge section for wind tunnel tests.

31.1 Introduction

In conventional high lift configurations, devices on leading and trailing edges with open slots are used to achieve additional lift. Because of the gap between the slat and the main wing when deployed, the air flow at conventional high lift devices is disturbed and causes transition to turbulent flow immediately after the slat gap. Thus, new high lift devices have to be developed [ACARE (Addendum to the Strategic Research Agenda, 2008), Horstmann (TELFONA, Contribution to Laminar Wing Development

M. Kintscher (✉) · M. Wiedemann
Institute of Composite Structures and Adaptive Systems,
German Aerospace Center DLR, Lilienthalplatz 7, 38108, Braunschweig,
Germany
e-mail: markus.kintscher@dlr.de

M. Wiedemann
e-mail: martin.wiedemann@dlr.de

M. Wiedemann and M. Sinapius (eds.), *Adaptive, Tolerant and Efficient Composite Structures*, Research Topics in Aerospace, DOI: 10.1007/978-3-642-29190-6_31,

for Future Transport Aircraft, 2006)]. Additionally, the construction space in the next generation of high aspect ratio wings is limited due to the employment of slim profiles. Furthermore the slots have been identified as a major source of noise during landing approach [4–6]. Thus smart seamless and gapless high lift devices especially at the wing's leading edge are a mandatory enabler for future wings to significantly increase aerodynamic efficiency and to reduce acoustic emission.

The considered application is an alternative to the slat or a conventional droop nose device. The benefit of a smart droop nose device comes from a smooth surface without gaps and steps leading to the reduction of parasite drag and an improved noise characteristic. The most important requirements in this scenario are the contour accuracy in cruise flight and the curvature on the upper wing side in drooped configuration for a maximum of aerodynamic performance. For compliance with the required contour accuracy in cruise the smart leading edge structure must fulfill the standard specifications for shape and waviness. The curvature on the upper wing side has to be as smooth as possible because of the negative influence of locally high curvature peaks on the distribution of pressure and boundary layer. For this reason the maximum droop angle is limited to about 30° to avoid a locally increasing curvature at the position of the attachment of the leading edge to the front spar.

A continuous deformable leading edge for the adaptation of the airfoil shape to the flight and flow conditions is required. A suitable skin has to provide elasticity for the deformation and stiffness at the same time. According to Campanile [7] a modal synthesis approach can be used for the design of smart structures with an inner mechanism. Within this approach the layout and the stiffness of the smart structure is "tuned" towards the desired change in shape when actuated. With this approach in mind a design and sizing procedure for the design of the smart leading edge aiming at a minimum restraint of the smart structure is developed. A continuous monolithic skin supported by omega shaped stringers at specific positions has been defined as the structural concept. The objective is a structural system with a flexible continuous skin and an activation concept in form of an "active rib". The main idea is to design a skin structure which is only charged by bending loads to limit the evolving strains in the skin. This can be achieved by an actuation concept which only guides the up and downward movement while the shape reached effectively is controlled by the stiffness design of the skin itself. Due to the unavailability of an ideal morphing skin, the approach of a monolithic skin supported by omega stringers requires a thin skin at critical locations to avoid high strains caused by structural deformation. The performance of the system therefore heavily depends on the optimization of the number and position of actuators and kinematics considering the stiffness of the skin.

31.2 Structural Design Process

Due to the low maturity of morphing skins [3] the challenge of high strains has to be solved by an adequate design and sizing process. The structural design process is based on a target shape which is derived from an aerodynamic optimization process.

Fig. 31.1 Design space with the undeformed part, the deformed one and front spar position

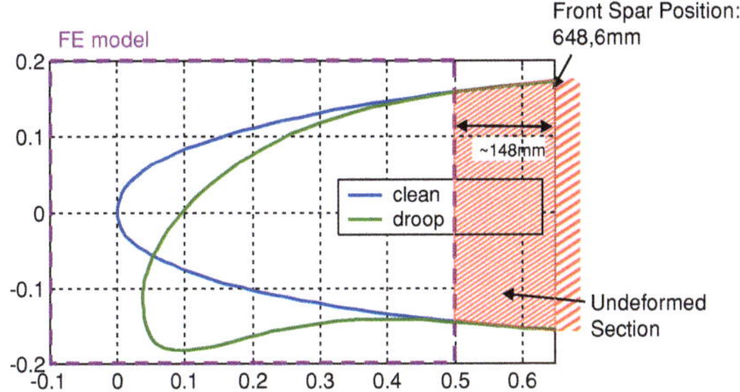

It is characterized by its smooth curvature on the upper surface of the airfoil. For the design of the smart leading edge device the front spar position is fixed. In Fig. 31.1 the position of the front spar, the undeformed and deformed part of the smart skin and the corresponding region of the FE model for the pre-design and sizing is shown.

For the adaptation of the structural stiffness to the aspired change of shape, the development of an automated design and sizing process which generates an optimal distribution of stiffness and configuration layout is necessary. A pre-design analysis is used for the estimation of initial values of design variables.

The initial distribution of skin thickness can be derived from the difference in curvature

$$\Delta\kappa = \kappa_1 - \kappa_0 \tag{31.1}$$

with

$$\kappa_i(x(s), y(s)) = \frac{x'(s)y''(s) - x''(s)y'(s)}{\left(x'(s)^2 + y'(s)^2\right)^{(3/2)}} \tag{31.2}$$

of the undeformed and the deformed shape for example.

Assuming that the skin structure is subjected only to bending loads, the maximum allowable skin thickness t over the normalized circumferential length of the leading edge from upper to lower skin s of the leading edge can be derived with a given maximum strain limit ε_{lim} by

$$t(s) = \frac{\varepsilon_{\text{lim}}}{^1\!/_2 \cdot \Delta\kappa(s)} . \tag{31.3}$$

A maximum of curvature difference is calculated especially at the nose tip from 35 to 55% of the profile. This is the most critical part of the structure. In this area the deformation leads to significant bending of the skin structure and therefore high strains and stresses. Assuming a minimum thickness of the laminat of 0.125 mm for reasons of manufacturing capability, the minimum skin thickness in this area is limited to one millimeter.

From the pre-design analysis the following conclusions can be drawn for the development of the automated design and sizing process:

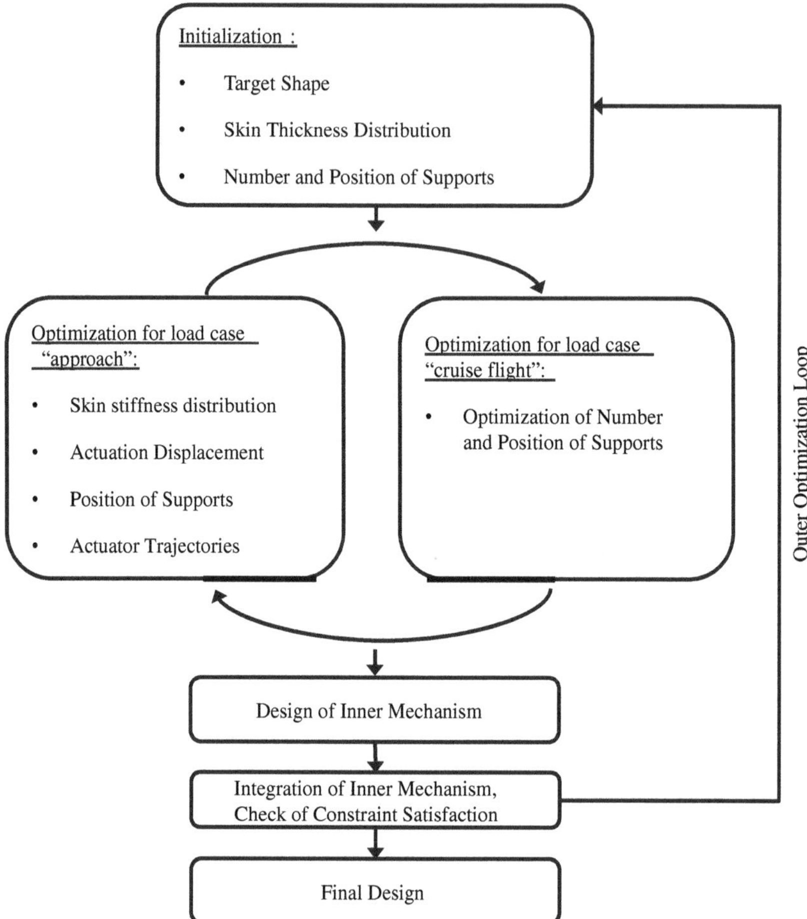

Fig. 31.2 Flowchart of the design and sizing process

- Intensive bending of the structure of the LE tip is necessary for the achievement of the specified aerodynamic target shape.
- Due to the bending, large strains are assumed in the outer fibers of the thin skin. Therefore a prepreg material has to be chosen which provides large strain capability.
- For reasons of manufacturing the design process has to consider requirements for the manufacturing of the fiber reinforced skin like a minimum skin thickness and a minimum distance for the ending of single layers as well as for the stacking sequence of layers.
- Due to the thin skin thickness an optimized setup of stiffeners and support stations is necessary to fulfill the requirements of the profile contour in cruise as well as in high-lift configuration.

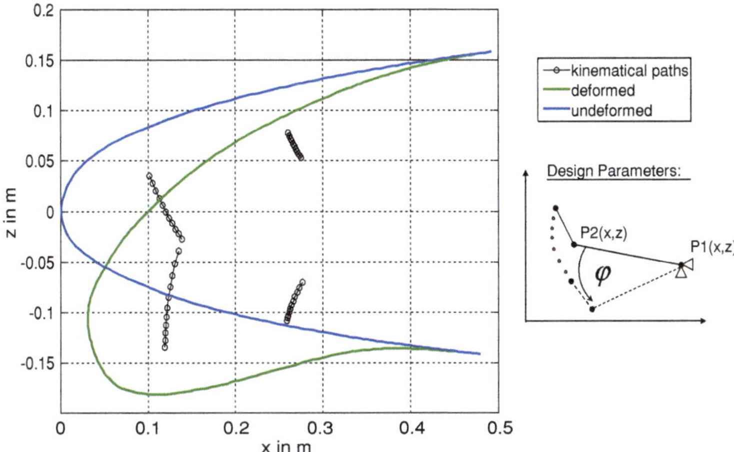

Fig. 31.3 Undeformed and deformed contour with kinematical trajectories of support stations

- The optimization of support stations requires a pre-design and integration of a simplified kinematical mechanism for the evaluation of simulation and design.

According to the major load cases cruise flight and landing approach the design and sizing process is divided into two optimization procedures which are implemented in an outer iterative loop (Fig. 31.2) This is necessary because of the bidirectional influence of each optimization approach on the global structural stiffness. Here the simplex search method of Lagarias [8] is used as a well-known and robust optimization approach.

The objective function for each optimization approach is to minimize the deviations from the given aerodynamic target and clean shape respectively.

$$obj = \min\left[\max\left(\sqrt{(x_i - x_{t\,arg\,et})^2 + (z_i - z_{t\,arg\,et})^2}\right)\right] \qquad (31.4)$$

The intention is a tailoring of the stiffness of the skin which leads to the desired deformation when actuated at specific points on the lower skin panel.

To design the simplified kinematical inner mechanism the displacement boundary conditions are used in the optimization procedures. Once the optimization of the skin thickness and the number and position of support stations is converged and finished, the kinematical trajectories of each support station in Fig. 31.3 can be used for the design of a simplified kinematical mechanism.

As assumption for the design of the kinematical mechanism trajectories are approximated assuming a circular motion of the load introduction points around a single pivot hinge point P2. With this assumption an optimization loop is started for each kinematical path with the objective to minimize the difference between the trajectory derived from the FE analysis and the path resulting from the assumption of a circular motion. The corresponding design parameters for each

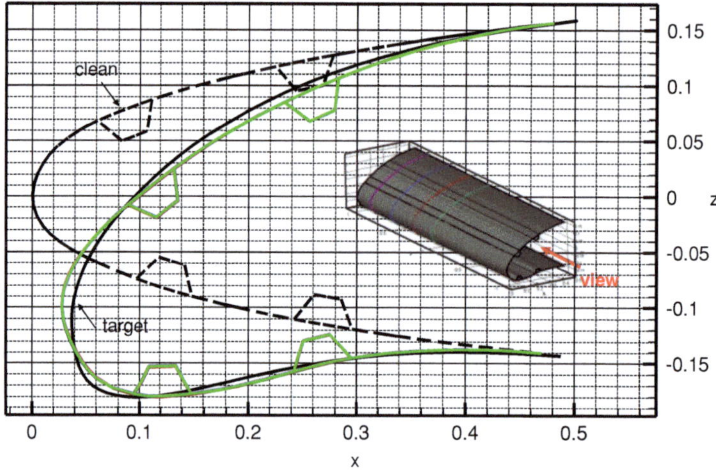

Fig. 31.4 Side-view of deformed slices at position $y_1 = 0.0$, $y_2 = 256.4$, $y_3 = 538.45$ and $y_4 = 820.5$ mm with clean- and target shape

support station in span are the coordinates of the points P1, P2 and the actuation angle φ.

The resulting mechanism on the one hand provides the desired motion for the deformation of the skin structure and on the other hand it provides sufficient support of the upper skin to guarantee the required profile contour under aerodynamic loading.

31.3 Evaluation of Results and Final Design

The final result of the optimization of the distribution of the skin thickness as well as the number and positions of support stations is presented in Fig. 31.4. The two stringers on the lower skin, i.e. at 22 and 35% circumferential length, are primarily used for load introduction during the change of shape. The stringers on the upper side (64 and 77%) serve as stiffeners in the span direction and support the upper skin against the aerodynamic loads.

As material a glass fiber prepreg with HexPly913 from Hexcel composites was chosen. The material is well- known from its application in the manufacturing of rotor blades. The material provides a good compromise concerning stiffness and large strain capability which is important for the stiffness under aerodynamic loading and the static strength for drooping the leading edge. The fiber layup and stacking sequence is chosen according to the maximum strains in the outer fibers of the elastic skin. The most critical material parameter is the tensile strength of the laminates matrix. Therefore the outer fiber layers are aligned in circumferential i.e. 0° direction of the skin. Providing stiffness in the span direction and at the same time torsional stiffness the fiber orientations of ±45 and $90°$ are used in layers which are nearer to the neutral surface of the laminate Due to the fact that the wind tunnel model is untapered and has no wing sweep the skin laminate setup is constant over the complete span of the model.

31.3.1 Performance in High-Lift Configuration

The evaluation of the performance in high-lift condition comprises mainly a check of the contour accuracy under aerodynamic loads and a check of the static strength of the structure. In Fig. 31.4 a side- view of the deformed profile is presented. The plot shows good agreement of the cross-sectional shapes according to FEM after optimization in the span direction which indicates sufficient stiffness and no buckling for example at the position of kinematics (y_2 and y_4) or in between. The target shape was derived by Kühn by an aerodynamic optimization [9].

As mentioned before the contour accuracy of the deformed configuration strongly depends on the interaction of the kinematics and the realized distribution of bending stiffness of the skin structure. Due to the deviation in the kinematic mechanism from the optimum paths a deviation in the size of the kinematics of around five millimetres is expected. Additionally, there is a deviation from the optimal distribution of stiffness of the skin due to the realization of the layer stacking sequence and the manufacturing constraints. The layer stacking sequence cannot be changed smoothly over the contour due to the balanced character of the laminate and the thickness of the discrete layers. Furthermore, minimum run-out distances of single layers have to be considered in order to avoid too rapid changes in stiffness and following stress concentrations. Finally, the minimum skin thickness at the leading edge tip is constrained to a minimum of one millimetre to be able to handle it even though a thinner skin would be sufficient to match the target shape.

In the evaluation of the structural strength the maximum strain criterion is used to indicate the failure of the structure. For a better understanding of the material and for the identification of the maximum strain limit, cyclic bending tests are conducted with glass fiber HexPly 913 prepreg, for which a three point bending test-setup is used [10].The results of the tests are shown in Fig. 31.5

In the cyclic bending tests a maximum bending strain of 3.5% of the material in unidirectional configuration has been found. This is far beyond the predicted and required strain level of around 1% in the outer fibers at critical locations in the leading edge structure. Even for the improper $0°/90°$ laminate design strains of up to 2.8% have been observed.

The tests were conducted according to DIN EN ISO 178. The thickness of the samples was chosen according to the most critical location in the leading edge structure. The cyclic tests were primarily used for a check of the thin skin structure under the mentioned loading conditions with their related large strain levels. The objective was to identify the maximum strain level for a limited number of 20 cycles. A starting damage in the laminate was assumed when a degradation of the

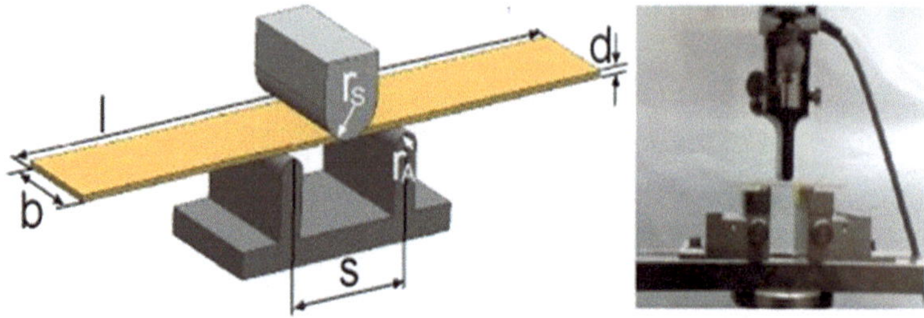

Fig. 31.5 Test setup for cyclic bending tests with HexPly 913

Table 31.1 Results of cyclic bending tests with HexPly913

Laminate	Degradation of e-modulus (%)	Maximum bending strain ε_b (%)
HexPly 913—0°/90°	0.789	2.8
HexPly 913—0°/±60°	0.514	3.1
HexPly 913—0°/±45°	0.386	3.2
HexPly 913—0°/±30°	0.285	3.5
HexPly 913—0°	0.395	3.5

Young's modulus greater than one percent could be measured. Fiber angles of ±30, ±45 and ±60° have been tested. A negative effect of increasing fiber orientation to the maximum bending strain has been found as reported in Table 31.1. Consequently the layer stacking sequence for the outer layers of the skin has to have an orientation in 0 (i.e. in circumferential direction) and ±45 and 90° layers have to be arranged closer to the neutral surface of the laminate.

Measurements of the critical strain in a deformed smart leading edge profile showed that the strain in the critical region of the leading edge tip in general stays below the strains predicted by the FE analysis. Local strain peaks which are predicted by the FE analysis due to the discrete character of the FE model in reality have less impact on the maximum strain values than calculated.

31.3.2 Performance in Cruise Configuration

The evaluation of the performance in cruise flight configuration focuses mainly on the accuracy of the profile contour under aerodynamic loading. For that purpose aerodynamic loads are interpolated and conservatively scaled from CFD calculations performed by Kühn [9]. Because the clean configuration is the initial shape of the leading edge structure, it is not pre-stressed in contrast to the drooped configuration. Therefore the unstressed clean configuration with its lower stiffness is much more susceptible to deformation under aerodynamic loading than the deformed configuration.

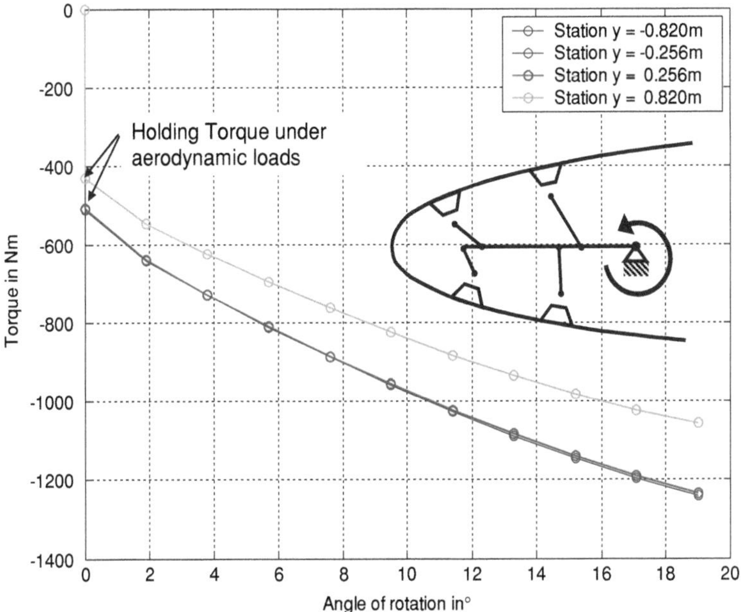

Fig. 31.6 Actuation torque for holding the clean position and for deformation

The maximum deformation in cruise flight occurs at the lower leading edge tip in the region with the minimum skin thickness of one millimetre where it measures about 1.3 mm. In the remaining part of the structure the overall deformation is below one millimetre which demonstrates sufficient stiffness of the system. For a more detailed evaluation of the performance of the structure in clean configuration as well as in high-lift configuration fluid–structure coupled calculations are necessary.

31.3.3 Estimation of the Actuation Torque

For the evaluation of the actuation torque necessary to hold the smart leading edge in clean position and for the calculation of the additional torque for the subsequent deformation of the structure the aerodynamic loads of the drooped configuration are used. In the calculations the variation of the aerodynamic loads during the deployment of the smart leading edge has not been considered.

The loads are applied as static loads in the first load step at zero degree droop angle. The actuation angle is subsequently increasing to the maximum droop angle of 20°. The results are presented in Fig. 31.6. The given values for the torque are derived directly from the reaction moments in the finite element model without the consideration of devices for an improved transmission of the moments.

31.4 Summary and Conclusion

The developed design and sizing process is well suited for the development of the chosen structural concept of a smart droop nose device. It comprises a flexible end-to-end skin without gaps and steps including a kinematical mechanism in style of an active rib for the downward deformation of the droop nose.

Large differences in the required curvature between the undeformed and the deformed shape indicate a large local deformation of the structure and high strains in the outer fiber due to bending. Due to limitations in the strength of the skin material and manufacturing constraints in the skin thickness it is not possible to match the given target shape completely by according to the design requirements. Therefore the developed design and sizing process lead to the solution fitting the best. For an effective design it is mandatory that the simulations of the structural deformation include the kinematics. The resulting deviations from the cruise shape and the drooped target shape are acceptable. By applying the cyclic 3-point bending tests the capability of the structure to sustain the required strain levels could be demonstrated.

Acknowledgments This work was carried out within/as part of the project SADE which is part of the 7th European Framework Program for research in aeronautics. The author is appreciative of the support in the pre-design of the kinematics by Dr. Lorkowski and Dr. Storm of EADS Innovation Works.

References

1. ACARE, : Addendum to the strategic research agenda, Advisory council for aeronautics research in Europe, 2008
2. Horstmann, K.H.: TELFONA, Contribution to laminar wing development for future transport aircraft, aeronautical days, Vienna, 19th-21st June 2006
3. Thill, C., Etches, J., Bond, I.P., Weaver, P.M., Potter, K.D.: Morphing skins—a review. Aeronaut. J. **112**(1129), 117–138 (2008)
4. Wild, J., Pott-Pollenske, M., Nagel, B.: An integrated design approach for low noise exposing high-lift devices, AIAA Paper (2006–2843) 3rd AIAA Flow Control Conference, San Francisco, 2006
5. Kreth, Stefan, König, Reinhard, Wild, Jochen: Aircraft noise determination of novel wing configurations. INTER-NOISE 2007, Istanbul Türkei (2007)
6. Pott-Pollenske, M., Wild, J.: Entwicklung und validierung eines verfahrens zur beurteilung der schallabstrahlung von hochauftriebssystemen, DAGA 2007, Stuttgart, 2007
7. Campanile, L.F.: Modal synthesis of flexible mechanisms for airfoil shape control. J. Intell. Mater. Syst. Struct. **19**, 779–789 (2008)
8. Lagarias, J.C., Reeds, J.A., Wright, M.H., Wright, P.E.: Convergence properties of the nelder-mead simplex method in low dimensions. SIAM J. Optim. **9**(1), 112–147 (1998)
9. Kühn, T.: Aerodynamic optimization of a two-dimensional two-element high lift airfoil with a smart droop nose device, EASN Workshop on Aerostructures, October 7 and 8th, Suresnes, 2010
10. Heintze, O., et al.: Die vorbereitung der faserverbundstruktur einer flexiblen und spaltfreien flügelvorderkante auf ihren ersten grossskaligen bodenversuch, Deutscher Luft- und Raumfahrtkongress 2010, Hamburg, 2010

Chapter 32
Experimental Investigation of an Active Twist Model Rotor Blade Under Centrifugal Loads

Peter Wierach, Johannes Riemenschneider, Steffen Opitz and Frauke Hoffmann

Abstract Individual Blade Control (IBC) for helicopter rotors promises to be a method to increase flight performance and to reduce vibration and noise. Quite a few concepts to realize IBC Systems have been proposed so far. Some of them have already been tested in wind tunnels or on real helicopters. A drawback of all systems that include discrete mechanical components like hinges, levers or gears is their vulnerability in a helicopter environment with high centrifugal loads and high vibration levels. That's why the idea of using smart materials that are directly embedded in the rotor blade structure is very attractive for this application. Operating as solid state actuators they can generate a twist deformation of the rotor blade without any friction and wear. In the common DLR-ONERA project "Active Twist Blade" (ATB), DLR designed and build a 1:2.5 mach scaled BO105 model rotor blade incorporating state of the art Macro Fiber Composite (MFC) Actuators. The design of the blade was optimized using a finite element code as well as rotor dynamic simulations to predict the benefits with respect to vibrations, noise and performance. Based on these tools a blade was designed that meets all mass and stiffness constraints. The blade has been intensively tested within some bench- and

P. Wierach (✉) · J. Riemenschneider · S. Opitz
Institute of Composite Structures and Adaptive Systems, German Aerospace Center DLR,
Lilienthalplatz 7, 38108, Braunschweig, Germany
e-mail: peter.wierach@dlr.de

J. Riemenschneider
e-mail: johannes.riemenschneider@dlr.de

S. Opitz
e-mail: steffen.opitz@dlr.de

F. Hoffmann
Institute of Flight Systems, German Aerospace Center DLR,
Lilienthalplatz 7, 38108, Braunschweig, Germany
e-mail: frauke.hoffmann@dlr.de

M. Wiedemann and M. Sinapius (eds.), *Adaptive, Tolerant and Efficient Composite Structures*, Research Topics in Aerospace, DOI: 10.1007/978-3-642-29190-6_32,

centrifugal tests. The mechanical properties of the blade obtained within the bench tests showed a good correlation between measured and calculated values. The centrifugal test comprised a measurement of the active twist performance at the nominal rotation speed of 1,043 RPM at different excitation frequencies from 2 up to 6/rev. It was proven, that also under centrifugal loads the predicted twist amplitudes can be achieved.

Nomenclature

Symbols	Variable (Unit)
a_∞	speed of sound (m/s)
c	blade chord (m)
C_n	normal force coefficient (–)
C_T	thrust coefficient, $C_T = T/(\rho\pi R^2(\Omega R)^2)$ (–)
F	force (N)
M	moment (Nm)
M_{tip}	Mach number at the blade tip, $M_{tip}= \Omega R/a_\infty$ (–)
N_b	number of blades (–)
QC_v	vibration quality criterion (–)
QC_n	noise quality criterion (–)
R	blade radius (m)
T	thrust (N)
V_∞	flight speed (m/s)
x, y, z	coordinates (m)
α_s	angle of attack of the rotor shaft (deg)
θ_{tw}	blade twist angle (deg)
$\theta_{a,n}$	control amplitude of active twist (deg)
μ	advance ratio, $\mu = V_\infty\cos\alpha_s/(\Omega R)$ (–)
ρ	air density (kg/m^3)
σ	rotor solidity, $\sigma = N_b c/(\pi R)$ (–)
Ω	rotor rotational speed (rad/s)
$\Psi_{a,n}$	control phase referred to the control input (deg)

Indices

a	due to active control
n	n-th harmonic
s	rotor shaft
tip	blade tip
tw	twist
0	baseline, without active control
∞	properties of undisturbed flow

32.1 Introduction

For helicopters in forward flight very complex flow conditions are apparent. The superposition of the flight speed of the helicopter to the rotational speed of the main rotor leads to asymmetric flow conditions between advancing and retreating side. This asymmetry of the flow conditions in the rotor disk is growing with increasing flight speed. At the advancing side the flow reaches transonic and supercritical velocity regimes resulting in local shock waves which are a major source of noise in high speed forward flight. In contrast, regions with flow separation respectively stall occur at the retreating side. Since the flow on the rotor blade strongly depends on rotor azimuth, flow separations as well as shock waves are of high dynamic nature and therefore cause vibrations.

In low speed flight as well as descent flight, rotor noise and vibration are mainly determined by blade vortex interactions (BVI). In contrast to fixed-wing aircrafts for which tip vortices are moving downstream away from the wing, those created by helicopters remain in the vicinity of the rotor for several revolutions. This causes multiple blade vortex interactions when rotor blades encounter previously generated tip vortices or pass them very close. Consequently, the velocity field around the blades is changing and the altered angle of attack generates unsteady airloads on the blades which originate noise and vibration.

A reduction of noise and vibration is most effective when the disturbing forces are attenuated at their origin e.g. by an individual control of the blades. A promising approach is the use of anisotropic piezoelectric strain actuators embedded in the rotor blade structure, capable of generating a direct twist deformation of the rotor blade. In comparison to approaches using flaps that generate an aerodynamic moment to deform the blade, the complexity of the actuation system is rather low. Since no moving parts are involved and there is no friction and wear. This is of special importance in a helicopter environment with high centrifugal forces. The same applies for the fact that no heavy mechanical components have to be installed inside the blade causing high loads and can also lead to a weakening of the rotor blade structure. In addition to that the active twist concept guarantees a very smooth surface of the blade whereas flaps have rough edges and therefore producing additional vortices and sources for noise and vibration.

32.2 State of the Art

The first active twist rotors using piezo-ceramic material to actuate the blade were presented by Chen and Chopra from the University of Maryland. From 1993 to 1996 they built and hover tested a series of 1/8th Froude scaled model rotors. The rotor blade skins incorporated piezoceramic plates using the transversal piezoelectric d31-effect [1].

In 1995 a team joining researchers from Boeing, Penn State University and MIT started investigating in the field of active rotors. After proving the concept with a 1/16th Froude scale model rotor, they investigated the capability of active twist of two 1/6th Mach scale rotors (Active Material Rotor AMR). Since at this moment it was not sure whether the generation of structural twist or the twist generation via flaps is favorable, the first phase of this project included the design, manufacturing and testing of both design principles. At the beginning of the project the active twist concept was rated as the high risk approach whereas the flap design was considered to have low risk. This direct comparison by the same team of engineers developing both concepts side-by-side pointed out many advantages of the active twist concept. Actually the active twist concept turned out to be the low risk approach. Because of the encouraging results the active twist concept was applied to a modern planform and airfoil rotor blade. The active twist blades were actuated by interdigitated piezo fiber composites integrated in the spar of the rotor blade [2].

In 1999 a joint venture from NASA, Army and MIT built and tested an active twist rotor (ATR) with a structural design similar to the Boeing model rotor. This rotor was conceived for the testing in heavy gas medium, in the NASA Langley Transonic Dynamics Tunnel. This rotor is the only one which is wind tunnel tested under forward flight conditions. The twist is generated via Active Fiber Composite piezoelectric actuators embedded into the rotor blade spar [3–5].

In 2004 Boeing investigated the possibility of scaling the results of the Mach scaled rotor to a full scale rotor blade. The main focus of this investigation was laid on production and manufacturing approaches concerning the incorporation of the piezoelectric actuators and a robust and reliable wiring to provide the necessary power to the actuators. A 1.8 m CH-47D blade section with 24 layers of active fiber composites embedded in the spar laminate was build and successfully tested. It was shown that a full scale active twist blade with a meaningful actuation capability and acceptable natural frequencies can be built within the weight limit of a passive blade [6].

Motivated by these promising results and the potential benefits, the goal of this work was the development and test of an active twist blade incorporating improved actuation technology and alternative structural concepts to bring this technology a further step forward. At this phase the test are limited to a verification of the actuation system and the structural concepts under centrifugal loads. The intention is to develop the necessary prerequisites for a successful wind tunnel campaign with an active twist model rotor.

32.3 Design and Manufacturing of the Active Twist Blade

In the design phase of the blade, several parameter studies were carried out, resulting in an optimal orientation for the ply lay-up of the skin as well as the actuation direction [7]. This optimum considers the twist deflection that can be achieved with respect to the torsional rigidity. The optimal configuration leads to a

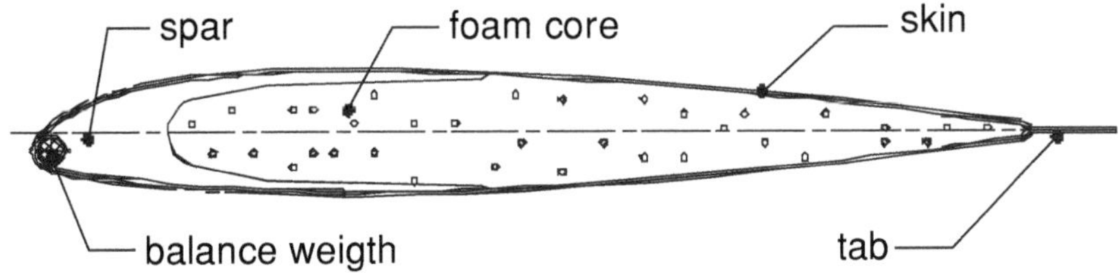

Fig. 32.1 Cross section of the original BO105 model rotor

blade with an anisotropic skin. With regard to these findings a model rotor blade was built. The main characteristics of this blade were taken from the well-known BO 105 model rotor blade. The BO-105 blade features a C-Spar made of unidirectional glass fiber, a glass fiber skin and a foam core (Fig. 32.1). The chord length of 121 mm and the radius of 2 m are in agreement with the original model rotor blade, whereas the profile was changed into a symmetrical NACA 0012, which does not really change the blade from a structural point of view. Because at this point no aerodynamic investigations were planned, it was not necessary to realize an aerodynamic effective blade and the manufacturing effort was reduced by building only one mold. For the same reason the blade was not pre-twisted. The actuator orientation was chosen to be $+40°$, whereas the skin was made of unidirectional glass fiber laminates with an orientation of $-30°$ (inner skin). The area surrounding the actuators was provided with additional unidirectional glass fiber layers in a $+40°$ direction in order to carry the loads of the actuators and to decrease the change in stiffness in the transition region between skin and actuators. The anisotropy of the skin allows the actuators to work in a relatively soft direction (approximately perpendicular to the fibers in the inner skin), whereas the complete blade still keeps its torsional stiffness by the shear stiffness perpendicular to the actuators.

A special focus was laid on the selection and design of the piezoelectric actuators. In all previous investigation (see Chap. 1) so called Active Fiber Composite (AFC) Actuators were used [8]. In this assembly a large number of PZT-fibers with typical diameters of 100–250 µm are aligned side-by side and are usually embedded in an epoxy resin (Fig. 32.2a). To benefit from the higher longitudinal effect (d33-effect) interdigitated electrodes (IDE-Electrodes) are used. This design is realized with two interlaced comb like electrodes symmetrically arranged on each side. With high voltages applied to the electrode fingers the polarization of the PZT material is established to fit to the electrode structure. An essential drawback of this concept is the very labor intensive manufacturing process. Up to know it is primarily handwork to place the many single fibers close to another. This causes quality problems, resulting in deviations of the actuator characteristics. Since the fibers have a circular cross section the electrode contacts the piezoelectric material only over a small area, leading to significant dielectric losses.

An alternative manufacturing process uses cheap PZT-wafers that are cut into ribbons (Fig. 32.2b). In this case the wafer is placed on a tacky film and cut using a

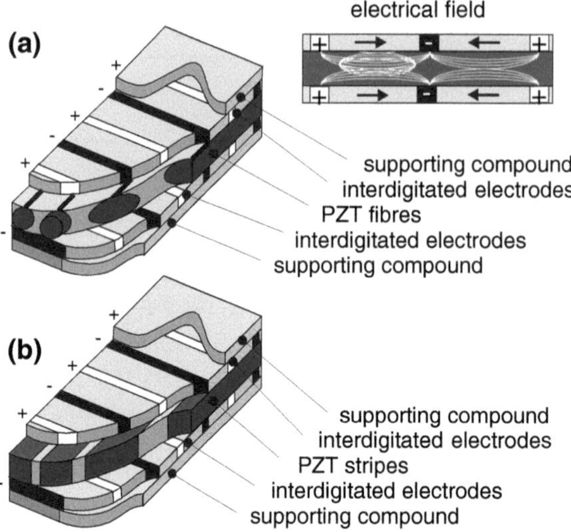

Fig. 32.2 **a** AFC design with fibers, **b** MFC design with ribbons

wafer saw, typically used within the production of computer chips. With this automated process the ribbons are aligned exactly in parallel. In the following manufacturing steps the gaps between the ribbons are filled with resin and polyimid films with etched IDE-electrodes are glued on the top and the bottom of the assembly. Due to the rectangular cross section of the ribbons the connectivity of the electrode to the piezoelectric material is considerably improved. This design has been developed by NASA [9] and is called Macro Fiber Composite (MFC).

To provide a good coverage with active material and to realize the desired actuation direction of 40°, special shaped MFC actuators were designed for the active twist blade (Fig. 32.3). Due to their particular electrode design these types of actuators have to be operated at relatively high voltages between −500 and +1,500 V to provide a sufficient electrical field. A total of six actuators per skin were implemented, resulting in a total active area of approx. 1,600 cm^2.

The manufacturing process of the upper and lower blade skin started with the placement of the MFC actuators into the mould followed by the glass fiber prepreg. Accordingly the strain gauge instrumentation and the complete wiring were positioned onto the uncured prepreg. In the next step the lay-up was put in a vacuum bag and cured in an autoclave at a temperature of 120°C and a pressure of 6 bars. Because actuation and instrumentation are entirely integrated into the rotor blade skins, it was possible to keep the internal design similar to that of conventional passive blades. The spar and the foam core were machined to the desired shape using unidirectional glass fiber composite and foam blocks, respectively. Balancing weights made of tungsten rods were added into the nose of the spar using a cold setting epoxy. Finally the upper and lower skin, the spar and the foam core were bonded together with an adhesive film and cured at 120°C (Fig. 32.4).

Fig. 32.3 Special shaped MFC actuators; height 60.2 mm; width 222 mm

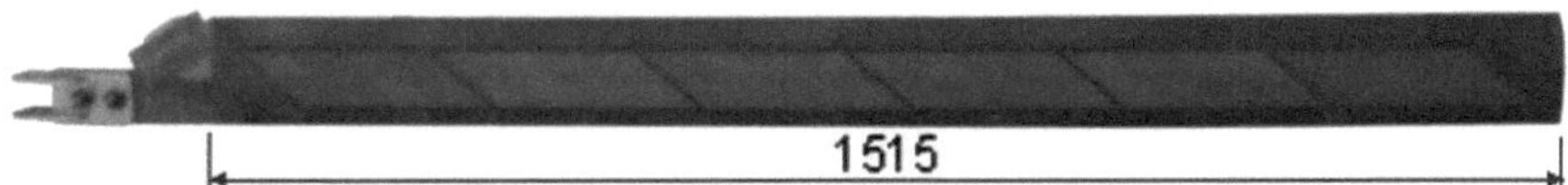

Fig. 32.4 Top view of the active twist blade

32.4 Experimental Test Setup

The objective of the test was to demonstrate the performance of the actuation system and the structural concept under centrifugal loads by showing that the expected twist deformation can be achieved at the nominal rotation speed and different actuation frequencies. For this purpose a test rig was installed at the DLR rotor tower in Braunschweig (Fig. 32.5). The test rig is driven by a 30 kW DC shunt-wound motor. A balancing weight is mounted on the opposite side to trim the blade. The direction of rotation is clockwise. To reduce the mechanical complexity the pitch links have been removed. Data transfer is realized by 24 slip rings and an additional telemetry system with 12 channels for strain gauge measurements (full bridge or half bridge) and 4 ICP channels for acceleration sensors. Four special designed high voltage slip rings transfer the required electrical power to the actuators in the blade. Depending of the excitation frequency the actuators were driven with up to three power amplifiers with a peak to peak voltage of +/−2,000 V and a maximum current of 400 mA. A camera, installed in the rotor tower, allows a permanent monitoring of the experiment from the control room.

The Active Twist Blade was equipped with nine sets of strain gauges. Six sets were implemented for measurement of torsion (one for each actuator) and three for flapping. The locations of the strain gauges are shown in (Fig. 32.6). As described before the strain gauges and the necessary wiring were embedded into the blade skin during the manufacturing process. For each torsion measurement point, two strain gauges were arranged on opposite sides of the upper and lower blade shell in an angle of +/−45° to measure the torsional deformation of the blade. The individual strain gauges were wired to a full bridge to compensate for any bending deformation of the blade, so that only torsional deformations were measured. Since it is not possible to directly measure the twist angle with strain gauges, two additional acceleration sensors were mounted at the leading and trailing edge of the blade tip respectively. As backup and to check the results of the strain gauge

Fig. 32.5 Test rig

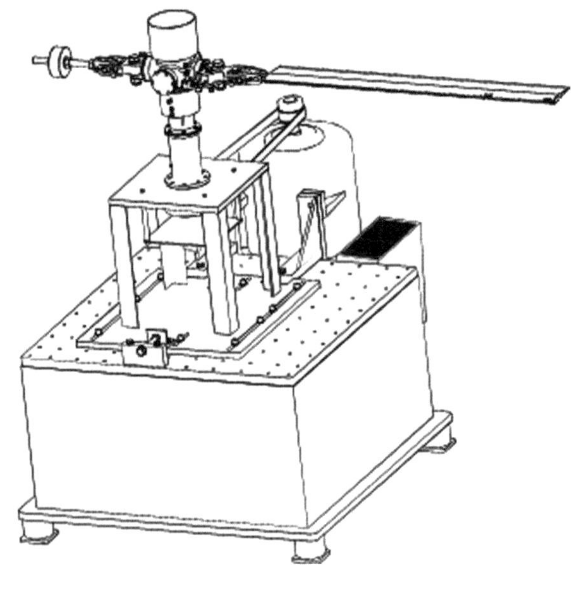

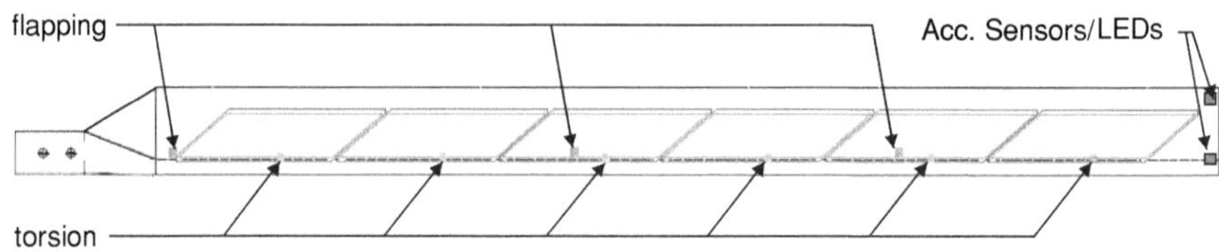

Fig. 32.6 Sensor positions

and acceleration sensor measurements, a supplementary optical measurement system was installed. The system consists of two LED's attached at the leading- and trailing edge of the rotor blade tip and a stationary high speed camera system. By properly triggering the camera the twist movement of the blade can be visualized by the two light dots of the LED's. In comparison to a system that needs a powerful stroboscope light, capable of sufficiently illuminate the blade tip, this solution is much cheaper and also facilitates the analysis of the blade tip motion. Because the LED's appear as clearly distinguishable points, standard image processing tools can be used to automatically determine the twist angle.

32.5 Experimental Results

Before the centrifugal test was conducted, the static properties of the Active Twist Blade were measured within a simple lab test configuration. Table 32.1 compares some of these results with the results of a finite element simulation and the

Table 32.1 Comparison of measured and calculated data of the AT1-blade

Parameter	Experiment	Simulation	Ref. (BO 105)
Flap bending stiffness	196 Nm^2	207 Nm^2	250 Nm^2
Torsional stiffness	194 Nm^2	189 Nm^2	160 Nm^2
Twist rate ($\Delta U = -500..+1,500$)	2.87°/m	2.7°/m	–

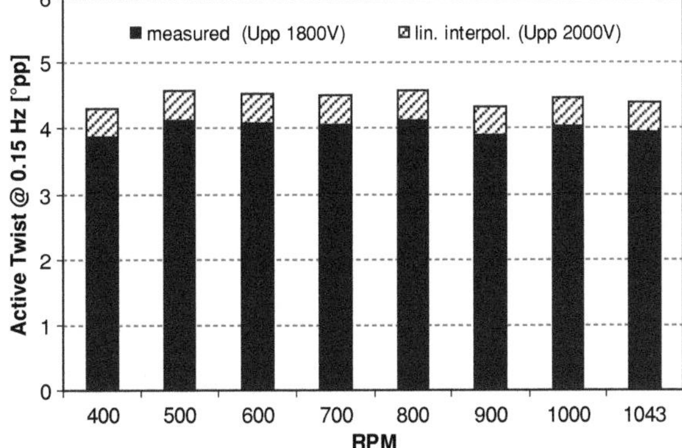

Fig. 32.7 Quasi static excitation of the blade at different rotation speeds

properties of the BO-105 reference rotor. There is a reasonable good correlation between the measured and simulated stiffness parameters and the active performance. In comparison with the reference rotor all parameters of the Active Twist Blade stay within the tolerated limits.

For the centrifugal tests a comprehensive test matrix was derived starting with the measurement of the static peak to peak twist displacement with increasing rotation speed starting from 400 RPM up to the nominal rotation speed of 1043 RPM (109 rad/s; 17.35 Hz) in steps of 100 RPM. All measurements were made using the optical measurement system. The actuators were driven within a voltage range of -500 to $+1,300$ V with a quasi-static excitation of 0.15 Hz. Though a maximum of $+1,500$ V is allowed for the MFC actuators the actuation voltage was reduced to 1,300 V to avoid any electrical overload. In Fig. 32.7 the result of the measurement with reduced actuation voltage are plotted as well as the linear interpolated twist deflection with a maximum voltage of 2,000 V$_{pp}$ ($-500...+1,500$ V).

For all rotation speeds the tip twist is in the order of 4° peak to peak with a deviation of $\pm0.1°$, which is due to the measurement accuracy. This measurement confirms, that the actuation system is capable of generating sufficient twist under centrifugal loads and that there is no performance decrease in comparison to the no rotating case.

The validity of the assumption of a linear correlation between twist deflection and actuation voltage is supported with the measurements plotted in Fig. 32.8. For different voltages the twist was measured at a rotation speed of 1,043 RPM and an excitation frequency of 69.5 Hz. Due to the material inherent nonlinearity of the piezoelectric actuators this assumption can only be an approximation but is reasonable for this case.

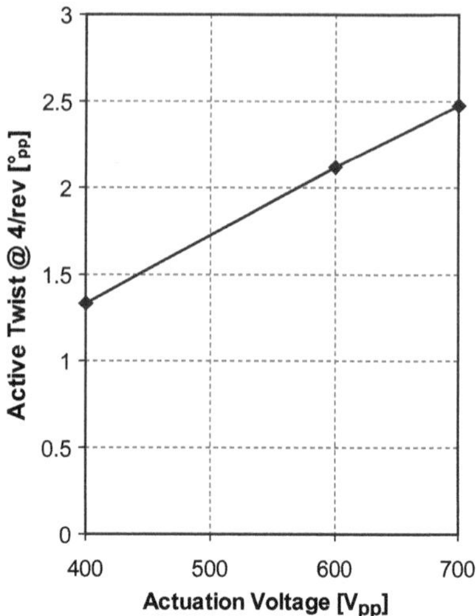

Fig. 32.8 Linear correlation between actuation voltage and twist deflection

The next test comprised a measurement of the tip twist at different excitation frequencies from 2/rev (34.8 Hz) to 6/rev (104.3 Hz). All measurements were done at the nominal rotation speed of 1,043 RPM using the optical measurement system. The results are shown in Fig. 32.9. The actuators were again only driven with a limited voltage of 1,800 V_{pp}. At a frequency of 2/rev the tip twist is equal to the quasi static twist. Because of resonance effects in the vicinity of the first torsional eigenfrequency the amplitude at 3 and 4/rev is even higher (>6°). A look at the rotor diagram (Fig. 32.10) for the first and second torsional eigenmodes explains why the measurement at 4/rev was only done with a much reduced voltage of 700 V_{pp}. Unfortunately the first torsional eigenmode is very close to 4/rev. Therefore the voltage was reduced to avoid any catastrophic resonance effects.

For the determination of the torsional eigenfrequencies only the upper and lower MFC actuators next to the blade root were excited with a low voltage sinusoidal sweep ranging from 0 to 400 Hz. Here the signals were acquired from strain gauges and acceleration sensors.

The reduced twist performance at 5 and 6/rev is mainly caused by the influence of second mode shape. In this case there is no steady increase of the twist deflection along the span width, and the actuators on both sides of the vibration node are operating counterproductive. To compensate for this problem an individual control of the MFC actuators along the span width was realized and investigated. In Fig. 32.11 the results of a measurement at 1,043 RPM, at an excitation frequency of 6/rev using a different number of activated actuator segments are plotted. The maximum twist deflection is achieved when only actuator segment three to six are activated. This indicates that there is a vibration node between segment three and two what is confirmed by strain gauge measurements. Ongoing activities are dealing with the development of optimized control laws to improve the twist deflection of the blade above the first torsional eigenfrequency.

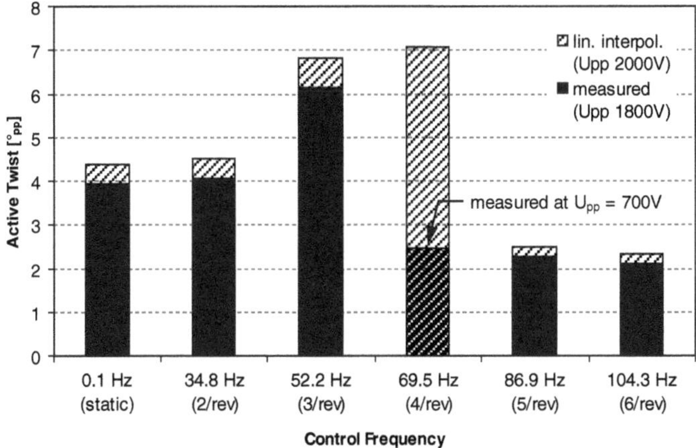

Fig. 32.9 Dynamic excitation at different control frequencies at 1,043 RPM

Fig. 32.10 Rotor diagram
for the 1st and 2nd torsional
eigenmode

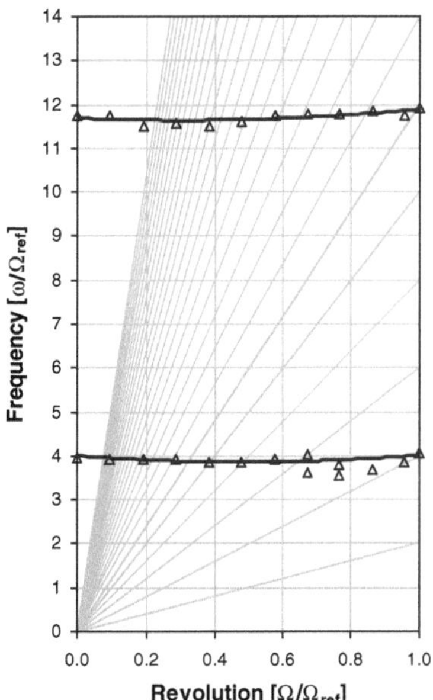

32.6 Rotor Simulation

In addition to the structural design and testing of the active twist blade also an
aeromechanical rotor simulation was run to evaluate the potential of the blade
concerning noise and vibration reduction in forward flight. For an analysis of the
rotor behavior regarding to noise, a descent flight was chosen as a typical flight

Fig. 32.11 Influence of a segmented excitation of the blade at 6/rev (104.3 Hz) and 1,043 RPM. **a** Noise quality criterion, $\psi_{\mathrm{a,n}} = 1.0°$. **b** Vibration quality criterion, $\psi_{\mathrm{a,n}}=0.5°$

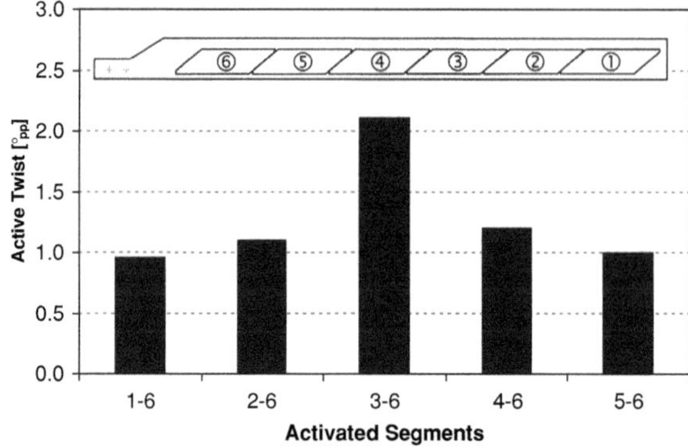

Table 32.2 Definition of flight cases for the analysis of the noise and vibration rotor characteristics

Parameter	Descent flight	High speed forward flight
μ	0.151	0.319
α_{sh}/deg	4.4	-8.8
C_{T}/σ	0.056	0.067
M_{tip}	0.64	0.63

Table 32.3 Rotor data of the AT2

Parameter	Definition
Airfoil	NACA23012
c	0.121 m
N_{b}	4
R	2 m
Scale factor	2.5
θ_{tw}	$-8°$
σ	0.077
Ω	109.27 rad/s

condition where very intensive BVI noise is radiated from the rotor. As described above, intensive vibrations are dominating in high speed forward flight. Thus, a high speed forward flight was chosen to examine the rotor characteristics concerning vibration. Table 32.2 summarizes the definitions for the noise and vibration flight case.

Rotor simulations were performed employing the aeromechanical simulation code S4. Within S4, aerodynamics are considered as unsteady. For the computation of the induced velocities, a prescribed wake model is included [10]. The wake model is taking into account additional deformation of the rotor wake due to higher harmonic lift components [11] where double vortex systems, which arise when the blade tip generates download, are implemented. Blade dynamics are represented by the Eigenmodes of the blade in flapping, lagging and torsion.

The model rotor is based on a BO105 rotor blade with the rotor specifications as given in Table 32.3. To be as close to reality as possible, the rotor data defined from the structural design process have been replaced by experimental data where available. Thus, the model rotor is specified from a semi-empirical data set. The active twist area extends from a radial position of 0.275 upto 0.96 R. The maximum control amplitude at the blade tip is used as input within the simulation program. As a simplification, the control amplitude is prescribed to rise linear from the initial radius up to the outer radial position of the active blade area independently of the control frequency.

32.7 Noise and Vibration Benefits

To obtain a general idea of the rotor behavior towards different active twist control inputs, frequency sweeps as well as phase sweeps were performed. Based on a cosine signal, control frequencies of 2–5 integer multiples of the rotor rotational frequency (1/rev) were applied, varying the control phase by increments of $30°$ from $\Psi_{a,n} = 0$–$360°$ (Fig. 32.12).

For the investigation of the noise characteristics, a control amplitude of $\theta_{a,n} = 1°$ was applied. From an analysis of the high-frequency components of the lift coefficients, a noise quality criterion has been calculated by S4. QC_n is computed as the sum of the $\partial C_n M^2 / \partial \Psi$ gradients in rotor areas where parallel BVI occurs since parallel BVI extends over a large radial area of the blade and thus originates intensive BVI noise. Those BVI areas are located between $\Psi = 20$–$100°$ at the advancing side and $\Psi = 270$–$330°$ at the retreating side. The noise criterion is weighted azimuthally taking into account the degree of BVI parallelism. Additionally, QC_n is weighted radially to consider the influence of the increasing Mach number going up to the blade tip. To obtain an overall criterion, the noise criterions for advancing and retreating side are accumulated.

Figure 32.12a illustrates the noise quality criterion QC_n referred to the quality criterion of the baseline case without active control $QC_{n,0}$. A 3 and 4/rev control frequency seem to be most efficient regarding to noise reductions. Minimum quality criteria are identified for control phases of $\psi_{a,3} = 150°/180°$ respectively $\psi_{a,4} = 240°$. It is noteworthy, that for all of the applied control frequencies, control phases with reduced QC_n compared to the baseline case could be detected.

For a study of the vibration behaviour of the rotor, a rather low control amplitude of $\theta_{a,n} = 0.5°$ has been introduced for taking into account the high sensitivity of rotor vibration towards active control. Rotor vibration is evaluated via the quality criterion QC_v which is computed by S4 (analogue to the calculation of a quality criterion for noise). The criterion is determined from the 4/rev force and moment components in the non-rotating frame since these components represent the source for vibration excitation of the helicopter airframe (1).

$$QC_v = \sqrt{\frac{F^2_{x,4/rev} + F^2_{y,4/rev} + F^2_{z,4/rev}}{N} + \frac{M^2_{x,4/rev} + M^2_{y,4/rev}}{Nm}}$$

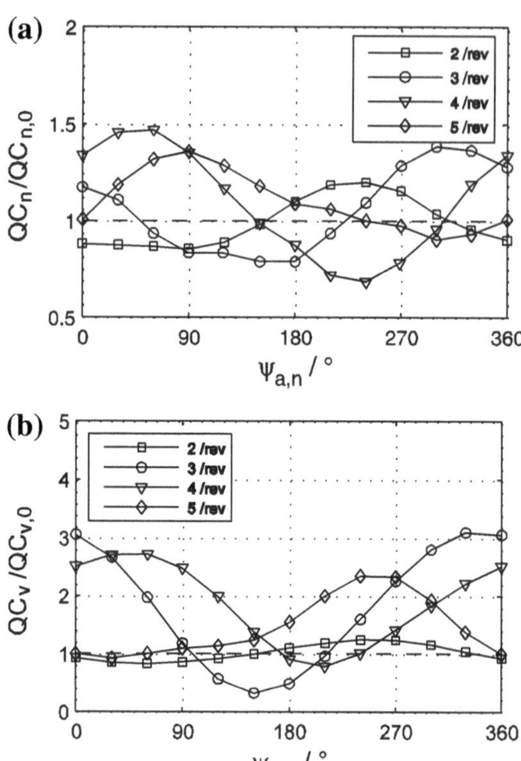

Fig. 32.12 Phase sweeps from 0 to 360° for 2–5/rev control frequencies. **a** Noise quality criterion, $\theta_{a,n} = 1.0$. **b** Vibration quality criterion, $\theta_{a,n} = 0.5°$

In high speed forward flight, control frequencies of 2–4/rev affect the vibration criteria such that for several control phases, the criteria can be reduced below the baseline value (Fig. 32.12b). The minimum noise criterion can be found for a 3/rev control with a control phase of $\psi_{a,3} = 150°$. Further improvements in noise and vibration reduction are expected from an operation of optimized control laws.

Based on Fig. 32.13, the physical background of the impact of active control on rotor noise can be discussed. The illustration shows the high-pass filtered lift distribution over the rotor disk without and with an active twist control ($\theta_{a,3} = 3.0°$). Since BVI are events of high-frequency and alter the pressure and lift distribution on the blade, strong fluctuations of the normal force coefficient $C_n M^2$ indicate the appearance of BVI.

As explained before, in descent flight noise mainly is emerged from those rotor areas where parallel BVI takes place. For the baseline case without active control, large areas of blade parallel BVI can be identified in Fig. 32.13a for both, advancing and retreating side. By introducing an active twist control, the number and extension of parallel BVI areas is decreased at both sides of the rotor disk (Fig. 32.13b). It can be seen that for the advancing side, BVI areas are moved more inboard. Because of the lower Mach number at inner radial positions, the generated BVI is less intensive.

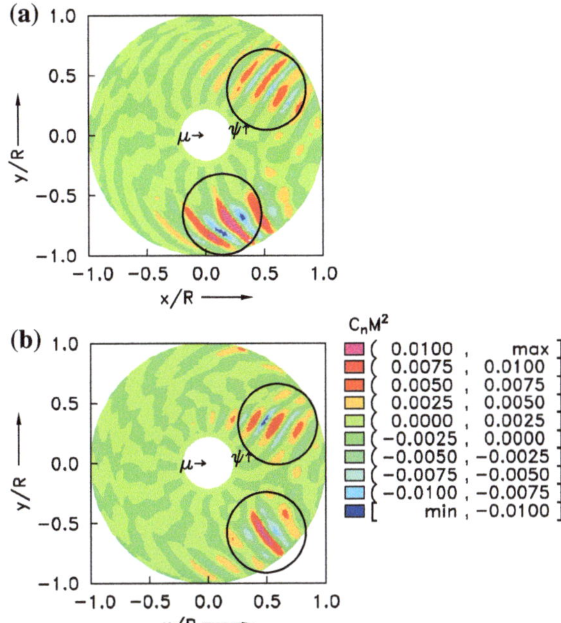

Fig. 32.13 12/rev high-pass-filtered CnM2; circles indicate locations of BVI. **a** BL. **b** 3/rev control, $\psi_{a,3} = 180°$, $\theta_{a,3} = 3.0°$

A more detailed analysis of the impact of different control amplitudes, phases and frequencies on the vibration characteristics of the rotor is given in Fig. 32.14. Here, the cosine and sine response of the 4/rev force and moment components in the non-rotating system from which the vibration quality criterion QC_v is derived, are plotted. The responses are shown for a 3/rev control frequency that provided best results in vibration reduction for the phase sweeps shown in Fig. 32.12b. Control amplitudes of $\theta_{a,3} = 0.5$–$2.0°$ are considered for this analysis. The baseline 4/rev forces and moments are illustrated by filled circles and marked as 'BL'. Filled symbols define the $\psi_{a,3} = 0°$ control phase with the direction of increasing phase denoted by an arrow. For all of the 4/rev force and moment components, beneficial control authority is gained. Already low amplitudes between approximately $\theta_{a,3} = 0.5$–$1.0°$ offer the opportunity to eliminate the 4/rev components. It is noteworthy, that for the lateral 4/rev force F_y hardly a reduction of the component is required. The sine as well as the cosine response of F_y are located near to the point of origin (Fig. 32.14b) which means that the force component already is close to zero for the baseline case. Optimum control phases of approximately $\psi_{a,3} = 120°$ are in good agreement between the longitudinal 4/rev force F_x and the roll moment M_y. In contrast, for the vertical 4/rev force F_z as well as the pitch moment M_x best results are obtained employing a control phase of approximately $\psi_{a,3} = 180°$. Additionally, the influence of further increased control amplitudes up to $\theta_{a,3} = 2.0°$ is demonstrated. It can be seen that improper control amplitudes generate more vibration than the baseline case itself.

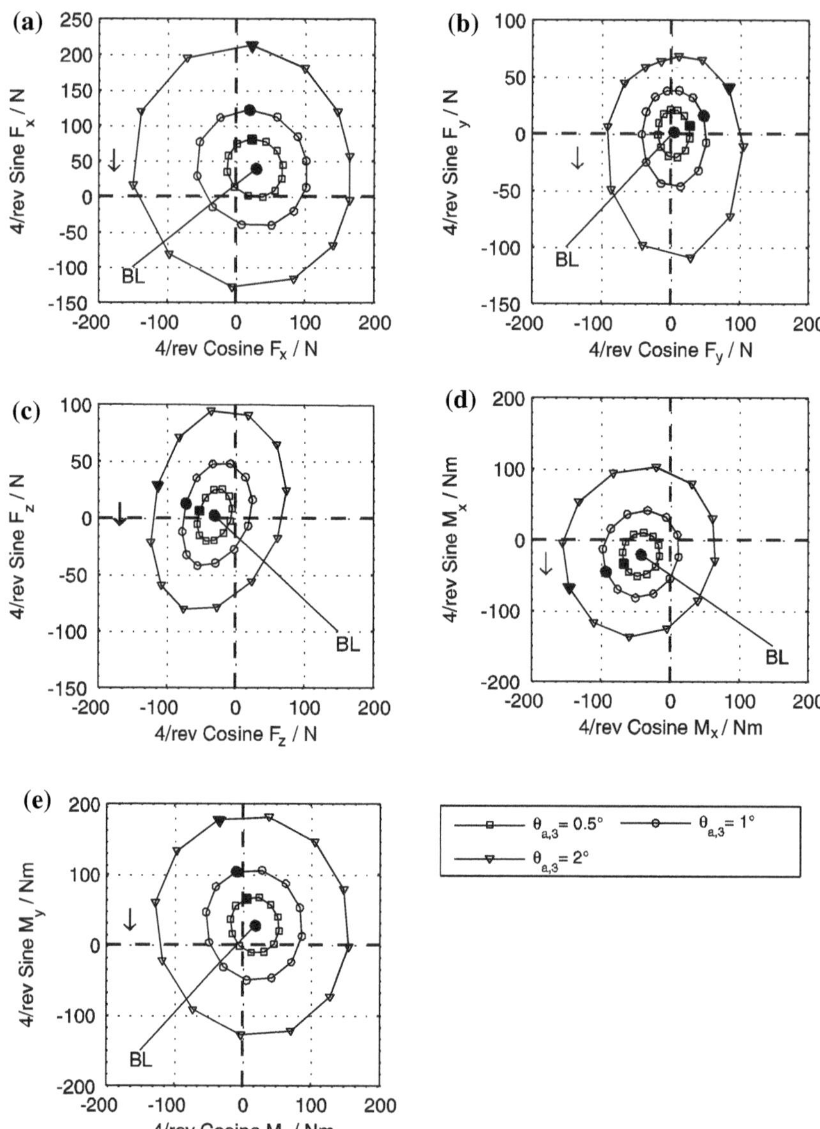

Fig. 32.14 Force and moment response in the non-rotating frame for a 3/rev control input. **a** Longitudinal force. **b** Lateral force. **c** Vertical force. **d** Roll moment. **e** Pitch moment

References

1. Chopra, I.: Status of application of smart structures technology to rotorcraft systems. J. Am. Helicopter. Soc **45**(4), 228–252 (2000)
2. Derham, R., Weems, D.B., Mathew, M.B., Bussom, R.C.: The Design Evolution of an Active Materials Rotor. American Helicopter Society, Washington (2001)
3. SangJoon, S., Carlos, E.C., Steven, R.H.: Closed-loop Control test of the NASA/ARMY/MIT active twist rotor for vibration reduction. In: 59th AHS Annual Forum 2003, Phoenix, Arizona, USA, 6–8. May 2003
4. Wilbur, M.L., Yeager, W.T., Wilkie, W.K., Cesnik, C.E.S., Shin, S.: Hover Testing of the NASA/ARMY/MIT Active Twist Rotor Prototype Blade. American Helicopter Society, Washington (2000)
5. Wilbur, M.L., Mirick, P.H., Yeager, W.T., Langston, C.L., Cesnik, C.E.S., Shin, S.: Vibratory Loads Reduction Testing of the NASA/ARMY/MIT Active Twist Rotor. American Helicopter Society, Washington (2001)
6. Douglas, B.W., David M.A., Mathew B.M., Richard C.B.: A large-scale active-twist rotor. In: 60th AHS Annual Forum, Baltimore, MD, USA, 7–10 June 2004
7. Riemenschneider, J., Keye, S., Wierach, P., Mercier des Rochettes, H.: Overview of the common DLR/ONERA project active twist blade. In: 30th European Rotorcraft Forum, Marseille, (2004)
8. Bent, A.A., Hagood, N.W., Rodgers, J.P.: Anisotropic Actuation with Piezoelectric Fiber Composites. J. Intell. Mater. Syst. Struct **6**, 338–349 (1995)
9. [9] Wilkie, W., High, J., Mirick, P., Fos, R., Little, B., Bryant, R., Hellbaum, R., Jalink, A.: Low cost piezocomposite actuator for structural control applications. In: Proceedings SPIE's 7th International Symposium on Structures and Materials, Newport Beach, California, 5–9 March 2000
10. Beddoes, T.S.: A wake model for high resolution airloads. In: Proceedings of the International Conference on Helicopter Basic Research, Army Research Office, Research Triangle Park, NC, USA, 1985
11. van der Wall, B.G.: Simulation of HHC on helicopter rotor BVI noise emission using a prescribed wake method. In: 26th European Rotorcraft Forum, The Hague, Netherlands, 2000

Chapter 33
Noise and Vibration Reduction with Hybrid Electronic Networks and Piezoelectric Transducers

Martin Pohl and Martin Wiedemann

Abstract Vibrations are problems encountered at almost every technical application when there are moving parts or fluids included. Upon the need for lightweight structures, especially in aerospace applications or electric mobility, conventional damping concepts are insufficient because of their extra-weight and low performance at low frequencies. Measures for active noise and vibration reduction have the potential to solve the drawbacks of passive systems. However, a limitation of these concepts arises from the need for complex system models for deriving the controllers. This leads to another possibility to reduce vibrations consisting of piezoelectric transducers attached to the structure and connected to hybrid electric networks. Within this paper, the basic principles of shunt damping, two hybrid electric networks circuits for shunted damping and the experimental validation of a damped system will be shown. Finally—as a challenging example—a circular saw blade is equipped with piezoelectric transducers and negative capacitance hybrid electric networks to reduce the vibration and noise amplitude excited by the cutting process.

33.1 Piezoelectric Shunt Damping

"Piezoelectric shunt damping" subsumes all concepts, where an oscillating structure is damped with attached piezoelectric actuators connected to hybrid electric networks. The basic principle can be seen in Fig. 33.1. Hybrid electronic

M. Pohl (✉) · M. Wiedemann
Institute of Composite Structures and Adaptive Systems, German Aerospace Center DLR, Lilienthalplatz 7, 38108, Braunschweig, Germany
e-mail: martin.pohl@dlr.de

M. Wiedemann
e-mail: martin.wiedemann@dlr.de

M. Wiedemann and M. Sinapius (eds.), *Adaptive, Tolerant and Efficient Composite Structures*, Research Topics in Aerospace, DOI: 10.1007/978-3-642-29190-6_33,
© Springer-Verlag Berlin Heidelberg 2013

Fig. 33.1 Principle of piezoelectric shunt damping

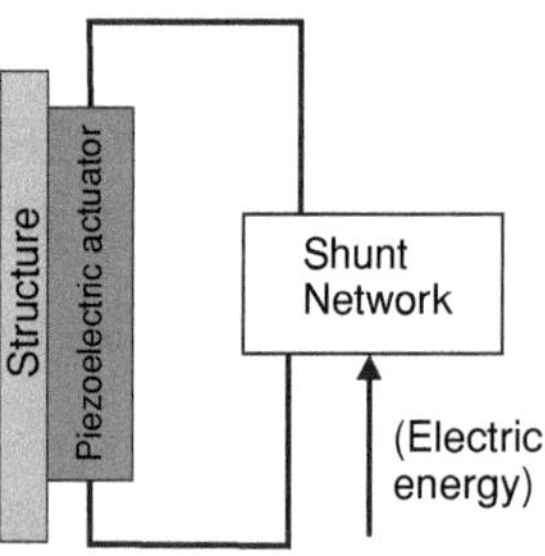

networks may represent any circuit consisting of different active and passive component. In most cases, this reduces to impedances consisting of capacitive, inductive and resistive parts. Then these networks are called shunt networks. According to the use of active and passive components, shunt networks can be divided into "passive" networks when there is no need for external electric energy or into "active" ones if external electric energy is required. In contradiction to active vibration control, no separate sensors, controllers or actuator amplifiers are needed. The electrical network incorporates all of these components and is called "semi-active" for this reason.

Shunt damping combines an oscillating structure with electric networks and piezoelectric actuators. The effect of damping is realized by an impedance matching between the structural and the electrical dynamics with the piezoelectric transducer as coupling element.

In the easiest way, this can be a simple resistor dissipating the electric charge created on the actuator by vibration energy. Because of the low effect this is an impractical solution. If an inductor is added to the resistor [1], as shown in Fig. 33.2 on the left, it is possible to tune this circuit to one structural eigenfrequency. Then, also an electrical resonance occurs which enlarges the energy transportation via the piezoelectric transducer and by this the damping of the vibration amplitude. In this case, there are two coupled resonant systems. This so-called resistive–capacitive–inductive (RCL) absorber can be compared to a mechanical tuned mass–damper–spring vibration absorber. The eigenfrequency and damping of the electrical system have to be fitted to the characteristics of the mechanical system.

This system can only be used for one single eigenfrequency. For damping more than one eigenfrequency, multiple mode RCL networks can be designed with filtered branches for every mode [2–4]. A common disadvantage of all passive RCL networks lies in the fixed parameters, which does not allow an adaptation to changing eigenfrequencies. For rotating structures as for example a saw blade, the centrifugal acceleration alters the eigenfrequencies, so that adaptive damping concepts are prerequisite.

A solution can be found in adaptive RL absorbers with tunable active inductors [5–7]. In this case synthetic inductors are used and tuned by external circuitry.

The effect of the electrical networks is based on an impedance matching between mechanical and electrical system. If the reactance of the actuator's

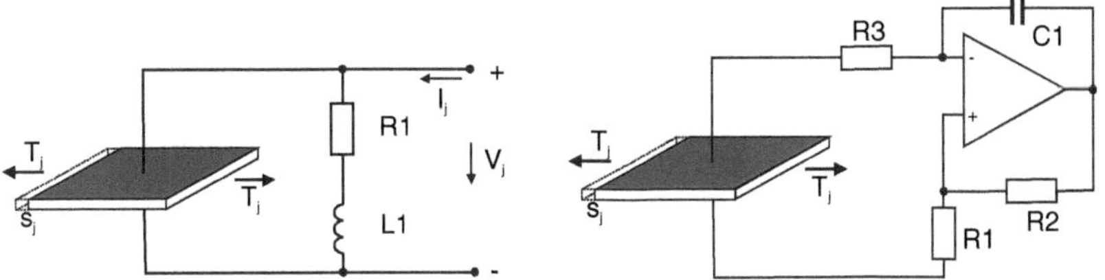

Fig. 33.2 Resistive–inductive and negative capacitance network

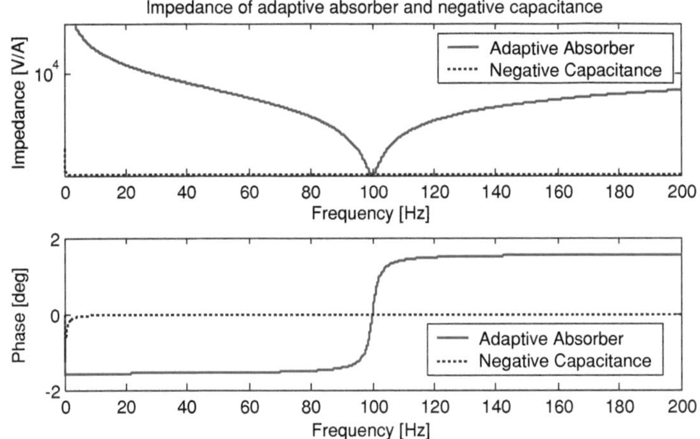

Fig. 33.3 Impedance of adaptive RL absorber and negative capacitance

capacitance is compensated by the reactance of the inductor in the RL absorber, maximum energy dissipation is possible. Due to the frequency dependence of both reactances this is only valid at a single frequency.

If it is possible to compensate the capacitive reactance of the actuator over a wider frequency range, damping would also be possible. In fact, a negative capacitance fulfills this requirement. A circuit with a negative impedance converter can be seen in Fig. 33.2 on the right.

It has the same reactive behavior over the frequency as the actuator's capacitance, but with opposite sign. Therefore, it is able to compensate its reactance. Figure 33.3 shows the impedance curves of a negative capacitance compared to a RL absorber tuned to a frequency of 100 Hz. The impedance of the negative capacitance does not show remarkable changes in amplitude and phase over the shown frequency range. In [8] a clamped beam is equipped with a negative capacitance network for vibration damping. An effect for multiple modes with a single network could be demonstrated. This makes this concept feasible for the use on structures with varying, rotation speed dependent eigenfrequencies like a circular saw blade. Also modern

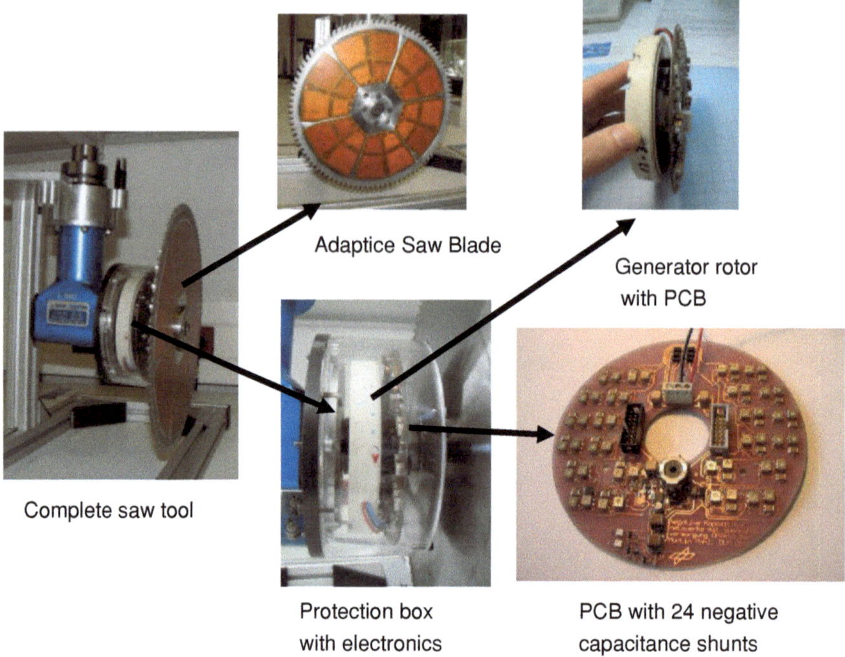

Fig. 33.4 Adaptive circular saw head

lightweight fiber composites with high sensitivity for structural vibration and noise emissions are able to be damped with shunt networks.

33.2 Design of Autonomous Saw Head Tool

Basic intention in the design of the saw tool is the usability without any change at the machine tool. This means that the saw head itself must contain all required components including power supply and electronic hardware. This is an extreme example for function integration, although not yet done within a composite structure. To include all the needed components a modified design of a commercial saw head tool is required. In [9] the basic concept of such a saw head is explained. A generator, derived from a permanently excited synchronous brushless motor, which also carries the circuit board (PCB) with the electric networks, is used for power supply. By this a rotating electrical connection between the rotating system and the saw head can be avoided. The whole set up of the tool is shown in Fig. 33.4.

The test blade consists of a commercially available 300 mm diameter circular saw blade with 96 teeth at its circumference. The actuators are applied on the

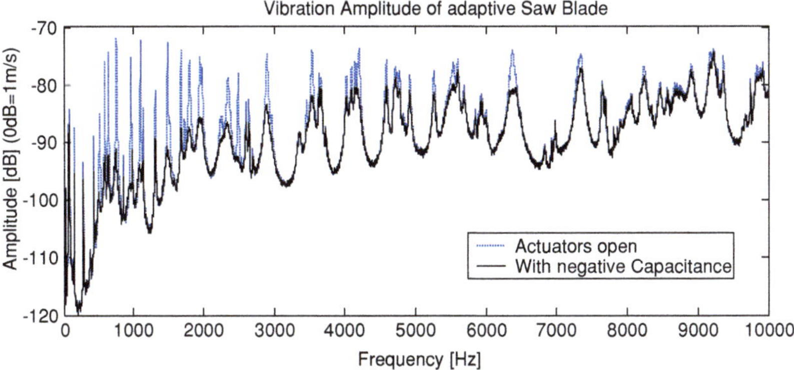

Fig. 33.5 Mean vibration amplitude of adaptive saw blade

surface of the saw blade by attaching them with an epoxy based resin. For an optimum damping efficiency 24 actuators are placed on the saw blade, 12 sectors with two actuators each. Every actuator is connected to a separate semi-active shunt network to avoid charge flow between different actuators due to opposite strain signs which would lower damping performance. With a simulation based on the modal data of the saw blade the diameter of the separation between the two actuator rows was determined [9].

33.3 Experimental Validation

For experimental validation sound and vibration amplitude measurements are used with a rotating and a non-rotating saw blade. Sound measurements are done according to DIN EN ISO 3744 with six microphones. For vibration measurement, a laser vibrometer and an accelerometer are used. In rotation the laser beam is self-tracked to the measurement location to obtain vibration velocity data as described in [6].

33.3.1 Results in the Non-Rotating System

For determining the effectiveness of the negative capacitance shunt network vibration tests were done in the non-rotating system with a laser scanning vibrometer. It could be shown that, depending on the mode shape and eigenfrequency, reductions of up to 20 dB in vibration amplitude can be achieved as shown in the mean vibration spectrum in Fig. 33.5. Damping is effective to more than 5 kHz without the need for tuning of the electrical network.

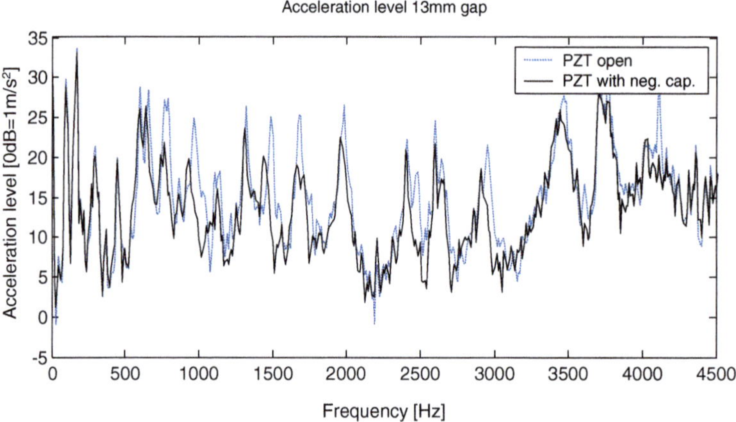

Fig. 33.6 Vibration spectrum of adaptive rotating saw blade

It can be seen that damping ratio decreases with rising frequency. This arises from the fact that higher frequencies have lower wavelengths. By that, node lines and strains with opposite sign within the area of an actuator occur. Charge equalization within the actuator lowers the output signal and damping performance. More detailed explanations can be found in [9] and [10].

33.3.2 Results in the Rotating System

Upon the good vibration attenuation in the non rotating system the adaptive saw blade was also investigated in rotation. Tests were performed while free-running as well as while cutting gaps into medium dense fiberboard (MDF) panels.

In rotation the amplitude reduction is lower compared to the results in the non rotating system. In Fig. 33.6 an exemplary acceleration spectrum of the adaptive saw blade is shown. In this case, the excitation arises from cutting a 13 mm deep gap into a MDF panel. Amplitude reductions of up to 5 dB in the eigenfrequencies could be demonstrated. The overall vibration amplitude reduction within a frequency band of 0–4.5 kHz was determined to about 1.6 dB.

In the rotating system two differences are present compared to the non-rotating blade. First, a dynamic stiffening of the blade occurs, which raises the eigenfrequency of the vibration modes. Due to the low frequency dependence of the damping effect of negative capacitance networks this can be ignored. Second, an additional excitation influences the blade at the relatively low rotation frequency leading to higher harmonic signals. Due to the higher voltage requirements of a negative capacitance shunt in low frequency vibration the electric devices are vulnerable to saturation of the output. This reduces the tolerable adjustment of the hybrid network and by this the effectiveness.

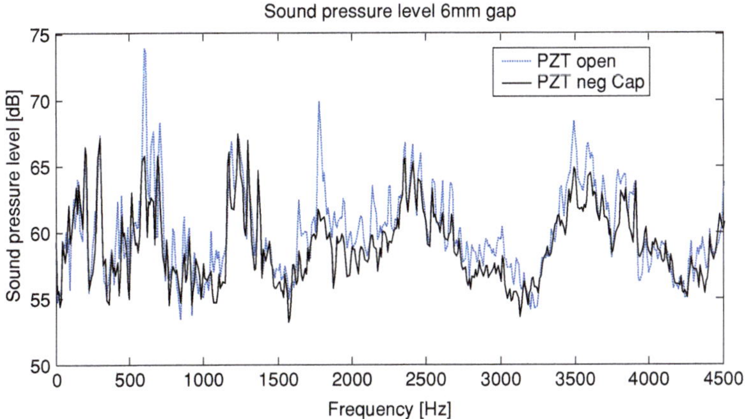

Fig. 33.7 Adaptive saw blade with and without negative capacitance shunt

Results of the sound measurement are comparable to the results of vibration measurement. Maximum amplitude reduction of sound pressure takes place in the eigenfrequencies with a maximum value of approx. 8 dB. The reduction of sound pressure in 1 m distance from the saw blade was measured to about 1.19 dB within a frequency range of 0–4.5 kHz. An exemplary sound pressure spectrum is shown in Fig. 33.7.

A reason for the low broadband effect lies in the measurement of multiple sound sources. When the saw blade is interacting with the workpiece, sound is emitted not only by the blade itself, but also by the driver, gears, by the contact of the teeth with the workpiece and in the end by the workpiece itself. From these multiple sources only the noise from the blade itself can be addressed by the piezoelectric actuators. This takes place at the eigenfrequencies of the saw blade and can clearly be seen in the spectrum of Fig. 33.7.

33.4 Conclusion

A vibration damping system based on piezoelectric actuators and negative capacitance networks has been developed and tested on a circular saw blade. Experimental investigations showed a remarkable reduction of vibration amplitude in the non rotating system of up to 20 dB. In the rotating system noise reduction with up to 8 dB in sound pressure and vibration amplitude reduction in the eigenfrequencies up to 5 dB could be demonstrated. Thus, the approach of hybrid electronic network shows to be promising for noise and vibration reduction in various applications.

Acknowledgments This project was funded by German Research Foundation (DFG) within the Priority Program SPP1156 "Adaptronics for Machine Tools".

References

1. Hagood, N.W., von Flotow, A.H.: Damping of structural vibrations with piezoelectric materials and passive electrical networks. J. Sound Vib. **146**(2), 243–268 (1991)
2. Hollkamp, J.J.: Multimodal passive vibration suppression with piezoelectric materials and resonant shunts. J. Intell. Mater. Syst. Struct. **5**, 49 (1994)
3. Wu, S.Y.: Method for multiple mode piezoelectric shunting with single PZT transducer for vibration control. J. Intell. Mater. Syst. Struct. **9**, 991 (1998)
4. Behrens, S., Moheimani, S.O.R., Fleming, A.J.: Multiple mode current flowing passive piezoelectric shunt controller. J. Sound Vib. **266**, 929–942 (2003)
5. Hollkamp, J.J., Starchville, T.F.: A self-tuning piezoelectric vibration absorber. J. Intell. Mater. Syst. Struct. **5**, 559 (1994)
6. Pohl, M.: Noise and Vibration Attenuation of a Circular Saw Blade with Applied Piezoceramic Patches and Negative Capacitance Shunt Networks, pp. 411–424. ISMA Leuven, Belgium (2010)
7. Torrez-Torres, J., Zornemann, M., Rose, M.: Damping of structural vibrations on a circular saw-blade dummy using an adaptive hybrid electromechanical network with piezoelectric devices, Adaptronic Congress, Göttingen, 2007
8. Park, C.H., Park, H.C.: Multiple-mode structural vibration control using negative capacitance shunt damping. KSME Int. J. **17**(11), 1650–1658 (2003)
9. Pohl, M., Rose, M.: Vibration and noise reduction of a circular saw blade with applied piezoceramic patches and semi-active shunt networks, pp. 85–91. Proceedings Adaptronic Congress, Berlin (2009)
10. Pohl, M., Rose, M., Breitbach, E.: Lärm- und Schwingungsreduktion eines Kreissägeblattes mit flächigen Piezokeramiken und hybriden elektrischen Netzwerken, Shaker (2010)

Chapter 34
Reduction of Turbulent Boundary Layer Noise with Actively Controlled Carbon Fiber Reinforced Plastic Panels

Stephan Algermissen, Malte Misol and Oliver Unruh

Abstract The turbulent boundary layer (TBL) is one of the dominant external noise sources in high subsonic aircrafts. Especially in modern aircrafts where common materials for fuselages are currently substituted by carbon-fiber-reinforced-plastics (CFRP), it is essential to avoid a decrease of passenger comfort as a result of an inferior transmission loss of the new materials. To increase the transmission loss of CFRP panels they are equipped with active noise reduction systems. In this paper the results of an experimental study in the aeroacoustic wind tunnel of the German Aerospace Center (DLR) are presented. An active panel excited by a TBL is tested at flow speeds up to Mach 0.16. The CFRP panel ($500 \times 800 \times 2.7$ mm^3) is equipped with five piezo-ceramic patch actuators and ten accelerometers. Active structural acoustic control (ASAC) and active vibration control (AVC) are used to reduce the broadband TBL noise transmission in the bandwidth from 1 to 500 Hz. Feedforward (FF) and feedback (FB) control algorithms are applied in the experiments and show high performance even in presence of plant uncertainties. To improve control results the generalized plant framework of robust control is utilized for global feedback control. Finally, an overview of the achieved results is given.

S. Algermissen (✉) · M. Misol · O. Unruh
Institute of Composite Structures and Adaptive Systems, German Aerospace Center DLR,
Lilienthalplatz 7, 38108, Braunschweig, Germany
e-mail: stephan.algermissen@dlr.de

M. Misol
e-mail: malte.misol@dlr.de

O. Unruh
e-mail: malte.misol@dlr.de

M. Wiedemann and M. Sinapius (eds.), *Adaptive, Tolerant and Efficient Composite Structures*, Research Topics in Aerospace, DOI: 10.1007/978-3-642-29190-6_34,
© Springer-Verlag Berlin Heidelberg 2013

34.1 State of Technology

The reduction of turbulent boundary layer (TBL) induced noise in modern aircrafts has been studied for many years. In contrast to propeller noise, TBL noise is typically broadband [1]. Moreover, pressure fluctuations in the turbulent boundary layer have small correlation lengths [2]. Common feed-forward techniques cannot be applied, due to lack of reference signals with adequate correlation. Therefore, mostly feedback controllers and their application were focused in the past.

Only few studies of the reduction of TBL induced noise led to practical implementations and experiments. Gibbs and Cabell [3] studied active control techniques on a TBL excited aircraft sidewall consisting of multiple aluminum panels. Local and global control concepts were tested in wind tunnel experiments with wind speeds of Mach 0.125. A maximum narrow-band reduction of the radiated sound power of -7.5 dB could be achieved. Nevertheless, most results of TBL induced noise reduction base on simulations of the entire system. This includes the work of Schiller and Fuller [4], Rohlfing and Gardonio [5], Thomas and Nelson [6] and Maury et al. [7].

In this paper experiments with a CFRP panel are presented. By means of feed-forward (FF) and feedback (FB) control strategies the TBL induced radiated sound is reduced. Experiments were carried out in the aeroacoustic wind tunnel of DLR in Braunschweig with a maximum wind speed of Mach 0.16. Measurements with a laser-scanning vibrometer (LSV) prove the performance of the different concepts.

34.2 Active Panel

In order to derive a lightweight-compliant solution for improved low-frequency transmission loss of modern fuselage structures, a passive panel is equipped with appropriate sensors, actuators and control. The CFRP panel is quasi-isotropic, has a clearance of 500×800 mm^2 and thickness of 2.7 mm. Figure 34.1 shows the panel during the fabrication process. An aluminum frame for fast mounting is bonded to the panel. When mounted in the test section, the actuator side of the panel will radiate sound into the anechoic chamber of the wind tunnel.

For control actuating and sensing the panel is equipped with five DuraAct® P-876.A15 piezo patch actuators (Dim: $61 \times 35 \times 0.8$ mm^3) and ten PCB® 352A24 accelerometers. The actuators were bonded in vacuum to ensure excellent strain transmission. Actuator's maximum voltage is 1000 V.

34.3 Actuator Placement

Optimized positions of the actuators are derived based on an estimation of radiated sound power in open- (deactivated) and closed-loop (activated optimal FF control). The optimization routines are implemented in a so-called ASAC pre-design tool

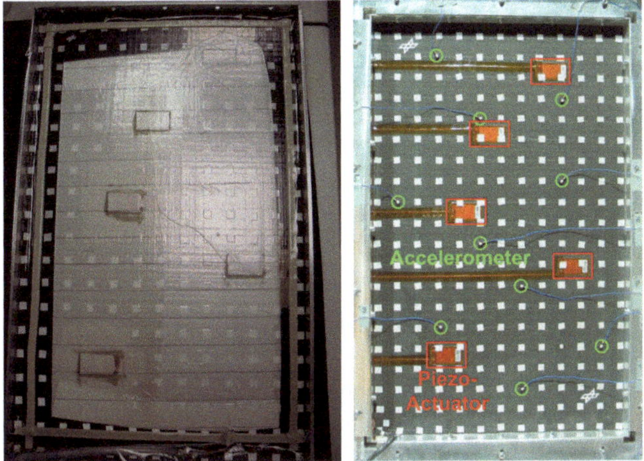

Fig. 34.1 Application of piezo-actuators in vacuum (*left*) and smart CFRP panel equipped with actuators and sensors in optimized positions (*right*)

[8] which simulates the complete sound-transmission process. It makes use of a synthetic turbulence model [9, 10], a modal model of the smart structure and a model of the vibro-acoustic coupling based on spatially discretized sound radiation resistance and normal surface velocities. The sensor positions are chosen for good observability of all eigenfrequencies in the control bandwidth.

34.4 System Identification

An accurate description of the controlled plant, from actuators to sensors, must be available for a good control performance. Models from finite element calculations lack precise description of vibration behavior due to inaccurate parameters. Especially phase response is characterized poorly. Therefore, models for control synthesis are generated by experimental system identification using experimental data. Here, multi-reference excitation is applied for system identification: The five piezo patch actuators excite the panel structure with uncorrelated random noise simultaneously and accelerometers record the structural responses. Measurement takes 16.4 s only. From these 15 time signals a subspace based identification algorithm [11] generates a discrete state-space model of the controlled plant. A maximum number of 90 states ensure a good accuracy of the model. The bandwidth of the system identification and control is set to 500 Hz.

For global vibration control of the panel, it is necessary to observe normal velocities at certain points at the panel's surface, to include these point velocities as additional outputs into the model of the controlled plant and to generate a so-

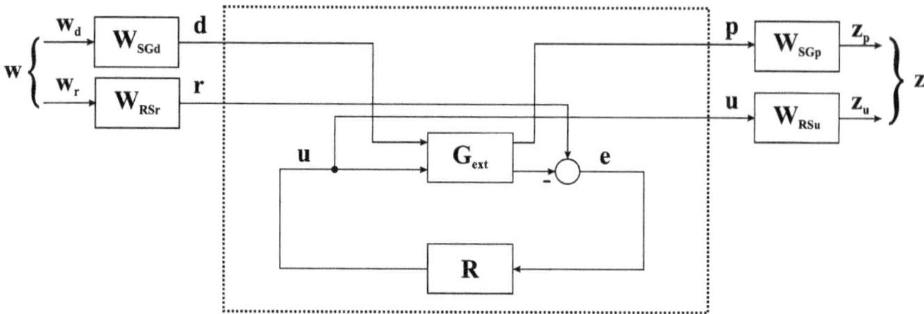

Fig. 34.2 Weighting scheme for FB-controller synthesis

called extended plant. The point coordinates are defined by a measuring grid of
13×20 points with a spacing of 40 mm. The grid is projected onto the panel and
a high reflective dot is placed at each point location. For system identification, a
laser scanning vibrometer measures the point velocities while the panel is excited
by a single actuator. Five measuring cycles are necessary to determine the FRF-
matrix $\mathbf{H}(j\omega_k)$ from all actuators to all scan points. System identification with fixed
$\mathbf{A}$ and $\mathbf{B}$ matrices of the controlled plant generates new $\mathbf{C}_v$ and $\mathbf{D}_v$ matrices for the
extended plant $\mathbf{G}_{ext}$:

$$\mathbf{x}(k+1) = \mathbf{A}\,\mathbf{x}(k) + \mathbf{B}\,\mathbf{u}(k)$$
$$\begin{bmatrix} \mathbf{y}_v(k) \\ \mathbf{y}(k) \end{bmatrix} = \begin{bmatrix} \mathbf{C}_v \\ \mathbf{C} \end{bmatrix} \mathbf{x}(k) + \begin{bmatrix} \mathbf{D}_v \\ \mathbf{D} \end{bmatrix} \mathbf{u}(k) \tag{34.1}$$

where the input- and state vectors are $\mathbf{u}$ and $\mathbf{x}$, and the velocity- and acceleration
outputs $\mathbf{y}_v$ and $\mathbf{y}$. By means of this so-called extended plant it is possible to
generate an observer for all point velocities.

34.5 Control Concepts

In the experiments, both FB- and FF-control concepts where applied. For FB
control a robust H_∞ controller was chosen whereas the FF control uses a filtered-x
least mean square (FxLMS) algorithm for the adaptation of the finite impulse
response (FIR) filters.

In H_∞ control the objectives have to be formulated in terms of so-called
weighting schemes. For realization of the control concept, the weighting scheme in
Fig. 34.2 was selected. Four weighting functions $\mathbf{W}$ formulate the constraints for
the controller synthesis. Their calculation is automated and depends only on two
parameters concerning disturbance rejection and control energy limitation [12, 13].
A control synthesis algorithm optimizes the controller such that the H_∞ norm of
the transfer function $\mathbf{T}_{zw}$ from disturbance input $\mathbf{w}$ to performance output $\mathbf{z}$ is less
than a specified value γ. The transfer matrix of the entire system in Fig. 34.2 reads:

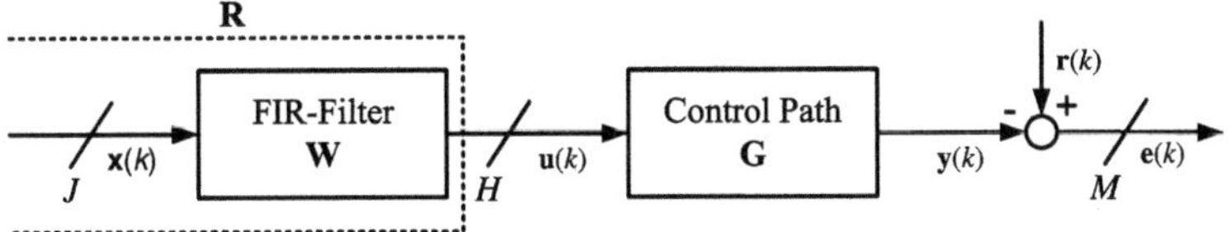

Fig. 34.3 Block diagram of a multiple reference FF-controller

$$\begin{bmatrix} \mathbf{z}_p \\ \mathbf{z}_u \end{bmatrix} = \begin{bmatrix} \mathbf{W}_{\text{SGp}}\mathbf{G}_{12}\mathbf{RSW}_{\text{RSr}} & \mathbf{W}_{\text{SGp}}(\mathbf{G}_{11} - \mathbf{G}_{12}\mathbf{RSG}_{21})\mathbf{W}_{\text{SGd}} \\ \mathbf{W}_{\text{RSu}}\mathbf{RSW}_{\text{RSr}} & -\mathbf{W}_{\text{RSu}}\mathbf{RSG}_{21}\mathbf{W}_{\text{SGd}} \end{bmatrix} \begin{bmatrix} \mathbf{w}_r \\ \mathbf{w}_d \end{bmatrix} \quad (34.2)$$

With plant extended plant $\mathbf{G}_{\text{ext}}$ and sensitivity $\mathbf{S}$:

$$\begin{bmatrix} \mathbf{p} \\ \mathbf{v} \end{bmatrix} = \mathbf{G}_{\text{ext}} \begin{bmatrix} \mathbf{d} \\ \mathbf{u} \end{bmatrix} = \begin{bmatrix} \mathbf{G}_{11} & \mathbf{G}_{12} \\ \mathbf{G}_{21} & \mathbf{G}_{22} \end{bmatrix} \begin{bmatrix} \mathbf{d} \\ \mathbf{u} \end{bmatrix} \quad \text{and} \quad \mathbf{S} = [\mathbf{E} + \mathbf{G}_{22}\mathbf{R}]^{-1}$$

The transfer function $(\mathbf{G}_{11} - \mathbf{G}_{12}\mathbf{RSG}_{21})$ corresponds to the disturbance transfer function, whose $\mathbf{H}_\infty$ norm has to be small compared to that of the plant to gain good disturbance rejection.

The FF-control scheme does not permit the formulation of an independent performance output. Furthermore the FF-control requires the real-time acquisition of reference signals which are highly correlated with the disturbance sources.

Figure 34.3 shows the block diagram of a FF-controller with J reference, H control and M sensor signals.

The time-domain formulation of an optimal and causal FF-controller with FIR-filters of length L is given in Eqs. (34.3)–(34.5). The theoretical background and further details can be found in [14, 15]. The performance metric $\xi(k)$ for the optimization of the filter weights is chosen as the expectation $E[\bullet]$ of the dot product of the error signal vector $\mathbf{e}(k)$.

$$\xi(k) = E\left[\mathbf{e}^T(k)\mathbf{e}(k)\right] = E\left[\left(\mathbf{r}(k) - \hat{\mathbf{X}}(k)\mathbf{w}(k)\right)^T \left(\mathbf{r}(k) - \hat{\mathbf{X}}(k)\mathbf{w}(k)\right)\right]$$

$$= E\left[\mathbf{r}^T(k)\mathbf{r}(k)\right] - E\left[\mathbf{r}^T(k)\hat{\mathbf{X}}(k)\mathbf{w}(k)\right] - E\left[\mathbf{w}^T(k)\hat{\mathbf{X}}^T(k)\mathbf{r}(k)\right]$$

$$+ E\left[\mathbf{w}^T(k)\hat{\mathbf{X}}^T(k)\hat{\mathbf{X}}(k)\mathbf{w}(k)\right] \quad (34.3)$$

The matrix of filtered reference signals $\hat{\mathbf{X}}(k)$ and the FIR-filter coefficient vector $\mathbf{w}(k)$ are defined as:

$$\hat{\mathbf{X}}(k) = \begin{bmatrix} \hat{\mathbf{x}}_1^T(k) & \hat{\mathbf{x}}_1^T(k-1) & \dots & \hat{\mathbf{x}}_1^T(k-L+1) \\ \vdots & \ddots & & \vdots \\ \hat{\mathbf{x}}_M^T(k) & \hat{\mathbf{x}}_M^T(k-1) & \cdots & \hat{\mathbf{x}}_M^T(k-L+1) \end{bmatrix} \in \mathbb{R}^{(M,HJL)} \quad (34.4)$$

$$\mathbf{w}(k) = \begin{bmatrix} \mathbf{w}_{11}(k) & \mathbf{w}_{12}(k) & \dots & \mathbf{w}_{1J}(k) & \mathbf{w}_{21}(k) & \cdots & \mathbf{w}_{HJ}(k) \end{bmatrix} \in \mathbb{R}^{HJL} \quad (34.5)$$

Because of the low spatial coherence of the flow-induced structural vibration, a multiple reference FF-control scheme with five structural reference sensors was

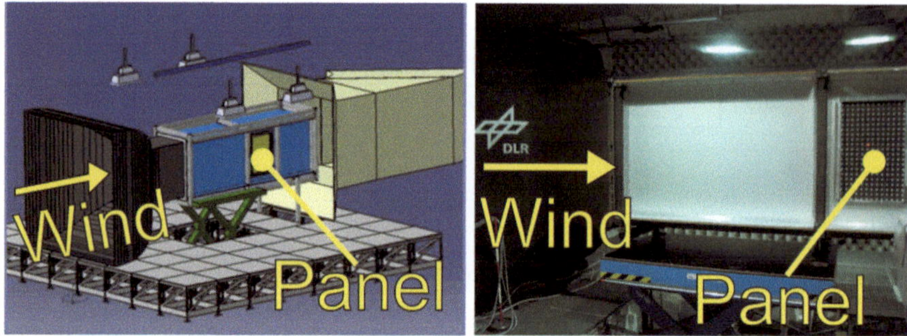

Fig. 34.4 Closed test section in the aeroacoustic wind tunnel: construction design (*left*); realization (*right*)

implemented. The strong actuator feedback to the reference sensors required the integration of a feedback-compensation filter which was implemented using the identified state-space model. Due to the lack of coherent disturbance source information, the use of pressure based reference signals was infeasible. Therefore, no time-advanced disturbance data where available and the causality constraints limited the performance of the FF-controller to the structural resonances.

34.6 Experiments

The aeroacoustic wind tunnel of DLR Braunschweig has an open test section. The nozzle has a cross-section of 1.2×0.8 m^2 and the maximum wind speed is 60 m/s or Mach 0.17. The test section is enclosed by an anechoic chamber to enable acoustic measurements. For realization of TBL experiments, a closed test section has been designed and built. In Fig. 34.4 the construction design and a picture of the realization can be seen. The test section is a rectangular channel that is placed right behind the nozzle. For vibration decoupling, the test section is mounted on shock absorbers. The CFRP panel is placed in a side wall of the test section. The thickness of the turbulent boundary layer is steadily growing in the closed test section to app. 41 mm at Mach 0.17 until it reaches the panel.

The vibro-acoustic experiments where performed in the acoustic wind tunnel with a flow velocity of Mach 0.16. The acquisition of normal surface velocity data of the CFRP panel by means of a LSV provided the database for the evaluation of structural vibration and active sound power in open- and closed-loop.

Active sound power was calculated by filtering the measured normal surface velocity through an analytically derived radiation resistance matrix (radiation filter).

To assure the correctness of this approach, a validation against sound pressure level (SPL)-based methods was performed in the absence of background noise. In this case, the structural excitation was realized using a shaker mounted at the backside of the plate. The deviations in calculated active sound power relative to the sound intensity probe data were less then 3 dB. This was considered as an argument

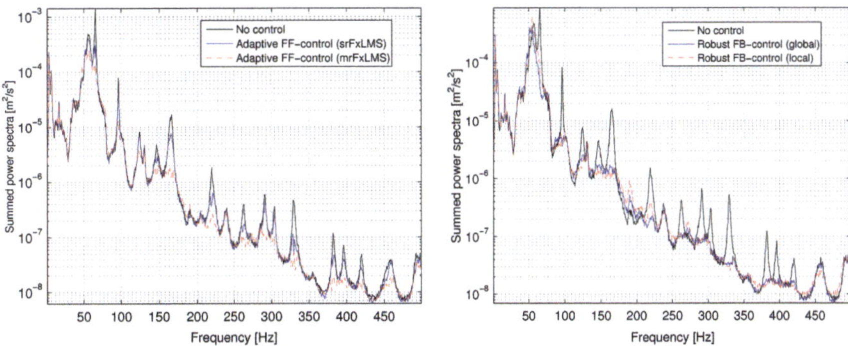

Fig. 34.5 Summed auto spectra of 260 LSV-scan points with single- and multiple-reference adaptive FF-control (*left*) and with local and global robust FB-control (*right*)

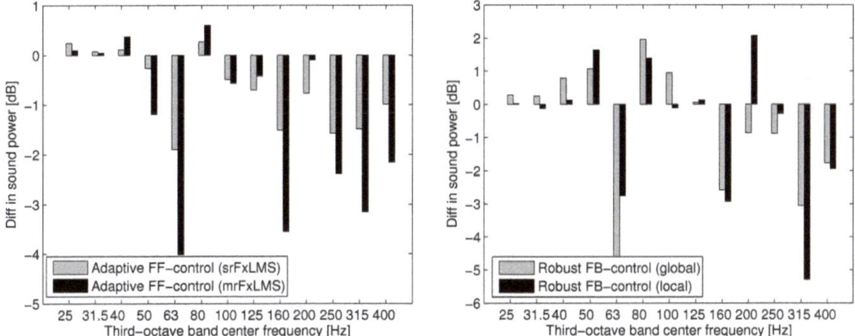

Fig. 34.6 Relative sound power reductions with single- and multiple-reference adaptive FF-control (*left*) and with local and global robust FB-control (*right*)

for the correctness of the implemented radiation filter. Figure 34.5 shows the effect of FF- and FB-control on the structural vibration of the CFRP panel. The left part of Fig. 34.5 depicts the results of adaptive FF-control using a single- (srFxLMS) and a multiple-reference (mrFxLMS) filtered-x least mean squares (FxLMS)—algorithm. The superior control performance of the mrFxLMS is achieved at the expense of a significantly higher computational complexity. According to Eq. (34.5), the implemented FF-controller with five references ($J = 5$), five actuators ($H = 5$) and 100 FIR-filter coefficients ($L = 100$) requires a total number of 2500 filter taps. The right part of Fig. 34.5 shows the results of local (accelerometer grid) and global (70 points of LSV-grid) FB-control. The control performance is comparable to the mrFxLMS with slight degradations at frequencies around the anti-resonances. Design flexibility, robustness and numerical efficiency (only 100–200 states) of the FB-controller are advantageous compared to the FF-controller.

Figure 34.6 shows the influence of active control on sound power radiation of the smart CFRP panel. The mrFxLMS achieved a total sound power reduction of

2.88 dB within the control bandwidth from 1 to 500 Hz. Narrowband sound power level reductions of more than 10 dB were achieved at some controllable structural resonances. The limited control authority around the efficiently radiating fundamental structural mode prevented the achievement of a higher total sound power reduction.

34.7 Achieved Results

A flat and non-stiffened CFRP panel of size $500 \times 800 \times 2.7$ mm^3 was designed, manufactured and equipped with actuators and sensors. The design was accomplished by means of a numerical pre-design tool which simulates the noise transmission through the smart structure in frequency domain. The smart CFRP panel was flush mounted in the closed test section which directed the turbulent inflow across the panel surface. A structure-based description of active sound power was implemented and experimentally validated. Measurement data of turbulent wall pressure fluctuations, structural response and sound power radiation were captured. Due to the influence of active control, the total radiated sound power in the bandwidth 1–500 Hz was reduced by 2.88 dB. The maximum reduction in third-octave band sound power level was above 5 dB. Due to the low predictability of the TBL excitation, the effect of active control was mainly present around the structural resonances. At these eigenfrequencies, reductions of sound power >10 dB where achieved.

References

1. Wilby, J.F.: Aircraft interior noise. J. Sound Vib. **190**(3), 545–564 (1996)
2. Corcos, G.M.: The structure of the turbulent pressure field in boundary-layer flows. J. Fluid Mech. **18**, 353–378 (1964)
3. Gibbs, G.P., Cabell, R.H.: Active control of turbulent boundary layer induced sound radiation from multiple aircraft panels. In: Proceedings of 8th AIAA/CEAS Aeroacoustics Conference, Breckenridge, CO (2002)
4. Schiller, N.H., Fuller, C.R.: A high-authority/low-authority control strategy for coupled aircraft-style bays. In: Proceedings of ACTIVE 2006, Adelaide, Australia (2006)
5. Rohlfing, J., Gardonio, P.: Homogeneous and sandwich active panels under deterministic and stochastic excitation. J. Acoust. Soc. Am. **125**(6), 3696–3706 (2009)
6. Thomas, D.R., Nelson, P.A.: Feedback control of sound radiation from a plate excited by a turbulent boundary layer. J. Acoust. Soc. Am. **98**(5), 2651–2662 (1995)
7. Maury, C., Gardonio, P., Elliott, S.J.: Model for active control of flow-induced noise transmitted through double partitions. AIAA J. **40**, 1113–1121 (2002)
8. Heintze, O., Rose, M., Algermissen, S., Misol, M.: Development and experimental application of a pre-design tool for active noise and vibration reduction systems. In: Proceedings of ACTIVE 2009, Ottawa, Canada (2009)

9. Ewert, R.: Rpm—the fast random particle-mesh method to realize unsteady turbulent sound sources and velocity fields for caa applications. In: Proceedings of 13th AIAA/CEAS Aeroacoustics Conference. http://elib.dlr.de/52932/ (2007)
10. Siefert, M., Ewert, R. (2009), Anisotropic synthetic turbulence with sweeping generated by random particle-mesh method. In: Peinke, J., Oberlack, M., Talamelli, A. (eds.) Progress in Turbulence III: Proceedings of the iTi Conference in Turbulence, p. 143
11. Katayama, T.: Subspace Methods for System Identification. Springer, London (2005)
12. Algermissen, S., Rose, M., Keimer, R., Monner, H.P., Breitbach, E.: Automated synthesis of robust controllers for smart-structure applications in parallel robots. In: Proceedings of AIAA/ASME/AHS Adaptive Structures Conference, Honolulu, USA (2007)
13. Algermissen, S., Rose, M., Keimer, R., Sinapius, M.: Robust gain-scheduling for smart-structures in parallel robots. In: 16th Annual International Symposium on Smart Structures and Materials, SPIE, San Diego (2009)
14. Elliott, S.: Signal Processing for Active Control. Academic Press, London (2001)
15. Kuo, S.M., Morgan, D.R.: Active Noise Control Systems. John Wiley & Sons, New York (1996)

Chapter 35
Active Structure Acoustic Control for a Truck Oil Pan

Olaf Heintze and Michael Rose

Abstract The oil pan of large diesel engine trucks is a significant contributor of external noise radiation. Especially at lower frequencies below 500 Hz, this undesired broadband noise cannot be treated effectively by passive measures due to weight and size restrictions. Augmenting such systems with an Active Structural Acoustic Control (ASAC) system is a promising way to effectively damp the sound radiation at critical frequency ranges. Such a system was to be realized within the European Union (EU) project "Intelligent materials for Active Noise Reduction" (InMAR) for the oil pan of a Volvo MD13 truck engine. Piezoceramic patch actuators have been used in a laboratory test stand to alter the vibrations in a broadband noise reduction manner. This chapter discusses the actuator placement strategy and how to obtain an estimation of the broadband sound power minimization capability. Finally the chosen actuator layout was validated by experimental observations of a serial production oil pan.

35.1 Motivation

Noise is one of the major environmental problems in modern societies and induces enormous financial losses as well as serious health problems. Very often large lightweight vibrating structures are the primary noise source as they occur in air, road or shipping traffic vehicles. In the European Union (EU), the steadily increasing road traffic raises the corresponding noise emissions. In the years from

---

O. Heintze (✉)
Invent Gmbh, Christian-Pommer-Straße 34, 38112, Braunschweig, Germany
e-mail: olaf.heintze@invent-gmbh.de

M. Rose
Institute of Composite Structures and Adaptive Systems, German Aerospace Center DLR,
Lilienthalplatz 7, 38108, Braunschweig, Germany
e-mail: michael.rose@dlr.de

M. Wiedemann and M. Sinapius (eds.), *Adaptive, Tolerant and Efficient Composite Structures*, Research Topics in Aerospace, DOI: 10.1007/978-3-642-29190-6_35,
© Springer-Verlag Berlin Heidelberg 2013

1975 to 2005 the noise level in Germany along highways increased by 2.5 dB and in cities by 3 dB at nighttime [1]. To counteract this trend, a stepwise lowering of the permissible noise emission per different vehicle type by legislative actions encourages car manufactures to continuously develop cars performing ahead of prior restrictions. Within the last 30 years, these measures already induced 8 dB reductions in cars and 11 dB reductions in trucks and busses.

But the future challenges will require even more lightweight structures employed in next vehicle generations due to increased requirements with respect to fuel consumption and CO_2 emission. These structures are more susceptible to vibration and greater noise emissions, which cannot be effectively treated by classic strategies and therefore needs a new technology. It is outlines in the road map for future noise reduction technologies [2] that road traffic noise technologies significantly lower vehicle noise emissions until the year 2020. The investigation on such technologies related to active noise reduction for traffic related applications was done in the EU-funded project InMAR (Intelligent Materials for Active Noise Reduction). One aspect of this research was the treatment of the entire pass-by noise of a heavy-duty truck, typically powered by a large diesel engine. About 43% of this noise is caused by its engine and related parts [3]. Especially the oil pan was identifies as a major contributor to the noise radiation. Due to the significant noise content in the low frequency range, the application of passive damping materials is ineffective and therefore no option. In addition, using sound absorbing material would prohibit the necessary engine cooling.

The noise emission of a vibrating oil pan directly exposed to ambient air can be attenuated by alteration of this vibration through the application of piezoceramic actuator modules onto the surface of the structure and suitable electrical activation. Artificial dephased vibration generated by the piezoceramic actuators interferes with the inherent vibration thereby reducing the resulting noise. This principle is referred to as "Active Structural Acoustic Control" (ASAC), which is proven successful [4–11] in reducing broadband noise up to 500 Hz and even above. Typically, such an ASAC system consists of sensors in addition to piezoceramic actuators, which both belong to an electronic network including controller and power supply unit.

In this section such an ASAC system in combination with the suitable placement of the corresponding actuators [12] will be discussed, leading to an estimation of its broadband noise reduction capability. As a prerequisite, the natural mode shapes, the corresponding eigenfrequencies and the structural response of the oil pan to a typical excitation were determined by the finite-element (FE) software ANSYS and exported to the matrix based numerical software environment MATLAB to obtain suitable actor positions, capable of addressing structural vibrations over a frequency range up to 500 Hz. The objective function for the underlying optimization of the actuator activation was given by the minimization of the radiated sound power. The latter was estimated by numerically derived so-called radiation modes using the boundary element software SYSNOISE. Radiation modes are a set of principle velocity patters for the sound radiating surface, obtained by a singular value decomposition of the underlying quadratic sound power functional, and therefore the filtering through these patterns delivers independent contributions to the radiated sound power. Finally, due to simulation and parameter uncertainties, some experimental observations of a serial production oil pan were conducted to validate the initially derived actuator placement for that specific oil pan device.

35.2 Structural Dynamics and Suitable Actuator Positions

The main parts of an ASAC system are given by the underlying structure, the applied actuators and sensors, the control unit and the power supply hardware. The placement of actuators and sensors is crucial with respect to controllability and observability of the structural vibrations. Especially the selected piezoceramic patch actuators are most efficient at locations of great strain and therefore such locations have to be determined.

The oil pan consists of glass-fiber reinforce plastics and has dimensions of approximately one meter in length and 0.3 m in width and height. There are various ways to mount it to the supporting structure. A modal analysis was conducted for two mount configurations: The decoupled mounting was achieved by string-loaded bolts and a rubber gasket in contrast to the fixed shoulder mounting with locking of all degrees of freedom (DOF) along the oil pan's shoulder. In the selected frequency range up to 500 Hz, 56 mode shapes were found for the decoupled support in contrast to 32 for the stiffer fixed shoulder type. Since the structure can be totally controlled within this frequency range by the control of each of these natural modes, each mode shape should have at least one dedicated actuator, which is positioned in a high-strain area of the mode in order to ensure high structural authority. To achieve such a practical compromise in determining suitable actuator positions, the equivalent modal strains

$$\varepsilon_{eq}^i(\mathbf{x}_s) = \frac{1/\sqrt{2}}{1 + v'} \left[\sum_{ab=xy,yz,zx} \left((\varepsilon_{ai} - \varepsilon_{bi})^2 + \frac{3}{2}\gamma_{abi}^2 \right) \right]^{1/2} \tag{35.1}$$

based on the von Mises Eq. (35.2) (taken from [17]) were computed for each structural mode shape i. Here $\varepsilon_{\alpha i}$ and $\gamma_{\alpha\beta i}$ denote the appropriate component strains depending on the position vector $\mathbf{x}_s$ and v' is an effective Poisson's ratio. These computed equivalent strains were subsequently exported to MATLAB, normalized and superposed through the relations

$$\bar{\varepsilon}_{eq}^i(\mathbf{x}_s) = \varepsilon_{eq}^i(\mathbf{x}_s) \bigg/ \max_{\mathbf{x}_s} \varepsilon_{eq}^i(\mathbf{x}_s), \quad \tilde{\varepsilon}_{eq}(\mathbf{x}_s) = \frac{1}{N} \sum_{i=1}^{N} \bar{\varepsilon}_{eq}^i(\mathbf{x}_s), \tag{35.2}$$

where N denotes the total number of considered structural mode shapes. In order to achieve significant structural authority, all structural modes in Eq. (35.2) were equally weighted due to the high modal density. Figure 35.1 illustrates the results of these calculations for the two different support types. The high-strain spots mark

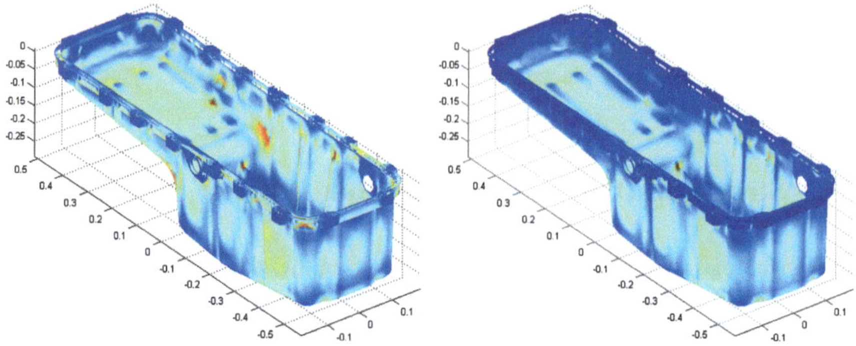

Fig. 35.1 FE calculation of superposed equivalent modal strains for two oil pan supports (*left* decoupled, *right* fixed shoulder)

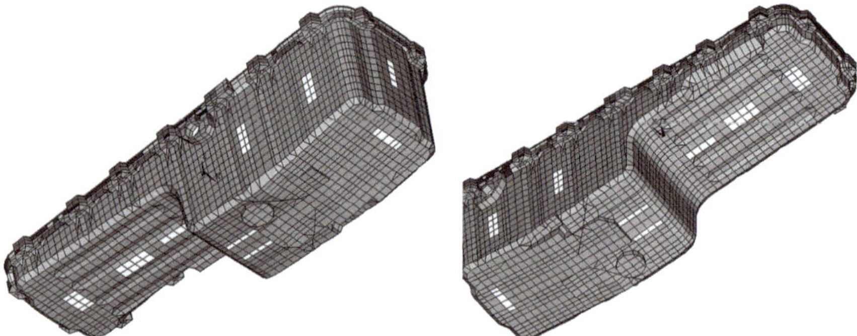

Fig. 35.2 FE-mesh of oil pan augmented with 11 piezoceramic patch actuators

potential actuator locations that represent a compromise for broadband contrallability and observability.

Typical areas of high-strain are located between stiffeners and in the rear flat part of the oil pan. To be able to use a moderate number of actuators (much less than the considered number of mode shapes), the placement strategy is maximization of the actuators modal coverage. In addition, very small areas of large strain do not permit actuator placement. Around thirty proposed positions with the greatest equivalent strains were identified as potential locations. Finally eleven actuators were applied at these locations as depicted in Fig. 35.2, which also shows the FE mesh of the layered shell elements. The piezoceramic actuators were modelled in ANSYS by employing the method of temperature analogy, because standard shell elements do not provide piezo-electric coupling.

The method uses the thermal expansion coefficient (CTE) to incorporate the actuator stress tensor of the modules by interpreting the temperature as the actuators driving voltage. A linear relation is assumed between the applied electric field and the resulting actuator strain tensor. This is reasonable as long as the maximum amplitudes of the driving voltage induce minimal hysteretic behavior of the module. The corresponding CTE can be computed directly from the piezoceramic material parameters.

35.3 Vibroacoustic Coupling and ASAC Efficiency Estimation

It is well known from literature [4–11] that in the absence of resonance the structures vibration is given by an operational deflection shape as a superposition of several mode shapes. To obtain a suitable control strategy for minimization of the sound radiation within a wide frequency range, it is therefore necessary to identify a relation between structural vibration and radiated sound. To reduce complexity, extra environmental details, road surface and engine compartment have been disregarded from the computation. Instead the active sound power as a measure of the acoustic energy transported to the far field was chosen as the acoustic quantity to be minimized through altering the inherent structural vibration. Several methods are discussed in literature to describe this sound power. Elliott and Johnson [5] substitute in the expression for the active sound power

$$\overline{W} = \frac{1}{2}\mathrm{Re}\left(\int_S v_{ns}^*(\mathbf{x}_s)p_s(\mathbf{x}_s)dS\right) = \frac{S_E}{2}\mathrm{Re}\left(\mathbf{v}_{ns}^*\mathbf{p}_s\right), \quad \mathbf{x}_s \in S \qquad (35.3)$$

the pressure p_S on the radiating surface S by the impedance relation between that pressure and the normal surface velocity v_{ns}. The second term in (35.3) results from the first term by using elementary radiators of uniform area S_E to discretize the structures surface. The superscript * denotes complex conjugated and transposed vectors or matrices. The impedance relation is given by $\mathbf{p}_S = \mathbf{Z}\mathbf{v}_{ns}$ with the acoustic impedance matrix $\mathbf{Z}$. Substituting this into Eq. (35.3), an expression

$$\overline{W} = v_{ns}^*\mathbf{R}v_{ns} \text{ with } \mathbf{R} = S_E/2\,\mathrm{Re}(\mathbf{Z}) \qquad (35.4)$$

for the active sound power results. The acoustic impedance matrix $\mathbf{Z}$ can be analytically described for plane and simply curved radiators [4]. But arbitrary shaped objects like the oil pan need sophisticated numerical procedures for this task. Kuijpers [13] and Visser [14] outlined methods to derive $\mathbf{Z}$ using the boundary element method (BEM). Instead of using the near field impedance relation, more accurate results can be obtained by expressing the active sound power in the far field. Photiadis [15] finds such a relation based on the expression

$$p_f(\mathbf{x}_f) = \int_S g(\mathbf{x}_f, \mathbf{x}_s)v_{ns}(\mathbf{x}_s)\,dS, \quad \mathbf{p}_f = \mathbf{G}\mathbf{v}_{ns}. \qquad (35.5)$$

The second term is again the discretized version of the first term. The elements of the matrix $\mathbf{G}$ are the values of the evaluated Green's function $g(\mathbf{x}_f, \mathbf{x}_s)$ at

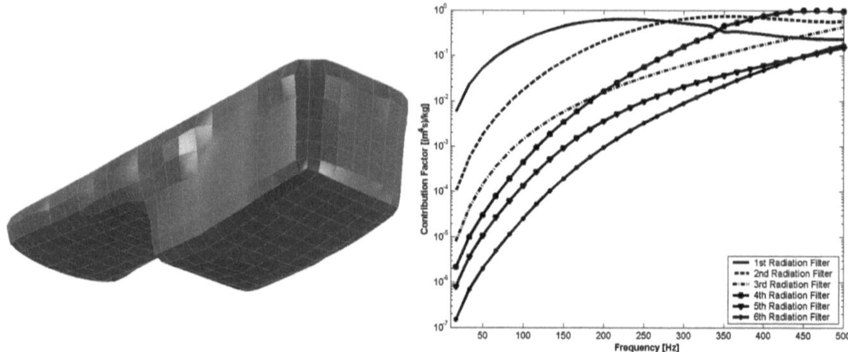

Fig. 35.3 Boundary element mesh derived from FE mesh (*left*) and radiation efficiency of the first six radiation modes in the frequency range of 17–500 Hz (*right*)

specified locations, weighted by an appropriate element area. The singular value decomposition (SVD) of $\mathbf{G} = \mathbf{U}\mathbf{S}\mathbf{V}^*$ delivers the radiation efficiencies from the real and positive singular values in the diagonal matrix $\mathbf{S}$, as well as the real radiation modes from the columns of the unitary matrix $\mathbf{V}$. The unitary matrix $\mathbf{U}$ is not needed for the sound power calculation due to cancelation in

$$\begin{aligned}
\overline{W}_f &= \tfrac{1}{2}\int_{S_f} v_{nf}^*(\mathbf{x}_f)p_f(\mathbf{x}_f)\,dS \\
&= \frac{S_{Ef}}{2\rho_0 c_0}\mathbf{p}_f^*\mathbf{p}_f \\
&\approx \frac{S_{Ef}}{2\rho_0 c_0}\mathbf{v}_{ns}^*\mathbf{V}\Lambda\mathbf{V}^*\mathbf{v}_{ns}.
\end{aligned} \tag{35.6}$$

Here v_{nf} denotes the particle velocity in the fluid with fluid density ρ_0 and speed of sound c_0, and S_{Ef} is the surface area of one element in an ISO 3744 microphone mesh. The matrix $\Lambda = \mathbf{S}\mathbf{S}$ is a diagonal matrix that contains the squared singular values, which implies that the radiation modes contribute to the sound power independently with an efficiency that is proportional to its respective squared singular value.

The BEM software SYSNOISE can be used to calculate the matrix $\mathbf{G}$. Figure 35.3 (left) depicts the boundary element (BE) mesh that was derived from the finite-element mesh presented above. Furthermore, a microphone mesh in the fluid region was set up according to ISO 3744. The computed matrix $\mathbf{G}$ was subsequently exported for each considered frequency to MATLAB to perform the SVD. A sorting algorithm as described by Johnson [16] had to be implemented to correctly assign the radiation modes to the continuously varying curves of radiation efficiencies with respect to frequency (for details compare with [12]). Figure 35.3 (right) presents the contribution factors of the first six radiation modes in the frequency range up to 500 Hz. The first radiation mode dominates the sound power radiation up to 250 Hz, which corresponds to a Helmholtz number $He = ka$

of 2.8 with k being the acoustic wave number and a as the typical radiator dimension. With increasing frequency, the radiation modes of higher order become more efficient. This leads to a contribution of all considered radiation modes with the same order of magnitude at the maximum frequency of 500 Hz, which is represented by a Helmholtz number of approximately 5.6. The contribution factors of Fig. 35.3 (right) directly reflect the significance of the corresponding radiation modes at each frequency.

As shown in [12], the first radiation mode represents a uniform normal surface velocity distribution that pulses in-phase with the inside and outside acting like a monopole radiator. In the low frequency range, this radiation mode is the most efficient radiator. In contrast, the second and third radiation modes exhibit a dipole character. With increasing order more complex spatial oscillations occur such that the fourth radiation mode exhibits characteristics of a quadrupole radiator. Multipole radiators are obtained for even higher radiation modes corresponding to the observations made by Visser [14] for a sphere.

Fortunately, the shape variation of the radiation modes is moderate with respect to frequency for a large part of the considered frequency range. To reduce the complexity of the control strategy, a set of the first four radiation modes at 250 Hz was selected to represent the entire frequency range. The sound power band level was estimated up to a frequency of 500 Hz employing these four modes in conjunction with frequency dependent radiation efficiencies. These band levels differ by 0.25 dB compared to the ones computed with all radiation modes and this set is applied from here on.

The dynamic model of the structure augmented with the actuators is used to obtain the frequency response matrix functions $\mathbf{H}_d$ and $\mathbf{H}_u$ for the disturbances and the actuator control voltages to the surface velocities respectively. Under the linearity assumptions, the vector of surface velocities is given by

$$\mathbf{v}_{ns} = \mathbf{H}_d\mathbf{f}_d + \mathbf{H}_u\mathbf{u} \tag{35.7}$$

in terms of the disturbance forces $\mathbf{f}_d$ and the actuator driving voltages $\mathbf{u}$. Using the plant Eq. (35.7) in combination with the quadratic functional (35.4), the control signal in the frequency domain can be calculated (with and without a penalty term for the control voltage amplitudes) by minimizing the implied quadratic equation (details are again presented in [12]). Here, a structural vibration is provided that reliably counters the inherent vibration in a manner which minimizes sound power over a wide range of frequencies. Although, such an optimally performing feedforward control algorithm can rarely be employed in realistic applications, the method provides an estimation of an upper performance limit for the noise reduction system.

Eighteen representative three dimensional structural excitations were simultaneously applied to all 18 oil pan mount points. The resulting active sound power in the frequency range up to 500 Hz was computed and its spectrum is depicted by the dark gray (upper) data set in Fig. 35.4 (left). Additionally, the significantly lowered sound power through the ASAC system is shown here as the light gray

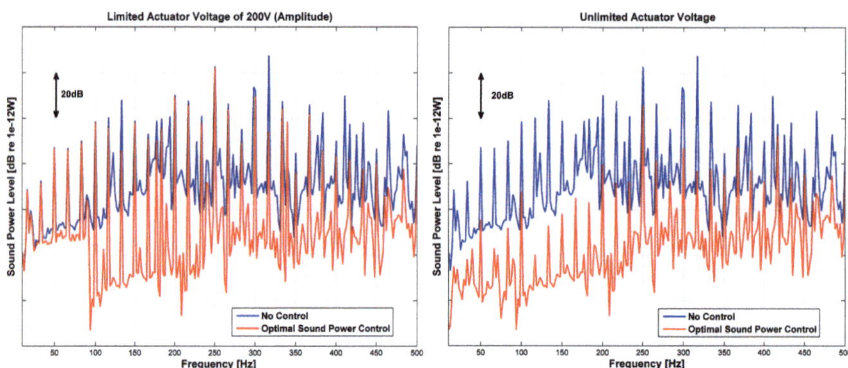

Fig. 35.4 Level of active sound power of oil pan with typical structural excitation without (*blue, upper graph*) and with (*red, lower graph*) control (*left* unlimited actuatorvoltage, *right* limited actuator voltage of 200 V amplitude)

(lower) curve. Inspection of the band level for the frequency range up to 500 Hz reveals a broadband sound power reduction of 23 dB compared to the uncontrolled sound power band level. This reduction was achieved assuming an optimally performing control which results in control voltages partially exceeding the limits of typical power supply unit and actuator specifications. Therefore, a provision was implemented in the voltage computation algorithm, which limits the maximally permissible voltage amplitude for actuator and power supply of 200 V. Even for this more realistic configuration with relatively large voltage restriction, a significant reduction in the sound power band level of about 6 dB was achieved as depicted in Fig. 35.4 (right).

35.4 The Serial Production Oil Pan Demonstrator

Within the project InMAR, an oil pan demonstrator was set up to validate numerical simulations. The test stand assembly consists of several parts. The oil pan itself is mounted to an aluminum frame using bolts and a gasket from serial production to demonstrate realistic mounting conditions. The oil pan's interior is occupied with sound absorbing open-cell foam. Figure 35.5 (left) illustrates this assembly without the sound absorbing foam, which is supported through suspension cords attached to a support frame. The structural excitation for a subsequent ASAC test was planned to be provided through a shaker device acting on the upper cross bar of the frame, where the oil pan is mounted. This ensures compatibility to the test facilities in the laboratories of DLR, providing a test environment as realistic as possible in comparison to actual operation conditions in a truck.

Due to geometric differences between the FE-model and the serial production oil pan, the actuator placement resulting from the simulations discussed above had to be validated for the serial production oil pan. Therefore, an experimental

Fig. 35.5 Oil pan test stand without sound absorbing foam from different views (*left*), oil pan equipped with piezoceramic patch actuators (*right*)

observation was conducted to fine tune and to validate the piezoceramic patch positions. This experimental observation was started with the locations proposed by the simulation by marking and numbering the feasible positions for patch actuators of a given size, 30 by 95 mm. For these locations, the mobility or frequency response functions (FRF) due to a shock hammer excitation at the shaker attachment point and an accelerometer (bonded to the structure with wax for easy repositioning) at the identified potential locations were measured.

The results of the conducted tests provide insight in the frequency content observable and controllable at the observed locations. All FRF's are an average of ten directly succeeding measurements for each of the 30 considered accelerometer locations. As a result of this experimental validation, the 11 chosen actuator positions are consistent with the FE simulation despite the geometric deviation of the serial production oil pan. All of the most significant peaks are covered by the actuator set and due to the observability of multiple frequency peaks a great modal coverage is given. Figure 35.5 (right) depicts the fully equipped oil pan.

35.5 Conclusions

In this work, a large truck engine oil pan has been investigated to establish an active structural acoustic control system. The focus has been set on a piezoceramic actuator placement strategy as well as on the estimation of the systems sound power reduction capabilities. For two different types of boundary conditions, the first natural mode shapes have been computed through FEM simulations. In both support types the modal density of the oil pan is too high to permit individual mode shape control. Hence, the actuator locations have been determined by the evaluation of superposed modal equivalent strains to grant a good multi-modal coverage. The primary goal of the ASAC system is to alter the inherent noise radiating vibration over a broad frequency range and to reliably minimize the sound power. Therefore, principle velocity patterns have been computed numerically, to allow for efficient radiated active sound power estimation and suitable real-time control implementation. Four dominating radiation modes have been used for the computation of the actuator voltages of a typical oil pan excitation due to the engine. A principle reduction of 23 dB in the active sound power band up to 500 Hz has been obtained by an ASAC simulation with an underlying optimal control scheme. More realistic operational conditions of the actuators were considered by limiting the maximum permissible voltage to 200 V. Even with this significant restriction, a broadband sound power reduction of approximately 6 dB through the ASAC system has been computed.

In order to demonstrate the ASAC system capabilities, a test stand with reasonably realistic mounting conditions have been designed for this particular oil pan in the DLR laboratories. Due to geometrical differences between the FEM mesh and the serial production oil pan, the numerically derived actuator placement has been adjusted by a set of experimental observations. Finally, the oil pan has been augmented with piezoceramic patch actuators, which are manufactured as modules by the DLR, featuring high durability, robustness, and performance.

On the basis of the work presented in this paper, the succeeding steps are the measurement of the actual sound radiation of the oil pan and the implementation of a control algorithm. Once the hardware implementation of the controller is performed, its capability to reduce the oil pan noise must be determined and compared to the simulation results presented here.

Acknowledgments The authors acknowledge gratefully the funding for this research through the EU project InMAR. Sincere thanks are given to VOLVO Technology Corporation and especially to Carl Fredrik Hartung for the profound technical support.

References

1. Affenzeller, J., Rust, A.: Road traffic noise—a topic for today and the future. In: VDA—Technical Congress, 2005
2. Redaelli, M., Manzoni, S., Cigada, A., Wimmel, R., Siebald, H., Fehren, H., Schiedewitz, M., Wolff, K., Lahey, H.-P., Nussmann, Ch., Nehl, J., Naake, A.: Different techniques for active and passive noise cancellation at powertrain oil pan. In: Adaptronic Congress 2007
3. Calm Network : Research for a quieter Europe in 2020. In: European Commision Research Directorate-General, www.calm-network.com, 2004
4. Bevan, J.S.: Piezoceramic actuator placement for acoustic control of panels, Technical Report NASA/CR-2001-211265, 2001
5. Elliott, S.J., Johnson, M.E.: Radiation modes and the active control of sound power. J. Acoust. Soc. Am. **94**(4), 2194–2204 (1993)
6. Fuller, C.R., Elliott, S.J., Nelson, P.A.: Active Control of Vibration. Academic, San Diego (1996)
7. Gibbs, G.P., Clar, R.L., Cox, D.E., Vipperman, J.S.: Radiation model expansion for acoustic control. J. Acoust. Soc. Am. **107**(1), 332–339 (2000)
8. Hansen, C.H., Snyder, S.D.: Active Control of Noise and Vibration. E and FN Spon, London (1997)
9. Nijhuose, M.H.H.O.: Analysis tools for the design of active structural acoustic control systems. PhD Thesis, University of Twente (2003)
10. Scors, T.C., Elliott, S.J.: Volume velocity estimation with accelerometer arrays for active structural acoustic control. J. Sound Vib. **258**(5), 867–883 (2002)

11. Weyer, T., Breitbach, E., Heintze, O.: Self-tuning active electromechanical absorbers for tonal noise reduction of a car roof. In: InterNoise07, 2007
12. Heintze, O.l., Rose, M.: Active structural acoustic control for a truck oil pan: actuator placement and efficiency estimation. Noise Control Eng. J., 58(3), 292–301 (2009)
13. Kuijpers, A.H.W.M.: Acoustic modeling and design of MRI scanners. PhD Thesis, TU Eindhoven, Eindhoven, The Netherlands (1999)
14. Visser, R.: A boundary element approach to acoustic radiation and source identification. PhD Thesis, University of Twente, Enschede, The Netherlands (2004)
15. Photiadis, D.M.: The relationship of singular value decomposition to wave-vector filtering in sound radiation problems. J. Acoust. Soc. Am. $\mathbf{88}$(2), 1152–1159 (1990)
16. Johnson, W.M.: Structural acoustic optimization of a composite cylindrical shell. PhD Thesis, Georgia Institute of Technology (2004)
17. ANSYS Inc.: POST1 and POST 26—Interpretation of equivalent strains. In: Release 10.0 Documentation for ANSYS, Chap. 19.12, 2005

Chapter 36
Experimental Study of an Active Window for Silent and Comfortable Vehicle Cabins

Malte Misol, Stephan Algermissen and Hans Peter Monner

Abstract The poor sound insulation of windows especially at low frequencies constitutes a severe problem, both in transportation and in the building sector. Due to additional constraints on vehicles or aircrafts regarding energy efficiency and lightweight construction, the demand of light-weight-compliant noise-reduction solutions is amplified in the transportation industry. Simultaneously, in order to satisfy the customer demands on visual comfort and modern design, the relative size of glazed surfaces increases in all sectors. The experimental study presented below considers the feasibility of actively controlled windows for noise reduction in passenger compartments by using the example of an automobile windshield. The active windshield consists of the passive windshield, augmented with piezoceramic actuators and sensors. The main focus of the subsequent work was the development and evaluation of feedforward and feedback control strategies with regard to interior noise reduction. The structural excitation of the windshield was realized by an electrodynamic exciter (shaker) applied at the roof brace between the A-pillars. By this choice it was possible to emulate the structural excitation of the windshield through the car body, induced by coasting and motor-force harmonics. The laboratory setup does not permit the consideration of hydrodynamic and acoustic loads, which might be important as well. However, the experimental results indicate the high noise reduction potential of active structural acoustic control of structure-borne sound that radiates into a cavity.

M. Misol (✉) · S. Algermissen · H. P. Monner
Institute of Composite Structures and Adaptive Systems, German Aerospace Center DLR,
Lilienthalplatz 7, 38108, Braunschweig, Germany
e-mail: malte.misol@dlr.de

S. Algermissen
e-mail: stephan.algermissen@dlr.de

H. P. Monner
e-mail: hans.monner@dlr.de

M. Wiedemann and M. Sinapius (eds.), *Adaptive, Tolerant and Efficient Composite Structures*, Research Topics in Aerospace, DOI: 10.1007/978-3-642-29190-6_36,
© Springer-Verlag Berlin Heidelberg 2013

36.1 State of Technology

The reduction of noise in passenger compartments of vehicles by means of active electronic devices has been a research issue for more than fifty years. In 1953, Olson and May [1] proposed the application of an electronic sound absorber in the vicinity of the passenger's head in order to reduce unwanted sound. More recent publications deal with the reduction of road noise using the car's audio system as actuator for the active noise control (ANC) system. Both feedforward and feedback algorithms are investigated. Sutton et al. [2] developed an active sound control system for automobiles that uses interior loudspeakers to counteract the low-frequency rumble of road noise when driving on typical road surfaces. In a publication of Oh et al. [3] the development of an active feedforward control system for the reduction of road booming noise is described. Four accelerometers were attached to the suspension system in order to detect reference vibration signals, and two loudspeakers were used for the attenuation of the noise near the headrests of the two front seats. Sano et al. [4] developed an active control system for low-frequency road noise in automobiles with the emphasis on cost efficiency. A recent paper by de Oliveira et al. [5] aims to evaluate, both numerically and experimentally, the effect of a feedback controller on the sound quality of a vehicle mock-up excited with engine noise. The controller performance is evaluated in terms of specific loudness and roughness. All of the mentioned publications implement an active noise reduction system by means of anti-sound. However, not many research results have been published in the field of active structural acoustic control (ASAC) for automobile application. Dehandschutter and Sas [6] propose the application of vibration absorbers at the car body. In Weyer et al. [7] active electromechanical absorbers (AEMA) are applied to a car roof for the realization of an ASAC system. The windshield, however, which constitutes an important factor regarding interior noise, has not been considered in former publications dealing with ASAC. The current paper considers the reduction of windshield-vibration-induced interior noise in an automobile passenger compartment by means of ASAC. The achieved results can be considered to be of generic importance also for other traffic carriers such as trains or aircraft.

36.2 Real-Time Control System of the Active Windshield

The sampling rate was set to 1 kHz. Apart from the microphone signal, all in- and outgoing signals were conditioned by low-pass filters of the type Kemo 21 M with a cut-off frequency set to 240 Hz. The piezo-amplifiers of the type E-471 and P-270 from PI were sufficiently powerful to supply the piezoelectric d31-patches with a maximum voltage of 400 V up to the cut-off frequency. For vibration sensing, six accelerometers of the type PCB 356A18 were placed at heuristically optimized positions on the windshield. The sound pressure level (SPL) at different

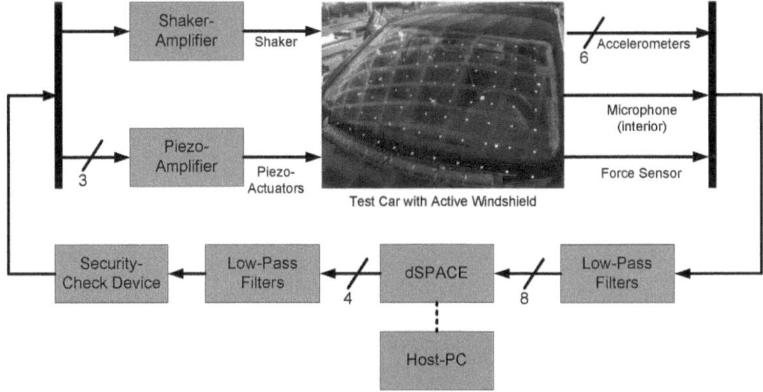

Fig. 36.1 Block-diagram of the real-time control system of the active windshield

Fig. 36.2 Definition of disturbance force (F), actuators (A) and sensors (S)

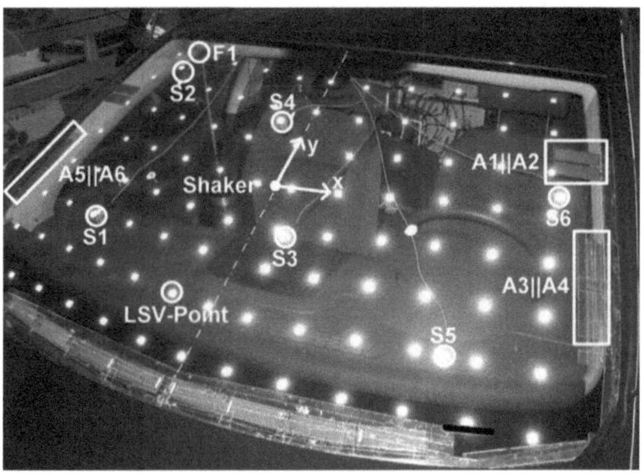

positions in the interior of the car was sensed by a PCB 377B02 microphone in combination with a PCB TMS426A01 amplifier. The reference signal for the adaptive feedforward controller was generated from a force sensor placed at the excitation point of the shaker at the roof brace (Fig. 36.1).

36.3 Definition of Sensors and Actuators

Figure 36.2 shows the accelerometer positions (S1–S6) as well as the selected piezo-actuators (A1–A6). The number and position of the sensors was chosen and optimized heuristically by using results from modal analysis and the principle of maximum modal observability. The selection process of the control actuators was guided by the evaluation of control-path frequency-response-functions (FRF). In order to achieve a reasonable trade-off between model complexity and control

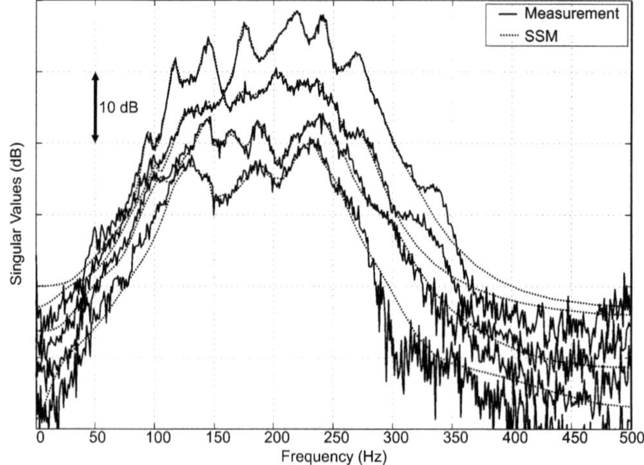

Fig. 36.3 Singular values of control- and disturbance-path FRF-matrix

authority, the number of actuator channels had been restricted to three. A further increase in control authority is obtained by operating adjacent actuators in parallel (Ax‖Ay).

In general, it can be stated that the lower actuators 3, 4, 5 and 6 have more authority below 150 Hz and the upper actuators 1 and 2 perform better above 150 Hz. This behavior can be explained by the mode shapes of the windshield system and the corresponding distribution of maximum modal strains. The final choice of actuators was A1‖A2, A3‖A4 and A5‖A6 because in this configuration the singular values of the control path FRF-matrix were largest throughout the controller bandwidth from 0 to 240 Hz.

36.4 Multi-Reference Test and System Identification

A high-precision model of the control system constitutes an important prerequisite for the successful design and implementation of a state-feedback or adaptive feedforward controller. In this study, the SLICOT toolbox [8] was used to identify a discrete-time state-space-model (SSM) for the coarse accelerometer grid (4 inputs, 6 outputs) by means of multiple-reference test data. In order to obtain the global system dynamics in terms of a fine, so-called SLDV-grid (101 points), a subsequent least-squares fit was performed by using the obtained state-space model and measurement data from the scanning laser doppler vibrometer (SLDV). The final result was an augmented state-space system of the same order, but with 101 outputs.

Figure 36.3 compares the singular values of the identified system model with a MATLAB® frequency response data (FRD)-model that had been calculated from measurement data. The agreement of the identified and measured singular values

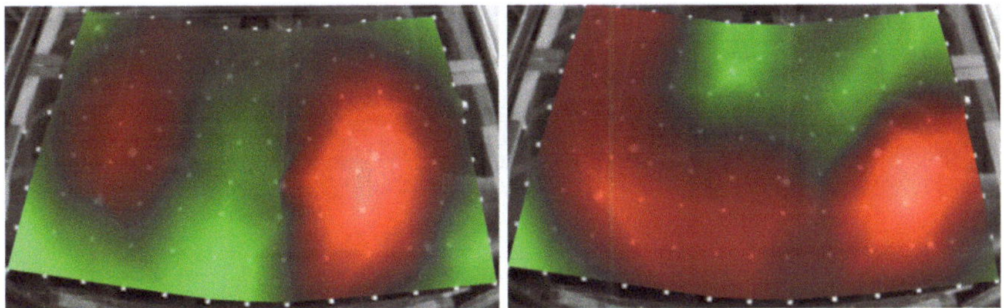

Fig. 36.4 Measured operational velocity shapes at 116 Hz (*left*) and 145 Hz (*right*) without control

within the control bandwidth proves that a numerically efficient and accurate modeling had been achieved with the use of only 60 states.

36.5 Implementation and Evaluation of the Control Algorithms

Modern control strategies are based on the concept of a generalized plant [9]. This method is very generic and provides great flexibility in control strategy formulation. In principle, the designed controllers are broadband within the control bandwidth. However, results of acoustic investigations obtained from road trials and roller test bench experiments emphasize the importance of the second and third eigenfrequency of the windshield system. Due to its importance for the low-frequency interior acoustics, the subsequent discussion is focused on these two structural resonances. In order to allow comparability, the color coding of the operational velocity shapes amplitudes was kept constant for all vibration measurements. The amplitudes are color coded from light green (0 m/s) to light red (9e-4 m/s at 116 Hz and 5e-4 m/s at 145 Hz) (Fig. 36.4).

36.5.1 State-Feedback Control

Figure 36.5 shows the schematic employed for the control design. This formulation follows the generalized plant concept with plant $\mathbf{G_{ext}}$, controller $\mathbf{R}$ and appropriately chosen weighting filters $\mathbf{W_{XYZ}}$. The zero-mean, white-noise processes $\mathbf{w_d}$ and $\mathbf{w_r}$ serve as input (process and measurement noise) whereas $\mathbf{z}$ represents the performance output of the generalized plant. The scheme is expressed mathematically in Eqs. 36.1 and 36.2. If $\mathbf{p}$ and $\mathbf{v}$ are identical, the controller minimizes a local performance metric. The global controller is designed with regard to the fine SLDV-grid. In order to enhance the control performance, a model of the disturbance path $\mathbf{G_{11}}$ was included into the generalized plant.

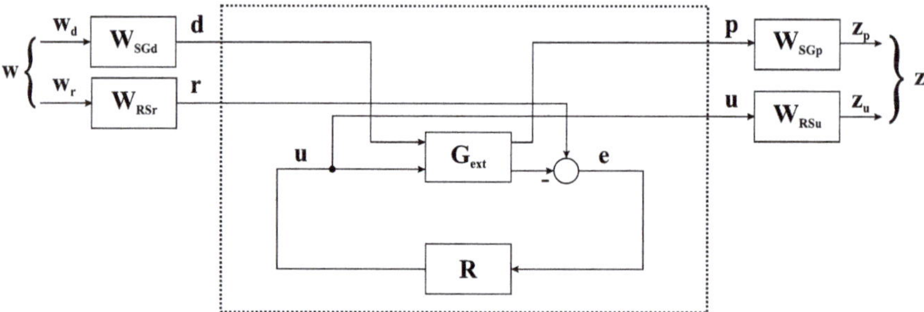

Fig. 36.5 Weighting scheme for FB-controller synthesis

Fig. 36.6 Measured operational velocity shapes at 116 Hz (*left*) and 145 Hz (*right*) with local (*above*) and global (*below*) state-feedback control

$$\begin{bmatrix} \mathbf{z}_p \\ \mathbf{z}_u \end{bmatrix} = \begin{bmatrix} \mathbf{W}_{SGp}\mathbf{G}_{12}\mathbf{RSW}_{RSr} & \mathbf{W}_{SGp}(\mathbf{G}_{11} - \mathbf{G}_{12}\mathbf{RSG}_{21})\mathbf{W}_{SGd} \\ \mathbf{W}_{RSu}\mathbf{RSW}_{RSr} & -\mathbf{W}_{RSu}\mathbf{RSG}_{21}\mathbf{W}_{SGd} \end{bmatrix} \begin{bmatrix} \mathbf{w}_r \\ \mathbf{w}_d \end{bmatrix} \quad (36.1)$$

With extended plant $\mathbf{G}_{ext}$ and sensitivity $\mathbf{S}$:

$$\begin{bmatrix} \mathbf{p} \\ \mathbf{v} \end{bmatrix} = \mathbf{G}_{ext} \begin{bmatrix} \mathbf{d} \\ \mathbf{u} \end{bmatrix} = \begin{bmatrix} \mathbf{G}_{11} & \mathbf{G}_{12} \\ \mathbf{G}_{21} & \mathbf{G}_{22} \end{bmatrix} \begin{bmatrix} \mathbf{d} \\ \mathbf{u} \end{bmatrix} \text{ and } \mathbf{S} = [\mathbf{E} + \mathbf{G}_{22}\mathbf{R}]^{-1} \quad (36.2)$$

According to Fig. 36.6, the vibration-reduction performance of local and global feedback control is very similar. One reason for this is the relatively large number of accelerometers compared to the structural wavelengths. This leads to a global vibration reduction even with a local control scheme. The global control would be

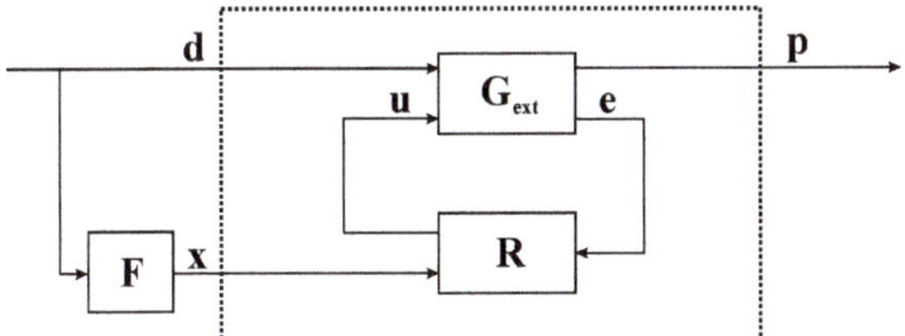

Fig. 36.7 Generalized plant framework for adaptive feedforward control

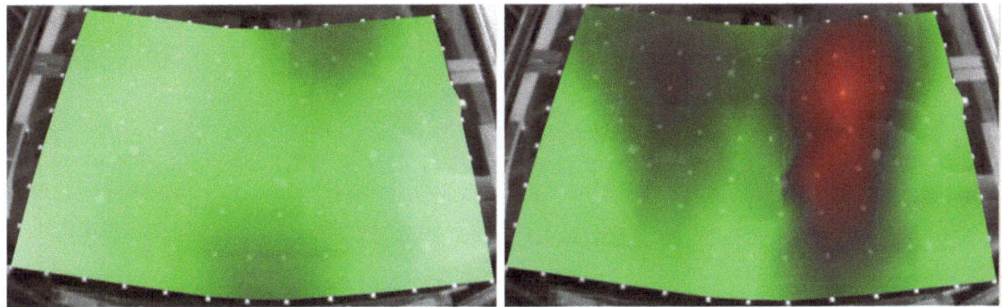

Fig. 36.8 Measured operational velocity shapes at 116 Hz (*left*) and 145 Hz (*right*) with adaptive feedforward control

advantageous if the ratio of structural sensors to control bandwidth was significantly reduced or if a sound power related performance metric was needed. The vibration level reductions of the local feedback controller averaged over the SLDV-grid are 7.54 dB at 116 Hz and 4.36 dB at 145 Hz.

36.5.2 Adaptive Feedforward Control

The implemented feedforward controller is based on finite impulse response (FIR) filters whose coefficients are adapted by means of the well-known filtered-x least mean squares algorithm (FxLMS) [10]. In contrast to the state-feedback control scheme, the FxLMS algorithm performs no post-processing of the sensor signals and hence can only process local information based on the coarse sensor grid ($\mathbf{p} \equiv \mathbf{e}$) (Fig. 36.7).

Adaptive filtering was performed with 200 FIR-filter taps for each control channel, a leakage-factor $V = 1$ and a normalized step size μ of 0.1 % times the theoretical maximum value. The adaption of the FIR-filter-weights for a single-reference FxLMS with M sensors and K actuators is described by Eq. 36.3.

$$\mathbf{w}(n+1) = \mathbf{w}(n) + \mu \mathbf{X}'(n)\mathbf{e}(n) \tag{36.3}$$

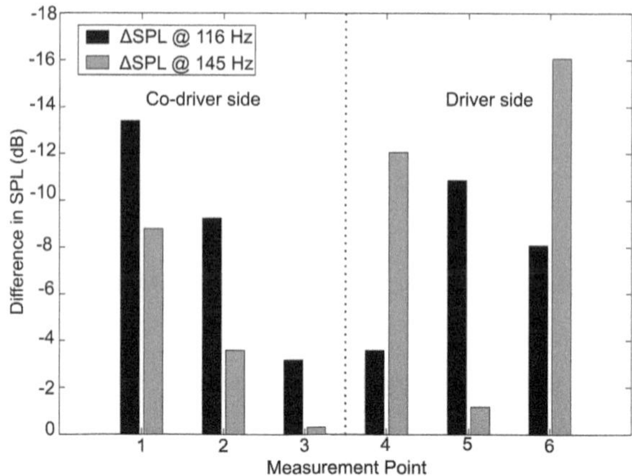

Fig. 36.9 Attenuation of interior sound pressure level (SPL) due to feedforward control

The dimensions of weight-vector **w**, filtered-reference matrix **X'** and error-vector **e** follow from Eq. 36.4. Matrix **X'** is formed by the Kronecker product convolution of impulse response matrix $\widehat{\mathbf{S}}$ and reference signal vector **x**.

$$\mathbf{w}(n) \equiv \left[\mathbf{w}_1^T(n) \quad \mathbf{w}_2^T(n) \quad \cdots \quad \mathbf{w}_K^T(n) \right]^T \in \mathbb{R}^{KL}$$
$$\mathbf{e}(n) \equiv \left[e_1(n) \quad e_2(n) \quad \cdots \quad e_M(n) \right]^T \in \mathbb{R}^{M} \tag{36.4}$$
$$\mathbf{X}'(n) = \widehat{\mathbf{S}}^T(n) \otimes \mathbf{x}(n) \in \mathbb{R}^{KLxM}$$

Figure 36.8 shows the vibration amplitude with the adaptive feedforward controller. The relative sound pressure levels shown in Fig. 36.9 were measured at six points equally spaced along the x-axis at $\mathbf{z} = -0.1$ m (interior).

36.6 Conclusion

Different feedback and feedforward control strategies for the active reduction of interior noise in vehicle cabins have been developed and experimentally approved. A maximum global vibration reduction of 7.5 dB was achieved in the acoustically relevant frequency band containing the second and third eigenfrequency of the windshield system (100–150 Hz). The interior SPL was reduced up to 16 dB.

References

1. Olson, H., May, E.: Electronic sound absorber. J. Acoust. Soc. Am. **25**, 829 (1953)
2. Sutton, T.J., Elliott, S.J., McDonald, A.M., Saunders, T.J.: Active control of road noise inside vehicles. Noise Control Eng. J. **42**(4), 137–147 (1994)
3. Oh, S.H., Kim, H.S., Park, Y.J.: Active control of road booming noise in automotive interiors. J. Acoust. Soc. Am. **111**(1), 180–188 (2002)

4. Sano, H., Inoue, T., Takahashi, A., Terai, K., Nakamura, Y.: Active control system for low-frequency road noise combined with an audio system. IEEE Trans. Speech Audio Process. **9**(7), 755–763 (2001)
5. De Oliveira, L.P.R., Janssens, K., Gajdatsy, P., Van der Auweraer, H., Varoto, P.S., Sas, P., Desmet, W.: Active sound quality control of engine induced cavity noise. Mech. Syst. Signal Process. **23**(2), 476–488 (2009)
6. Dehandschutter, W., Sas, P.: Active control of structure-borne road noise using vibration actuators. J. Vib. Acoust. Trans. ASME **120**(2), 517–523 (1998)
7. Weyer, T., Breitbach, E., Heintze, O.: Self-tuning active electromechanical absorbers for tonal noise reduction of a car roof. In: INTER-NOISE—International Congress and Exhibition on Noise Control Engineering, Istanbul, Turkey. http://elib.dlr.de/52319/ (2007)
8. Favoreel, W., Sima, V., Van Huffel, S., Verhaegen, M., De Moor, B.: Subspace model identification of linear systems in slicot, Technical report, European Community BRITE-EURAM III Thematic Networks Programme NICONET. SLICOT Working Note 1998-6 (1998)
9. Clark, R., Saunders, W., Gibbs, G.: Adaptive structures: dynamics and control. **28**(2) (1998)
10. Kuo, S., Morgan, D.: Active Noise Control Systems: Algorithms and DSP Implementations. John Wiley & Sons Inc., New York (1995)

Chapter 37
Structural Health Monitoring Based on Guided Waves

Daniel Schmidt, Wolfgang Hillger, Artur Szewieczek and Michael Sinapius

Abstract Structural Health Monitoring (SHM) based on ultrasonic guided waves, so-called Lamb waves, is a promising method for in-service inspection of composite structures without time consuming scanning like conventional ultrasonic techniques. Lamb waves are able to propagate over large distances and can be easily excited and received by a network of piezoelectric actuators and sensors. In principle different kinds of structural defects can be detected and located by analyzing the sensor signals. The chapter describes recent research activities at DLR on Structural Health Monitoring. The research are focused on the visualisation of Lamb wave propagation fields based on air-coupled ultrasonic technique, the simulation of virtual sensors, mode selective actuators as well as manufacturing of actuator and sensor networks. Additionally, the chapter present the development of a SHM system for impact detection in a helicopter tailboom (Eurocopter—EC 135).

D. Schmidt (✉) · W. Hillger · A. Szewieczek · M. Sinapius
Institute of Composite Structures and Adaptive Systems, German Aerospace Center DLR, Lilienthalplatz 7, 38108, Braunschweig, Germany
e-mail: daniel.schmidt@dlr.de

W. Hillger
e-mail: wolfgang.hillger@dlr.de

A. Szewieczek
e-mail: artur.szewieczek@dlr.de

M. Sinapius
e-mail: michael.sinapius@dlr.de

M. Wiedemann and M. Sinapius (eds.), *Adaptive, Tolerant and Efficient Composite Structures*, Research Topics in Aerospace, DOI: 10.1007/978-3-642-29190-6_37,
© Springer-Verlag Berlin Heidelberg 2013

37.1 Introduction

Structural Health Monitoring (SHM) based on ultrasonic guided waves, so-called Lamb waves, is a promising method for in-service inspection of aerospace structures without time consuming scanning like conventional ultrasonic techniques. The implementation of SHM systems into aerospace applications enhances reliability, safety and maintenance performance as well as economic aspects. Lamb waves are disperse guided waves which propagate between two parallel surfaces, e.g. the upper and lower plate surface. Due to the disperse characteristic, the wave velocity depends on the product of excitation frequency and plate thickness. At a given frequency-thickness product a finite number of different Lamb wave modes (symmetric, S_0, S_1, S_2,..., and anti-symmetric, A_0, A_1, A_2,..., modes) exist [1, 2].

Lamb waves are able to propagate over long distances in thin-walled structures with low attenuation and are highly sensitive to a variety of structural damages [3–6].

Lamb waves are best excited and received using a network of piezoelectric actuators and sensors which are permanently attached on the structure. Piezo-polymers, like polyvinylidene fluoride (PVDF), or piezoceramics (PZT) are commonly used as actuator and sensor materials for Lamb wave based Structural Health Monitoring. Piezopolymers are flexible and can be applied on curved structures. They are lightweight as well as cheaper and easier to manufacture than piezoceramics [7]. However, piezopolymers possess a much lower Young's modulus and actuation force in comparison to piezoceramics. Thus piezopolymers are less suitable as actuators. Moreover, piezopolymers are inappropriate for aerospace applications due to their limited temperature range of -40 to $+110°C$. Hence, the presented work is focused on piezoceramic based Structural Health Monitoring.

Different kinds of structural defects can be detected in principle and located by analysing the sensor signals of guided Lamb waves [8, 9]. However, the presence of at least two Lamb wave modes at any given frequency, their dispersive characteristic and their interference at structural discontinuities produce complex wave propagation fields and sensor signals which are difficult to evaluate. Furthermore, discrete sensors provide only point information. An understanding of the complete wave propagation is not derivable out of these sensor signals. Therefore it is necessary to visualize the wave propagation in order to get a profound understanding of the propagation of different Lamb wave modes and their individual interaction with defects. Usually this is done by scanning laser vibrometry [10]. DLR has developed a method using air-coupled ultrasonic scanning technique, which is presented in the next section. The third section describes the design and positioning of virtual sensors based on the scanned wave propagation field. By using this method different sensor layouts and positions can be compared and evaluated in order to optimize SHM systems. Mode selective actuators are able to excite a particular Lamb wave mode to

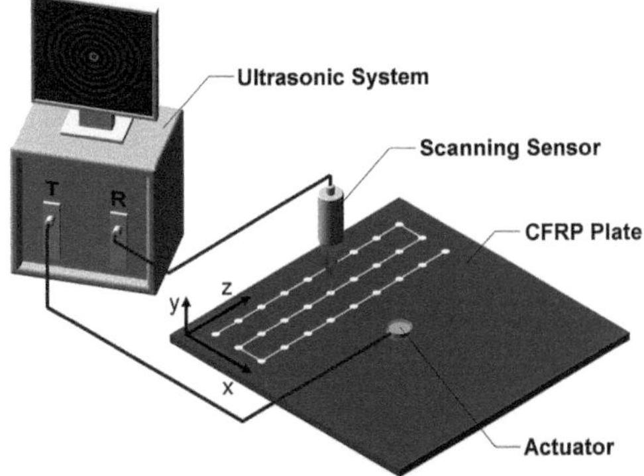

Fig. 37.1 Experimental set-up for lamb wave investigations based on air-coupled ultrasonic technique

reduce the complexity of the wave propagation field. Section 37.4 describes the development and the manufacturing of mode selective actuators. The experimental validation of the mode selective actuators is elucidated thru experiments on a CFRP (Carbon Fiber Reinforced Plastics) plate. Finally, a SHM system for impact detection in a helicopter tailboom (Eurocopter—EC 135) is proposed in Sect. 37.5.

37.2 Visualisation of the Lamb Wave Propagation Field

Air-coupled ultrasonic technique is a well-established method at DLR for the inspection of CFRP structures. This technique has been combined with Lamb wave investigations in order to visualize the Lamb wave propagation field and its interaction with defects. Figure 37.1 shows the experimental set-up used for visualization of Lamb wave propagation fields.

A piezoceramic actuator is applied in the set-up on the lower surface of the specimen and operates as transmitter. A rectangle burst signal is used for excitation. Due to the harmonics of the rectangular signal the receiver signal is filtered in order to provide a narrow band signal. An air-coupled ultrasonic sensor is moved in a meander-shaped track by a portal scanner for receiving the out-of-plane component of the Lamb wave propagation field. Broadband capacitive sensors providing a frequency range from 10 to 100 kHz can also be utilized in addition to ultrasonic sensors having a relative narrowband characteristic with centre frequencies of e.g. 120, 200, 300 and 400 kHz. The amplitudes of the receiver signal are very low due to the air-coupling of the sensors and the mode conversion into longitudinal waves. In consequence, ultra-low noise preamplifiers and a band pass filter on the receiver side are required. The amplitude signal as a function of time $[A = f(t)]$ is measured during the scanning process at each point in x- and y-direction of the scanning grid. Depending on the x- and y-coordinates the amplitude signals are recorded into a so-called volume data file $[A = f(x,y,t)]$.

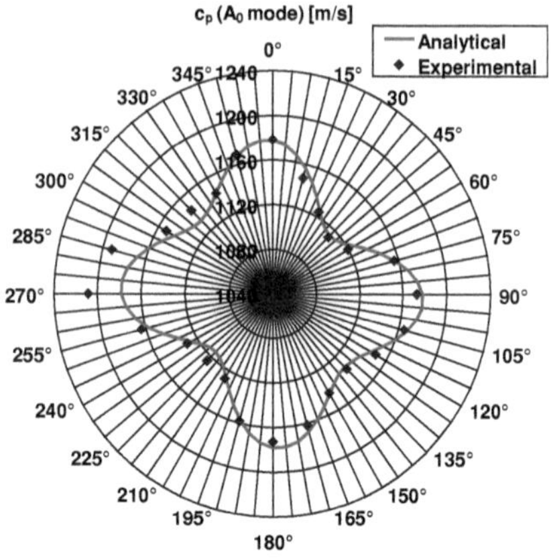

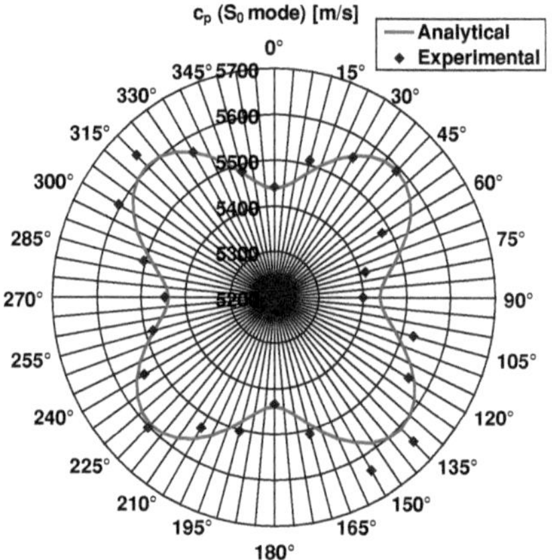

Fig. 37.2 Phase velocity of A_0 mode in x–y plane, 131 kHz, 2 mm thick CFRP plate

Fig. 37.3 Phase velocity of S_0 mode in x–y plane, 131 kHz, 2 mm thick CFRP plate

This volume data file contains the entire information of the Lamb wave propagation field so that several analysis tools can be calculated:

- A-scan, amplitude signal as a function of time at specific x–y-positions
- B-scan, amplitude signal as a function of time along an axis in x–y-plane
- C-scan, two dimensional image of maximal amplitudes
- D-scan, two dimensional image of time-of-flight
- Video animation of the wave propagation

Further analysis tools provide the measurement of specific Lamb wave modes regarding wavelength, phase and group velocity as well as signal attenuation. This measurement can be performed in different angles of the x–y-plane which is fundamental for the characterisation of anisotropic structures, such as CFRP.

Within the investigations the phase velocities of the A_0 and S_0 mode in a CFRP plate are experimentally determined by air-coupled ultrasonic technique. The

CFRP plate have dimensions of $1000 \times 1000 \times 2$ mm and are made of [7] plies in a $[(0/90)_f/+45/-45/(0/90);_f]_S$ configuration. The experimentally determined phase velocities of A_0 and S_0 mode at 131 kHz are shown in Figs. 37.2 and 37.3. In these figures the experimental measurements are compared with the theoretical phase velocities which are determined by the method of guided waves in multiple layers proposed in [9] and [11]. In addition, the phase velocities of the CFRP plates were measured using scanning laser vibrometry which was performed by the Otto von Guericke University of Magdeburg-Germany [12]. The scanning laser vibrometry as well as the ultrasonic technique are appropriate for measurement the dispersion diagram with the same accuracy. Figures 37.2 and 37.3 show that the phase velocity also depends on the direction of propagation in the x–y plane. It can be seen, that the phase velocity pattern is different between A_0 and S_0 mode. This is caused by the lay-up of the CFRP plate which leads to varying stiffnesses in different directions and thicknesses. The varying stiffness of the CFRP plate in combination with the complex displacement field of the Lamb wave modes results in different phase velocity pattern. For instance, the displacement field across the thickness of the A_0 mode resemble for low frequencies (fd $\rightarrow$ 0) a conventional flexure wave. In contrast, the displacement field of the S_0 mode resemble for low frequencies (fd $\rightarrow$ 0) a conventional axial plate wave [8].

37.3 Virtual Design and Evaluation of Sensors

The volume data file described in chap. 37.2 includes the entire information of the wave propagation field for a chosen excitation frequency. This volume data can be used in combination with structure information to simulate the signal of a sensor with known dimensions and behaviour as if it would be bonded on the structure surface. In this way it is possible to optimize the sensor design and network configuration (number of sensors and their positions) with higher time and cost efficiency [13].

After defining the coordinated of the virtual sensor within a wave propagation snapshot, the software calculates the anticipated signal of an equivalent sensor by analysing the volume data file. Every imaginable layout can be simulated. The software considers different coupling techniques and sensor layouts. Figure 37.4 shows an investigation of simulated sensors with an interdigital transducer layout for mode selection 0, [14]. An aluminium plate with the dimensions of $1000 \times 1000 \times 2$ mm was used for experimental verification. A circular piezoceramic actuator having a diameter of 10 mm and a thickness of 0.5 mm was bonded on the plate centre. The actuator is excited by a rectangle burst signal with a frequency of 120 kHz. The Lamb wave propagation field is recorded with the air-coupled ultrasonic scanning technique, as presented above. The wave propagation field is shown in Fig. 37.4 for two different points of time (152.2 and 200 μs). The wavelengths of the excited Lamb wave modes are

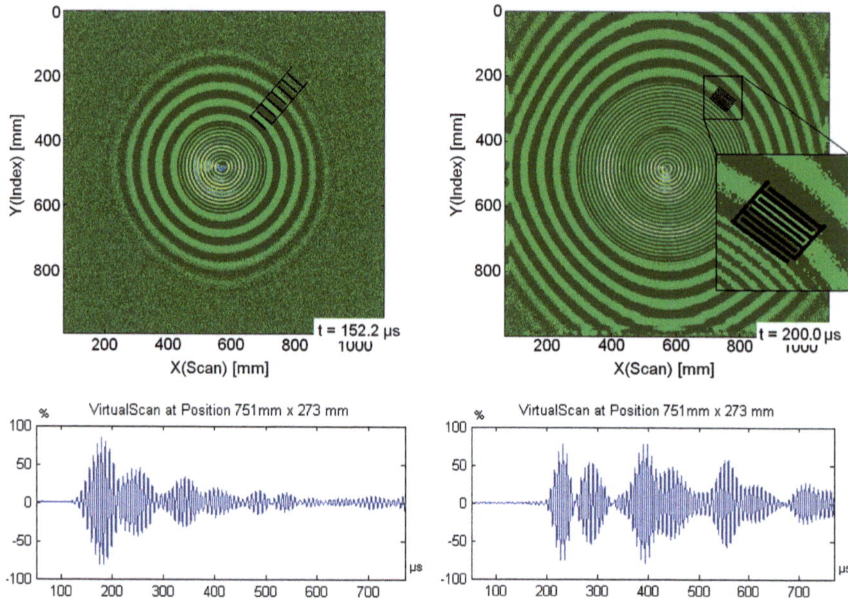

Fig. 37.4 Snapshots of lamb wave propagation field in an aluminium plate with different. virtual sensors (*left* S$_0$ mode, *right* A$_0$ mode) and their calculated signals

measured with the in-house developed analysis software. The procedure of the wavelength measurement is detailed described in [12]. The results show that the S$_0$ mode propagates with a wavelength of 49 mm and the A$_0$ mode propagates with a wavelength of 13 mm.

Two different virtual sensors are designed in the simulation according to the wavelengths of S$_0$ mode (Fig. 37.4—left) and A$_0$ mode (Fig. 37.4—right). The width of each sensor is 60 mm. As a result of the simulation the calculated signals in Fig. 37.4 show a high separation of both Lamb wave modes at the time of arriving at the sensor positions.

37.4 Mode Selective Actuator Design and Manufacturing Process

The generation of a particular Lamb wave mode can be achieved by controlling the frequency as well as the wavelength (λ) of the desired mode within the excitation. An appropriate technical solution is to use piezoelectric substrate with applied interdigitated electrode pattern, so-called interdigital transducers [7, 14, 15]. Such transducers are widespread in telecommunication systems as surface acoustic wave filters (SAW) for frequency selection [16]. The electrode configuration is made of two comb-like electrodes with opposite polarity. The electrode distance

Fig. 37.5 Schematic design of an IDT with apodization

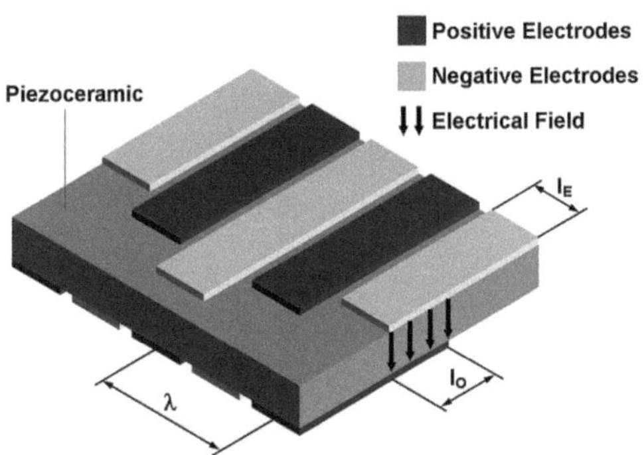

corresponds to the half-wavelength of the desired Lamb wave mode which will be excited at a frequency in the thin-walled structure. The bandwidth of the frequency response function of the excited Lamb wave can be controlled by the number of electrodes. Furthermore, the frequency response function can be modified by apodization which is known from the theory of surface acoustic wave filters. Apodization means that the overlaps (l_O) of each electrode pair are varying along the length of the transducer (Fig. 37.5). By suitable dimensioning of the overlaps the transducer can be in principle designed to a specific frequency response. These possibilities of modifying the frequency response function can be utilized to enhance the effectiveness of the actuator regarding mode selectivity.

Starting point for the manufacturing of interdigital transducer is a commercial available piezoceramic plate with a typical thickness of 0.2–0.5 mm. This piezoceramic plate is already provided with uniform electrodes on the upper and lower surface and polarized by the manufacturer. In a first step the piezoceramic plate is additionally metallised with gold by a sputtering process. This procedure ensures wrap-around electrodes and thus the electrical connection from the upper side of the final actuator. In a second step the electrode structure is made by a laser ablation process. The laser parameters are adjusted to remove only the metallised layer and avoid mechanical damages of the piezoceramic. Due to the polarisation direction and the electrical field, which is generated through the thickness of the piezoceramic, the transducer is working in the piezoelectric d_{31}-effect. In this case a positive electrical field causes in-plane contraction of the piezoceramic material which is used for Lamb wave excitation. A typical example of such an interdigital transducer is shown in Fig. 37.6.

The main problems of piezoceramics are their inherent brittleness and the failures of electrical contacts which lead to insufficient reliability. A promising alternative to conventional piezoceramics is to utilize the piezocomposite technology to increase the reliability of brittle piezoceramics. Piezocomposites consist of piezoceramic materials embedded in a ductile polymer. Further components like electrodes, electrical contacts or insulators are also embedded into the composite. Within the embedding process the polymer is typically cured in a temperature range of 120–180°C. Due to the different coefficients of thermal expansion of the

Fig. 37.6 Mode selective lamb wave actuator based on interdigital transducer design with apodization

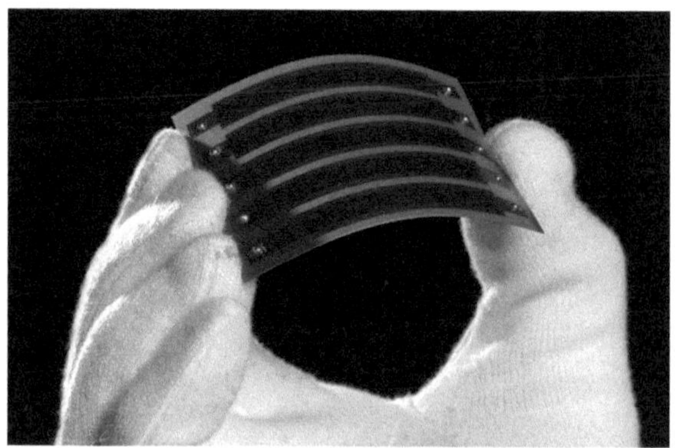

Fig. 37.7 Mode selective lamb wave actuators based on piezocomposite technology

polymer and the piezoceramic material as well as to the shrinking of the polymer during curing, the piezoceramic material is provided with a mechanical pre-compression. This pre-compression protects the brittle piezoceramic material and allows bending loads on the piezocomposite which is essential for the application on curved structures. Further advantages of piezocomposites are reliable electrical contacts, electrical insulation as well as high durability under variable environments. In recent years different piezocomposite configurations have been designed and manufactured [17].

One possible solution of manufacturing mode selective actuators based on the piezocomposite technology is the embedding of the piezoceramic with interdigitated electrode pattern, which is shown in Fig. 37.6, into the polymer. Another solution is to arrange individual piezoceramic elements in a distance of half-wavelength within the piezocomposite (Fig. 37.7). The piezoceramic elements work in the piezoelectric d_{31}-effect.

For experimental tests a mode selective actuator is designed and manufactured. The actuator is designed to attenuate the S_0 mode and thus to amplify the A_0 mode at a frequency of 40 kHz. At this frequency only the lowest order of symmetric

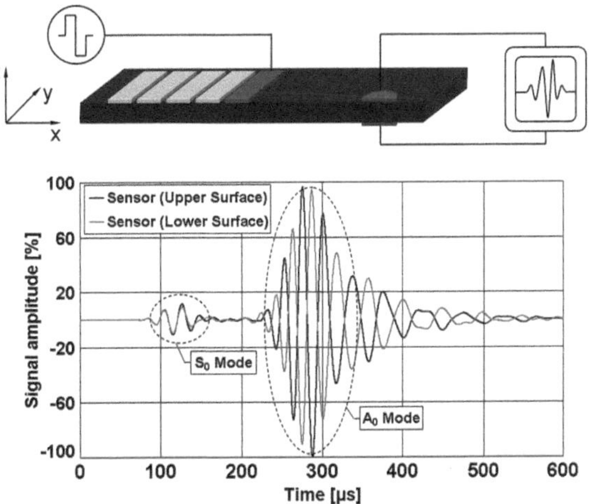

Fig. 37.8 Experimental set-up and sensor signal by driving the first element of the actuator

and anti-symmetric modes exists which is a first reduction of the complex Lamb wave propagation field. Due to the dispersion diagram of the CFRP plate the wavelength of the A_0 mode at 40 kHz is approx. 20 mm. The actuator is manufactured as piezocomposite and consists of five piezoceramic elements which have respective dimensions of $50 \times 8 \times 0.2$ mm. The distance between the elements corresponds to the half-wavelength of 10 mm. The final actuator that is used for the experimental tests is shown in Fig. 37.7. The actuator is bonded in $0°$-direction on the upper surface of the CFRP plate which is presented in Sect. 37.2. As bonding layer a cyanoacrylate adhesive is used. Due to the low viscosity of the uncured adhesive very thin and uniform bonding layers can be ensured. Each actuator element is excited by a rectangle burst signal with 3 bipolar pulses, whereby the signal of adjacent elements has a $180°$ phase difference. The voltage of the excitation signal is 15 V (peak to peak). In order to distinguish the S_0 mode from the A_0 mode a pair of circular piezoceramic sensors is collocated bonded on the upper and lower plate surface. In case of the symmetric S_0 mode both sensors show equal amplitude signals over time without a phase shift. But in case of the anti-symmetric A_0 mode the amplitude signals of the upper and lower sensor show a $180°$ phase shift. The sensors are piezocomposites with an embedded circular piezoceramic which has a diameter of 10 mm and a thickness of 0.2 mm. The distance between the actuator and the sensors is set to 200 mm. The sensor signals are filtered using a 12th order band pass of 20–60 kHz. The amplitudes are measured building the absolute value of the signal of each Lamb wave mode. In a first setup only the first element of the actuator is driven. The sensor signals in Fig. 37.8 shows that the A_0 and the S_0 mode are generated in an amplitude ratio of 100 to 11%.

In a second setup all elements of the actuator are driven. The resulting amplitude ratio is 100% of the A_0 mode to 1.7% of the S_0 mode. The amplitude ratio can be improved by applying an apodization. This is realized by controlling the voltage of the excitation signal of each actuator element using adjustable

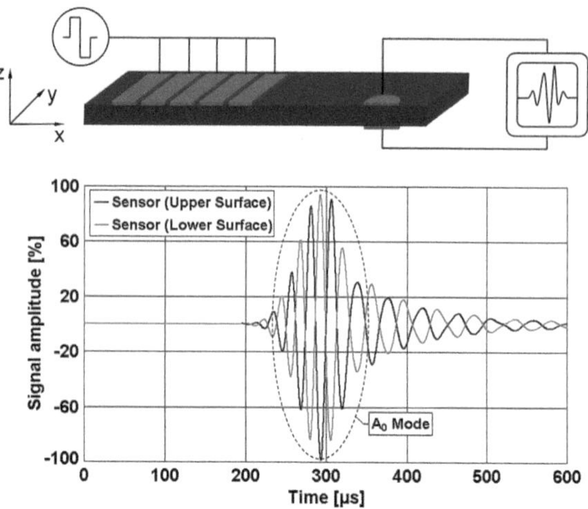

Fig. 37.9 Experimental set-up and sensor signal by driving all elements of the actuator with applied apodization

resistors. The apodization is set in such a way that the S_0 mode shows minimal amplitudes over the frequency bandwidth of the sensor signal. The aim of the apodization is to modify the frequency response functions of the S_0 mode as well as the A_0 mode so that a broadband reduction of the S_0 mode is achieved. Figure 37.9 shows the signal amplitudes of the apodized actuator. The signal amplitude of the S_0 mode is reduced to 0.2% in contrast to the amplitude of the A_0 mode of 100%. As a result, the experimental tests show that he designed actuator can be sufficiently attenuated the S_0 mode in a CFRP plate. Furthermore, an apodization modifies the frequency response of the actuators in such a way that the mode selectivity can be improved.

37.5 Concept of Damage Detection in a Helicopter Tailboom

Bridging the gap between basic research and application DLR's research on SHM covers the investigation of a SHM system on a half-shell of the EC 135 helicopter tailboom. The research is focused on the ability of defect detection in the helicopter tailboom. Figure 37.10 shows the half-shell of the tailboom (3.49 × 0.57 m).

The tailboom consists of a honeycomb sandwich structure (skin thicknesses 1.0 and 0.5 mm) with copper mesh in the outer skin for lightning protection and different inserts. The wave propagation in such sandwich structure is quite complex. Therefore the wave propagation is visualized using the air-coupled ultrasonic scanning technique which is presented in Sect. 37.2.

Initial experiments reveal that only excitation at frequencies below 30 kHz generate wave modes which propagate in both skins and in the sandwich core. This aspect is important because the actuator and sensor system is applied on the inner skin and must also be able to detect damages in the outer skin as well as in the core.

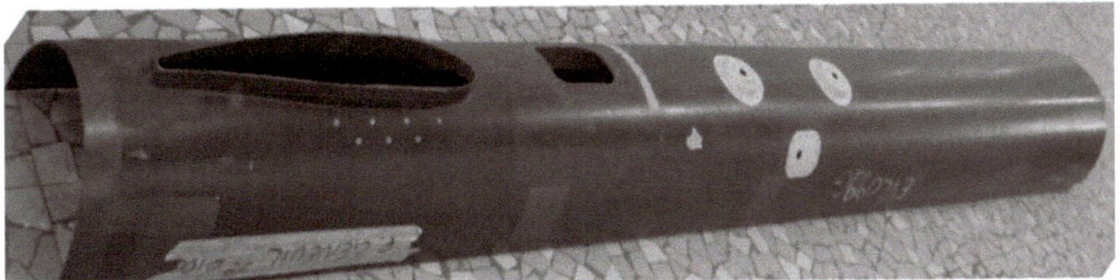

Fig. 37.10 Tailboom (half-shell) of Eurocopter EC 135

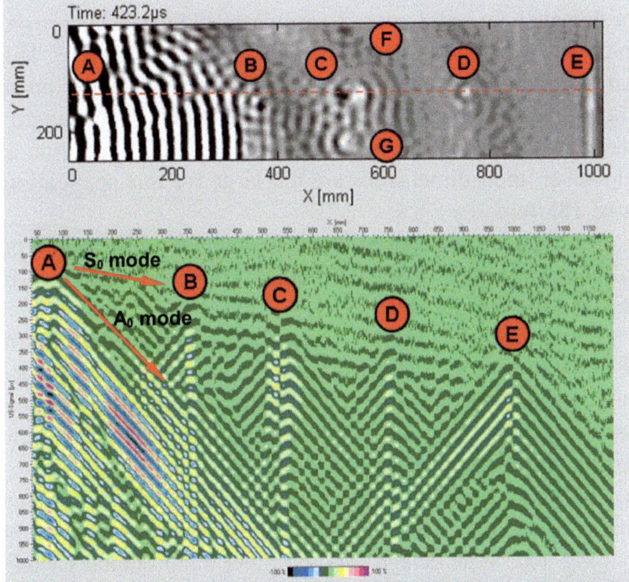

Fig. 37.11 Snapshot of the wave propagation in the tailboom (*top*), B-scan along the red dashed line in the snapshot (*bottom*)

In order to study the wave interaction with an impact and applied piezoceramic sensors, the wave propagation in the tailboom has been recoded over a length of 1 m. Figure 37.11 shows the wave propagation in the tailboom. The actuator is situated at point (A) and excites the S- and the A-mode. At position (B), (D), (F) and (G) circular piezoceramic sensor (PIC 255, Ø 10 × 0.2 mm, PI Ceramic GmbH) are bonded on the inner skin. At position (C) an impact with energy of 10 J is applied on the outer skin. Position (E) indicates a separation of the sandwich core. It can be seen that every stiffness change, like sensors, impacts and core separations, produce mode conversion from the S-mode into the A-mode. Based on the wave propagation these stiffness changes behave like virtual A-mode sources. This means the more sensors required for precise defect detection and localisation, the more additional mode conversions appear.

Fig. 37.12 Air-coupled
sensors are used for the
sensor network

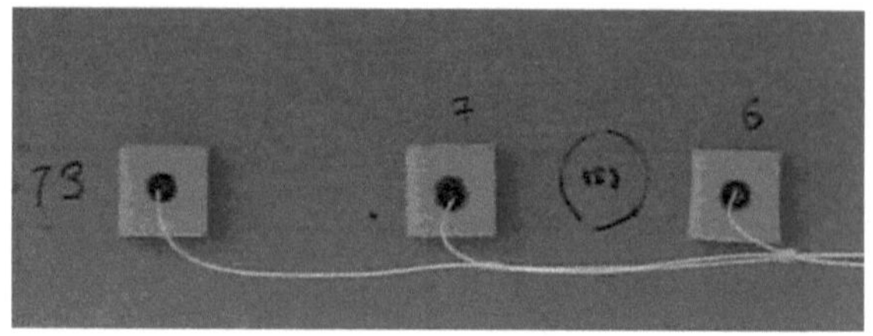

Based on the results of the wave propagation a concept of damage detection in the tailboom has been developed. This concept uses mode conversion as an indicator for damages. In order to have a minimal additional distortion of the wave propagation only few actuators should be applied on the structure. This is possible because Lamb waves propagate over large distances with low attenuation. The actuators should be optimised for the S0-mode with a wavelength 185 mm at 22 kHz. Interdigital actuators, as is presented in Sect. 37.4, provide a wavelength selection. However, their dimensions are too large for this application requiring a wavelength of 185 mm.

Because of the mode conversions at glued actuators they should be position at "natural stiffness changes" of the tailboom. Those positions are internal core bondings in a distance of about 1 m.

In order to avoid additional mode conversion air-coupled sensor networks are used. These kinds of sensors are sensitive to the out-of-plane component of Lamb waves. Low frequencies between 20 and 30 kHz generate Lamb waves with higher out-of-plane components of the A-mode than the S-mode which is another advantage of the air-coupled system (Fig. 37.12).

Integrated SHM networks underlie different kinds of electromagnetic perturbation. A high signal to noise ratio of the measurements requires an optimized signal processing and cable design. On the other hand complex cabling of entire networks accompanies with a high system weight. A weight-saving solution is a tree-like organisation of multiplexer entities. Local groups of sensors can be connected to a multiplexer. Different multiplexers can be arranged together in dependence of the network layout for minimal cable usage. The aim is a solution as simple as possible with a minimum of (shielded) cables from the SHM system to the tailboom. A standard VGA monitor cable contains three shielded coax cables (75 Ω) and ten control lines and is easily deliverable up to a length of 10 m. The actuators are multiplexed by relays so that only one internal coax cable is required. The 64 sensors are segmented in eight sensor arrays, each with eight selected sensors (Fig. 37.13). The eight sensors of an array are connected to the sensor multiplexer. Its output is amplified by a low noise pre-amplifier and connected to an array multiplexer so that only one coax cable connection to the SHM system is necessary. A DC coupled technique between sensors, multiplexers and amplifiers is used so that only a single supply is necessary.

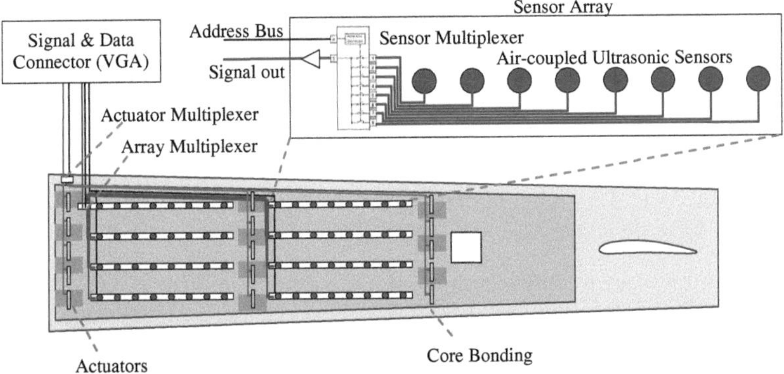

Fig. 37.13 SHM system of the helicopter tailboom

37.6 Conclusion

Structural Health Monitoring based on Lamb waves is a promising technology for in-service inspection of aerospace structures. For the design of SHM systems in respect to different application requirements the visualisation of the wave propagation field is essential. As presented, the air-coupled ultrasonic scanning technique is a suitable methodology for this task, especially the recording of the complete wave propagation field by volume data files. These data files in combination with different analysis tools provide the basis for the design of actuators, sensors, signal processing and algorithms for damage detection. It has been shown that the signal of virtual sensors can be simulated using the scanned propagation field. It could be demonstrated that mode selective actuators are able to excite a particular Lamb wave mode in a CFRP plate. The design and manufacturing of reliable actuator and sensor networks are presented. Finally, a concept of damage detection in a helicopter tailboom has been shown.

References

1. Viktorov, I.A.: Rayleigh and Lamb Waves: Physical Theory and Applications. Plenum Press, New york (1967)
2. Achenbach, J.D.: Wave Propagation in Elastic Solids. North-Holland Publishing Company, Amsterdam (1973)
3. Boller, C., Chang, F.-K., Fujino, Y.: Encyclopedia of Structural Health Monitoring. John Wiley & Sons, New Jersey (2009)
4. Staszewski, W., Boller, C., Tomlinson, G.: Health Monitoring of Aerospace Structures, Smart Sensor Technologies and Signal Processing. John Wiley & Sons, New Jersey (2004)
5. Raghavan, A., Cesnik, C.E.S.: Review of guided-wave structural health monitoring. The Shock Vib. Digest. **39**, 91–114 (2007)

6. Su, Z., Ye, L., Ye, L.: Guided lamb waves for identification of damage in composite structures: a review. J. Sound Vib. **295**, 753–780 (2006)
7. Monkhouse, R.S.C., Wilcox, P.D., Cawley, P.: Flexible interdigital PVDF transducers for the generation of lamb waves in structures. Ultrasonics **35**, 489–498 (1997)
8. Giurgiutiu, V.: Structural Health Monitoring with Piezoelectric Wafer Active Sensors. Academic Press—Elsevier, Amsterdam (2008)
9. Rose, J.L.: Ultrasonic Waves in Solid Media. Cambridge University Press, Cambridge (2004)
10. Staszewski, W.J., Lee, B.C., Mallet, L., Scarpa, F.: Structural health monitoring using scanning laser vibrometry: I. Lamb wave sensing. Smart Mater. Struct. **14**(2), 251–260 (2004)
11. Nayfeh, A.L.: Wave propagation in anisotropic media, North Holland Series. In: Applied Mathematics and Mechanics, vol. 39 (1995)
12. Pohl, J., Mook, G., Szewieczek, A., Hillger, W., Schmidt, D.: Determination of lamb wave dispersion data for SHM, 5th European Workshop on Structural Health Monitoring, Italy (2010)
13. Hillger, W., Szewieczek, A.: Verfahren zur Optimierung eines Sensornetzwerks (Procedure for optimization of a sensor network). Patent No. 10 2009 019 243.3, German Patent and Trade Mark Office (2009)
14. Monkhouse, R.S.C., Wilcox, P.W., Lowe, M.J.S., Dalton, R.P., Cawley, P.: The rapid monitoring of structures using interdigital lamb wave transducers. Smart Mater. Struct. **9**, 304–309 (2000)
15. Veidt, M., Liu, T., Kitipornchai, S.: Modelling of lamb waves in composite laminated plates excited by interdigital transducers. NDT&E Int. **35**, 437–447 (2002)
16. Matthews, H.: Surface Wave Filters—Design, Construction, and Use. John Wiley & Sons, New Jersey (1977)
17. Wierach, P.: Elektromechanisches Funktionsmodul, German Patent DE 10051784 C1 (2002)

MIX
Papier aus verantwortungsvollen Quellen
Paper from responsible sources
FSC® C105338

If you have any concerns about our products,
you can contact us on
ProductSafety@springernature.com

In case Publisher is established outside the EU,
the EU authorized representative is:
Springer Nature Customer Service Center GmbH
Europaplatz 3, 69115 Heidelberg, Germany

Printed by Libri Plureos GmbH
in Hamburg, Germany